Lecture Notes in Physics

Springer-Verlag Berlin Heidelberg GmbH

The Editorial Policy for Proceedings

The series Lecture Notes in Physics reports new developments in physical research and teaching – quickly, informally, and at a high level. The proceedings to be considered for publication in this series should be limited to only a few areas of research, and these should be closely related to each other. The contributions should be of a high standard and should avoid lengthy redraftings of papers already published or about to be published elsewhere. As a whole, the proceedings should aim for a balanced presentation of the theme of the conference including a description of the techniques used and enough motivation for a broad readership. It should not be assumed that the published proceedings must reflect the conference in its entirety. (A listing or abstracts of papers presented at the meeting but not included in the proceedings could be added as an appendix.)
When applying for publication in the series Lecture Notes in Physics the volume's editor(s) should submit sufficient material to enable the series editors and their referees to make a fairly accurate evaluation (e.g. a complete list of speakers and titles of papers to be presented and abstracts). If, based on this information, the proceedings are (tentatively) accepted, the volume's editor(s), whose name(s) will appear on the title pages, should select the papers suitable for publication and have them refereed (as for a journal) when appropriate. As a rule discussions will not be accepted. The series editors and Springer-Verlag will normally not interfere with the detailed editing except in fairly obvious cases or on technical matters.
Final acceptance is expressed by the series editor in charge, in consultation with Springer-Verlag only after receiving the complete manuscript. It might help to send a copy of the authors' manuscripts in advance to the editor in charge to discuss possible revisions with him. As a general rule, the series editor will confirm his tentative acceptance if the final manuscript corresponds to the original concept discussed, if the quality of the contribution meets the requirements of the series, and if the final size of the manuscript does not greatly exceed the number of pages originally agreed upon. The manuscript should be forwarded to Springer-Verlag shortly after the meeting. In cases of extreme delay (more than six months after the conference) the series editors will check once more the timeliness of the papers. Therefore, the volume's editor(s) should establish strict deadlines, or collect the articles during the conference and have them revised on the spot. If a delay is unavoidable, one should encourage the authors to update their contributions if appropriate. The editors of proceedings are strongly advised to inform contributors about these points at an early stage.
The final manuscript should contain a table of contents and an informative introduction accessible also to readers not particularly familiar with the topic of the conference. The contributions should be in English. The volume's editor(s) should check the contributions for the correct use of language. At Springer-Verlag only the prefaces will be checked by a copy-editor for language and style. Grave linguistic or technical shortcomings may lead to the rejection of contributions by the series editors. A conference report should not exceed a total of 500 pages. Keeping the size within this bound should be achieved by a stricter selection of articles and not by imposing an upper limit to the length of the individual papers. Editors receive jointly 30 complimentary copies of their book. They are entitled to purchase further copies of their book at a reduced rate. As a rule no reprints of individual contributions can be supplied. No royalty is paid on Lecture Notes in Physics volumes. Commitment to publish is made by letter of interest rather than by signing a formal contract. Springer-Verlag secures the copyright for each volume.

The Production Process

The books are hardbound, and the publisher will select quality paper appropriate to the needs of the author(s). Publication time is about ten weeks. More than twenty years of experience guarantee authors the best possible service. To reach the goal of rapid publication at a low price the technique of photographic reproduction from a camera-ready manuscript was chosen. This process shifts the main responsibility for the technical quality considerably from the publisher to the authors. We therefore urge all authors and editors of proceedings to observe very carefully the essentials for the preparation of camera-ready manuscripts, which we will supply on request. This applies especially to the quality of figures and halftones submitted for publication. In addition, it might be useful to look at some of the volumes already published. As a special service, we offer free of charge LaTeX and TeX macro packages to format the text according to Springer-Verlag's quality requirements. We strongly recommend that you make use of this offer, since the result will be a book of considerably improved technical quality. To avoid mistakes and time-consuming correspondence during the production period the conference editors should request special instructions from the publisher well before the beginning of the conference. Manuscripts not meeting the technical standard of the series will have to be returned for improvement.

For further information please contact Springer-Verlag, Physics Editorial Department II, Tiergartenstrasse 17, D-69121 Heidelberg, Germany

Julius Wess Vladimir P. Akulov (Eds.)

Supersymmetry and Quantum Field Theory

Proceedings of the D. Volkov Memorial Seminar
Held in Kharkov, Ukraine, 5–7 January 1997

Springer

Editors

Julius Wess
Sektion Physik
Universität München
Theresienstrasse 37
D-80333 München, Germany

Vladimir P. Akulov
Institute for Theoretical Physics
National Science Centre (NSC)
Kharkov Institute of Physics and Technology
1, Akademicheskaya str.
310108 Kharkov, Ukraine

Cataloging-in-Publication Data applied for.

Die Deutsche Bibliothek - CIP-Einheitsaufnahme

Supersymmetry and quantum field theory : proceedings of the D. Volkov Memorial Seminar, held in Kharkov, Ukraine, 5 - 7 January 1997 / Julius Wess ; Vladimir P. Akulov (ed.).
(Lecture notes in physics ; Vol. 509)
ISBN 978-3-662-14200-4 ISBN 978-3-540-69217-1 (eBook)
DOI 10.1007/978-3-540-69217-1

ISSN 0075-8450
ISBN 978-3-662-14200-4

Originally published by Springer-Verlag Berlin Heidelberg New York in 1998
Softcover reprint of the hardcover 1st edition 1998

Typesetting: Camera-ready by the authors
Cover design: *design & production* GmbH, Heidelberg
SPIN: 10644165 55/3144-543210 - Printed on acid-free paper

Preface

This volume is dedicated to the memory of Dmitrij Volkov (03.07.1925 – 05.01.1996), a prominent scientist, who made an outstanding contribution to the theoretical physics of elementary particles and quantum field theory. The volume contains the Proceedings of the D. Volkov Memorial Seminar "Supersymmetry and Quantum Field Theory", which was held in Kharkov, Ukraine, during January 5–7, 1997, and invited contributions. These are divided in two topics: supermembranes, supersymmetry and field theory; and quantum groups. It also contains Volkov's pioneering works in spin and statistics, supersymmetry, supergravity, and superstrings, and his last interview.

Kharkov, Munich
20th February, 1998

V.Akulov, J.Wess

Contents

Last Interview with D.V. Volkov 1

Part I Supermembranes, Supersymmetry and Quantum Field Models

p-Brane Chemistry
K.S. Stelle 11

Intersecting Branes and Supersymmetry
M. de Roo 18

Properties of Intersecting p-Branes in Various Dimensions
I. Ya. Aref'eva, M.G. Ivanov and O.A. Rytchkov 25

Super D-Branes
E. Bergshoeff 42

Volkov–Akulov Theory and D-Branes
R. Kallosh 49

Ten to Eleven: It Is Not Too Late
M.J. Duff 59

Aspects of Superembeddings
P.S. Howe, E. Sezgin and P.C. West 64

Superbrane Actions and Geometrical Approach
I. Bandos, P. Pasti, D. Sorokin and M. Tonin 79

The M Theory Five-Brane and the Heterotic String
J.H. Schwarz 92

A Linear Representation for the Topological Extensions of the Poincaré Superalgebra in $d = 11$
A.A. Deriglazov and A.V. Galajinsky 103

On the Construction of Global Duality Maps in Strings and Supermembrane Theories
I. Martin and A. Restuccia 109

Planckian Energy Scattering of *D*-Branes and M(atrix) Theory in Curved Space
I.V. Volovich 116

Self-Duality in Nonlinear Electromagnetism
M.K. Gaillard and B. Zumino 121

Progress Toward A Classical (SUSY)^2 4D, $N = 1$ Green–Schwarz σ-Model Action
S.J. Gates, Jr. 130

$N = 4$ Supersymmetric Integrable Systems
E.A. Ivanov 136

On Some Puzzles in $N = 2$ Supersymmetric Gauge Theory
L. O'Raifeartaigh 150

Alternative Formulations of $N = 2$ Supesymmetric Gauge Theory in Harmonic Superspace
B.M. Zupnik 157

Lie-Algebraic Characterization of 2*D* (Super-)Integrable Models
F. Toppan 161

Universal Hidden Supersymmetry in Classical Mechanics and Its Local Extension
E. Gozzi 166

The Hamiltonian Structure of the "Bosonic" and "Fermionic" Extensions of $N = 2$ KdV Hierarchy
L. Bonora and S. Krivonos 173

Mass Generation in the Supersymmetric Nambu–Jona–Lasinio Model in an External Magnetic Field
I.A. Shovkovy 182

On Extension of Minimality Principle in Supersymmetric Electrodynamics
V.V. Tugai and A.A. Zheltukhin 187

Stochastic Wess–Zumino–Witten Models
R. Léandre 196

On $(k \oplus l|q)$-Dimensional Supermanifolds
A. Konechny and A. Schwarz 201

Let the Spin and the Charges Unify
N.M. Borštnik 207

Kerr Spinning Particle and Superparticle Models
A. Burinskii 214

Spinons and Parafermions in Fermion Cosets
D.C. Cabra 220

Exact Solutions in Einstein–Yang–Mills Theories
I.P. Volobuev 230

Higher Massless Irreducible Spins in the BRST Approach
A. Pashnev and M. Tsulaia 237

Sonoluminescence and Black Holes as Sources of Squeezed Light
Yu.P. Stepanovsky 246

Remark Concerning Integrable Hamilton Systems
V.A. Soroka 252

Part II Quantum Symmetries and q-Deformed Groups

q-Deformed Heisenberg Algebra
J. Wess 259

Supersymmetric Reflection Matrices
M. Moriconi and K. Schoutens 265

Deformed Oscillator Algebras and Higher-Spin Gauge Interactions of Matter Fields in 2+1 Dimensions
M.A. Vasiliev 274

Universality of the R-Deformed Heisenberg Algebra
M.S. Plyushchay . . . 289

The Dual Algebra of the Jordanian $GL_{g,h}(2)$
B.L. Aneva, V.K. Dobrev and S.G. Mihov . . . 298

Supertraces on Some Deformations of Heisenberg Superalgebra
S.E. Konstein . . . 305

Harish–Chandra Embedding and q-Analogues of Bounded Symmetric Domains
S. Sinel'shchikov and L. Vaksman . . . 312

q-Differential Calculus and Deformed Light–Cone
V. P. Akulov, V.V. Chitov and S. Duplij . . . 317

σ-Models on the Quantum Group Manifolds $SL_q(2,R)$, $SL_q(2,R)/U_h(1)$, $C_q(2|0)$ and Infinitesimal transformations
V.D. Gershun . . . 324

Integrating a Generic Algebra
R. Casalbuoni . . . 332

Part III Selected Works and List of Main Publications of D.V. Volkov

On the Quantization of Half-Integer Spin Fields
D.V. Volkov . . . 341

S-Matrix in the Generalized Quantization Method
D.V. Volkov . . . 350

Regge Poles in Nucleon–Nucleon and Nucleon–Antinucleon Scattering Amplitudes
D.V. Volkov and V.N. Gribov . . . 357

$SU(3) \times SU(3)$ Symmetry and the Baryon–Meson Coupling Constants
D.V. Volkov . . . 368

Phenomenological Lagrangian for Spin Waves
D.V. Volkov, A.A. Zheltukhin, and Yu.P. Bliokh . . . 370

Possible Universal Neutrino Interaction
D.V. Volkov, and V.P. Akulov 383

Higgs Effect for Goldstone Particles with Spin 1/2
D.V. Volkov and V.A. Soroka 386

Gauge Fields on Superspaces with Different Holonomy Groups
V.P. Akulov, D.V. Volkov, and V.A. Soroka 389

Spontaneous Compactification of Subspace Due to Interaction of the Einstein Fields with the Gauge Fields
D.V. Volkov and V.I. Tkach 393

Hamiltonian Systems with Even and Odd Poisson Brackets: Duality of Their Conservation Laws
D.V. Volkov, A.I. Pashnev, V.A. Soroka and V.I. Tkach 396

List of Main Publications of D.V. Volkov 399

List of Participants

Akulov V.P.	akulov@juno.com
Aref'eva I.	arefeva@class.mian.su
Bandos I.	bandos@kipt.kharkov.ua
Berezovoj V.	berezovoj@kipt.kharkov.ua
Borstnik N.	Norma.S.Mankoc@ijs.si
Bugrij E.	ebugrij@itp.kiev.ua
Burinskij A.	grg@ibrae.msk.su
Cabra D.	cabra@venus.fisica.unlp.edu.ar
Casalbuoni R.	casalbuoni@fi.infn.it
de Roo M.	deroo@th.rug.nl
Deriglazov S.	deriglaz@phys.tsu.tomsk.su
Duplij S.	duplij@physik.uni-kl.de
Fedoruk S.	aigumen@kipt.kharkov.ua
Fomin P.	pfomin@gluk.apc.org
Galajinski A.	galajin@phys.tsu.tomsk.su
Gershun V.	gershun@kipt.kharkov.ua
Gozzi E.	gozzi@trieste.infn.it

Gumenchuk A.	alex-gumenchuk@geocities.com
Ivanov E.	eivanov@thsun1.jinr.dubna.su
Kapustnikov A.	alexandr@ff.dsu.dp.ua
Konshtein S.	konstein@td.lpi.ac.ru
Krivonos S.	krivonos@thsun1.jinr.dubna.su
Lapshin V.	lapshin@kipt.kharkov.ua
Leandre R.	Remi.Leandre@antares.iecn.u-nancy.fr
Nurmagambetov A.	ajn@kipt.kharkov.ua
Pashnev A.	pashnev@thsun1.jinr.dubna.su
Pasti P.	PASTI@padova.infn.it
Plyushchay M.	mikhail@canastra.fisica.ufjf.br
Rekalo A.	rekaloa@kipt.kharkov.ua
Restuccia A.	arestu@usb.ve
Schoutens K.	kjs@phys.uva.nl schouten@itp.ucsb.edu
Shovkovy I.	eppaitp@gluk.apc.org
Sitenko Y.	yusitenko@gluk.apc.org
Skalozub V.	skalozub@phd.dp.ua
Soroka V	vsoroka@kipt.kharkov.ua
Sorokin D.P.	dsorokin@kipt.kharkov.ua
Stelle K.	k.stelle@ic.ac.uk
Stepanovski Yu.	aigumen@kipt.kharkov.ua
Tolstoy V.	tolstoy@anna19.npi.msu.su

Tonin M.	TONIN@padova.infn.it
Toppan F.	TOPPAN@padova.infn.it
Vaksman L.	vaksman@ilt.kharkov.ua
Volobuev I.	volobuev@ira.npi.msu.su
Volovich I.	volovich@class.mian.su
Zheltukhin A.A.	zheltukhin@kipt.kharkov.ua
Zima V.	aigumen@kipt.kharkov.ua
Zupnik B.	zupnik@thsun1.jinr.dubna.su

Last Interview with D.V. Volkov

My conversation today [1] is with the renowned scientist and academician Dmitry Vasiljevitch Volkov, head of the theoretical laboratory of the Kharkov Physical-Technical Institute (KhFTI).

QUESTION: Dmitry Vasiljevitch, could you tell me, please, what was your path into theoretical science, how did you become a theoretician, and was this accidental or was there any specific cause.

ANSWER: I was born in Leningrad. When I was 16 years old and when I was studying in the 8th class, the Great Patriotic War began. I was evacuated from Leningrad. These were very difficult years for young people. In this period I came to work on a collective farm and in a military factory. After that I was drafted into the army and took part as a soldier in the war on the Karelian front, above the polar circle. When the war began with Japan, I participated in military action on the Far Eastern front. I want to say that the war had a considerable influence on my attitude to life. In my generation the war created a feeling of responsibility for the country. After the war we carried over the same ideology into civilian life. When the question of a choice of profession arose, many of us thought about how we might be useful to our country. During all the war years I dreamed about going into science, because already in school I was attracted especially to the exact sciences: mathematics and physics. After the demobilisation I entered Leningrad State University, in the faculty of physics. At that time prominent scientists such as V.A. Fok and T.P. Kravets were teaching there. The lectures of T.P. Kravets were distinguished by his ability to link the study material with personal moments. He taught us that physics is created by living people and he spoke much about his teacher Lebedev. From the first days Kravets infected us with a deep love for science, and for physics in particular. Aside from that, I listened to the lectures of V.I. Smirnov, whose widely known multi-volume works were specially intended for theoretical physicists , and, in fact, formed the basis of our whole education. We learned a lot from other mathematicians of his school: O.A. Ladyzhenskaya, M.I. Petrashen. In the final year I received a profound training in the specialisation of theoretical physics thanks to the excellent teacher L.E. Gurevitch. There were also other teachers, which I remember to this time with thankfulness.

Unfortunately, a great tragedy happened in Leningrad when I was studying there, and this had an impact on my further destiny. It is now commonly known that at the end of the forties took place the so-called Leningrad trial, as a result of which prominent party leaders and representatives of the scientific and cultural community of the city were repressed, and among their

[1] Questions by Yu.N. Ranyuk

number were leading scientists of Leningrad University. Therefore, the group in which I was studying was dissolved and the scientific line on which I was working ceased to exist there. Here one should give credit to the Kharkov scientists. From the very beginning, the Kharkov scientists had close contacts with their colleagues in Leningrad. In particular, it is known, that our Institute, KhFTI, was organised thanks to the arrival in Kharkov of the group of scientists headed by the well-known physicists Kirill Dmitrievitch Sinelnikov, Anton Karlovitch Walter and others. One of the major concerns at the foundation of the Institute was how to attract young people so that in the future the Institute would be staffed by highly qualified scientists. Therefore a special representative came to Leningrad and our students were offered the possibility to move to Kharkov and to continue their studies there. I agreed. Together with me came the already well-known scientists E.V. Inopin, K.N. Stepanov, V.F. Alexin and some other students.

QUESTION: Dmitry Vasiljevitch, in which year of your studies did this happen?

ANSWER: This was after the fourth year. I finished the fourth year in Leningrad, and in the fifth year I was already studying at Kharkov University. Education here was also carried out on a high level. Lectures were given by outstanding lecturers and popularisers of science, such as, e.g. Alexander Iljitch Akhiezer. The lectures of L.N. Rosenzweig made a special impression upon me. He was a man of sparkling wit, and, besides that, he was aware of the latest developments, which were taking place in physics. And at this time, physics was receiving an extremely interesting boost connected with the discovery of new elementary particles, and, in fact, the physics of elementary particles was going through its birth period. This was a very interesting period. Lipa Natanovitch was fascinated by this new physics and imagine that while we were still students, he came to a lecture, and, instead of reading to us material according to the programme, took a fresh article from the "Physical Review" and summarised it to us. Aside from that, he would even give these articles to us young students to learn from and to discuss in seminars.

QUESTION: Dmitry Vasiljevitch, when did you move to Kharkov, and how did your further scientific interests develop?

ANSWER: I moved to Kharkov in 1951. Now, about my scientific interests and their development. During my student years, the mathematical aspects of science attracted me even more than the physical ones. At that time I had already developed a deep feeling for the general theory of relativity. Quantum field theory was just beginning. Just at this same time the development of physics received some sharp impacts. On the one hand, quantum electrodynamics was effectively reborn thanks to the theory of renormalization, and, on the other hand, the physics of new particles arose. At this time the pi-meson was observed and the neutral pi-0-meson was discovered and discussions on the definition of spin and parity were going on. This was a real revolution in science. Later on, after graduation from university, I began my graduate training. My advisor was A.I. Akhiezer. He organised a small group of graduate students, including R. Polovin, P. Fomin, V. Alexin and me. All of us actively studied quantum electrodynamics. This was especially important for me, because I continued to work in this area and quantum electrodynamics became for me a sort of initial example of the theories that I work on now.

I would like to discuss in some more detail the directions of science that interested me most and on which I worked later. I don't know why - somehow intuitively - but already in my years of study my favourite subject was connected with the theory of symmetry groups. At that time, the theory of symmetry groups was already being adopted in physics, but not widely enough. Later on, the application of this theory came into a full flowering, especially when many new elementary particles were discovered and the question of systematising them on the basis of symmetry groups became very important. When I started to work, there were no powerful group-theoretical methods. The first pieces of work that I consider to be somehow connected to what I am doing now were concerned with the properties of particles with high spin. I would like to emphasise that questions that were also related to supersymmetry interested me already at the earliest stage of my activity. How do particles with different spin differ one from another? Why are there bosons; why are there fermions? I remember, in part, that just when I learned about the group SU(3), it amazed me that the multiplets of this group contained simultaneously particles with integral and with half-integral isospin. I was already trying to do something in this direction, replacing isotopic spin with ordinary spin, that is, actually, the idea that now lies at the basis of supersymmetry. An important role in the conception of the idea of supersymmetry was played by work on phenomenological Lagrangians, in which the interactions of Goldstone particles in the presence of spontaneous symmetry breaking are practically uniquely described by geometrical group-theoretical methods. I actively participated in the development of this line; it fascinated me very much. At the same time, the ideology of gauge fields started to penetrate actively into physics and I returned to the old questions. It amazed me that all these particles, the Goldstone particles and the gauge fields, were bosons, but the fermions somehow were not involved at all. Here, a kind of inequality appeared: why were some particles - bosons - selected, but others - fermions - not included in this group? This was a key moment, because the very thought that fermions can also be Goldstone particles or gauge fields contained in itself somehow the answer: if it were clear how to build a general scheme of Goldstone particles with integral spin, then upon making the transition to fermions one should replace the corresponding operators with operators firstly carrying spin one-half and secondly being anticommuting, in accordance with their fermionic nature; and in fact I did this. After that, when this was done, I, together with my co-authors V. Akulov and V. Soroka, considered first the global properties of supersymmetric theories and their local properties.

I would like to say that I worked not only on the theory of symmetry groups, but I also had pieces of work on nuclear physics and even on the theory of accelerators, but, nevertheless, the theory of symmetry groups has always been my favourite subject. And certainly, my main work was connected with the application of this theory to the physics of elementary particles.

By main, I mean my work on the establishment of parastatistics, work on the discovery of the conspiracy of Regge poles, and on the application of symmetry groups to the classification of hadronic resonances, that is, to baryonic and mesonic resonances. But my most important result, which is the most widely-known one now in the world, is certainly the discovery of a new symmetry group - supersymmetry - and the extension of this to aspects of the general theory of relativity, the so-called supergravity. I can speak about this in some more detail. First, a couple of words about how I came to the discovery of supersymmetry and supergravity. The starting point was the idea of W. Heisenberg. Let me tell you what this idea was. At the end of the 60's, Heisenberg proposed the idea that all elementary particles can be described by a uniform theory on the basis of the nonlinear equation formulated by him. He started this work together with W. Pauli, but sometime later Pauli dissociated himself from this theory and even sharply criticised Heisenberg for his assumptions. But, nevertheless what did Heisenberg contribute to the discovery of supersymmetry? Heisenberg tried to explain the spectrum of all elementary particles on the basis of his theory. And thanks to his great intuition, he guessed the places of the particles and predicted how the photon should appear in this theory. Moreover, he found a place for the neutrino within his theory and assumed that the neutrino emerges as a result of spontaneous symmetry breaking. This was a very unusual assumption, because all known Goldstone particles emerging as a result of spontaneous symmetry breaking had spin zero in all theories known at that time. This idea of Heisenberg was revolutionary, because he was the first to formulate that the thought that there might exist Goldstone particles in nature with spin one-half. To tell the truth, he found this particle by an incorrect method, but nonetheless it was an idea that had a strong impact on me and from that time I often although not constantly would think about whether this idea could be realised. And when I started to consider how such a particle might appear in the theory, I understood that this requires an extension of the usual physical groups, which is the basis for all relativistic processes. That means an extension of the Lorentz group and an extension of the Poincare group so that new operators would be present, which would correspond to a quantum number of the neutrino. Thus, the main result that we obtained was that we managed to create such an extension of the Poincare group, which is now called the Poincare supergroup.

Later on, we encountered certain difficulties and, if one speaks about the direct application to the neutrino, this idea nonetheless did not work. Why? Because the group that we were considering contained the Poincare group and it is known that the Poincare group leads eventually to the general theory of relativity, if one considers the local transformations of this group. That means that the same local transformations of the supergroup would lead to the emergence of certain superpartners of the ordinary gravity field, and these superpartners would totally absorb the Goldstone particles. And thus, when I understood such ideas, I, together with my collaborators, proposed an extension even of Einstein's general theory of relativity, which would include just what we call supersymmetry. So, in fact, certain superpartners would emerge. Moreover, in the extension of this theory arose not only superpartners - particles with spin 3/2 - but also particles with spin 1, with spin 1/2 and with spin zero. That is, there arose the idea that all elementary particles might be included into the system of supergravity. Now, many scientists all over the world are working on it. Different variants of supergravity are being considered and also different variants of superstring theory, which are based on a certain extension of supergravity that takes into account the fact that elementary particles can be not simply point objects but also extended objects. This direction is now the main direction in theoretical physics and proposes that there is a unified theory of all elementary particles, based on one common principle - namely on local supersymmetry.

Heisenberg's idea was, if one can say so, a physical idea, but in order to give this idea shape in a new precise mathematical theory, a certain mathematical formalism was required. And here I am very thankful to J. Schwinger, whom I personally knew, and who in a number of cases could have discovered supersymmetry. Moreover, I have recently seen an article by Schwinger, in which he writes why he didn't discover supersymmetry in spite of the fact that he was completely ready for it.

Why did I think that Schwinger could have discovered supersymmetry? Firstly, he introduced the concept of certain anticommuting variables, which played the roles of physical variables in field theory. He was the first to do this. Secondly, when I was saying that in supergravity the superpartner of the gravity field appears, well, these superpartners are particles of spin 3/2, and the theory of such spin 3/2 particles was first developed by Schwinger. It is even called the Rarita-Schwinger theory.

Recently, I received this article by Schwinger in which he writes about the reasons why he did not discover supersymmetry. Schwinger also says that he was very close to the discovery of supersymmetry and explains why. At the same time, in response to a general question as to why he didn't do it, which he received from the audience in an auditorium, he answered that this is a philosophical question and that this philosophy of discoveries is discussed in a number of monographs and that he could give only a general answer to the question.

The next moment that played an immense role in the mathematical formulation of supersymmetry, this was the work of the greatest French mathematician Elie Cartan. He created a special formalism of differential geometry, which seems to be specially appropriate for the system that I was considering. This particular circumstance, that I knew this work very well, helped me to create the mathematical formalism of supersymmetry. So, I think that there were actually three sources that helped me to develop the theory of supersymmetry and the theory of supergravity: the idea of Heisenberg, the idea of Schwinger and the mathematical idea of the great mathematician Elie Cartan.

QUESTION: Dmitry Vasiljevitch, I am not a theoretician, and it is very difficult for me to assess your accomplishments. Could you tell me in popular terms how your work is recognised and about its place in theoretical science.

ANSWER: It's very easy to answer this question, mostly because there are many references to our work. The number of references is more than a thousand. This is a very large indicator for scientific works. Besides that, testimony to the fact that this work is recognised in the world is the fact that in 1994 I gave a talk at a very important conference where the results of scientific development for the past 50 years were summarised and where the speakers were authors who had made fundamental contributions to the development both of theoretical ideas and in the acquisition of new experimental data.

QUESTION: Where was this conference?

ANSWER: It was in Erice, in Italy. It was called the "International Conference on the History of Original Ideas and Basic Discoveries in Particle Physics." I was invited to speak on supergravity, and it was a great honour for me. And so I gave my talk at this conference.

QUESTION: Dmitry Vasiljevitch, who else was studying the same problems that you were working on at that time?

ANSWER: At that time, in fact there was one more work that was done at the same time as ours. This was the work by Yu. Gol'fand and Lichtman, who introduced formulations of the theory of supersymmetry, but for completely different reasons. They tried to explain on the basis of supersymmetry the breaking of parity that exists in nature. Unfortunately, Gol'fand passed away in 1994. There was no further work. If one talks about supergravity, our first work was done in 1973. The next work in this field appeared in the west only in 1976.

QUESTION: Dmitry Vasiljevitch, I know that you have worked at CERN for some time. Could you say something about your work at CERN, and its influence on your scientific activity and, if possible, about your own influence on the theoreticians at CERN?

ANSWER: CERN has a very close importance for me. I was at CERN five times: three times for conferences and two times for work - once for three months and one time for a month. Certainly, the work at CERN stimulated my scientific activity. CERN in general plays an important role in the development of world science, and this is not only my opinion. I have talked to Americans and to other scientists. They consider that CERN is the most ideal place both for theoreticians and for experimentalists. It is the biggest centre, uniting mostly European physicists but also scientists from America continually working there as visitors. I was lucky, for instance, to work together with M. Gell-Mann, a famous Nobel-Prize laureate, and had the possibility to discuss many idea with Eitiro Nambu and other no-less-famous physicists. Therefore, of course, the fact that I was at CERN and had the chance to be in contact with outstanding scientists from Europe and America was very important for me.

QUESTION: Dmitry Vasiljevitch, could you please tell us about your scientific contacts with your colleagues from the countries of the CIS and with the scientists of KhFTI?

ANSWER: At KhFTI, when I was starting my work, there was a group studying quantum electrodynamics. Later on, I worked practically alone. But I had very good contacts with physicists from Moscow and Leningrad. I would like in particular to stress the role of Isaac Yakovlevich Pomeranchuk, who was working at that time in Moscow.

QUESTION: Did you know him personally?

ANSWER: I knew him personally. Every time I came to Moscow, I always visited Pomeranchuk. We had lengthy discussions together and it was interesting that we had the same way of thinking, because we usually discussed the conceptual part of work without going into formulas. We didn't even write formulas, but clarified what connections are possible between phenomena. I would like to tell about the following episode as an interesting example. It was in 1962 and I had been sent as a member of a delegation from the Soviet Union to CERN. I was lucky to share a hotel room with Isaac Yakovlevich Pomeranchuk. This was a stressful time for me, because just then I had finished some work together with V.N. Gribov about the conspiracy of Regge poles, which I spoke about earlier. This was the most important subject for me at that time. So, I remember lying on a hotel bed and that I couldn't fall asleep. It was already two o'clock in the morning and a certain idea came to me. I shook Isaac Yakovlevich awake, he awoke and said "What!" and I told him about this idea. After that, we discussed this idea for a couple of hours without turning on the light and we fell asleep only at dawn. This certainly characterises a person, if he can discuss a topic that interests him at any time of day or night. Such a person was Isaac Yakovlevich Pomeranchuk. I was also very impressed by meetings with D.I. Blokhintsev. We often spoke about science too, but he always amazed me mostly by the firmness of his conviction, which had a very high philosophical level. He was a man of multiple interests: he studied art and many questions of philosophy. He also had his own views on different fields of science and culture. His character seemed very integral to me. I learned many things about him only after his death from conversations with his students, and I regretted very much that I did not profit more from the possibilities of having contact with him.

Now, about Kharkov. At first I worked in relative solitude. Then I managed to form a small group of young capable scientists. The personnel of my laboratory consist of about 8 people, and I continue working with them.

QUESTION: Dmitry Vasiljevitch, could you tell us please what you and your laboratory are working on now?

ANSWER: Now, I am working on a problem that I consider very interesting. Up until recent times, if we take, for example, the textbooks by L.D. Landau, it has been stated that the requirements of relativistic invariance lead to the fact that elementary particles can only be point objects. Now, a new theory has been created which shows that elementary particles can also be extended. They can look like strings or membranes. At the present time, this line is developing very actively. It is interesting, for example, that when speaking about extended objects it has been proved that such extended objects can exist and the theory can be consistently formulated only in the case when supersymmetry is required. Without this requirement, these theories become senseless. Therefore I and my collaborators continue working in this area. I consider that we have found some quite interesting results, which I will include in a talk at an upcoming international conference.

QUESTION: Dmitry Vasiljevitch, I would like to remind you that the record of our conversation will be kept in the American Institute of Physics' Niels Bohr Library for the history of physics. I think that it will be interesting to historians of physics to learn about the contribution of our institute to the development of science, and in particular the contribution of our theoreticians. And of course, my conversation with you will useful to them. I thank you for agreeing to this interview.

ANSWER: I thank you too, for the interesting questions that you asked and which I have tried to answer as best I could.

Translated by K. Stelle

Part I

Supermembranes, Supersymmetry and Quantum Field Models

p-Brane Chemistry

K.S. Stelle

The Isaac Newton Institute for Mathematical Sciences, University of Cambridge, Cambridge CB3 0EH, UK and The Blackett Laboratory, Imperial College, Prince Consort Road, London SW7 2BZ, UK **

Abstract. A brief account is given of the spectrum of extended-object solutions to supergravity theories that saturate Bogomol'ny-Prasad-Sommerfield bounds, and which therefore are candidates for exact states in the underlying string theory. These are organized into a coherent scheme by a pattern of vertical and diagonal dimensional reduction trajectories on a plot of worldvolume versus spacetime dimension. These intersecting reduction trajectories are related by duality symmetries at the intersection points, and these duality symmetries govern also the multiplicities of the p-branes at each (worldvolume, spacetime) dimension.

A Reminiscence of Dimitry Vassiljevich Volkov

Dimitry Vassiljevich Volkov made fundamental contributions to many subjects in theoretical physics, including in particular the early development of supersymmetry and also the development of the theory of non-linear realizations of space-time symmetries. Currently, these subjects are becoming linked in a way that makes particularly clear the importance of his pioneering work, with the discovery that states characterized by a partial breaking of supersymmetry may provide an essential clue to the non-perturbative structure of superstring theory. It is a remarkable tribute to his insight that he chose to focus on these specific subjects, which are now potentially the key to a major unification of theoretical physics. It is thus very fitting to remember him in this volume of papers devoted to the elaboration of ideas that originally flowed from his original insights.

The p-Brane Spectrum of Supergravity

One of the fascinations of superstring theory is the way in which it repeatedly seems to give us glimpses of its inner secrets by leaving hints in places we have already come to find familiar. Thus, much of the current hope for discovering the non-perturbative structure of string theory is based in observations of the patterns of solutions to supergravity theories that are about twenty years old now. Among these solutions, we have a class that achieves a partial breaking of supersymmetry, *i.e.* the so-called Bogomol'ny-Prasad-Sommerfield, or

** Research supported in part by the Commissionof the European Communities under contract ERBFMRX-CT96-0045.

BPS solutions, which saturate bounds on their mass densities that are determined by the charges they carry. These charges, known generically as Page charges following their original discussion in the context of $D = 11$ supergravity,(Page (1983)) are of importance primarily because they occur in the relevant supersymmetry algebras. For example, in the $D = 11$ theory one has

$$\{Q,Q\} = C(\Gamma^A P_A + \Gamma^{AB} U_{AB} + \Gamma^{ABCDE} V_{ABCDE}) \ , \qquad (1)$$

where the supercharge $Q = \int_{\partial\Sigma} \Gamma^{ABC}\Psi_C dS_{AB}$ is given as usual by a Gauss' law integral over the boundary of a spatial hypersurface Σ, while the "electric" and "magnetic" Page charges U and V are given as integrals over the boundaries of spatial *sub*manifolds $\mathcal{M}_8$ and $\tilde{\mathcal{M}}_5$ respectively. Given the various orientations that these can take in their embeddings into the spatial hypersurface Σ, the resulting charges may be labeled with two-form and five-form indices as in (1); for particular choices of $\mathcal{M}_8$ and $\tilde{\mathcal{M}}_5$, they are given by

$$U = \int_{\partial\mathcal{M}_8} ({}^*F_{[4]} + \tfrac{1}{2}A_{[3]} \wedge F_{[4]}) \ , \qquad V = \int_{\tilde{\mathcal{M}}_5} F_{[4]} \ . \qquad (2)$$

As with the more familiar $D = 4$ electric and magnetic charges, the U and V charges differ in that conservation of the U charges is guaranteed by the equations of motion for the $A_{[3]}$ gauge field, while the V charges are conserved by virtue of the Bianchi identity for the corresponding field strength $F_{[4]}$.

The indices carried by the Page charges show that they are not "central" in the mathematical sense, as is the case for the permitted extensions to the supersymmetry algebra in $D = 4$, but they nonetheless play a similar rôle. Upon dimensional reduction to $D = 4$, these indices become "internal," and then the corresponding charges are indeed central, *i.e.* they become Lorentz scalars. For the reduction to $D = 4$, these internal indices range over 7 values, so the reductions of the U and V charges each give rise to $7 \cdot 6/2 = 21$ charges in $D = 4$. These are then accompanied by an additional 7 electric and magnetic charges associated to the Kaluza-Klein vectors that arise from the metric during the dimensional reduction procedure, giving in the end $28 + 28$ electric + magnetic charges in the reduced $D = 4$, $N = 8$ supergravity theory. These together form a **56** under the $N = 8$ theory's linearly realized automorphism symmetry SU(8). One of the dramatic surprises of this maximal $D = 4$ supergravity was the appearance of a hidden nonlinear $E_{(7,+7)}$ symmetry,(Cremmer, Julia (1979)) linearly realized on $SU(8)/Z_2$. While the scalar fields of the theory transform nonlinearly in the corresponding $133 - 63 = 70$ dimensional coset space, the 56 electric and magnetic charges in fact transform linearly under E_7.

Now consider the implications of the above structure for extended-object solutions in supergravity theories, *i.e.* p-branes. The most elementary class of these (those carrying a single Page charge) may be derived from a simplified action obtained from that for a full supergravity theory by a consistent truncation, that is, one in which the solutions of the reduced theory are automatically solutions of the unreduced theory as well. Here, it is sufficient to consider the D-dimensional action

$$I = \int \mathrm{d}^D x \sqrt{-g}(R - \tfrac{1}{2}(\partial\phi)^2 - \tfrac{1}{2n!}\mathrm{e}^{a\phi}F_{[n]}^2) \quad , \tag{3}$$

where the coefficient a governs the interaction of the dilatonic scalar field ϕ with the n-form field strength $F_{[n]}$. One considers a general setting such as this, with the coefficient a and the rank of the field strength left undetermined, because the dimensional reductions of supergravity theories give rise to a large number of such cases with differing (a, n). For the subsystem (3), two types of extended-object solution arise. Firstly, by a direct, "electric" coupling of the gauge potential $A_{[n-1]}$ to the worldsheet of an extended object, analogous to the coupling of a Maxwell potential to a particle's worldline, one expects to find the occurrence of a p-brane solution with $p = n - 2$. This solution in fact exists, and generalizes the extreme Reissner-Nordstrom black hole solution of Einstein-Maxwell theory. It also carries an electric-type charge that one may recognize as one of the electric Page charges U (or as one of the supplementary Kaluza-Klein electric charges) in the original supergravity theory. The second type of extended-object solution displays a magnetic charge. Accordingly, one may anticipate its spatial dimension by replacing the rank n of $F_{[n]}$ by that of its Hodge dual in D dimensions, $D - n$, thus finding a spatial dimensionality $\tilde{p} = D - p - 4$ for the expected magnetic solution; the corresponding worldvolume dimensions $d = p + 1$ are related by $\tilde{d} = D - d - 2$.

The electric and magnetic solutions are all characterized by a Poincaré$_d$ $\times$ SO$(D - d)$ set of isometries; they may be written together as (Lü, Pope, Sezgin, Stelle (1996))

$$\begin{aligned} \mathrm{d}s^2 &= H^{\frac{-4\tilde{d}}{\Delta(D-2)}}\mathrm{d}x^\mu \mathrm{d}x^\nu \eta_{\mu\nu} + H^{\frac{4d}{\Delta(D-2)}}\mathrm{d}y^m \mathrm{d}y^m \\ \mathrm{e}^\phi &= H^{\frac{2a}{\epsilon\Delta}} \qquad H = 1 + \frac{k}{r^{\tilde{d}}} \qquad r = \sqrt{y^m y^m} \end{aligned} \tag{4}$$

where $\epsilon = \pm 1$ for the (electric, magnetic) cases, and k is the integration constant that determines the mass density of the solution. The x^μ coordinates $(\mu = 0, 1, \ldots d - 1)$ correspond to the worldsheet of the p-brane, and the y^m coordinates correspond to the space transverse to the worldsheet. The function H occurring in (4) is a harmonic function in the transverse y^m space, arising as the solution to a Laplace equation that emerges from the coupled gravity-scalar-antisymmetric tensor equations upon insertion of the simplifying assumptions of the p-brane ansatz. The antisymmetric tensor field strength $F_{[n]}$ for the p-brane solutions takes two different forms, related by duality, in the electric and magnetic cases. In the electric case, one has

$$F_{m\mu_1\ldots\mu_{n-1}} = \epsilon_{\mu_1\ldots\mu_{n-1}}\partial_m e^{C(r)}$$
$$e^{C(r)} = \frac{2}{\sqrt{\Delta}}H^{-1} \quad . \tag{5}$$

In the magnetic case, one has

$$F_{m_1\ldots m_n} = \lambda\epsilon_{m_1\ldots m_n s}\frac{y^s}{r^{n+1}}, \tag{6}$$

where the charge constant λ is given by $\lambda = -2k\tilde{d}/\sqrt{\Delta}$ and the power of r in the denominator is determined by the requirement that the Bianchi identity be satisfied. The quantity Δ appearing in (4, 5) is given in terms of the parameter a appearing in (3) by

$$a^2 = \Delta - \frac{2d\tilde{d}}{(D-2)} \quad . \tag{7}$$

Unlike the parameter a, Δ has the property of being preserved under dimensional reduction (Lü, Pope, Sezgin, Stelle (1996)) in all couplings of dilatonic scalars to field strengths $F_{[n']}$ arising in the dimensional reduction of (3). Thus, the corresponding electric and magnetic p-branes solutions to the dimensional reductions of (3) will all be characterized by the same value of Δ, even though the corresponding a values differ. Δ is thus something like a "principal quantum number," governing multiplets of p-branes in supergravity theories.

The p-brane solutions are tailor-made for interpretation in a Kaluza-Klein context, owing to their translational symmetries in the worldvolume directions. By taking the Kaluza-Klein reduction coordinate z to coincide with one of the x^μ coordinates, on which the solution does not depend, one naturally can reinterpret a p-brane in D spacetime dimensions as a $(p-1)$-brane in $(D-1)$ dimensions. This process thus may be considered to be a "diagonal" dimensional reduction of p-branes.(Lü, Pope, Stelle (1996a))

Another type of dimensional reduction is also available to relate the p-brane solutions, relying on more specific properties of this class of solutions. All the static p-brane solutions involve finding a solution H to a Laplace equation derived from the ϕ equation. The single-centered p-brane solutions may be generalized to multi-center solutions, by simply generalizing H to a multi-center solution of the Laplace equation, $\partial_m\partial_m H = 0$. Physically, the existence of static multi-center solutions may be interpreted as being due to a cancellation of attractive metric and scalar-field forces against repulsive antisymmetric-tensor forces. Taking the limit of a densely-packed "stack" of p-branes, in which case the sum tends to an integral, one obtains a translationally-invariant solution in a transverse direction; this may then be reconsidered as an isotropic p-brane solution in one fewer spacetime dimension. Because the worldvolume dimension is not changed in this process, it may be considered to be a "vertical" dimensional reduction of the p-branes (Lü, Pope, Stelle (1996a)). Both the diagonal and the vertical dimensional reductions preserve the value of Δ associated to a solution and also the amount of unbroken supersymmetry.

The overall pattern of p-branes is shown in the Figure, which plots worldvolume versus spacetime dimensions, showing the diagonal and vertical reduction trajectories. At the top of a given trajectory one finds a p-brane solution that cannot be "oxidized" into a p-brane solution in a higher dimension, either for lack of an appropriate supergravity theory for the oxidized solution to belong to, or because the higher-dimensional solution does not possess the full set of Poincaré$_d$ × SO$(D-d)$ isometries of a static p-brane. Such solutions may be called "stainless;" they constitute the originating solutions of a reduction trajectory, characterized by a particular value of Δ.

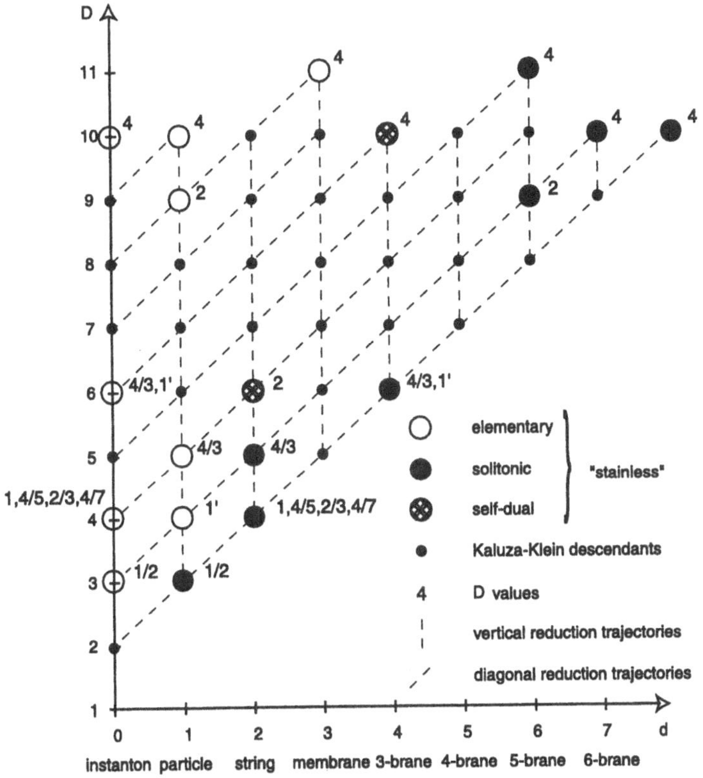

Fig.1 Brane-scan of supergravity p-brane

The spectral pattern of p-branes lying below the $p = D - 3$ trajectory is rather different from the pattern above this trajectory,(Cowdall, Lü, Pope, Stelle, Townsend (1996)) and is not appropriate to show them together on the same plot. Attempting to make a vertical dimensional reduction from $(D-3)$-branes to $(D-2)$-branes, one runs into the apparent difficulty that the $(D-3)$ brane solutions are not really asymptotically flat, but only asymptotically locally flat, owing to the existence of a deficit angle at infinity, proportional to the mass density of the $(D-3)$ brane. The attempt to create a "domain wall" solution by stacking up such asymptotically locally flat solutions within a standard supergravity theory fails because the deficit angle grows beyond 2π. Nonetheless, one may still press on blindly with the vertical reduction procedure, and obtain a solution after the stacking-up procedure with a zero-form gauge potential (*i.e.* a scalar) that contains a term with linear dependence on the compactification coordinate z. Normally, retaining any dependence on the compactification coordinate would seem to prevent making a consistent Kaluza-Klein reduction, so this would seem to be the Kaluza-Klein analogue of the global difficulties with the asymptotic space.

In certain circumstances, however, such a reduction does make sense: when the zero-form potential in question appears in the field equations only in differentiated form. In that case, it is in terms of the 1-form field strength that the reduction is effectively done, and this field strength does not itself have any z dependence. What one obtains from this generalization of the Kaluza-Klein procedure is in fact an instance of Scherk-Schwarz dimensional reduction,(Scherk, Schwarz (1979)) except that in this case supersymmetry is not broken by the procedure. The Scherk-Schwarz reduction in this case does, however, spontaneously break a number of the bosonic symmetries of the theory, giving rise to Higgs effects and thus to mass terms for various antisymmetric tensor fields. A distinct Scherk-Schwarz reduction of this type is available for each of the "axion" fields of the theory, *i.e.* for each of the scalar fields that can be arranged to be covered everywhere by derivatives. The number of such axionic scalars turns out to be equal to the dimension of the supergravity isotropy group H in all cases. Each of the theories obtained by Scherk-Schwarz reduction is distinct, with differing patterns of the masses obtained and with differing residual symmetries.(Cowdall, Lü, Pope, Stelle, Townsend (1996)) Each such reduced theory does, however, have a domain-wall solution of the general form (4), but now with the field strength considered to be a zero-form, equal in value to the mass parameter m of the Scherk-Schwarz procedure. In contrast to the "one theory – many solutions" character of the $(p \le D-3)$ branes, these domain walls may instead be characterized as "one solution – many theories." The consequences of this structure for the relations between supergravity theories and string theories remain to be worked out.

Finally, we come to the question of how to count the multiplicities of a set of p-branes of the same dimension, characterized by a given value of Δ. It now becomes relevant to consider also the requirements of the charge quantization condition for p-branes (Nepomechie (1985), Teitelboim (1986)), which has the effect of restricting the allowed electric and magnetic charges to lie on a "charge lattice," and also of restricting (Hull, Townsend (1995)) the supergravity symmetry group G (*i.e.* E_7 in the $D=4$, $N=8$ case) to a discrete subgroup $G(\mathbb{Z})$. The classification of solutions is made using the linearly realized subgroup H of G (*i.e.* SU(8) in the $D = 4$, $N = 8$ case), which is also the isotropy group for the scalar moduli of the theory, *i.e.* of the asymptotic values of the scalar fields at infinity. Different sets of scalar moduli give rise to different embeddings of H in G that are related by conjugation, $H \to gHg^{-1}$, where $g \in G$ is the group element that moves the moduli. The embedding of $G(\mathbb{Z})$, which is determined by the charge quantization condition, also moves by conjugation, but in the opposite way, *i.e.* $G(\mathbb{Z}) \to g^{-1}G(\mathbb{Z})g$. The quantum classification symmetry for p-branes is given by $G(\mathbb{Z}) \cap H$ for a given set of moduli; owing to the different behaviors under change of scalar moduli, this symmetry is sensitive to the values of these moduli, and so special points in the scalar modulus space appear at the quantum level. This symmetry $G(\mathbb{Z}) \cap H$ acts on solutions with the same value of mass per unit p-volume, and is the analogue of the classification symmetry that commutes with the Hamiltonian in ordinary gauge theories (*i.e.* the group of rigid gauge transformations). At the special point in modulus space where all the scalar asymptotic values vanish, this group is maximal, and becomes the Weyl group (Lü, Pope, Stelle (1996), Lü, Pope, Stelle (1996a)) for the supergravity symmetry group G. This identification arises because the couplings of the theories' dilatonic scalars are controlled by "dilaton vectors" generalizing the single parameter a appearing above in (3), and these dilaton vectors may be associated to the fundamental weights of the symmetry group G. The maximal $G(\mathbb{Z}) \cap H$ symmetry preserves the dot products of these dilaton vectors, just as the Weyl group preserves the dot products of fundamental weight vectors.

References

Cowdall, P.M., Lü, H., Pope, C.N., Stelle, K.S. and Townsend, P.K. (1996) "Domain Walls in Massive Supergravities," preprint `hep-th/9608173`.

Cremmer, E. and Julia, B. (1979): m Nucl. Phys. B **159**, 141 (.)

Hull, C.M. and Townsend, P.K. (1995): Nucl. Phys. B **438**, 109 (.)

Lü, H., Pope, C.N., Sezgin, E. and Stelle, K.S. (1996): Nucl. Phys. B **456**, 669 (.)

Lü, H., Pope, C.N. and Stelle, K.S. (1996): "Weyl Group Invariance and p-brane Multiplets," Nucl. Phys. B **476**, 89 (.)

Lü, H., Pope, C.N. and Stelle, K.S. (1996a): "Vertical Versus Diagonal Dimensional Reduction for p-branes," Nucl. Phys. B **481**, 313 (.)

Nepomechie, R. (1985): Phys. Rev. D **31**, 1921 (;)
Teitelboim, C. (1986): Phys. Lett. B **67**, 63, 69 (.)

Page, D.N. (1983): Phys. Rev. D **28**, 2976 (.)

Scherk, J. and Schwarz, J.H. (1979): Phys. Lett. B **82**, 60 (.)

Intersecting Branes and Supersymmetry

M. de Roo

Institute for Theoretical Physics, Nijenborgh 4, 9747 AG Groningen, The Netherlands

Abstract. We consider intersecting M-brane solutions of supergravity in eleven dimensions. Supersymmetry turns out to be a powerful tool in obtaining such solutions and their generalizations.

1 Introduction

The revival of the concept of strong-weak coupling duality has drastically changed our view of string theories. The five apparently different ten-dimensional superstring theories are now interpreted as different limits of a single theory, the conjectured M-theory. The study of extended objects, which by duality must manifest themselves in each of the descendents of M-theory, has been a decisive factor in establishing this picture of a united string theory[1].

Of particular interest are those extended objects (p-branes, where p is the dimension of the spatial extension) which satisfy a BPS-bound and preserve partial supersymmetry. Such objects can satisfy a "no-force" condition, implying that static configurations of several such objects can exist due to a cancellation of the gravitational and gauge forces between them. Several authors have contributed to the rather complete picture that now exists of these intersecting p-brane configurations (Papadopoulos and Townsend (1996a), Tseytlin (1996a), Behrndt et al. (1996), Gauntlett et al. (1996), Tseytlin (1996b), Costa (1996), Bergshoeff et al. (1996b)). Here I would like to report on the work done in (Bergshoeff et al. (1996b)), where a classification of multiple intersections in $D = 10$ and $D = 11$ was obtained. I will limit myself to our results in eleven dimensions, and, in the spirit of this meeting, I would like to discuss in particular how supersymmetry can be helpful in obtaining intersections of M-branes. In particular, we will find that supersymmetry is a useful guide in constructing the intersections of the $M2$- and $M5$-brane, and it shows that these should be extended to include objects with 1, 6, and 9 spatial extensions.

2 Pair Intersections

The basic solutions in $D = 11$ are the $M2$-brane (Duff and Stelle (1991)):

[1] For a recent review of these developments, see, e.g., (Townsend (1996))

$$\mathrm{d}s^2 = H^{-2/3}\, \mathrm{d}x^2_{(0-2)} - H_2^{1/3}\, \mathrm{d}x^2_{(3-10)}, \qquad F_{012i} = \partial_i H^{-1} \ , \tag{1}$$

where H is harmonic on the eight-dimensional space transverse to the membrane, and the $M5$-brane solution[2]: (Güven (1992)):

$$\mathrm{d}s^2 = H^{-1/3}\, \mathrm{d}x^2_{(0-5)} - H^{2/3}\, \mathrm{d}x^2_{(6-10)}, \qquad F_{012345i} = \partial_i H^{-1} \ . \tag{2}$$

In this case H is harmonic on the five-dimensional transverse space.

For our purposes it is useful to represent the metric for these solutions pictorially as

$$\mathrm{d}s^2 = \underbrace{\times\ \times\ \ldots\ \times}_{p+1}\ \overbrace{-\ -\ \ldots\ -}^{10-p}\ , \tag{3}$$

where $\times$ indicates a worldvolume coordinate of, $-$ a direction transverse to the p-brane. In this notation, the basic intersections (Papadopoulos and Townsend (1996a), Tseytlin (1996a), Gauntlett et al. (1996)) of the $M2$- and $M5$-brane can be represented by[3]

$$(0|M2, M2) = \left\{ \begin{array}{c|cccccccccc} \times & \times & \times & - & - & - & - & - & - & - & - \\ \times & - & - & \times & \times & - & - & - & - & - & - \end{array} \right. , \tag{4}$$

$$(1|M2, M5) = \left\{ \begin{array}{c|cccccccccc} \times & \times & \times & - & - & - & - & - & - & - & - \\ \times & \times & - & \times & \times & \times & \times & - & - & - & - \end{array} \right. , \tag{5}$$

$$(3|M5, M5) = \left\{ \begin{array}{c|cccccccccc} \times & \times & \times & \times & \times & \times & - & - & - & - & - \\ \times & \times & \times & \times & - & - & \times & \times & - & - & - \end{array} \right. , \tag{6}$$

$$(1|M5, M5) = \left\{ \begin{array}{c|cccccccccc} \times & \times & \times & \times & \times & \times & - & - & - & - & - \\ \times & \times & - & - & - & - & \times & \times & \times & \times & - \end{array} \right. . \tag{7}$$

Each intersection is determined by two harmonic functions, H_1 and H_2. We distinguish between overall worldvolume directions (both rows have an $\times$, the harmonic functions are in all cases independent of these directions), relative transverse directions (only one row has an $\times$), and overall transverse directions (both rows have a $-$). In (4-6) either both H_i must depend on the overall transverse directions, or one H must depend on overall transverse, the other on relative transverse directions. In (7) the dependence of the H_i must be on the relative transverse directions only.

The metric for these basic pairs is easily constructed. In general, in the intersection of type $(q|q+r, q+s)$ the form of the metric is

$$\begin{aligned} \mathrm{d}s^2 = H_1^{\alpha_1} H_2^{\alpha_2} \big\{ &\mathrm{d}x^2_{(0-q)} - H_1 \mathrm{d}x^2_{(q+1,q+s)} \\ &H_2 \mathrm{d}x^2_{(q+s+1,q+s+r)} - H_1 H_2 \mathrm{d}x^2_{(q+r+s+1,10)} \big\} \ . \end{aligned} \tag{8}$$

[2] Supergravity in $D = 11$ is formulated in terms of a three-form gauge field. For the solutions considered here the contribution of the Chern-Simons term to the equations of motion, which depends on the three-form gauge field, does not contribute. In that case it is possible to represent the fivebrane in terms of a six-form gauge field, the field strength $F_{012345i}$ being the dual of F_{jklm}.

[3] We denote the intersection of a p_1- and a p_2-brane over a common $q+1$ dimensional spacetime by $(q|p_1, p_2)$.

Here α is $-2/3$ for $M2$, $-1/3$ for $M5$. The curvature tensors F for the basic pairs correspond to the sum of the curvatures of the separate branes, except for (7), where a slight modification is required (Gauntlett et al. (1996), Behrndt et al. (1996)).

The basic rule in constructing intersections of $N > 2$ fundamental objects is, that each pair among the N objects must be one of the above pairs. This leads to configurations with a maximum of nine branes (Bergshoeff et al. (1996b)). In the next section, we will discuss the role of supersymmetry in obtaining multiple intersections.

3 Supersymmetry

The BPS $M2$-and $M5$-brane each preserves 1/2 of the $D = 11$ supersymmetry. The supersymmetry transformation of the gravitino reads:

$$\delta\psi_\mu = \partial_\mu\epsilon - \tfrac{1}{4}\omega_\mu{}^{ab}\epsilon - \tfrac{i}{576}\left(\Gamma_\mu\Gamma^{abcd} - 3\,\Gamma^{abcd}\Gamma_\mu\right)\epsilon\, F_{abcd}\ . \tag{9}$$

Supersymmetry is partially preserved, if the configuration is such that $\delta\psi_\mu$ vanishes for some ϵ. For $M2$ and $M5$ a simple calculation leads to the following conditions:

$$M2\ :\quad \epsilon = H^{-1/6}\eta,\ \eta \text{ constant with } P_2\eta = \eta,\ \text{ where } P_2 = i\Gamma^{012}\ , \tag{10}$$
$$M5\ :\quad \epsilon = H^{-1/12}\eta,\ \eta \text{ constant with } P_5\eta = \eta,\ \text{ where } P_5 = \Gamma^{012345}. \tag{11}$$

So η is algebraically restricted by a product of Γ-matrices corresponding to the worldvolume directions.

Given the supersymmetry preserving conditions (10, 11), the obvious question is how to formulate the preservation of supersymmetry for pairs of M-branes. If η must satisfy two conditions, then compatibility requires that the corresponing P_p must *commute*. For a pair consisting of a p_1 and a p_2 brane, intersecting over a common worldvolume of dimension $d_{12} + 1$, one can derive the following rule:

– If both p_1 and p_2 are even, d_{12} must be even, otherwise d_{12} must be odd.

Such a pair will preserve 1/4 of the $D = 11$ supersymmetry. For M_2 and M_5 this condition leads precisely to the four possibilities given in (4-7).

Once intersections of three or more fundamental branes have been obtained, there is a simple method to add additional branes which do not lead to further supersymmetry breaking. Consider a triple p_1, p_2 and p_3 satisfying the above conditions, i.e., such that the P_{p_i} commute. Then the product $P_{p_4} \equiv P_{p_1}P_{p_2}P_{p_3}$ clearly commutes with each P_i, and a brane with spatial extension p_4 can be added to the configuration. Note that this calculation also determines the orientation of the p_4-brane.

For any allowed triple of $M2$ and $M5$, one finds that p_4, calculated as above, is always one of the numbers $1, 2, 5, 6, 9$, i.e., p_4 is of the form $4k + 1$

or $4k+2$. More precisely, we find the following: Let p_1, p_2 and p_3 form an intersecting triple with $1/8$ supersymmetry, then

- If either one or three p_i are of the form $4k+1$, then so is p_4, otherwise p_4 is of the form $4k+2$.

It now becomes interesting to extend the intersecting pairs of Sect. 2 to the case of M-branes with spatial dimensions $1, 2, 5, 6, 9$. As we have seen above, the allowed pairs are determined by supersymmetry. The result is given in the Table 1. In this table we have left out intersections of the form $(p|p, p)$, where the two intersecting branes overlap completely. These are still expressed in terms of a single harmonic function and preserve $1/2$ of supersymmetry. In the table the numbers d_{12}, p_1 and p_2 are therefore restricted by $d_{12} < \max(p_1, p_2)$. The fact that the configuration must fit in ten spatial dimensions implies $p_1 + p_2 - d_{12} \leq 10$.

Table 1. Basic pair intersections $(d_{12}|p_1, p_2)$ **in** $D = 11$**.** The table indicates the possible values of d_{12} for each pair p_1 and p_2. The 2- and 5-branes are discussed in Sect. 2, the nature of 1-, 6- and 9-branes in Sect. 4.

p_i	1	2	5	6	9
1	–	1	1	1	1
2	1	0	1	0, 2	1
5	1	1	1, 3	1, 3	5
6	1	0, 2	1, 3	2, 4	5
9	1	1	5	5	–

We have seen that supersymetry determines the pair intersections, and is helpful in obtaining, for a given configuration, an additional brane which does not lead to further supersymmetry breaking. For the last point we used triple configurations with $1/8$ supersymmetry. A further use of supersymmetry arises for the pair intersections themselves. Consider a pair $(d_{12}|p_1, p_2)$. By taking the product of P_{p_1} and P_{p_2} we obtain a matrix $\Gamma^{(p_1+p_2-2d_{12})}$, where (p) stands for a set of p spatial indices. The indices correspond to the relative transverse coordinates of the pair. This matrix does not involve Γ^0, so the worldvolume is spacelike and cannot be used to define an additional brane. But in $D = 11$ the matrix $i\Gamma^{012\ldots10} = 1$. Therefore $\Gamma^{(p_1+p_2-2d_{12})} = i\Gamma^{0(10-p_1-p_2+2d_{12})}$, which does define a suitable worldvolume. Note that if p_1 and p_2 are both of the form $4k+1$ or $4k+2$, then so is $10 - p_1 - p_2 + 2d_{12}$. In this way we can obtain configurations of three branes with $1/4$ supersymmetry, which have no overall transverse directions. However, one has to be careful with the way the harmonic functions are allowed to depend on the coordinates. Following the rules for intersecting pairs, one finds that only in a few cases a nontrivial solution arises. There is only one example involving only $M2$ and $M5$. This arises from the pair $(1|5, 5)$ (see (7)), to which we can add an $M2$, such that the triplet has a common string direction (see also Tseytlin (1997), Gauntlett et al. (1997)).

4 The 1-, 6- and 9-Brane

In Table 1 we find the pairs (4-7) as a subset. Now we must discuss the nature of the branes of extension 1, 6 and 9. For the first two cases we have obvious candidates. The $M1$-brane can be interpreted as the Brinkmann wave in $D = 11$:

$$ds^2 = (2 - H)dt^2 - Hdz^2 + 2(1 - H)\,dtdz - (dx_2^2 + ... + dx_{10}^2) \ , \tag{12}$$

where H is a harmonic function in the variables $t + z, x_2, \ldots, x_{10}$. Its interpretation as an $M1$-brane makes sense, since it indeed preserves $1/2$ supersymmetry, and its direct dimensional reduction to $D = 10$ gives the fundamental string solution. The double dimensional reduction gives the $D0$-brane in $D = 10$.

Also the $M6$-brane allows a natural interpretation. It must be the Kaluza-Klein monopole (Sorokin (1983), Gross and Perry (1983)), with metric ($i = 1, 2, 3$)

$$ds^2 = dt^2 - dx_1^2 - ... - dx_6^2 - H^{-1}(dz + A_i dy_i)^2 - Hdy_i^2 \ , \tag{13}$$

where H and A_i depend on y_i, and the relation between H and A_i is

$$F_{ij} \equiv \partial_i A_j - \partial_j A_i = \epsilon_{ijk}\partial_k H \ . \tag{14}$$

Direct dimensional reduction to $D = 10$ gives a $D6$-brane, double dimensional reduction the solitonic fivebrane in $D = 10$. Recently we have extended our results on $M2$- and $M5$-branes (Bergshoeff et al. (1996b)) to include also the wave (12) and the monopole (13) (Bergshoeff et al. (in prep.)). Interestingly, the intersections of pairs of waves and monopoles with $M2$ and $M5$, and with themselves, are precisely as given in Table 1. This, and the results on multiple intersections (Bergshoeff et al. (1996b)), gives us some confidence that supersymmetry may indeed be used to predict the allowed configurations of intersecting branes. According to this point of view, the construction of a multiple intersections involving N basic objects is the same as the construction of N commuting matrices $\Gamma^{0(p_i)}$, $i = 1, \ldots N$, where (p_i) denotes the spatial orientation of the worldvolume of the p_i-brane.

There is no known 9-brane solution of $D = 11$ supergravity. Nevertheless, the above results indicate that we should seriously consider the existence of such an object[4]. There are also other indications that a 9-brane should exist. In $D = 10$ there is an $D8$-brane solution (Polchinski and Witten (1995, 1996), Bergshoeff et al. (1996a)), and, according to the M-theory interpretation of string theories, it should have an eleven-dimensional counterpart. However, the $D8$-brane requires the massive extension of $D = 10$ IIA supergravity (Romans (1986)), which we do not know how to lift to $D = 11$.

[4] The $D = 11$ 9-brane has been discussed before. See remarks in 9Bergshoeff et al. (1996a), Howe, Sezgin (1996), Papadopoulos and Townsend (1996b), Polchinski (1996), Duff (1996)).

Our analysis does not tell us what the conjectured 9-brane solution is. But, assuming that it preserves 1/2 supersymmetry, and that the condition of preservation of supersymmetry is of the standard form, its pair intersections with the known solutions of $D = 11$ supergravity are determined (see Table 1). For instance, this analysis tells us that the 9-brane can occur in configurations of n $M5$-branes for $n \leq 7$. Such configurations would reduce in $D = 10$ to an intersection of n $D4$-branes with the $D8$-brane, which is known to be a solution of massive $D = 10$ IIA supergravity.

Acknowledgements

It is a pleasure to thank the Organising Comittee of this meeting for their invitation, and the participants from the Ukraine and abroad for the pleasant atmosphere during this meeting. The work described here was done in collaboration with Eric Bergshoeff, Eduardo Eyras, Bert Janssen and Jan Pieter van der Schaar. This work is also supported by the European Commission TMR programme ERBFMRX-CT96-0045, in which I am associated to the University of Utrecht.

References

Behrndt, K., Bergshoeff, E., Janssen, B. (1996): *Intersecting D-branes in ten and six dimensions*, hep-th/9604168, to appear in Phys. Rev. D.

Bergshoeff, E., de Roo, M., Green, M.B., Papadopoulos, G. and Townsend, P.K. (1996a): Nucl. Phys. **B470**, 113, hep-th/9601150

Bergshoeff, E., de Roo, M., Eyras, E., Janssen, B. and van der Schaar, J.P. (1996b): *Multiple intersections of D-branes and M-branes*, hep-th/9612095, to appear in Nucl. Phys. B.

Bergshoeff, E., de Roo, M., Eyras, E., Janssen, B. and van der Schaar, J.P. (in prep.): *Intersections involving monopoles and waves in eleven dimensions*, in preparation

Costa, M. (1996): *Composite M-branes*, hep-th/9609181.

Duff, M.J. (1996): *Supermembranes*, TASI lectures, Boulder 1996, hep-th/9611203

Duff, M.J. and Stelle, K.S. (1991): Phys. Lett. **B253**, 113.

Gauntlett, J., Kastor, D. and Traschen, J. (1996): Nucl. Phys. **B478**, 544, hep-th/9604179.

Gauntlett, J.P., Gibbons, J.W., Papadopoulos, G. and Townsend, P.K. (1997): *Hyper-Kähler manifolds and multiply intersecting branes*, hep-th/9702202

Güven, R. (1992): Phys. Lett. **B276**, 49.

Howe, P.S. and Sezgin, E. (1996): *Superbranes*, hep-th/9607227

Papadopoulos, G. and Townsend, P.K. (1996a): Phys. Lett. **B380**, 273, hep-th/9603087.

Papadopoulos, G. and Townsend, P.K. (1996b): *Kaluza-Klein on the Brane*, hep-th/9609095

Polchinski, J. (1996): *TASI-lectures on D-branes*, hep-th/9611050

Polchinski, J. and Witten, E. (1995, 1996): Nucl. Phys. **B460**, 525, hep-th/9510169
Romans, L. (1986): Phys. Lett. **169B**, 374
Sorokin, R.D. (1983): Phys. Rev. Lett. **51**, 87;
Gross, D.J. and Perry, M.J. (1983): Nucl. Phys. **B226**, 29.
Townsend, P.K. (1996): *Four lectures on M-theory*, to appear in the proceedings of the 1996 ICTP Summer School in High Energy Physics and Cosmology, Trieste, hep-th/9612121
Tseytlin, A.A. (1996a): Nucl. Phys. **B475**, 149, hep-th/9604035.
Tseytlin, A.A. (1996b): *'No force' condition and BPS combinations of p-branes in 11 and 10 dimensions*, hep-th/9609212.
Tseytlin, A.A. (1997): *Composite BPS configurations of p-branes in 10 and 11 dimensions*, hep-th/9702163

Properties of Intersecting p-Branes in Various Dimensions

I. Ya. Aref'eva[1], M.G. Ivanov[2] and O.A. Rytchkov[3]

[1] Steklov Mathematical Institute,Gubkin 8, 117966, Moscow, Russia
[2] Moscow Institute of Physics and Technology, Russia
[3] Moscow State University, Moscow 119 899, Russia

Abstract. General properties of intersecting extremal p-brane solutions of gravity coupled with dilatons and several different d-form fields in arbitrary space-time dimensions are considered. It is shown that heuristically expected properties of the intersecting p-branes follow from the explicit formulae for solutions. In particular, harmonic superposition and S-duality hold for all p-brane solutions. Generalized T-duality takes place under additional restrictions on the initial theory parameters.

1 Introduction

There has been recently considerable progress in the study of classical p-brane solutions (for review see Stelle (1997) and references therein) of higher dimensional gravity coupled with matter. p-brane solutions in 10 and 11 dimensions play a key role for probing the duality conjectures (Hull, Townsend (1995), Witten (1995), Townsend (1995), Schwarz (1996, 1997), Sen (1996)) which relate five known superstrings and M-theory. A study of p-brane intersections is a subject of growing interest because when enough branes intersect one gets, as a rule, solutions with a regular horizon. The microscopic interpretation of the Bekenstein-Hawking entropy within string theory (Strominger, Vafa (1996)) has also stimulated investigations of the intersecting (composite) p-brane solutions. Several composite p-brane solutions in $D = 10$ and $D = 11$ have been obtained (Tseytlin (1996)-Tseytlin (1997)).

Heuristic scheme of constructing of p-brane intersections was based on string theory representation of the branes, duality and supersymmetry. This scheme involves the harmonic function superposition rule for the intersecting p-branes. This rule was formulated (Tseytlin (1996b)) in $D = 11$ and $D = 10$. Using T-duality and the supersymmetry requirements intersections of p-branes (more exactly M-branes and D-branes) have been recently classified in (Bergshoeff,de Roo, Eyras, Janssen, Schaar (1997)).

Other approach to the problem was elaborated in the papers (Volovich (1995, 1997)-Hambli (1997)). In these papers p-brane intersection rules were found from the equations of motion. The aim of this paper is to summarize these results, and to show that heiristically expected properties of the intersecting p-branes follow from the explicit formulae for solutions.

The paper is organized as follows. In Sect. 2 we remind single p-brane solutions and introduce graphic representations for them. In Sect. 3 the main steps of finding p-brane solutions are sketched and general composite p-brane solutions are presented. In Sect. 4 we collect the formulae for the entropy and ADM mass. In the section 5 we discuss S-duality, which is a specific property of our solutions. Harmonic function rule is generalized on arbitrary space-time dimensions in Sect. 6. T-duality transformations are considered in Sect. 7. In Sect. 8 we modify our results for the case of an arbitrary space–time signature. In Sect. 9 we analyze supersymmetry in the special case of 11D supergravity.

2 Single p-Brane Solutions

Let us consider the theory with the following action

$$I = \frac{1}{2\kappa^2} \int \mathrm{d}^D X \sqrt{-g} \left(R - \frac{1}{2} (\nabla\phi)^2 - \frac{\mathrm{e}^{-\alpha\phi}}{2(d+1)!} F_{d+1}^2 \right) , \tag{1}$$

where F_{d+1} is a $d+1$ differential form, $F_{d+1} = \mathrm{d}A_d$, ϕ is a dilaton.

This action admits the elementary p-brane solution (Dubholkar et al. (1990))

$$\mathrm{d}s^2 = H^\tau(x) \sum_{\mu,\nu=0}^{d-1} \eta_{\mu\nu} \mathrm{d}y^\mu \mathrm{d}y^\nu + H^\rho(x) \sum_{\gamma=1}^{D-d} \mathrm{d}x^\gamma \mathrm{d}x^\gamma , \tag{2}$$

$$A_d = 2\Delta^{-1/2} H^{-1}(x) \mathrm{d}y^0 \wedge \ldots \wedge \mathrm{d}y^{d-1} , \tag{3}$$

where $H(x)$ is a harmonic function.

Solution (2), (3) generalizes the well-known Majumdar-Papapetrou solutions (Majumdar (1947), Papapetrou (1947)). The exponents are defined by the parameters of the theory

$$\tau = -\frac{4(D-2-d)}{\Delta(D-2)}, \quad \rho = \frac{4d}{\Delta(D-2)}, \quad \text{where} \quad \Delta = \alpha^2 + \frac{2d(D-2-d)}{D-2} . \tag{4}$$

In the p-brane terminology (Stelle (1997)) the solution (2),(3) describes an electrically charged p-brane, where y-coordinates correspond to the world-volume directions and x-coordinates to directions transverse to the brane. It is convenient to represent every y^k-coordinate by "×" and every x-coordinate by "–" (Bergshoeff,de Roo, Eyras, Janssen, Schaar (1997)). One has the following representation of the metric

$$\mathrm{d}s^2 = \underbrace{\times\ \times \cdots \times}_{d}\ \underbrace{-\ - \cdots -}_{D-d} . \tag{5}$$

It is also convenient to present the gauge field (3) as a row with d circles "o"

∘ ∘ ··· ∘	

Fig.1

where the circles correspond to indices of the non-zero component.

The action (1) admits also a solitonic p-brane solution (Callan, Harvey, Strominger (1991))

$$ds^2 = U^{-\rho}(x) \sum_{\mu,\nu=0}^{D-d-3} \eta_{\mu\nu} dy^{\mu} dy^{\nu} + U^{-\tau}(x) \sum_{\gamma=1}^{d+2} dx^{\gamma} dx^{\gamma} \ , \tag{6}$$

$$F = 2\Delta^{-1/2} * dU \ , \tag{7}$$

where $*$ is a Hodge dual on $\mathbb{R}^{d+2}$, $U(x)$ is a harmonic function. Using the p-brane interpretation one says that the solution (6), (7) describes the magnetically charged p-brane lying in the y-directions. One can present the metric in the following way

$$ds^2 = \underbrace{* \ * \cdots \ *}_{D-d-2} \underbrace{- \ - \cdots -}_{d+2} \ , \tag{8}$$

where "$*$" denote the worldvolume directions. Note that we use the symbol "$\times$" for the electric p-brane and the symbol "$*$" for the magnetic one.

For the gauge field we have a different picture with $d+2$ circles "$\bullet$"

	• • ··· •

Fig.2

where by "$\bullet$" we denote subspace on which a Hodge dual acts. For the more complicated ansatzes we have pictures with more then one rows. For example,

∘	∘	∘						
∘			∘	∘				
	•		•			•	•	•
		•		•		•	•	•

Fig.3

We will call such pictures for the antisymmetric field as incidence tables. In the p-brane terminology the corresponding metric could be presented as

$$ds^2 = \left\{ \begin{array}{c|c|c|c|c|ccc|c|c|c|} \times & \times & \times & - & - & - & - & - & - & - & - \\ \times & - & - & \times & \times & - & - & - & - & - & - \\ * & - & * & - & * & * & * & * & - & - & - \\ * & * & - & * & - & * & * & * & - & - & - \end{array} \right. . \tag{9}$$

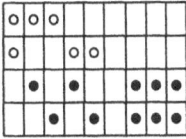

We will call such pictures for the metric as *brane incidence tables.* The intersections of the p-branes are not arbitrary and governed by characteristic equations which are considered below.

3 Composite p-Brane Solutions

The method (Volovich (1995, 1997), Aref'eva, Vismanathan, Volovich (1997), Aref'eva, Volovich (1996), Aref'eva, Rytchkov (1996)) of finding the intersecting p-brane solutions involves direct solving the equations of motion for the theory (1). The method consists of the following steps.

Step 1. We assume the metric in the special form (see (10)). In order to simplify the Ricci tensor we also assume that the form of the metric satisfies the Fock–De Donder gauge condition (see (12) below).

Step 2. In order to describe an ansatz for the antisymmetric field we introduce the *incidence* matrices (see (15)).

Step 3. Using assumed form for the antisymmetric field we calculate components of the energy-momentum tensor. They could be crucially simplified if one assumes so-called "no-force" conditions. Their origin is simple: they eliminate exponents from the components of the energy-momentum tensor.

Step 4. From Maxwell's equations and Bianchi identity we conclude that scalar functions specifying our ansatzes should be harmonic ones.

Step 5. We check that Einstein equations and equation of motion for the dilaton are fulfilled if the incidence matrices satisfy *characteristic* equations (see (25), (26), (27)).

Let us briefly demonstrate the realization of this program. We assume the metric in the following form

$$\mathrm{d}s^2 = \mathrm{e}^{2A(x)} \sum_{\mu,\nu=0}^{q-1} \eta_{\mu\nu} \mathrm{d}y^\mu \mathrm{d}y^\nu + \sum_{i=q}^{D-s-3} \mathrm{e}^{2F_i(x)} \mathrm{d}y^i \mathrm{d}y^i + \\ + \mathrm{e}^{2B(x)} \sum_{\gamma=D-s-2}^{D-1} \mathrm{d}x^\gamma \mathrm{d}x^\gamma \ , \tag{10}$$

where $\eta_{\mu\nu}$ is a flat Minkowski metric, A, B and F_i are functions of x. Using the notations $F_\mu(x) = A(x)$ and $F_\gamma(x) = B(x)$ the above metric can be rewritten as

$$\mathrm{d}s^2 = \sum_{L=0}^{D-1} \mathrm{e}^{2F_L(x)} \eta_{KL} \mathrm{d}X^K \mathrm{d}X^L \ . \tag{11}$$

If we assume the Fock–De Donder gauge condition, which for metric (10) takes the form

$$\sum_{L=0}^{D-1} F_L - 2B = 0 \ , \tag{12}$$

then the Ricci tensor has the following components

$$R_{KL} = -\sum_{N=0}^{D-1} \partial_K F_N \partial_L F_N + 2\partial_K B \partial_L B - \mathrm{e}^{2F_L-2B}\eta_{KL}\Delta F_L \ . \qquad (13)$$

For the d-form A_d we consider a class of ansatzes corresponding to E electric and M magnetic charges,

$$F = \sum_a^E \mathrm{d}A_a^{(\mathcal{E})} + \sum_b^M F_b^{(\mathcal{M})} \ . \qquad (14)$$

We will refer to the different terms in (14) as to the different branches of electric and magnetic fields.

To describe electric and magnetic configurations $A_a^{(\mathcal{E})}$ and $F_b^{(\mathcal{M})}$ we introduce electric and magnetic *incidence* matrices

$$\begin{aligned} &\Delta = (\Delta_{aL}), \quad \Lambda = (\Lambda_{bL}), \\ &a = 1, \ldots, E, \quad b = 1, \ldots, M, \quad L = 0, \ldots, D-1 \ , \end{aligned} \qquad (15)$$

respectively. Their rows correspond to independent branches of the electric (magnetic) gauge field and columns refer to the space-time indices. The entries of the incidence matrices are equal to 1 or 0. Furthermore the electric (magnetic) incidence matrix has an equal number of units in each row, and there are no rows which coincide. Here we don't consider Euclidean p-branes so we assume $\Delta_{a0} = 1$, $\Delta_{a\alpha} = 0$, $\Lambda_{b0} = 0$ $\Lambda_{b\alpha} = 1$ for all a and b (see section 8). In a graphic representation we draw "∘" (for the electric incidence matrix) and "•" (for the magnetic incidence matrix) instead of 1. Empty space denotes 0 in both cases. Fig.1 and Fig.2 are nothing but a graphic representation of the incidence matrices for the single elementary and solitonic p-brane solutions. The incidence table Fig.3 is an example representing electric and magnetic incidence matrices in a more complicated case.

In the terms of the incidence matrices an electric field is assumed to have the form

$$A_a^{(\mathcal{E})} = h_a \mathrm{e}^{C_a(x)} \bigwedge_{\{L|\Delta_{aL}=1\}} \mathrm{d}X^L \ , \qquad (16)$$

where we use the notation $\bigwedge_{i=1}^n \mathrm{d}X^i = \mathrm{d}X^1 \wedge \cdots \wedge \mathrm{d}X^n$.

In the terms of the incidence matrices a magnetic field is assumed in the form

$$F_b^{(\mathcal{M})\, K_1 \ldots K_{d+1}} = \frac{1}{\sqrt{-g}} v_b \mathrm{e}^{\alpha\phi} \epsilon^{K_1 \ldots K_{d+1}\beta} \partial_\beta \mathrm{e}^{\chi} \ , \qquad (17)$$

where we take K_i such that $\Lambda_{bK_i} = 1$.

We will use Einstein equations in the form $R_{KL} = G_{KL}$, there G_{KL} is related with the stress-energy tensor T_{KL} as $G_{KL} = T_{KL} - \frac{g_{KL}}{D-2} T_P{}^P$. For the considered ansatz the tensor G is

$$G_{KL} = \frac{1}{2}\partial_K\phi\partial_L\phi \tag{18}$$
$$+\sum_{a=1}^{E}\frac{h_a^2}{2}e^{2F_L-2B+\mathcal{F}_a}\left(-\partial_K C_a\partial_L C_a - \eta_{KL}\left\{\Delta_{aL} - \frac{d}{D-2}\right\}(\partial C_a)^2\right)$$
$$+\sum_{b=1}^{M}\frac{v_b^2}{2}e^{2F_L-2B+\mathcal{F}_b}\left(-\partial_K\chi_b\partial_L\chi_b + \eta_{KL}\left\{\Lambda_{bL} - \frac{d}{D-2}\right\}(\partial\chi_b)^2\right),$$

$$\mathcal{F}_a = -\alpha\phi - 2\sum_{N=0}^{D-1}\Delta_{aN}F_N + 2C_a,$$
$$\mathcal{F}_b = \alpha\phi + 2\sum_{N=0}^{D-1}\Lambda_{bN}F_N - 4B + 2\chi_b\ . \tag{19}$$

In order to guarantee the above form of the G-tensor (the absence of the (ij)-, $(\mu\nu)$-, (μi)-, $(\mu\alpha)$- and $(i\alpha)$-components, where $i \neq j$, $\mu \neq \nu$) we have to impose additional restrictions on the incidence matrices. Namely, each two rows assumed to have differences in more then two columns. Furthermore, the difference between each row of the electric incidence matrix and each row of the magnetic incidence matrix have to be in more then four columns.

According to step 3 we assume the following "no-force" conditions

$$\mathcal{F}_a = 0,\ a = 1,\ldots,E, \qquad \mathcal{F}_b = 0,\ b = 1,\ldots,M\ . \tag{20}$$

The LHS of these conditions are nothing but exponents which enter in the G tensor. Using (20) the magnetic field (17) could be rewritten in the following form

$$F_b^{(\mathcal{M})} = v_b e^{-\chi_b(x)} * d\chi_b(x) \bigwedge_{\{i|\Lambda_{bi}=1\}} dy^i\ , \tag{21}$$

where $*$ is a Hodge dual on the x-subspace.

From Maxwell's equations and from Bianchi identity for F under conditions (20) we get

$$\Delta C_a = (\partial C_a)^2, \qquad \Delta\chi_b = (\partial\chi_b)^2\ , \tag{22}$$

therefore, $H_a = e^{-C_a}$, $a = 1,\ldots,E$ and $U_b = e^{-\chi_b}$, $b = 1,\ldots,M$ are harmonic functions.

In order to solve the Einstein equations and the equation of motion for the dilaton we suppose

$$F_L = \sum_{a=1}^{E} h_a^2 C_a\left(\frac{1}{2}\Delta_{aL} - u\right) - \sum_{b=1}^{M} v_b^2\chi_b\left(\frac{1}{2}\Lambda_{bL} - u\right),$$
$$\phi = \frac{\alpha}{2}\left[\sum_{a=1}^{E} h_a^2 C_a - \sum_{b=1}^{M} v_a^2\chi_b\right], \tag{23}$$

where

$$u = \frac{d}{2(D-2)} \ . \tag{24}$$

Substituting (23) in (12) one can check that equation (12) is fulfilled. Under assumption of the independence of C_a and χ_b a substitution of (23) in the "no force" conditions gives three types of *characteristic* equations on the incidence matrices

$$\frac{\alpha^2}{2} - \frac{\mathrm{d}^2}{D-2} + \sum_{L=0}^{D-1} \Delta_{aL}\Delta_{a'L} = 0, \quad a \neq a', \quad a, a' = 1, \dots, E, \tag{25}$$

$$\frac{\alpha^2}{2} - \frac{\mathrm{d}^2}{D-2} + \sum_{L=0}^{D-1} \Lambda_{bL}\Lambda_{b'L} - 2 = 0, \ b \neq b', \quad b, b' = 1, \dots, M, \tag{26}$$

$$\frac{\alpha^2}{2} - \frac{\mathrm{d}^2}{D-2} + \sum_{L=0}^{D-1} \Delta_{aL}\Lambda_{bL} = 0, \quad a = 1, \dots, E, \quad b = 1, \dots, M \ , \tag{27}$$

and $h_a^2 = v_b^2 = \sigma$, where

$$\sigma = \frac{1}{dt + \alpha^2/4}, \qquad t = \frac{D-d-2}{2(D-2)} \ . \tag{28}$$

Note that $\sum_{L=0}^{D-1} \Delta_{aL}\Delta_{a'L}$ is a number of common columns of a and a' branches in the incidence table. Characteristic equations admit solutions only for quantized values of scalar coupling parameter that is in accordance with (Bergshoeff, de Roo, Panda (1997), Volovich (1995, 1997), Aref'eva, Volovich (1996)).

Note that if the dilaton is absent, then the p-brane solutions are given by the above formula with $\alpha = 0$. Corresponding characteristic equations are very restrictive since they have to be solved for integers and they admit non-trivial solutions only for special D. In particular, they are $D = 6$, 10, 11, 18, 20, 26.

One can check that if the incidence matrices satisfy the characteristic equations then the Einstein equations hold. Therefore, the form of the metric which solves the theory is

$$\mathrm{d}s^2 = (H_1 H_2 \cdots H_E)^{2u\sigma} (U_1 U_2 \cdots U_M)^{2t\sigma} \times \left\{ \sum_{K,L=0}^{D-s-3} \left(\prod_a H_a^{\Delta_{aL}} \prod_b U_b^{1-\Lambda_{bL}} \right)^{-\sigma} \eta_{KL} \mathrm{d}y^K \mathrm{d}y^L + \sum_\gamma \mathrm{d}x^\gamma \mathrm{d}x^\gamma \right\}, \tag{29}$$

where H_a and U_b are harmonic functions, u, t and σ are given by (24) and (28).

The approach considered above could be obviously generalized to the actions with k antisymmetric fields and several dilatons

$$I=\frac{1}{2\kappa^2}\int \mathrm{d}^D X\sqrt{-g}\left(R-\frac{1}{2}(\nabla\phi)^2-\sum_{I=1}^{k}\frac{\mathrm{e}^{-\alpha^{(I)}\phi}}{2(d_I+1)!}F^{(I)2}_{d_I+1}\right). \quad (30)$$

In this case we introduce $2k$ incidence matrices. Electric and magnetic fields are assumed in the form

$$A_a^{(I)}=h_a^{(I)}\mathrm{e}^{C_a^{(I)}(x)}\bigwedge_{\{L|\Delta_{aL}^{(I)}=1\}}\mathrm{d}X^L\ , \quad (31)$$

$$F_b^{(I)\ K_1\ldots K_{d_I+1}}=\frac{1}{\sqrt{-g}}v_b^{(I)}\mathrm{e}^{\alpha^{(I)}\phi}\epsilon^{K_1\ldots K_{d+1}\beta}\partial_\beta\mathrm{e}^{\chi^{(I)}}\ , \quad (32)$$

where $I=1,\ldots,k$ and we take K_i such that $\Lambda_{bK_i}^{(I)}=1$. Instead of (25), (26) and (27) one has the following characteristic equations

$$(1-\delta_{IJ}\delta_{aa'})\left\{\frac{\alpha^{(I)}\alpha^{(J)}}{2}-\frac{d_Id_J}{D-2}+\sum_{L=0}^{D-1}\Delta_{aL}^{(I)}\Delta_{a'L}^{(J)}\right\}=0\ , \quad (33)$$

$$(1-\delta_{IJ}\delta_{bb'})\left\{\frac{\alpha^{(I)}\alpha^{(J)}}{2}-\frac{d_Id_J}{D-2}+\sum_{L=0}^{D-1}\Lambda_{bL}^{(I)}\Lambda_{b'L}^{(J)}-2\right\}=0\ , \quad (34)$$

$$\frac{\alpha^{(I)}\alpha^{(J)}}{2}-\frac{d_Id_J}{D-2}+\sum_{L=0}^{D-1}\Delta_{aL}^{(I)}\Lambda_{b'L}^{(J)}=0\ . \quad (35)$$

The constants $h_a^{(I)}$ and $v_b^{(I)}$ are given by

$$h_a^{(I)^2}=v_b^{(I)^2}=\sigma^{(I)}\ , \quad (36)$$

where $\sigma^{(I)}=\frac{1}{t^{(I)}d_I}+\frac{1}{4}\,\alpha^{(I)^2}$, $t^{(I)}=\frac{D-2-d_I}{2(D-2)}$.

Let us present the metric in the form where the overall conformal factor which multiplies the transverse x-part is separated

$$\mathrm{d}s^2=\prod_{I=1}^{k}\left(H_1^{(I)}H_2^{(I)}\cdots H_{E_I}^{(I)}\right)^{2u^{(I)}\sigma^{(I)}}\left(U_1^{(I)}U_2^{(I)}\cdots U_{M_I}^{(I)}\right)^{2t^{(I)}\sigma^{(I)}}$$
$$\times\left\{\sum_{L=0}^{D-s-3}\prod_{I=1}^{k}\left(\prod_a H_a^{(I)\Delta_{aL}^{(I)}}\prod_b U_b^{(I)1-\Lambda_{bL}^{(I)}}\right)^{-\sigma^{(I)}}\eta_{KL}\mathrm{d}y^K\mathrm{d}y^L\right.$$
$$\left.+\sum_\gamma \mathrm{d}x^\gamma\mathrm{d}x^\gamma\right\}\ , \quad (37)$$

where

$$u^{(I)}=\frac{d_I}{2(D-2)}\ . \quad (38)$$

Under assumption $s > 0$ we take the harmonic functions in the form

$$H_a^{(I)} = 1 + \sum_c \frac{Q_{ac}^{(I)}}{|x - x_{ac}|^s}, \qquad U_b^{(I)} = 1 + \sum_c \frac{P_{bc}^{(I)}}{|x - x_{bc}|^s} \ . \tag{39}$$

The algebraic method (Volovich (1995, 1997), Aref'eva, Vismanathan, Volovich (1997), Aref'eva, Volovich (1996), Aref'eva, Viswanathan, Volovich, Volovich (1997)) can be also used for finding solutions with depending harmonic functions.

4 ADM Mass and Area of Horizon

The representation (37) is convenient for calculating the ADM mass and the entropy. Let the harmonic functions have the form (39) and all these functions have the same centers ($Q_{ac}^{(I)}, P_{bc}^{(I)} > 0$ for all a, b, c, I).

The ADM mass has the form

$$M = \frac{L^{d-1}\omega_{s+1}s}{2\kappa^2} \sum_I \sigma^{(I)} \left(\sum_{ac} Q_{ac}^{(I)} + \sum_{bc} P_{bc}^{(I)} \right) , \tag{40}$$

where ω_{s+1} is a volume of the $s+1$-dimensional sphere, and L is a period of all y_i, $i = 1, \ldots, D - s - 3$.

Under the condition

$$\frac{s}{2} \sum_I \sigma^{(I)} (E_I + M_I) = s + 1 \tag{41}$$

the area of horizon has the form

$$\mathcal{A}_{D-2} = \omega_{s+1} L^{D-s-3} \sum_c \prod_I \left(\prod_a Q_{ac}^{(I)} \prod_b P_{bc}^{(I)} \right)^{\frac{\sigma^{(I)}}{2}} , \tag{42}$$

One can also get non-trivial entropy if $q \geq 2$ and instead of (41) the following condition is assumed

$$\frac{s}{2} \sum_I \sigma^{(I)} (E_I + M_I) = \frac{s}{2} + 1 \ . \tag{43}$$

To get non-trivial entropy we make a boost $-\mathrm{d}y_0^2 + \mathrm{d}y_1^2 \longrightarrow \mathrm{d}u\mathrm{d}v + K(x)\mathrm{d}u^2$, where

$$u = y_1 + y_0, \ v = y_1 - y_0, \ K(x) = \sum_c \frac{Q_c}{|x - x_c|^s}, \ x_c = x_{ac} \ . \tag{44}$$

In this case the area of gorizon is given by

$$\mathcal{A}_{D-2} = \omega_{s+1} L^{D-s-3} \sum_c Q_c^{\frac{1}{2}} \prod_I \left(\prod_a Q_{ac}^{(I)} \prod_b P_{bc}^{(I)} \right)^{\frac{\sigma^{(I)}}{2}} \ . \tag{45}$$

5 S-Duality

In order to demonstrate S-duality let us consider a new action, which could be obtained from the action (30) by replacing an antisymmetric field $F^{(I)}_{d_I}$ by another field $F^{(I)}_{\tilde{d}_I}$ and changing the signs of the corresponding dilaton coupling constants on the opposite ones:

$$\tilde{d}_I = D - 2 - d_I, \qquad \tilde{\alpha}^{(I)} = -\alpha^{(I)} . \tag{46}$$

S-duality transforms the solutions of the theory (30) into the solutions of the theory with a new action. The corresponding transformations of the incidence matrices are

$$\Delta^{(I)}_{aL} \to \tilde{\Delta}^{(I)}_{bL} = 1 - \Lambda^{(I)}_{bL}, \qquad \Lambda^{(I)}_{bL} \to \tilde{\Lambda}^{(I)}_{aL} = 1 - \Delta^{(I)}_{aL} . \tag{47}$$

One can check that the new incidence matrices also satisfy the characteristic equations.

Also one can perform S-duality transformation (47) only for some branches of the fields. In this case the dual theory may have more fields in comparison with the initial one.

S-duality becomes evident if we present our results in the p-brane terminology. In order to consider the electric and magnetic p-branes together we introduce the *brane incidence matrix* Υ_{PL}, where $L = 0, \dots, D-1$; $P = 1, \dots, B$, B is a whole number of p-branes. The brane incidence matrix is constructed in the following way

$$\Upsilon_{PL} = \begin{pmatrix} \Delta^{(1)}_{aL} \\ \vdots \\ \Delta^{(k)}_{aL} \\ 1 - \Lambda^{(1)}_{bL} \\ \vdots \\ 1 - \Lambda^{(k)}_{bL} \end{pmatrix} . \tag{48}$$

The entries of the brane incidence matrix are equal to 1 or 0. The rows of this matrix correspond to p-branes. Electrically charged p-branes occupy the upper rows of the matrix, magnetically charged p-branes occupy the down rows. The columns of the matrix Υ_{PL} correspond to the space-time indices (similar to Δ and Λ). This matrix could be represented as the brane incidence table (see example (9)). We denote by "×" electrically charged p-branes and by "*" magnetically charged ones. For Υ_{PL} we have only one characteristic equation instead of three ones. Namely, the equations (33), (34) and (35) in the terms of Υ_{PL} could be rewritten in the following form

$$\varsigma_R \varsigma_{R'} \frac{\alpha^{(R)}\alpha^{(R')}}{2} - \frac{d_R d_{R'}}{D-2} + \sum_{L=0}^{D-1} \Upsilon_{RL}\Upsilon_{R'L} = 0, \; R \neq R' , \tag{49}$$

where ς_R is -1 for the electric p-branes and $+1$ for the magnetic ones, $\alpha^{(R)}$ are dilatons coupling constants connected with the p_R-brane, $d_R = p_R + 1$. This unified form of the characteristic equations is a manifestation of the S-duality.

6 Harmonic Function Rule

Our solution (37) has a very simple structure. This becomes obvious if one rewrites the metric in the following form:

$$g_{KK} = \prod_I \prod_a g_{KK}^{(I)a} \prod_b g_{KK}^{(I)b} , \tag{50}$$

where

$$\begin{aligned} g_{KK}^{(I)a} &= \left(H_a^{(I)\Delta_{aK}}\right)^{\tau^{(I)}} \left(H_a^{(I)(1-\Delta_{aK})}\right)^{\rho^{(I)}} , \\ g_{KK}^{(I)b} &= \left(U_b^{(I)\Lambda_{bK}}\right)^{-\tau^{(I)}} \left(U_b^{(I)(1-\Lambda_{bK})}\right)^{-\rho^{(I)}} . \end{aligned} \tag{51}$$

According to (36) and (38) the exponents are given by:

$$\tau^{(I)} \equiv -2t^{(I)}\sigma^{(I)} = -\frac{4(D-2-d_I)}{\Delta^{(I)}(D-2)}, \quad \rho^{(I)} \equiv 2u^{(I)}\sigma^{(I)} = \frac{4d_I}{\Delta^{(I)}(D-2)} , \tag{52}$$

where $\Delta^{(I)}$ is a generalization of (4)

$$\Delta^{(I)} = \alpha^{(I)2} + \frac{2d_I(D-2-d_I)}{D-2} . \tag{53}$$

For given incidence matrices and values of $\tau^{(I)}$ and $\rho^{(I)}$ (50), (51) gives the following rule for constructing a metric. For each space-time direction the coefficient in the metric is a product of functions H_a and U_b in an appropriate power. Namely, we put ${H_n^{(I)}}^{\tau^{(I)}}$ $({U_n^{(I)}}^{-\rho^{(I)}})$ if the corresponding direction belongs to the n-th $(d-1)-$ electric $((D-d-3)-$ magnetic) brane, and we put ${H_n^{(I)}}^{\rho^{(I)}}$ $({U_n^{(I)}}^{-\tau^{(I)}})$ if the corresponding direction is transverse to $(d-1)-$ electric $((D-d-3)-$ magnetic) brane. Note that $\tau^{(I)}$ and $\rho^{(I)}$ are the same as in the corresponding single brane. Certainly, one has to assume that the incidence matrices satisfy the characteristic equations (33)–(35).

There is another point of view, based on the metric representation (37). The overall conformal factor in (37) can be rewritten as

$$\begin{aligned} &\prod_{I=1}^k \left(H_1^{(I)}H_2^{(I)}\cdots H_{E_I}^{(I)}\right)^{2u^{(I)}\sigma^{(I)}} \left(U_1^{(I)}U_2^{(I)}\cdots U_{M_I}^{(I)}\right)^{2t^{(I)}\sigma^{(I)}} \\ &= \prod_{I=1}^k \left(H_1^{(I)}H_2^{(I)}\cdots H_{E_I}^{(I)}\right)^{\frac{d_I}{D-2}\sigma^{(I)}} \left(U_1^{(I)}U_2^{(I)}\cdots U_{M_I}^{(I)}\right)^{\frac{D-d_I-2}{D-2}\sigma^{(I)}} . \end{aligned} \tag{54}$$

Since the magnetic p-brane is connected with the $(D-p-3)$-form, the conformal factor could be unified (see eq.(55) below). This factor is a product of all harmonic functions in an appropriate power which is independent on the p-brane charge and is determined by the p-brane dimension. For each space-time dimension the coefficient in the metric (in brackets) is also a product of harmonic functions in power $\sigma^{(R)}$. Namely, we put $H_R^{-\sigma^{(R)}}$ if the corresponding direction belongs to the R-th brane. So the form of the metric doesn't depend on the type of p-brane charges and could be constructed in terms of the brane incidence matrix (which describes p-brane intersections only)

$$ds^2 = \prod_{R=1}^{B} H_R^{\frac{p_R+1}{D-2}\sigma^{(R)}} \left\{ \sum_{L=0}^{D-s-3} \left(\prod_{R=1}^{B} H_R^{\Upsilon_{RL}} \right)^{-\sigma^{(R)}} \eta_{KL} \mathrm{d}y^K \mathrm{d}y^L + \sum_{\gamma} \mathrm{d}x^\gamma \mathrm{d}x^\gamma \right\} . \tag{55}$$

Therefore the formula (55) gives a D-dimensional generalization of the "harmonic function rule" found before for $D = 10,\ 11$ (Tseytlin (1996b)).

7 T-Duality

Let us consider generalized T-duality transformations. T-duality transforms solutions for the action (30) with one set of fields into solutions of the action (30) with another set of fields. We perform T-duality transformation along the direction corresponding to y_{i_0} coordinate, $q \leq i_0 \leq D-s-3$. T-duality acts on the brane incidence matrix Υ_{RL} as follows. We select the i_0-th column, change 1 into 0 and vice versa and obtain a new brane incidence matrix. This matrix satisfies the characteristic equation if we simultaneously change dilaton coupling constants. More precisely, new dilaton coupling constants β_R are connected with the old ones $\boldsymbol{\alpha}_R$ in the following way

$$\begin{aligned} \frac{\beta_R\beta_{R'}}{2} = \frac{\alpha_R\alpha_{R'}}{2} - 1 + \Upsilon_{Ri_0} + \Upsilon_{R'i_0} + \frac{1}{D-2}[(1-2\Upsilon_{Ri_0})(1-2\Upsilon_{R'i_0}) \\ +(1-2\Upsilon_{Ri_0})d_{R'} + (1-2\Upsilon_{R'i_0})d_R] \ . \end{aligned} \tag{56}$$

In particular cases these relations give rather restrictive conditions on the initial theory parameters. For example, let us consider the case of one dilaton and one antisymmetric field and let us deal with the electrically charged p-branes. We make the mapping

$$I(D,d,\alpha) \longmapsto I(D,d-1,d+1,\alpha_1,\alpha_2) \tag{57}$$

and call it as a generalized T-duality transformation. T-duality (Hull, Townsend (1995), Witten (1995), Townsend (1995), Schwarz (1996, 1997), Sen

(1996), Bergshoeff, Hull, Ortin (1995)) in the case of IIA superstring transforms solutions with non-zero 2-form into solutions with non-zero 1- and 3-forms (Bergshoeff,de Roo, Eyras, Janssen, Schaar (1997)). After performing T-duality we have a new brane incidence matrix Υ'_{RL} which is effectively a composition of the brane incidence matrices for $(d-2)$- and d-branes. In the case $E' > 1,\ E'' > 1$ the characteristic equations (56) are consistent only if

$$\alpha^2 = \frac{(2d - D + 2)^2}{2(D-2)} \tag{58}$$

(for $E' = 1$ or $E'' = 1$ the characteristic equations are less restrictive). Only under condition (58) T-duality in the form (57) takes place. Using (29) and (37) one can check that the values of σ which specify the relative transverse components of the metrics corresponding to the theories related by T-duality are the same.

8 The Case of Arbitrary Signature

p-brane solutions also exist in the case of an arbitrary space-time signature: $\eta = \mathrm{diag}(\pm 1, \ldots, \pm 1)$. In this case one deals with a modified action

$$I = \frac{1}{2\kappa^2} \int \mathrm{d}^D X \sqrt{|g|} \left(R - \frac{1}{2}(\nabla\phi)^2 - \sum_{I=1}^{k} \frac{s_I e^{-\alpha^{(I)}\phi}}{2(d_I+1)!} F^{(I)2}_{d_I+1} \right) , \tag{59}$$

where $s_I = \pm 1$. Instead of old restrictions for the incidence matrices one has new ones. In the unified notation we may rewrite these restrictions as

$$s_I \prod_{L=0}^{D-1} (\eta_{LL})^{\Delta_{RL}} = \varsigma_R , \tag{60}$$

where Δ_{RL} is the *matter incidence matrix*, which contain rows corresponding to both types of p-branes, electric and magnetic. The characteristic equations remain the same.

Under the S-duality transformation in addition to (46) one has also to perform a change

$$s_I \to \tilde{s}_I = -\det(\eta_{KL}) s_I . \tag{61}$$

9 Supersymmetry in 11D Supergravity

Let us recall that some p-brane solutions were obtained using requirements of supersymmetry. Known supersymmetric p-brane solutions admit an existence of Killing spinors. In this section we are going to examine relations between the problem of finding Killing spinors and our scheme of finding solutions taking as example 11D supergravity. The bosonic sector of $D = 11$

supergravity consists of a metric and a three-form potential. Killing spinors ε satisfy the following equations

$$\tilde{D}_L\varepsilon = 0, \tag{62}$$

$$\tilde{D}_L = \partial_L + \frac{1}{4}\omega_L{}^{AB}\Gamma_{AB} - \frac{1}{288}(\Gamma^{PQRS}{}_L + 8\Gamma^{PQR}\delta^S_L)F_{PQRS} \ , \tag{63}$$

where P, Q, R, S, L are 11D world indices. Γ_A are the 11D Dirac matrices

$$\{\Gamma_A, \Gamma_B\} = 2\eta_{AB}, \tag{64}$$

$$\Gamma_{AB\cdots C} = \Gamma_{[A}\Gamma_B \cdots \Gamma_{C]} \ . \tag{65}$$

Indices A, B, C are 11D vielbein indices, related with the orthonormal vielbein.

If we assume the ansatz for the metric (10) and for the matter field (16), (17), then in the Fock–De Donder gauge the covariant derivative is

$$\begin{aligned}\tilde{D}_L = {} & \partial_L - \frac{1}{2}\Gamma_L{}^K\partial_K F_L + \sum_a \frac{h_a}{6}\mathrm{e}^{\mathcal{F}_a/2}\partial_L C_a\Gamma(a) - \sum_b \frac{v_b}{12}\mathrm{e}^{\mathcal{F}_b/2}\partial_L\chi_b\Gamma(b) \\ & + \Gamma_L{}^K\sum_a \frac{h_a}{2}\left\{\Delta_{aL} - \frac{1}{3}\right\}\mathrm{e}^{\mathcal{F}_a/2}\partial_K C_a\frac{\Gamma(a)}{2} \\ & + \Gamma_L{}^K\sum_b \frac{v_b}{2}\left\{\Lambda_{bL} - \frac{1}{3}\right\}\mathrm{e}^{\mathcal{F}_b/2}\partial_K\chi_b\frac{\Gamma(b)}{2} \ ,\end{aligned} \tag{66}$$

where

$$\begin{aligned}\Gamma(a) &= \frac{1}{3!}E^{\mu_1^a\mu_2^a\mu_3^a}\Gamma_{\mu_1^a\mu_2^a\mu_3^a} = \prod_{\{A|\Delta_{aA}=1\}}\Gamma_A \ , \\ \Gamma(b) &= \frac{1}{5!}E^{\mu_1^b\mu_2^b\mu_3^b\mu_4^b\mu_5^b}\Gamma_{\mu_1^b\mu_2^b\mu_3^b\mu_4^b\mu_5^b} = \prod_{\{A|\Lambda_{bA}=1\}}\Gamma_A \ ,\end{aligned} \tag{67}$$

and

$$E^a_{\mu_1^a\cdots\mu_d^a} = \sqrt{-\prod_{L=0}^{D-1} g_{LL}^{\Delta_{aL}}}\,\epsilon_{\mu_1^a\cdots\mu_d^a}, \quad E^b_{\mu_0^b\cdots\mu_d^b\alpha} = \sqrt{\prod_{L=0}^{D-1} g_{LL}^{\Lambda_{bL}}}\,\epsilon_{\mu_0^b\cdots\mu_d^b\alpha} \tag{68}$$

are forms of d and $d+2$ dimensional volume, $\mu^a \in \{L|\Delta_{aL} = 1\}$, $\mu^b \in \{L|\Lambda_{bL} = 1\}$. For definition of $\mathcal{F}_a$ and $\mathcal{F}_b$ see (19).

To analyze the consequences of the supersymmetry it is convenient to rewrite the covariant derivative as

$$\begin{aligned}\tilde{D}_L = {} & \partial_L - \frac{1}{2}\Gamma_L{}^K\partial_K F_L + \sum_R \frac{1-3\varsigma_R}{24}h_R\mathrm{e}^{\mathcal{F}_R/2}\partial_L C_R\Gamma(R) \\ & + \Gamma_L{}^K\sum_R \frac{h_R}{2}\left\{\Delta_{RL} - \frac{1}{3}\right\}\mathrm{e}^{\mathcal{F}_R/2}\partial_K C_R\frac{\Gamma(R)}{2} \ .\end{aligned} \tag{69}$$

The notations in (69) are obvious

$$C_R = \begin{cases} C_a, \ R = a \\ \chi_b, \ R = b \end{cases}, \qquad h_R = \begin{cases} h_a, \ R = a \\ v_b, \ R = b \end{cases}.$$

Definition of the matrix Δ_{RL} one can find after equation (60). Note that in the expression (69) for the covariant derivative we do not assume that the metric solves the Einstein equations.

To construct the Killing spinor it is natural to assume that all $\mathcal{F}_R$ vanish. These conditions coincide with "no force" conditions (20).

Now we can see that if we assume that there exist spinors ε_0 such that

$$\Gamma_L{}^K \left[-\frac{1}{2}\partial_K F_L + \sum_R \frac{h_R}{2} \left\{ \Delta_{RL} - \frac{1}{3} \right\} \partial_K C_R \frac{\Gamma(R)}{2} \right] \varepsilon_0 = 0 \ , \tag{70}$$

then the Killing spinors can be found in the form $\varepsilon = \varepsilon_0 f(x)$, where a scalar function $f(x)$ solves the following equations

$$\left[\partial_L + \sum_R \frac{1 - 3\varsigma_R}{24} h_R \partial_L C_R \Gamma(R) \right] \varepsilon_0 f(x) = 0 \ . \tag{71}$$

In order to solve the equation (70), let us assume that F_L are linear combinations of the functions C_R (a "relax" harmonic superposition rule), for simplification of our calculation we shall write $F_L = -1/2 \sum l_{RL} h_R \{\Delta_{RL} - 1/3\} C_R$ (we assume that $h_R \neq 0$). Under this assumption and taking into account the independence of functions C_R we get

$$[l_{RL} + \Gamma(R)]\varepsilon_0 = 0 \ . \tag{72}$$

Therefore constants l_{RL} do not depend on subscript L, i.e.

$$[l_R + \Gamma(R)]\varepsilon_0 = 0 \ . \tag{73}$$

This equation admits non-trivial solutions only for $|l_R| = 1$. Note that substituting the relax harmonicity conditions in "no force" conditions one can find a relation $l_R h_R = \varsigma_R$ and the characteristic equation. Since $|l_R| = 1$, we get $h_R^2 = 1$, $l_R = \varsigma_R h_R$. So under "no force" conditions the relax harmonic superposition rule coincides with the harmonic superposition rule and one can say that the requirement of the supersymmetry supports the harmonic superposition.

To guarantee that a configuration obtained as a result of intersecting of single branes is supersymmetric one has to study a compatibility conditions of all requirements (73). In order to analyse these requirements let us introduce S as a special set of signs $S = \{\pm, \ldots, \pm\}$, or $S(R) = \pm$. We search a set S which admits an existence of ε_0 such that

$$P_R^{S(R)} \varepsilon_0 = 0 \ , \tag{74}$$

for all R. Here $P_R^{S(R)}$ is a projector $P_R^{S(R)} = \frac{1+S(R)\Gamma(R)}{2}$. We will demonstrate, that one can find an appropriate S and ε_0, if the incidence matrices satisfy the characteristic equations. In this case of 11D supergravity they lead to the restrictions on the incidence matrix

$$\sum_{N=0}^{D-1} \Delta_{RN}\Delta_{R'N} = 1, \quad \text{or} \quad \sum_{N=0}^{D-1} \Delta_{RN}\Delta_{R'N} = 3, \qquad R \neq R' . \tag{75}$$

If the incidence matrix satisfies these conditions one can check that $[\Gamma(R), \Gamma(R')] = 0$. Using this relation one can introduce new projection operators $P^S = \prod_R P_R^{S(R)}$ with the following properties:

$$P^S P^{S'} = \delta_{SS'} P^S, \qquad \sum_S P^S = 1 . \tag{76}$$

Acting by $\sum_S P^S$ on an arbitrary spinor $\varepsilon_i \neq 0$ and using (76), one can obtain the existence of the set S_0, such that $P^{S_0}\varepsilon_i = \varepsilon_0 \neq 0$. Moreover for every R $P_R^{-S_0(R)}\varepsilon_0 = 0$, comparing this result with the equation (74) one can conclude, that $-\varsigma_R S_0(R)$ is supersymmetric signs set for h_R.

To summarize, using a simple algebraic method we have constructed the general D-dimensional intersecting p-brane solutions which satisfy the harmonic function superposition rule and possess S- and T-dualities. The intersections of p-branes are controlled by the characteristic equations. These equations have solutions only for quantized values of scalar coupling parameters. Some of these solutions in the cases of D=11 and D=10 provide the metrics with regular horizons and non-zero entropy.

We are grateful to I. V. Volovich for useful discussions. This work is partially supported by the RFFI grants 96-01-00608 (I.A. and O.R.) and 96-01-00312 (M.I.).

References

Arguirio, R., Englert, F. and Houart, L. (1997): Phys.Lett. **B398**, 61-68, hep-th/9701042

Aref'eva, I.Ya., Ivanov, M.G. and Volovich, I.V. (1997): hep-th/9702079

Aref'eva, I.Ya and Rytchkov, O.A. (1996): hep-th/9612236

Aref'eva, I.Ya., Viswanathan, K.S. and Volovich, I.V. (1997): Phys.Rev. **D55**, 4748-4755, hep-th/9609225

Aref'eva, I.Ya., Viswanathan, K.S., Volovich, A.I., Volovich, I.V. (1997): hep-th/9701092

Aref'eva, I.Ya. and Volovich, A.I. (1996): hep-th/9611026

Behrndt, K., Cardoso, G.L., de Wit, B., Kallosh, R., Lust, D. and Mohaupt, T. (1997): Nucl.Phys. **B488**, 236-260, hep-th/9610105

Behrndt, K., Bergshoeff, E. and Janssen, B. (1997): Phys. Rev. **D55**, 3785–3792, hep-th/9604168

Bergshoeff, E., Hull, C.M. and Ortin, T. (1995): Nucl. Phys. **B451**, 547, hep-th/9504081
Bergshoeff, E., Kallosh, R., Ortin, T. (1996): Nucl. Phys. **B478**, 156–180, hep-th/9605059
Bergshoeff, E., de Roo, M., Eyras, E., Janssen, B. and van der Schaar, J.P. (1997): hep-th/9612095
Bergshoeff, E., De Roo, M. and Panda, S. (1997): Phys. Lett. **B390**, 143–147, hep-th/9609056
Callan, C.G., Harvey, J.A. and Strominger, A. (1991): Nucl. Phys. **B 359**, 611
Callan, C. and Maldacena, J. (1996): Nucl. Phys. **B472**, 591–610, hep-th/9602043
Callan, C.G. Jr., Gubser, S.S., Klebanov, I.R. and Tseytlin, A.A. (1997): Nucl.Phys. **B489**, 65-94, hep-th/9610172
Dubholkar, A., Gibbons, G.W., Harvey, J.A., Ruiz Ruiz, F. (1990): Nucl. Phys. **340**, 33.
Emparan, R. (1997): Nucl.Phys. **B490**, 365-390, hep-th/9610170
Gauntlett, J.P., Kastor, D.A. and Traschen, J. (1996): Nucl. Phys **B478**, 544–560, hep-th/9604179
Hambli, N. (1997): hep-th/9703179
Horowitz, G.T. (1996): gr-qc/9604051
Hull, C. and Townsend, P. (1995): Nucl. Phys. B438, 109
Ivashchuk, V.D. and Melnikov, V.N. (1996): hep-th/9612089
Majumdar, S.D. (1947): Phys. Rev **72**, 930;
Papapetrou, A. (1947): Proc. R. Irish Acad. **A51**, 191
Klebanov, I.R. and Tseytlin, A.A. (1996): Nucl. Phys. **B475**, 179, hep-th/9604166
Ohta, N. (1997): hep-th/9702164
Papadopoulos, G. and Townsend, P.K. (1996): Phys. Lett. **B380**, 273–279, hep-th/9603087
Papadopoulos, G. (1996): Fortsch. Phys. **44**, 573–584, hep-th/9604068
Papadopoulos, G. and Townsend, P.K. (1997): Phys. Lett. **B393**, 59–64, hep-th/9609095
Sen, A. (1996): Mod. Phys. Lett. **A11** (1996) 827
Schwarz, J.H. (1996, 1997): Phys. Lett. **B367**, 97–103, hep-th/9510086; Nucl.Phys.Proc.Suppl. **55B**, 1-32, hep-th/9607201
Stelle, K.S. (1997): hep-th/9701088
Strominger, A. and Vafa, C. (1996): Phys. Lett. **B379**, 99–104, hep-th/9601029
Townsend, P. (1995): hep-th/9507048
Tseytlin, A.A. (1996): Mod. Phys. Lett. **A11**, 689–714, hep-th/9601177
Tseytlin, A.A. (1996b): Nucl. Phys. **B475**, 149, hep-th/9604035
Tseytlin, A.A. (1996c): gr-qc/9608044
Tseytlin, A.A. (1997): Nucl. Phys. **B487**, 141–154, hep-th/9609212
Volovich, A. (1995, 1997): hep-th/9608095, Nucl. Phys. **B492**
Witten, E. (1995): Nucl. Phys. B443, 85

Super D-Branes

E. Bergshoeff

Institute for Theoretical Physics Nijenborgh 4, 9747 AG Groningen,
The Netherlands

Abstract. We present a "kappa–symmetric" worldvolume action for all type II Dirichlet p–branes, $0 \leq p \leq 9$, in a general type II supergravity background, including massive backgrounds in the IIA case. The $p = 9$ case provides a supersymmetric $D = 10$ Born–Infeld action. We emphasize the existence, in the "kappa–transformations", of a projection operator, leading to a correct matching of Bose and Fermi degrees of freedom on the worldvolume. Some recent developments are sketched.

1 Introduction

It is a pleasure for me to write a contribution dedicated to the memory of D.V. Volkov. Since Volkov, together with Akulov, was one of the first to introduce the concept of supersymmetry in quantum field theory (Volkov, Akulov (1972, 1973)), it seems appropriate to discuss, as a special example of a supersymmetry, the so–called "kappa–symmetry". The work I will describe is based on a recent paper I wrote with P. Townsend (Bergshoeff, Townsend (1997)).

The kind of kappa–symmetry I will discuss is encountered when one considers the effective action of certain Ramond-Ramond (R-R) p–brane solutions in ten–dimensional $N = 2$ supergravity theories. These solutions have an interpretation as Dirichlet p–branes, or Dp–branes, of type II superstring theory (Polchinski (1995)). In general the worldvolume actions for p-brane solutions of supersymmetric field theories may be viewed as $(p+1)$-dimensional non-linear sigma-models with a superspace as the target space. A distinguishing feature of the worldvolume actions I will consider is that the worldvolume fields include a vector, the so-called Born-Infeld field. These actions are called D-brane actions. The complications introduced by this gauge field have, until recently, prevented the construction of the complete, supersymmetric, D-brane action.

For reasons explained elsewhere (see, e.g. (Bergshoeff, Sezgin, Townsend (1988))) super p-brane actions are required to have a fermionic gauge invariance, usually called 'kappa-symmetry', and super D-branes are no exception. Starting from the bosonic result, it is not difficult to guess what the super Dp-brane action is likely to be for general p, and in general supergravity backgrounds. The difficulty lies in verifying the correctness of this guess. The aim of this work is to elucidate some aspects of this verification. For more details, we refer to (Bergshoeff, Townsend (1997)).

2 The Bosonic Case

In this section we first briefly review the form of the bosonic D-brane actions. Our starting point is the following effective action of a bosonic Dp-brane $(0 \leq p \leq 9)$:

$$I_{\mathrm{Dp}} = -\int \mathrm{d}^{p+1}\xi \mathrm{e}^{-\phi}\sqrt{|\det(g_{ij} + \mathcal{F}_{ij})|} + C\mathrm{e}^{\mathcal{F}} + mI_{\mathrm{CS}} \ . \tag{1}$$

Here ϕ is the dilaton, g_{ij} is the induced metric

$$g_{ij} = \partial_i X^\mu \partial_j X^\nu g_{\mu\nu} \ , \tag{2}$$

($g_{\mu\nu}$ is the D=10 target space metric) and $\mathcal{F}_{ij}$ ($i = 1, \cdots (p+1)$) is the modified 2-form field strength

$$\mathcal{F} = F - B \ . \tag{3}$$

The curvature $F = \mathrm{d}V$ is the field-strength of the Born-Infeld vector V and B is the pull-back of the NS-NS 2-form field. The first term in (1) is usually called the kinetic term I_{DBI} and is of the Born-Infeld type. The second term in (1) is a Wess-Zumino (WZ) term I_{WZ} where

$$C = \sum_{r=0}^{10} C^{(r)} \tag{4}$$

is a formal sum of the RR potentials $C^{(r)}$ and where it is understood that after expanding the expotential only the (p+1)-form is retained. The last term is only present for even p (the IIA case). Its coefficient m is the IIA mass parameter.

We note that recently alternative forms of the Dp–brane actions (1) have been obtained that are quadratic in derivatives of X^μ and linear in $\mathcal{F}_{ij}$ by introducing an auxiliary metric which has both symmetric and antisymmetric parts, thereby generalising the simplification of the Nambu-Goto action for p–branes using a symmetric metric (Zeid and Hull (1997)). For earlier work in this direction, using only a symmetric auxiliary metric, see (Cederwall et al. (1997)).

3 The Supersymmetric Case

In the supersymmetric case the Dp-brane actions look the same with the understanding that the coordinates X^μ have been replaced by supercoordinates Z^M and the fields have been replaced by corresponding superfields, whose leading component in a θ-expansion is the original (bosonic) field. Furthermore the induced metric in the supersymmetric case is given by

$$g_{ij} = E_i^a E_j^b \eta_{ab} \ , \tag{5}$$

where

$$E_i^A = \partial_i Z^M E_M{}^A \ . \tag{6}$$

In the latter expression we have used the supervielbein $E_M{}^A$. We use a notation where the flat index A decomposes under the action of the Lorentz group into a Lorentz vector index a and a Lorentz spinor index α. The latter is a 32-component Majorana spinor in the IIA case and a pair of chiral Majorana spinors in the IIB case:

$$A = \begin{cases} (a, \alpha) & \text{IIA} \ , \\ (a, \alpha I) \quad I = 1, 2 & \text{IIB} \ . \end{cases} \tag{7}$$

4 Kappa-Symmetry

In (Bergshoeff, Townsend (1997), Aganagic, Popescu, Schwarz (1997), Cederwall et al. (1996)) it has been shown that the action (1) is invariant under so-called kappa-transformations which are worldvolume gauge transformations with a fermionic spinor parameter κ:

$$\begin{aligned} \delta Z^M E_M^a &= 0 \,, \\ \delta Z^M E_M^\alpha &= [\bar{\kappa}(1 + \Gamma)]^\alpha \,, \\ \delta V_i &= E_i^A \delta E^B B_{BA} \ . \end{aligned} \tag{8}$$

A similar derivation using actions with simultaneous worldvolume and spacetime supersymmetry has been given in (Howe, Sezgin (1997)). The expression for Γ is given by

$$\Gamma = \begin{cases} \frac{1}{\sqrt{|\det\ (1+X)|}} s e^{\frac{1}{2}\gamma^{ij}\mathcal{F}_{ij}\Gamma_{11}} \Gamma_{(0)} & \text{IIA} \ , \\ \frac{1}{\sqrt{|\det\ (1+X)|}} s e^{-\frac{1}{2}\gamma^{ij}\mathcal{F}_{ij}\otimes\sigma_3} \Gamma_{(0)} & \text{IIB} \ , \end{cases} \tag{9}$$

where $\Gamma_{(0)}$ is given by[1]

$$\Gamma_{(0)} = \begin{cases} (\Gamma_{11})^{\frac{p-2}{2}} \frac{1}{(p+1)!} \frac{1}{\sqrt{|g|}} \epsilon^{i_1 \cdots i_{p+1}} \gamma_{i_1 \cdots i_{p+1}} & \text{IIA} \ , \\ \frac{i}{(p+1)!} \frac{1}{\sqrt{|g|}} \epsilon^{i_1 \cdots i_{p+1}} \gamma_{i_1 \cdots i_{p+1}} \otimes (\sigma_3)^{\frac{p-3}{2}} \sigma_2 & \text{IIB} \ . \end{cases} \tag{10}$$

It is straightforward to verify that the matrix $\Gamma_{(0)}$ defined in (10) satisfies

$$\left(\Gamma_{(0)}\right)^2 = 1 \,, \qquad \text{Tr}\ \Gamma_{(0)} = 0 \ . \tag{11}$$

[1] Note that this $\Gamma_{(0)}$ slightly differs form the one given in (Bergshoeff, Townsend (1997)).

The symbol se in the above equations defines a "skew-symmetric exponential", which means that in the expansion as a usual exponential one only keeps the terms where the gamma-matrices are completely antisymmetrized, i.e.

$$se^{\gamma^{ij}X_{ij}} = \sum_{n=0}^{\infty} \gamma^{i_1 j_1 \cdots i_n j_n} X_{i_1 j_1} \cdots X_{i_n j_n} \quad . \tag{12}$$

The matrices γ_i are defined as

$$\gamma_i = E_i{}^a \Gamma_a \quad . \tag{13}$$

Finally, the matrix X is defined as

$$X^i{}_j = g^{ik} \mathcal{F}_{kj} \quad , \tag{14}$$

where g^{ij} is the inverse induced metric.

The actual proof that the super D-brane actions (1) are invariant under the kappa transformations (8) is rather involved. Here we give a brief sketch of the proof. For more details, see e.g. section 4 of (Bergshoeff, Townsend (1997)). For simplicity we restrict ourselves to bosonic backgrounds and take $m = 0$.

We first calculate the kappa variation of the super D-brane action (1) without actually using the form of δE^α. For this purpose we note that the kappa transformation of the Born-Infeld vector V implies that its curvature transforms "kappa-covariantly", i.e.

$$\delta_\kappa \mathcal{F}_{ij} = -E_{[i}^A E_{j]}^B \delta_\kappa E^C H_{CBA} \quad . \tag{15}$$

It also follows that the kappa transformation of the embedding metric is given by

$$\delta_\kappa g_{ij} = -2 E_i^a \delta_\kappa E^B E_j^C T_{CB}{}^b \eta_{ab} \quad . \tag{16}$$

Substituting in the above transformations the superspace constraints for the torsion T and the 3-form curvature H we can calculate the kappa variation of the kinetic term in (1). Similarly, we find that the kappa variation of the WZ term in (1) is given by

$$\delta_\kappa I_{\mathrm{WZ}} = \int_W \frac{1}{(n-1)!} E^{A_{p+2}} \cdots E^{A_2} \delta E^{A_1} R(C)_{A_1 A_2 \cdots A_n} \mathrm{e}^{\mathcal{F}} \quad , \tag{17}$$

where $R(C)$ is the curvature of the RR fields C (Green, Hull, Townsend (1996)):

$$R(C) = \mathrm{d}C - HC \quad . \tag{18}$$

Substituting the superspace constraints for $R(C)$ we can calculate the kappa variation of the WZ term.

We next combine the kappa variations of the kinetic and WZ term and find after some straightforward algebra the following result:

$$\delta_\kappa I_{\rm Dp} = -i\int \mathrm{d}^{p+1}\xi\, \mathcal{L}_{\rm DBI}\left(\delta_\kappa E K^i_{(p)} E_i\right) \ , \tag{19}$$

where

$$K^i_{(p)} = N^i + M^i_{(p)} \ , \tag{20}$$

with N^i and $M^i_{(p)}$ given by

$$N^i = \begin{cases} \left[(1-X^2)^{-1}\right]^i{}_j\gamma^j + \left[(1-X^2)^{-1}X\right]^i{}_j\gamma^j\,\Gamma_{11} & \text{(IIA)} \ , \\ \left[(1-X^2)^{-1}\right]^i{}_j\gamma^j\otimes \mathbf{1}_2 - \left[(1-X^2)^{-1}X\right]^i{}_j\gamma^j\otimes\sigma_3 & \text{(IIB)} \ , \end{cases} \tag{21}$$

and

$$M^i_{(p)} = \frac{1}{\sqrt{\det(1+X)}}\sum_{n=0}^{\infty}\frac{1}{2^n n!}\gamma^{ij_1k_1\dots j_nk_n} X_{j_1k_1}\dots X_{j_nk_n}\begin{cases}(-\Gamma_{11})^{n+(p-2)/2} & \text{(IIA)} \ , \\ (-\sigma_3)^{n+(p-3)/2} & \text{(IIB)} \ , \end{cases} \tag{22}$$

respectively.

The proof of kappa-symmetry of the Dp-brane actions (1) now boils down to showing that the following identity holds

$$(1+\Gamma)K^i_{(p)} \equiv 0 \ , \tag{23}$$

with Γ given in (9) and $K^i_{(p)}$ given above. This identity has sofar only been proven by brute force. This concludes our proof of kappa–invariance.

We note that the $p = 9$ case is equivalent to a supersymmetric $D = 10$ Born–Infeld action. The explicit supersymmetry rules, after gauge-fixing the kappa-symmetry and for a flat background, have been given in the second reference of (Aganagic, Popescu, Schwarz (1997)). A curious feature of the super Born-Infeld action in a curved background is that it makes essential use of an 11-form superspace field strength, which vanishes identically when restricted to space-time. The possibility of such superspace gauge fields has been explored in the past in (Gates Jr. (1981)).

5 New Developments

A crucial property of the kappa-rules is that they must eliminate half of the fermionic degrees of freedom. The fermionic coordinates θ^α therefore describe $32/(2\times 2) = 8$ degrees of freedom. This matches the 8 bosonic degrees of freedom described by the scalars X^μ (10–(p+1) d.o.f.) and the Born-Infeld vector V_i ((p–1) d.o.f.). Kappa-symmetry therefore leads to a correct Bose-Fermi matching on the worldvolume.

In order that the kappa-symmetry eliminates half of the fermionic degrees of freedom, the matrix Γ defined in (9) must satisfy the properties of a projection operator, i.e.

$$\Gamma^2 = 1\,, \qquad\qquad \mathrm{Tr}\,\Gamma = 0\ . \tag{24}$$

One recent development is that the proof of the first identity in (24) can be considerably simplified (Bergshoeff et al. (in preparation)). More explicitly, it can be shown that Γ can be rewritten in terms of an ordinary exponential as follows:

$$\Gamma = \begin{cases} \mathrm{e}^{\frac{1}{2}\gamma^{ij}\mathcal{F}_{ij}\Gamma_{11}}\,\Gamma_{(0)} & \mathrm{IIA}\ , \\ \mathrm{e}^{-\frac{1}{2}\gamma^{ij}\mathcal{F}_{ij}\otimes\sigma_3}\,\Gamma_{(0)} & \mathrm{IIB}\ , \end{cases} \tag{25}$$

where the matrices X and Y are related in the following way. First choose a Lorentz frame such that the non-zero components of X are given by (we give the IIB case, the IIA case goes in a similar way)

$$X_{0,n} = \tanh\frac{\theta_0}{2}\,, \qquad X_{\alpha,n+\alpha} = \tan\frac{\theta_\alpha}{2}\,, \quad \alpha = 1,\cdots,n\ . \tag{26}$$

In this Lorentz frame the non-vanishing matrix elements of Y are given by

$$Y_{0n} = \frac{\theta_0}{2}\,, \qquad Y_{\alpha,n+\alpha} = \frac{\theta_\alpha}{2}\,, \quad \alpha = 1,\cdots,n\ . \tag{27}$$

One thus establishes a relationship between kappa-symmetry and "rotated branes" (Berkooz et al. (1996), Balasubrananian, Leigh (1996)). For more details, see (Bergshoeff et al. (in preparation)).

Acknowledgements

The work described here was done in collaboration with Paul Townsend. I also report about work in preparation done in collaboration with Renata Kallosh, George Papadopoulos and Tomas Ortín. This work is supported by the European Commission TMR programme ERBFMRX-CT96-0045, in which I am associated to the University of Utrecht.

References

Abou Zeid, M. and Hull, C.M. (1997): `hep-th/9704021`.

Aganagic, M., Popescu, C. and Schwarz, J.H. (1997): Phys. Lett. B393, 311; `hep-th/9612080`.

Cederwall, M., von Gussich, A., Nilsson, B.E.W. and Westerberg, A. (1996): `hep-th/9610148`;
Cederwall, M., von Gussich, A., Nilsson, B.E.W., Sundell, P. and Westerberg, A. (1996): `hep-th/9611159`.

Cederwall, M., von Gussich, A., Mikovic, A., Nilsson, B.E.W. and Westerberg, A. (1997): Phys. Lett. B390, 148.

Bergshoeff, E., Sezgin, E. and Townsend, P.K. (1988): Phys. Lett. **189B**, 75; Ann. Phys. (NY) **185**, 330.

Bergshoeff, E. and Townsend, P.K. (1997): Nucl. Phys. B490, 145.

Bergshoeff, E., Kallosh, R., Papadopoulos, G. and Ortín, T. (in preparation).

Berkooz, M., Douglas, M.R. and Leigh, R.G. (1996): Nucl. Phys. B480, 296; Balasubramanian, V. and Leigh, R.G. (1996): hep-th/9611165.

Gates, S.J. Jr. (1981): Nucl. Phys. B184, 381.

Green, M.B., Hull, C.M. and Townsend, P.K. (1996): Phys. Lett. B382, 65.

Howe, P.S. and Sezgin, E. (1997): Phys. Lett. B390, 148.

Polchinski, J. (1995): Phys. Rev. Lett. **75**, 4724.

Volkov, D.V. and Akulov, V.P. (1972, 1973): JETP Lett. 16, 438; Phys. Lett. 46B (1973) 109; JETP Lett. 17, 261.

Volkov–Akulov Theory and D-Branes

R. Kallosh

Physics Department, Stanford University,
Stanford, CA 94305-4060, USA **

Abstract. The action of supersymmetric Born-Infeld theory (D-9-brane in a Lorentz covariant static gauge) has a geometric form of the Volkov-Akulov-type. The first non-linearly realized supersymmetry can be made manifest, the second worldvolume supersymmetry is not manifest. We also study the analogous 2 supersymmetries of the quadratic action of the covariantly quantized D-0-brane. We show that the Hamiltonian and the BRST operator are build from these two supersymmetry generators.

This article is dedicated to the memory of D. V. Volkov whose insights into the nature of supersymmetry and geometry proved to be enlightening for few generations of high-energy physicists. His ideas inspired the most active developments in theoretical physics over the last quarter of the century.

Extended objects with global supersymmetry have local κ-symmetry. This symmetry is difficult to quantize in Lorentz covariant gauges keeping finite number of fields in the theory. A revival of interest to κ-symmetric objects is due to the recent discovery of D-p-branes (Polchinski 1996), κ-symmetric non-linear effective actions and/or equations of motion for D-branes (Cederwall et al. 1997), (Aganagic et al. 1996), (Bergshoeff and Townsend 1997), (Howe and Sezgin 1997a), (Abou Zeid and Hull 1997), (Bergshoeff et al. 1997) and the M-5-brane action (Howe and Sezgin 1997b), (Pasti et al. 1997), (Bandos et al. 1997b), (Aganagic et al. 1997), complementing the κ-symmetric superstring and M-2-brane (Green and Schwarz 1984), (Bergshoeff et al. 1987). A nice review on worldvolume actions in a doubly supersymmetric geometric approach initiated by D. V. Volkov is presented in these Proceedings (Bandos et al. 1997a).

The new issues in quantization of D-branes have been analyzed recently in (Aganagic et al. 1996), (Bergshoeff et al. 1997), (Kallosh 1997).

The κ-symmetric D-brane action in the flat background geometry[1] consists of the Born-Infeld-Nambu-Goto term S_1 and Wess-Zumino term S_2:

$$S_{\rm DBI} + S_{\rm WZ} = T\left(- \int {\rm d}^{p+1}\sigma \sqrt{-\det(G_{\mu\nu} + \mathcal{F}_{\mu\nu})} + \int \Omega_{p+1} \right) \ . \quad (1)$$

Here T is the tension of the D-brane, $G_{\mu\nu}$ is the manifestly supersymmetric induced world-volume metric

** This work is supported by the NSF grant PHY-9219345.

[1] We use notation of (Aganagic et al. 1996).

$$G_{\mu\nu} = \eta_{mn} \Pi_\mu^m \Pi_\nu^n , \qquad \Pi_\mu^m = \partial_\mu X^m - \bar\theta \Gamma^m \partial_\mu \theta \ , \tag{2}$$

and $\mathcal{F}_{\mu\nu}$ is a manifestly supersymmetric Born-Infeld field strength (for p even) [2]

$$\mathcal{F}_{\mu\nu} \equiv F_{\mu\nu} - b_{\mu\nu} =$$
$$= \left[\partial_\mu A_\nu - \bar\theta \Gamma_{11} \Gamma_m \partial_\mu \theta \left(\partial_\nu X^m - \frac{1}{2} \bar\theta \Gamma^m \partial_\nu \theta\right)\right] - (\mu \leftrightarrow \nu) \ . \tag{3}$$

When p is odd, Γ_{11} is replaced by $\tau_3 \otimes I$. The action has global supersymmetry

$$\delta_\epsilon \theta = \epsilon, \qquad \delta_\epsilon X^m = \bar\epsilon \Gamma^m \theta \ . \tag{4}$$

and local κ-supersymmetry:

$$\delta X^m = \bar\theta \Gamma^m \delta\theta = -\delta\bar\theta \Gamma^m \theta, \qquad \delta\bar\theta = \bar\kappa (1 + \Gamma) \ , \tag{5}$$

and

$$\Gamma = \mathrm{e}^{\frac{a}{2}} \Gamma'_{(0)} \mathrm{e}^{-\frac{a}{2}} , \qquad a = \begin{cases} +\frac{1}{2} Y_{jk} \gamma^{jk} \Gamma_{11} & \text{IIA} \ , \\ -\frac{1}{2} Y_{jk} \gamma^{jk} \sigma_3 \otimes 1 & \text{IIB} \ . \end{cases} \tag{6}$$

Here $\Gamma'_{(0)}$ is the product structure, independent on the BI field, $(\Gamma'_{(0)})^2 = 1, \mathrm{tr}\, \Gamma'_{(0)} = 0)$. All dependence on BI field $\mathcal{F} =$ "tan"Y is in the exponent (Bergshoeff et al. 1997). The matrix Γ_{11} in IIA and $\sigma_3 \otimes 1$ in IIB theory anticommute with $\Gamma'_{(0)}$ and with Γ. Therefore in the basis where Γ_{11} and $\sigma_3 \otimes 1$ are diagonal, $\Gamma'_{(0)}$ and Γ are off-diagonal. We introduce a 16×16-dimensional matrix $\hat\gamma$ which does not depend on BI field.

$$\Gamma'_{(0)} = \begin{pmatrix} 0 & \hat\gamma \\ \hat\gamma^{-1} & 0 \end{pmatrix} , \qquad \Gamma = \begin{pmatrix} 0 & \hat\gamma \mathrm{e}^{\hat a} \\ (\hat\gamma \mathrm{e}^{\hat a})^{-1} & 0 \end{pmatrix} , \tag{7}$$

where

$$\hat a = \begin{cases} +\frac{1}{2} Y_{jk} \gamma^{jk} & \text{IIA} \ , \\ -\frac{1}{2} Y_{jk} \gamma^{jk} & \text{IIB} \ . \end{cases} \tag{8}$$

The fact that Γ is off-diagonal and that the matrix $\gamma \mathrm{e}^{\hat a}$ is invertible is quite important and the significance of this for covariant quantization of D-branes was already discussed in (Aganagic et al. 1996), (Bergshoeff et al. 1997), (Kallosh 1997). In particular this allows us to consider only irreducible κ-symmetry transformations by imposing a Lorentz covariant condition on $\bar\kappa$ of the form

$$\bar\kappa_1 = 0 \qquad \bar\kappa_2 \neq 0 \qquad \text{IIA} \tag{9}$$
$$\bar\kappa_2 = 0 \qquad \bar\kappa_1 \neq 0 \qquad \text{IIB} \ . \tag{10}$$

[2] We define spinors for even p as $\theta = \theta_1 + \theta_2$ where $\theta_1 \equiv \frac{1}{2}(1 + \Gamma_{11})\theta$ and $\theta_2 \equiv \frac{1}{2}(1 - \Gamma_{11})\theta$.

In this way we have an irreducible 16-dimensional κ-symmetry since the matrix $\hat{\gamma}e^{\hat{a}}$ is invertible, acting as

$$\delta\bar{\theta}_1 = \bar{\kappa}_2\hat{\gamma}e^{\hat{a}} \qquad \delta\bar{\theta}_2 = \bar{\kappa}_2 \qquad \delta X^m = -\bar{\kappa}_2\Gamma^m\theta_2 \qquad \text{IIA} \quad (11)$$

$$\delta\bar{\theta}_1 = \bar{\kappa}_1 \qquad \delta\bar{\theta}_2 = \bar{\kappa}_1(\hat{\gamma}e^{\hat{a}})^{-1} \qquad \delta X^m = -\bar{\kappa}_1\Gamma^m\theta_1 \qquad \text{IIB} \ . \quad (12)$$

Recently a covariant gauge fixing κ-symmetry of D-branes has been discovered (Aganagic et al. 1996). The fermionic gauge is of the form

$$\theta_2 = 0 \qquad \text{IIA} \qquad \theta_1 \equiv \lambda \qquad (13)$$

$$\theta_1 = 0 \qquad \text{IIB} \qquad \theta_2 \equiv \lambda \ . \qquad (14)$$

Note that our choice of irreducible κ-symmetry (which is not unique) was made here with the purpose to explicitly eliminate θ_2 (θ_1) in IIA (IIB) case using $\delta\bar{\theta}_2 = \bar{\kappa}_2$ ($\delta\bar{\theta}_1 = \bar{\kappa}_1$). The gauge-fixed action has one particularly useful property: the Wess-Zumino term vanishes in this gauge (Aganagic et al. 1996). We are left with the reparametrization invariant action:

$$S_{\kappa-\text{fixed}} = -\int \mathrm{d}^{p+1}\sigma\sqrt{-\det(G_{\mu\nu} + \mathcal{F}_{\mu\nu})} \ , \qquad (15)$$

$$G_{\mu\nu} = \eta_{mn}\Pi_\mu^m\Pi_\nu^n \ , \qquad \Pi_\mu^m = \partial_\mu X^m - \bar{\lambda}\Gamma^m\partial_\mu\lambda \ , \qquad (16)$$

$$\mathcal{F}_{\mu\nu} = [\partial_\mu A_\nu - \bar{\lambda}\Gamma_m\partial_\mu\lambda\left(\partial_\nu X^m - \frac{1}{2}\bar{\lambda}\Gamma^m\partial_\nu\lambda\right)] - (\mu \leftrightarrow \nu) \ . \qquad (17)$$

This action (15) has a local reparametrization symmetry and a 32-component global supersymmetry. The form of the action is such that it can be brought to the form close to the one discovered by Volkov-Akulov (Volkov and Akulov 1972). Consider for example a D-9-brane in a static gauge (Aganagic et al. 1996) $X^\mu = \sigma^\mu$:

$$S_{\text{g.f.}} = -\int \mathrm{d}^{10}\sigma\sqrt{-\det(G_{\mu\nu} + \mathcal{F}_{\mu\nu})} \ , \qquad (18)$$

where

$$G_{\mu\nu} = e_\mu{}^m e_\nu{}^n \eta^{mn} \qquad (19)$$

and

$$e_\mu{}^m = \delta_\mu{}^m - \bar{\lambda}\Gamma^m\partial_\mu\lambda \ . \qquad (20)$$

If we introduce the 1-forms depending on fermion fields $\lambda(\sigma)$

$$e^m = d\sigma^\mu e_\mu{}^m[\lambda(\sigma)] = d\sigma^m + \bar{\lambda}\Gamma^m d\lambda \qquad (21)$$

we can rewrite the supersymmetric Born-Infeld 9-brane action as

$$S = \int e^{m_0} \wedge e^{m_1} \wedge \cdots \wedge e^{m_9} \sqrt{-\det(\eta_{mn} + \mathcal{F}_{mn})} \ , \qquad (22)$$

where $\mathcal{F}_{mn} = e^{\mu}{}_m e^{\nu}{}_n \mathcal{F}_{\mu\nu}$. In absence of the two-form $\mathcal{F}_{mn}$ the supersymmetric Born-Infeld action is reduced to geometric action of Volkov-Akulov (Volkov and Akulov 1972), generalized to d=10:

$$S = \int e^{m_0} \wedge e^{m_1} \wedge \cdots \wedge e^{m_9} = \int \mathrm{d}^{10}\sigma \, \mathrm{dete}[\lambda(\sigma)] \ . \tag{23}$$

This action depends only on fermions $\lambda(\sigma)$ and has a non-linearly realized supersymmetry manifest, since the 1-forms e^m are supersymmetric. The second supersymmetry of the supersymmetric Born-Infeld 9-brane action (22) is not manifest. It explicit form can be obtained from the preservation of the gauge-fixing condition for the kappa-symmetry.

The gauge-fixed actions of extended supersymmetric objects in static gauge are known to lead to complicated non-linear actions. For example, the action of d=2 massive superparticle (Achúcarro et al. 1980)

$$S = -M \int \mathrm{d}t \left(\left[-(\dot{X}^m - \bar{\theta}\Gamma^m \dot{\theta})(\dot{X}^n - \bar{\theta}\Gamma^n \dot{\theta})\eta_{mn} \right]^{1/2} - 1 + \theta\Gamma^3\dot{\theta} \right)$$

$$m = 0, 1 \ . \tag{24}$$

quantized in the gauge $\Gamma^3\theta = \theta$ for κ-symmetry and the static gauge for reparametrization symmetry, $X^0 = t$, gives the kink effective action (Achúcarro et al. 1980)

$$S_{\mathrm{g.f.}} = -M \int \mathrm{d}t \left(\left[1 - \dot{\phi}^2 \right]^{1/2} - 1 + \frac{\mathrm{i}}{4M} \rho\dot{\rho} \right) \ . \tag{25}$$

Here the bosonic field is the remaining coordinate of the d=2 superparticle, $\phi = X^1$. The Hamiltonian associated with this action is also non-linear:

$$H = (p^2 + M^2)^{1/2} - M \ , \tag{26}$$

p is the momentum canonical conjugate to X^1.

Here we will perform a covariant quantization of the D-0-brane which is a d=10 generalization of the d=2 massive superparticle (Achúcarro et al. 1980). Instead of the static gauge, which belongs to a class of canonical gauges with the non-propagating ghosts, we will use a covariant gauge for reparametrization symmetry. Consider the κ-symmetric action of a D-0-brane (Cederwall et al. 1997), (Aganagic et al. 1996), (Bergshoeff and Townsend 1997), (Howe and Sezgin 1997a), (Abou Zeid and Hull 1997). D-0-brane action does not have Born-Infeld field since there is no place for an antisymmetric tensor of rank 2 in one-dimensional theory. The D-p-brane action for $p = 0$ case reduces to

$$S = -T \left(\int \mathrm{d}\tau \sqrt{-G_{\tau\tau}} + \int \bar{\theta}\Gamma^{11}\dot{\theta} \right) \ . \tag{27}$$

This action can be derived from the action of the massless 11-dimensional superparticle (Bergshoeff and Townsend 1997).

$$S = \int d\tau \sqrt{-g_{\tau\tau}} g^{\tau\tau} \left(\dot{X}^{\hat{m}} - \bar{\theta}\Gamma^{\hat{m}}\dot{\theta}\right)^2 , \qquad \hat{m} = 0, 1, \cdots, 8, 9, 10 . \quad (28)$$

We may solve equation of motion for $X^{\hat{10}}$ as $\mathbf{P}_{\hat{10}} = Z$, where Z is a constant, and use $\Gamma^{11} = \Gamma^{\hat{10}}$. From this one can deduce a first order action

$$S = \int d\tau \left(\mathbf{P}_m(\dot{X}^m - \bar{\theta}\Gamma^m\dot{\theta}) + \frac{1}{2}V(\mathbf{P}^2 + Z^2) - Z\bar{\theta}\Gamma^{11}\dot{\theta} + \bar{\chi}_1 d_2\right) . \quad (29)$$

We will show now that the D-0-brane action can be obtained from this one upon solving equations of motion for $\mathbf{P}_m$, V, χ_1, and d_2. Here $V(\tau)$ is a Lagrange multiplier, $Z = T$ is some constant parameter in front of the WZ term and $\mathbf{P}^2 \equiv \mathbf{P}^m \eta_{mn} \mathbf{P}^n$. The chiral spinors χ_1 and d_2 are auxiliary fields. They are introduced to close the gauge symmetry algebra off shell. To verify that this first order action is one of the D-p-brane family actions given in (1) we can use equations of motion for $\mathbf{P}_m$

$$\mathbf{P}_m = -\frac{1}{V}(\dot{X}^m - \bar{\theta}\Gamma^m\dot{\theta}) , \quad (30)$$

and for the auxiliary fields $\chi_1 = 0$ and $d_2 = 0$. The action (29) becomes

$$S = \int d\tau \left(-\frac{1}{2V}(\dot{X}^m - \bar{\theta}\Gamma^m\dot{\theta})^2 + \frac{1}{2}VZ^2 - Z\bar{\theta}\Gamma^{11}\dot{\theta}\right) . \quad (31)$$

Equation of motion for V is

$$V^2 = -\frac{1}{Z^2}(\dot{X}^m - \bar{\theta}\Gamma^m\dot{\theta})^2 , \quad (32)$$

and we can insert $V = -\frac{1}{Z}\sqrt{-(\dot{X}^m - \bar{\theta}\Gamma^m\dot{\theta})^2}$ back into the action (31) and get

$$S = -Z\left(\int d\tau \sqrt{-(\dot{X}^m - \bar{\theta}\Gamma^m\dot{\theta})^2} + Z\bar{\theta}\Gamma^{11}\dot{\theta}\right) . \quad (33)$$

This is the action (1) for D-0-brane at $T = Z$ as given in (27).

The action (29) is invariant under the 16-dimensional irreducible κ-symmetry and under the reparametrization symmetry. The gauge symmetries are (we denote $\Gamma^m \mathbf{P}_m = \not{\mathbf{P}}$):

$$\delta\bar{\theta} = \bar{\kappa}_2(\Gamma^{11}Z + \not{\mathbf{P}}) , \quad (34)$$

$$\delta X^m = -\eta\mathbf{P}^m - \delta\bar{\theta}\Gamma^m\theta - \bar{\kappa}_2\Gamma^m d , \quad (35)$$

$$\delta V = \dot{\eta} + 4\bar{\kappa}_2\dot{\theta} + 2\bar{\chi}_1\kappa_2 \quad (36)$$

$$\delta\bar{\chi} = \bar{\kappa}_2 \dot{\not{\mathbf{P}}} , \quad (37)$$

$$\delta d = [\mathbf{P}^2 + Z^2]\kappa_2 . \quad (38)$$

Here $\eta(\tau)$ is the time reparametrization gauge parameter and $\kappa_2(\tau) = \frac{1}{2}(1 - \Gamma^{11})\kappa(\tau)$ is the 16-dimensional parameter of κ-symmetry. The gauge symmetries form a closed algebra

$$[\delta(\kappa_2), \delta(\kappa_2')] = \delta(\eta = 2\bar{\kappa}_2 \not{P} \kappa_2') \ . \tag{39}$$

To bring the theory to the canonical form we introduce canonical momenta to θ and to V and find, excluding auxiliary fields

$$L = \mathbf{P}_m \dot{X}^m + P_V \dot{V} + \bar{P}_\theta \dot{\theta} + \frac{1}{2} V(\mathbf{P}^2 + Z^2) + P_V \varphi + (\bar{P}_\theta + \bar{\theta}(\not{P} + Z\Gamma^{11}))\,\psi \ . \tag{40}$$

We have primary constraints $\bar{\Phi} \equiv \bar{P}_\theta + \bar{\theta}(\not{P} + Z\Gamma^{11}) \approx 0$ and $P_V = 0$. The Poisson brackets for 32 fermionic constraints are

$$\{\Phi, \Phi\} = 2C(\not{P} + \Gamma_{11} Z) \ . \tag{41}$$

We also have to require that the constraints are consistent with the time evolution $\{P_V, H\} = 0$. This generates a secondary constraint

$$t = \mathbf{P}^2 + Z^2 \ . \tag{42}$$

Thus the Hamiltonian is weakly zero and any physical state of the system satisfying the reparamet-
rization constraint is a BPS state $M = |Z|$ since

$$\mathbf{P}^2 + Z^2 |\Psi\rangle = 0 \qquad \Longrightarrow \qquad Z^2|\Psi\rangle = -\mathbf{P}^2|\Psi\rangle = M^2|\Psi\rangle \ . \tag{43}$$

The 32×32 -dimensional matrix $C(\not{P} + \Gamma_{11} Z)$ is not invertible since it squares to zero when the reparametrization constraint is imposed. This is a reminder of the fact that D-0-brane is a d=11 massless superparticle. The 32 dimensional fermionic constraint has a 16-dimensional part which forms a first class constraint and another 16-dimensional part which forms a second class constraint. We notice that the Poisson brackets reproduce the $d = 10$, $N = 2$ algebra with the central charge which can also be understood as $d = 11$, $N = 1$ supersymmetry algebra with the constant value of $\mathbf{P}_{11} = Z$.

We proceed with the quantization and gauge-fix κ-symmetry covariantly by taking $\theta_2 = 0, \theta_1 \equiv \lambda$ and find

$$L^\kappa_{\text{g.f.}} = \mathbf{P}_m(\dot{X}^m - \bar{\lambda}\Gamma^m \dot{\lambda}) + \frac{1}{2} V(\mathbf{P}^2 + Z^2) \ . \tag{44}$$

The 16-dimensional fermionic constraint

$$\bar{\Phi}_\lambda \equiv (\bar{P}_\lambda + \bar{\lambda} \not{P}) \approx 0 \tag{45}$$

forms the Poisson bracket

$$\{\Phi^\alpha_\lambda, \Phi^\beta_\lambda\} = 2(\not{P} C)^{\alpha\beta} \ . \tag{46}$$

The matrix $\not{P}C$ is perfectly invertible as long as the central charge Z is not vanishing. The inverse to (46) is

$$\{\Phi^\alpha, \Phi^\beta\}^{-1}\,|_{t=0} = [2(\not{P}C)^{\alpha\beta}]^{-1} = \frac{(C\not{P})_{\alpha\beta}}{2\mathbf{P}^2} \ . \tag{47}$$

This proves that the fermionic constraints are second class and that the fermionic part of the Lagrangian

$$-\bar\lambda\not{P}\dot\lambda \equiv -\mathrm{i}\lambda^\alpha\Phi_{\alpha\beta}\dot\lambda^\beta \ , \qquad \Phi_{\alpha\beta} = -\mathrm{i}(C\not{P})_{\alpha\beta} \ , \tag{48}$$

is not degenerate in a Lorentz covariant gauge. None of this would be true for a vanishing central charge. Note that in the rest frame $\mathbf{P}_0 = M, \mathbf{P} = 0$, hence

$$\Phi_{\alpha\beta} = M\delta_{\alpha\beta} \ . \tag{49}$$

For D-0-brane one can covariantly gauge-fix the reparametrization symmetry by choosing the $V = 1$ gauge and including the anticommuting reparametrization ghosts b, c. This brings us to the following form of the action:

$$L^{\kappa,\eta}_{\mathrm{g.f.}} = \mathbf{P}_m\dot X^m - \bar\lambda\not{P}\dot\lambda + \frac{1}{2}(\mathbf{P}^2 + Z^2) + b\dot c \ . \tag{50}$$

Now we can define Dirac brackets

$$\{\lambda, \bar\lambda\}^* = \{\lambda, \bar\Phi\}\{\bar\Phi, \Phi\}^{-1}\{\Phi\,, \bar\lambda\} = \frac{\not{P}}{2\mathbf{P}^2} = -\frac{\not{P}}{2Z^2} \ . \tag{51}$$

The generator of the 32-dimensional supersymmetry is

$$\bar\epsilon Q = \bar\epsilon(\not{P} + \Gamma^{11}Z)\lambda \ . \tag{52}$$

It forms the following Dirac bracket

$$\begin{aligned}[\bar\epsilon Q\,, \bar Q\epsilon']^* &= \bar\epsilon(\not{P} + \Gamma^{11}Z)\frac{\not{P}}{2\mathbf{P}^2}(\not{P} + \Gamma^{11}Z)\epsilon' = \\ &= \bar\epsilon\Gamma^{\hat m}\mathbf{P}_{\hat m}\epsilon' = \bar\epsilon(\not{P} + \Gamma^{11}Z)\epsilon' \ .\end{aligned} \tag{53}$$

We can also rewrite it in d=11 Lorentz covariant form

$$[\bar\epsilon Q\,, \bar Q\epsilon']^* = \bar\epsilon\Gamma^{\hat m}\mathbf{P}_{\hat m}\epsilon' = \bar\epsilon\hat{\not{P}}\epsilon' \ , \qquad \hat m = 0, 1, \cdots, 8, 9, 10 \ ,$$

$$Z = \mathbf{P}_{\hat{10}} \ , \qquad \Gamma^{11} = \Gamma^{\hat{10}} \ . \tag{54}$$

This Dirac bracket realizes the d=11, N=1 supersymmetry algebra or, equivalently, d=10, N=2 supersymmetry algebra with the central charge Z.

One can also to take into account that the path integral in presence of second class constraints has an additional term with $\sqrt{\mathrm{Ber}\{\Phi_\lambda, \Phi_\lambda\}} \sim \sqrt{\mathrm{Ber}\,\Phi_{\alpha\beta}}$ (Fradkin 1973). It can be used to make a change of variables

$$S_\alpha = \Phi^{1/2}_{\alpha\beta}\,\lambda^\beta \ . \tag{55}$$

The action becomes

$$L = \mathbf{P}_m \dot{X}^m - \mathrm{i} S_\alpha \dot{S}_\alpha + b\dot{c} - H \tag{56}$$

$$H = -\frac{1}{2}(\mathbf{P}^2 + Z^2) . \tag{57}$$

The generators of global supersymmetry commuting with the Hamiltonian take the form

$$\bar{\epsilon}Q = \bar{\epsilon}(\not{P} + \Gamma^{11} Z)\Phi^{-1/2} S . \tag{58}$$

Taking into account that $\{S_\alpha, S_\beta\}^* = -\frac{i}{2}\delta_{\alpha\beta}$ we have again realized $d = 10$, $N = 2$ supersymmetry algebra in the form (53) or (54). The nilpotent BRST operator in this gauge where only reparametrization ghosts are propagating is given by

$$Q_{\mathrm{BRST}} = cH \qquad H = \{b, Q_{\mathrm{BRST}}\} \qquad (Q_{\mathrm{BRST}})^2 = 0 \tag{59}$$

and here we used the fact that $\{b, c\} = 1$. In turn, the Hamiltonian (and therefore the BRST operator) can be constructed from supersymmetry generators as follows:

$$H = \frac{1}{2}\{Q_\alpha, Q_\beta\}^* \{\bar{Q}^\alpha, \bar{Q}^\beta\}^* = \frac{1}{2}(C\hat{\not{P}})_{\alpha\beta}(\hat{\not{P}}C)^{\alpha\beta} =$$

$$= -\frac{1}{2}\hat{\not{P}}^2 = -\frac{1}{2}(\not{P}^2 + Z^2) . \tag{60}$$

The terms with anticommuting fields S_α can be rewritten in a form where it is clear that they can be interpreted as world-line spinors,

$$L = \mathbf{P}_m \partial_0 X^m + \bar{S}_\alpha \rho^0 \partial_0 S_\alpha + b\dot{c} - H . \tag{61}$$

Here $\bar{S}_\alpha = \mathrm{i} S_\alpha \rho^0$ and $(\rho^0)^2 = -1$, $\rho^0 = \mathrm{i}$ being a 1-dimensional matrix.

Thus, we have the original 10 coordinates X^m and their conjugate momenta $\mathbf{P}_m$, and a pair of reparametrization ghosts. There are also 16 anticommuting world-line spinors S, describing 8 fermionic degrees of freedom. The Hamiltonian is quadratic. The ground state with $M^2 = Z^2$ is the state with the minimal value of the Hamiltonian. Thus for the D-superparticle one can see that the condition for the covariant quantization is satisfied in the presence of a central charge which makes the mass of a physical state non-vanishing. The global supersymmetry algebra is realized in a covariant way, as different from the light-cone gauge.

Thus we have found that covariantly quantized D-0-brane has a quadratic action with the physical state being the BPS state $M = |Z|$. The resulting supersymmetry generator is $d = 10$ Lorentz covariant and the Dirac bracket of the quantized theory form $d = 10$, $N = 2$ supersymmetry algebra with a central charge.

References

Achúcarro, A., Gauntlett, J., Itoh, K. and Townsend, P.K., (1980): Nucl. Phys. **B314**, 129 .

Aganagic, M., Popescu, C. and Schwarz, J.H., (1996): D-Brane Actions with Local Kappa Symmetry, Phys. Lett. **B393**, 311 (1997), hep-th/9610249;
Aganagic, M., Popescu, C. and Schwarz, J.H.: Gauge Invariant and Gauge Fixed D-brane Actions, Report CALT-68-2088, hep-th/9612080;
Aganagic, M., Park, J., Popescu, C. and Schwarz, J.H.: Dual D-brane Actions, Report CALT-68-2099, hep-th/9702133.

Abou Zeid, M. and Hull, C.M.(1997):, Intrinsic geometry of D-branes, hep-th/9704021.

Aganagic, M., Park, J., Popescu, C. and Schwarz, J.H., (1997): Worldvolume Action of the M-Theory Five-Brane, hep-th/9701166.

Bandos, I., Pasti, P., Sorokin, D. and Tonin, M., (1997a): Superbrane Actions and Geometrical Approach, hep-th/9705064.

Bandos, I., Lechner, K., Nurmagambetov, A., Pasti, P., Sorokin, D. and Tonin, M., (1997b): Covariant Action for the Super-Five-Brane of M-Theory, hep-th/9701149.
Bandos, I., Lechner, K., Nurmagambetov, A., Pasti, P., Sorokin, D. and Tonin, M.: On the Equivalence of Different Formulations of the M-Theory Five-Brane, hep-th/9703127.

Bergshoeff, E., Sezgin, E. and Townsend, P.K., (1987): Supermembranes And Eleven-Dimensional Supergravity, Phys. Lett. **B189** 75.

Bergshoeff, E. and Townsend, P.K., (1997): Super D-Branes, Nucl. Phys. **B490**, 145; hep-th/9611173.

Bergshoeff, E., Kallosh, R., Ortín, T., and Papadopoulos, G., (1997): κ-symmetry, Supersymmetry and Intersecting branes, CERN–TH/97–87, hep-th/9705040.

Cederwall, M., von Gussich, A., Nilsson, B.E.W., and Westerberg, A., (1997): The Dirichlet Super Three Brane In Ten-Dimensional Type IIIB Supergravity, Nucl. Phys. **B490**, 163 ;
Cederwall, M., von Gussich, A., Nilsson, B.E.W., Sundell, P., and Westerberg, A., (1997): The Dirichlet Super P-Branes In Ten-Dimensional Type IIA And IIB Supergravity, Nucl. Phys. **B490**, 179 ;
Cederwall, M.: Aspects of D-brane actions, Report GOTEBORG-ITP-96-20, hep-th/9612153.

Fradkin, E. (1973): Acta Universitatis Wratislaviensis **207**. Proceedings of X-th Winter School in Karpach, 1973;
Fradkin, E. S. and Vilkovisky, G.A., (1975): Phys. Lett. **55B**, 224 ;
Batalin, I.A. and Vilkovisky, G.A., (1977): Phys. Lett. **69B**, 309;
Fradkin, E.S. and Fradkina, T.E., (1978): Phys. Lett. **72B**, 343.

Green, M.B. and Schwarz, J.H., (1984): Covariant Description of Superstrings, Phys. Lett. **136B** 367-370.

Howe, P.S. and Sezgin, E., (1997a): Superbranes, Phys. Lett. **B390** 133-142.

Howe, P.S. and Sezgin, E., (1997b): D=11, p=5, Phys. Lett. **B394** (1997) 62.
Howe, P.S. and Sezgin, E., West, P.C.,: Covariant Field Equations of the M-Theory Fivebrane, hep-th/9702008.
Howe, P.S. and Sezgin, E., West, P.C.,: The Six-Dimensional Self-Dual Tensor, hep-th/9702111.

Kallosh, R., (1997): Covariant quantization of D-branes, hep-th/9705056.
Pasti, P., Sorokin, D. and Tonin, M., (1997): Covariant Action for a D=10 Five-Brane with the Chiral Field, hep-th/9701037.
Polchinski, J., (1996): TASI Lectures on D-branes, hep-th/9611050.
Volkov, D.V., Akulov, V.P., (1972): JETP Lett. **16**, 438 ; Phys. Lett. **46B**, 109 (1973).

Ten to Eleven: It Is Not Too Late

M.J. Duff

Center for Theoretical Physics
Texas A&M University, College Station, Texas 77843 **

Abstract. Superunification underwent a major paradigm shift in 1984 when eleven-dimensional supergravity was knocked off its pedastal by ten-dimensional superstrings. This last year has witnessed a new shift of equal proportions: perturbative ten-dimensional superstrings have in their turn been superseded by a new non-perturbative theory called *M-theory*, which describes supermembranes and superfivebranes, which subsumes all five consistent string theories and whose low energy limit is, ironically, eleven-dimensional supergravity. In particular, six-dimensional string/string duality follows from membrane/fivebrane duality by compactifying M-theory on $S^1/Z_2 \times K3$ (heterotic/heterotic duality) or $S^1 \times K3$ (Type IIA/heterotic duality) or $S^1/Z_2 \times T^4$ (heterotic/Type IIA duality) or $S^1 \times T^4$ (Type IIA/Type IIA duality).

1 M-theory (The theory formerly known as strings)

The maximum spacetime dimension in which one can formulate a consisistent supersymmetric theory is eleven (Nahm 1978). For this reason in the early 1980's many physicists looked to $D = 11$ supergravity (Cremmer, Julia and Scherk 1978), in the hope that it might provide that superunification (Duff, Nilsson and Pope 1986(they were all looking for. Then in 1984 superunification underwent a major paradigm shift: eleven-dimensional supergravity was knocked off its pedastal by ten-dimensional superstrings (Green, Schwarz and Witten 1987), and eleven dimensions fell out of favor. This last year, however, has witnessed a new shift of equal proportions: perturbative ten-dimensional superstrings have in their turn been superseded by a new non-perturbative theory called *M-theory*, which describes (amongst other things) supersymmetric extended objects with two spatial dimensions (***supermembranes***), and five spatial dimensions (***superfivebranes***), which subsumes all five consistent string theories and whose low energy limit is, ironically, eleven-dimensional supergravity.

The reason for this reversal of fortune of eleven dimensions is due, in large part, to the 1995 paper by Witten (Witten 1995). One of the biggest problems with $D = 10$ string theory (Green, Schwarz and Witten 1987) is that there are *five* consistent string theories: Type I SO(32), heterotic SO(32), heterotic $E_8 \times E_8$, Type IIA and Type IIB. As a candidate for a unique *theory of everything*, this is clearly an embarrassment of riches. Witten put forward a

** Research supported in part by NSF Grant PHY-9411543.

convincing case that this distiction is just an artifact of perturbation theory and that non-perturbatively these five theories are, in fact, just different corners of a deeper theory. Moreover, this deeper theory, subsequently dubbed *M-theory*, has $D = 11$ supergravity as its low energy limit! Thus the five string theories and $D = 11$ supergravity represent six different special points[1] in the moduli space of M-theory. The small parameters of perturbative string theory are provided by $< e^{\Phi} >$, where Φ is the dilaton field, and $< e^{\sigma_i} >$ where σ_i are the moduli fields which arise after compactification. What makes M-theory at once intriguing and yet difficult to analyse is that in $D = 11$ there is neither dilaton nor moduli and hence the theory is intrinsically non-perturbative. Consequently, the ultimate meaning of M-theory is still unclear, and Witten has suggested that in the meantime, M should stand for "Magic", "Mystery" or "Membrane", according to taste.

The relation between the membrane and the fivebrane in $D = 11$ is analogous to the relation between electric and magnetic charges in $D = 4$. In fact this is more than an analogy: electric/magnetic duality in $D = 4$ string theory (Font, Ibanez, Lust and Quevedo 1990), (Rey 1991) follows as a consequence of string/string duality in $D = 6$ (Duff 1995). The main purpose of the present paper is to show how $D = 6$ string/string duality (Duff and Lu 1991), (Duff and Khuri 1994), (Duff and Lu 1994), (Duff and Minasian 1995), (Hull and Townsend, 1995), (Duff, Khuri and Lu 1995), (Witten 1995) follows, in its turn, as a consequence of membrane/fivebrane duality in $D = 11$. In particular, heterotic/heterotic duality, Type IIA/heterotic duality, heterotic/Type IIA duality and Type IIA/Type IIA duality follow from membrane/fivebrane duality by compactifying M-theory on $S^1/Z_2 \times K3$ (Duff, Minasian and Witten 1996), $S^1 \times K3$ (Duff, Liu, Minasian 1995), $S^1/Z_2 \times T^4$ and $S^1 \times T^4$, respectively.

2 String/string duality from M-theory

Let us consider M-theory, with its fundamental membrane and solitonic fivebrane, on $R^6 \times M_1 \times \tilde{M}_4$ where M_1 is a one-dimensional compact space of radius R and $\tilde{M}_4$ is a four-dimensional compact space of volume V. We may obtain a fundamental string on R^6 by wrapping the membrane around M_1 and reducing on $\tilde{M}_4$. Let us denote the fundamental string sigma-model metrics in $D = 10$ and $D = 6$ by G_{10} and G_6. Then from the corresponding Einstein Lagrangians

$$\sqrt{-G_{11}}R_{11} = R^{-3}\sqrt{-G_{10}}R_{10} = \frac{V}{R}\sqrt{-G_6}R_6 \tag{1}$$

[1] Some authors take the phrase *M-theory* to refer merely to this sixth corner of the moduli space. With this definition, of course, M-theory is no more fundamental than the other five corners. For us, *M-theory* means the whole kit and caboodle.

we may read off the strength of the string couplings in $D = 10$ (Duff, Minasian and Witten 1996)

$$\lambda_{10}{}^2 = R^3 \tag{2}$$

and $D = 6$

$$\lambda_6{}^2 = \frac{R}{V} \ . \tag{3}$$

Similarly we may obtain a solitonic string on R^6 by wrapping the fivebrane around $\tilde{M}_4$ and reducing on M_1. Let us denote the solitonic string sigma-model metrics in $D = 7$ and $D = 6$ by $\tilde{G}_7$ and $\tilde{G}_6$. Then from the corresponding Einstein Lagrangians

$$\sqrt{-G_{11}}R_{11} = V^{-3/2}\sqrt{-\tilde{G}_7}\tilde{R}_7 = \frac{R}{V}\sqrt{-G_6}R_6 \tag{4}$$

we may read off the strength of the string couplings in $D = 7$ (Duff, Minasian and Witten 1996)

$$\tilde{\lambda}_7^2 = V^{3/2} \tag{5}$$

and $D = 6$

$$\tilde{\lambda}_6^2 = \frac{V}{R} \ . \tag{6}$$

Thus we see that the fundamental and solitonic strings are related by a strong/weak coupling:

$$\tilde{\lambda}_6^2 = 1/\lambda_6{}^2 \ . \tag{7}$$

We shall be interested in $M_1 = S^1$ (in which case from (Duff, Howe, Inami and Stelle 1987) the fundamental string will be Type IIA) or $M_1 = S^1/Z_2$ (in which case from (Horava and Witten 1996) the fundamental string will be heterotic $E_8 \times E_8$). Similarly, we will be interested in $\tilde{M}_4 = T^4$ (in which case the solitonic string will be Type IIA) or $\tilde{M}_4 = K3$ (in which case from (Townsend 1995), (Harvey and Strominger 1995) the solitonic string will be heterotic). Thus there are four possible scenarios which are summarized in Table 1. (N_+, N_-) denotes the $D = 6$ spacetime supersymmetries. In each case, the fundamental string will be weakly coupled as we shrink the size of the wrapping space M_1 and the dual string will be weakly coupled as we shrink the size of the wrapping space $\tilde{M}_4$.

Table 1. String/string dualities

$(\mathbf{N_+, N_-})$	$\mathbf{M_1}$	$\mathbf{\tilde{M}_4}$	**fundamental string**	**dual string**
$(1,0)$	S^1/Z_2	$K3$	heterotic	heterotic
$(1,1)$	S^1	$K3$	Type IIA	heterotic
$(1,1)$	S^1/Z_2	T^4	heterotic	Type IIA
$(2,2)$	S^1	T^4	Type IIA	Type IIA

In fact, there is in general a topological obstruction to wrapping the fivebrane around $\tilde{M}_4$ provided by

$$\int K_4 = 2\pi m \ , \tag{8}$$

where K_4 is the 4-form field strength of $D = 11$ supergravity, because the fivebrane cannot wrap around a 4-manifold that has $m \neq 0$. This is because the anti-self-dual 3-form field strength T on the worldvolume of the fivebrane obeys

$$\mathrm{d}T = K_4 \tag{9}$$

and the existence of a solution for T therefore requires that K_4 must be cohomologically trivial. For M-theory on $R^6 \times S^1/Z_2 \times T^4$ this is no problem. For M theory on $R^6 \times S^1/Z_2 \times K3$, with instanton number k in one E_8 and $24 - k$ in the other, however, the flux of K_4 over $K3$ is (Duff, Minasian and Witten 1996)

$$m = 12 - k \ . \tag{10}$$

Consequently, the M-theoretic explanation of heterotic/heterotic duality requires $E_8 \times E_8$ with the symmetric embedding $k = 12$. This has some far-reaching implications. For example, the duality exchanges gauge fields that can be seen in perturbation theory with gauge fields of a non-perturbative origin (Duff, Minasian and Witten 1996).

The dilaton $\tilde{\Phi}$, the string σ-model metric $\tilde{G}_{MN}$ and 3-form field strength $\tilde{H}$ of the dual string are related to those of the fundamental string, Φ, G_{MN} and H by the replacements

$$\begin{aligned} \Phi &\rightarrow \tilde{\Phi} = -\Phi \\ G_{MN} &\rightarrow \tilde{G}_{MN} = \mathrm{e}^{-\Phi} G_{MN} \\ H &\rightarrow \tilde{H} = \mathrm{e}^{-\Phi} * H \ . \end{aligned} \tag{11}$$

In the case of heterotic/Type IIA duality and Type IIA/heterotic duality, this operation takes us from one string to the other, but in the case of heterotic/heterotic duality and Type IIA/Type IIA duality this operation is a discrete symmetry of the theory. This Type IIA/Type IIA duality is hardly ever discussed in the literature in these terms, but we can recognize this symmetry as subgroup of the $SO(5,5;\mathbb{Z})$ U-duality (Duff and Lu 1990), (Hull and Townsend, 1995) of the $D = 6$ Type IIA string.

References

Cremmer, E., Julia, B. and Scherk, J., (1978): Supergravity in theory in 11 dimensions, Phys. Lett. **76B**, 409.

Duff, M.J., (1995): *Strong/weak coupling duality from the dual string*, Nucl. Phys. **B442**, 47.

Duff, M.J. and Khuri, R. R., (1994): Four-dimensional string/string duality, Nucl. Phys. **B411**, 473.
Duff, M.J. and Lu, J.X., (1990): Duality rotations in membrane theory, Nucl. Phys. **B347**, 394.
Duff, M.J. and Lu, J.X.,(1991): Loop expansions and string/five-brane duality, Nucl. Phys. **B357**, 534.
Duff, M.J. and Lu, J. X., (1994): Black and super p-branes in diverse dimensions, Nucl. Phys. **B416**, 301.
Duff, M.J. and Minasian, R., (1995): Putting string/string duality to the test, Nucl. Phys. **B436**, 507.
Duff, M.J., Howe, P.S., Inami, T. and Stelle, K.S., (1987): Superstrings in $D = 10$ from supermembranes in $D = 11$, Phys. Lett. **B191**, 70.
Duff, M.J., Khuri, R.R. and Lu, J.X., (1995): String solitons, Phys. Rep. **259**, 213.
Duff, M.J., Liu, J.T., Minasian, R.,(1995): Eleven dimensional origin of string-string duality: a one-loop test, Nucl. Phys. **B452**, 261.
Duff, M.J., Minasian, R. and Witten, E., (1996): Evidence for heterotic/heterotic duality, Nucl. Phys. **B465**, 413.
Duff, M.J., Nilsson, B.E.W. and Pope, C.N. (1986): Kaluza-Klein supergravity, Phys. Rep. **130**, 1.
Font, A., Ibanez, L., Lust, D. and Quevedo, F., (1990): Strong-weak coupling duality and nonperturbative effects in string theory, Phys. Lett. **B249**, 35.
Green, M.B., Schwarz, J.H. and Witten, E., (1987): *Superstring Theory* (Cambridge University Press).
Harvey, J.A. and Strominger, A., (1995): The heterotic string is a soliton, Nucl. Phys. **B449**, 535.
Horava, P. and Witten, E., (1996): Heterotic and Type I string dynamics from eleven dimensions, Nucl. Phys. **B460**, 506.
Hull, C.M. and Townsend, P.K., (1995): Unity of superstring dualities, Nucl. Phys. **B438**, 109.
Nahm, W., (1978): Supersymmetries and their representations, Nucl. Phys. **B135**, 409.
Rey, S.-J., (1991): The confining phase of superstrings and axionic strings, Phys. Rev. **D43**, 526.
Townsend, P. K., (1995): String/membrane duality in seven dimensions, Phys. Lett. **B354**, 247.
Witten, E., (1995): String theory dynamics in various dimensions, Nucl. Phys. **B443**, 85.

Aspects of Superembeddings

P.S. Howe[1], E. Sezgin[2] and P.C. West[3]

Abstract. Some aspects of the geometry of superembeddings and its application to supersymmetric extended objects are discussed. In particular, the embeddings of the (3|16) and (6|16) dimensional superspaces into (11|32) dimensional superspace, corresponding to the supermembrane and superfivebrane in eleven dimensions, are treated in some detail.

1 Introduction

One of the many contributions that D.V. Volkov made to modern theoretical physics was the realisation that supersymmetric particles, moving in three or four dimensional spacetimes, can be described using a formalism which has both worldline and spacetime supersymmetries built in (Sorokin, Tkach, Volkov (1989), Sorokin, Tkach, Volkov, Zheltukhin (1989)). Up until then it had been thought that, although the superstring can be described with either worldsheet (Ramond (1971), Neveu, Schwarz (1971), Gliozzi, Scherk, Olive (1977)) or spacetime supersymmetry (Green, Schwarz (1984)), all other extended supersymmetric objects, including the superparticle (Brink, Schwarz (1981)), could only be written with manifest spacetime supersymmetry (Hughes, Liu, Polchinski (1986), Bergshoeff, Sezgin, Townsend (1987)). The spacetime supersymmetric formalism does involve a local fermionic symmetry, called κ-symmetry (Siegel (1983)), but the geometric nature of this symmetry remained obscure until the work of the Kharkov group showed that it can be derived from, and is equivalent to, local worldsurface supersymmetry. Subsequently the formalism has been developed by the Kharkov group and others, and has been applied to various other supersymmetric extended objects (see (Bandos, Sorokin, Tonin, Pasti, Volkov (1995)) for a full list of references). It also gradually became clear that the formalism can be understood in terms of the embedding of one superspace, the worldsurface, into another, the target superspace. Although this point of view was implicit in the early papers it was made explicit in a study of the heterotic string (Delduc, Galperin, Howe, Sokatchev (1993)) and was further developed in (Bandos, Sorokin, Tonin, Pasti, Volkov (1995)). More recently, in (Howe, Sezgin (1997)), it was shown that all supersymmetric extended objects can be understood in this way, including objects such as the Dirichlet branes of string theory which have additional physical worldsurface bosonic fields to the usual transverse coordinate fields. The latter are scalars on the worldsurface whereas the new fields are gauge fields; we shall refer to the two types of object as type I (scalars only) and type II (additional gauge fields). The formalism was applied in particular to construct the full equations of motion of the five-brane in eleven-dimensional superspace (Howe, Sezgin (1997), Howe, Sezgin, West (1997)), an object that plays an important role in M-theory. Moreover, it turns out that all branes are described by the same simple embedding condition which is extremely natural from the point of view of supergeometry (Howe, Sezgin (1997)). We refer the reader to the literature for discussions of the component (GS) approach to Dirichlet branes (Cederwall, von Gussich, Nilsson, Westerberg (1997), Aganagic, Popescu, Schwarz (1997), Aganagic, Popescu, Schwarz (1996), Cederwall, von Gussich, Nilsson, Sundell, Westerberg (1997), Bergshoeff, Townsend (1997)) and the eleven-dimensional five-brane (Townsend (1996), Aharony (1996), Bergshoeff, de Roo, Ortin (1996), Witten (1996), Perry, Schwarz (1997), Schwarz (1997), Pasti, Sorokin, Tonin (1997)).

Before describing superembeddings in more detail it is perhaps worthwhile recalling the problem that Volkov and his collaborators solved. Consider a superparticle moving on a superworldline parametrised by (even, odd) coordinates (t, τ) in flat $(3|2)$-dimensional superspace coordinatised by (x^a, θ^α). Expanding the supercoordinates describing the particle in τ we have

$$\begin{aligned} x^a(t,\tau) &= x^a(t) + \tau\lambda^a(t) , \\ \theta^\alpha(t,\tau) &= \theta^\alpha(t) + \tau u^\alpha(t) . \end{aligned} \tag{1}$$

The problem seems to be that there are two "wrong statistics" fields, λ^a and u^α. The solution to this problem found by the Kharkov group was to identify u as a twistor variable, in fact, as the "square-root" of the lightlike momentum of the particle, and to regard λ as an auxiliary field. This is summarised in the superspace equation

$$Dx^a - \frac{\mathrm{i}}{2} D\theta^\alpha (\gamma^a)_{\alpha\beta}\theta^\beta = 0 , \tag{2}$$

where $D = \frac{\partial}{\partial\tau} + \frac{\mathrm{i}}{2}\tau\frac{\partial}{\partial t}$ is the superworldline covariant derivative. The first component of 2 (in a τ-expansion) allows one to solve for λ while the second component relates $\dot{x}$ to u^2. Volkov and his team were able to find a somewhat unusual Lagrangian which gives rise to these conditions on the fields λ and u, but in what follows we shall not have much to say about actions, rather we focus on the dynamics of the extended objects directly and show how these can be understood from the perspective of superembeddings. It is important to notice that the geometrical interpretation of 2 is that, at any point on the superworldline, the odd tangent space of the superworldline is a subspace of the odd tangent space of the target superspace.

2 Flat Branes

We define a flat brane to be an embedding of a flat superspace, of dimension $(d|\frac{1}{2}D')$ in a flat superspace of dimension $(D|D')$. The existence of such objects determines in which dimensions one can have branes and the structure of the worldsurface multiplets can be obtained by considering small deformations. In fact, the allowed super-dimensions correspond to the points on the modified brane scan (Duff, Lu (1993), Howe, Sezgin (1997)). One could also consider branes which preserve fewer than half of the target space supersymmetries, but we shall not do so here. In order for flat branes to exist it is necessary that the Γ-matrices should decompose in the right way. If $(x^{\underline{a}}, \theta^{\underline{\alpha}})$, $\underline{a} = 0, 1, \ldots D-1$; $\underline{\alpha} = 1, \ldots D'$ are coordinates on the target superspace split into (x^a, θ^α), $a = 1, 0, \ldots d-1$; $\alpha = 1, \ldots \frac{1}{2}D'$ and $(x^{a'}, \theta^{\alpha'})$, $a' = d, \ldots D-1$; $\alpha' = \frac{1}{2}D' + 1, \ldots D'$, we require that the Γ-matrices split as follows:

$$(\Gamma^{\underline{a}})_{\underline{\alpha\beta}} \to (\Gamma^a)_{\alpha\beta},\ (\Gamma^{a'})_{\alpha\beta'} = (\Gamma^{a'})_{\beta'\alpha},\ (\Gamma^a)_{\alpha'\beta'}\ , \tag{3}$$

with all other components vanishing. If this is the case we have

$$[D_\alpha, D_\beta] = \mathrm{i}(\Gamma^a)_{\alpha\beta}\partial_a\ , \tag{4}$$

or, equivalently, there is a subalgebra of the supertranslational algebra of the required dimension. The covariant derivative is defined as usual to be

$$D_{\underline{\alpha}} = \partial_{\underline{\alpha}} + \frac{\mathrm{i}}{2}(\Gamma^{\underline{a}})_{\underline{\alpha\beta}}\theta^{\underline{\beta}}\partial_{\underline{a}}\ . \tag{5}$$

The brane itself is given by the embedding $(x^a, \theta^\alpha) \mapsto (x^a, \theta^\alpha; 0, 0)$, or, equivalently, as the solution of the equations $x^{a'} = \theta^{\alpha'} = 0$ in the target superspace. The condition that the Γ-matrices split as above tells us when branes preserving half-supersymmetry can exist.

We next consider a small deformation of the flat brane, for which the embedding becomes

$$(x, \theta) \mapsto (x, \theta; x'(x, \theta), \theta'(x, \theta))\ , \tag{6}$$

where x' and θ' are small, so that we only need to work to first order in these variables. The odd basis tangent vectors to the submanifold, collectively denoted E_α, are given as the image of the (worldsurface) D_α under the embedding,

$$E_\alpha = D_\alpha + D_\alpha\theta^{\beta'}D_{\beta'} + (D_\alpha X^{b'} - \mathrm{i}(\Gamma^{b'})_{\alpha\beta'}\theta^{\beta'})\partial_{b'}\ . \tag{7}$$

Note that D_α on the target space (which occurs on the right-hand side of the above equation) differs from D_α on the brane; the former, which is the α component of 5, includes a term involving θ' which is absent from the latter. The even basis vectors are

$$E_a = \partial_a + \partial_a X^{b'} + \partial_a\theta^{\beta'}D_{\beta'}\ , \tag{8}$$

where

$$X^{a'} = x^{a'} + \frac{i}{2}\theta^{\beta}(\Gamma^{a'})_{\beta\gamma'}\theta^{\gamma'} \ . \tag{9}$$

We now impose the requirement that the odd tangent space at any point of the embedded submanifold should be a subspace of the odd tangent space of the target space at that point. This condition is required for all supersymmetric extended objects and implies that we must impose

$$D_{\alpha}X^{a'} = i(\Gamma^{a'})_{\alpha\beta'}\theta^{\beta'} \ . \tag{10}$$

Computing the commutator of the odd tangent vectors (to first order in the transverse variables) one finds

$$[E_{\alpha}, E_{\beta}] = i(\Gamma^{a})_{\alpha\beta}E_{a} + i(2D_{(\alpha}\theta^{\gamma'}(\Gamma^{b'})_{\beta)\gamma'} - (\Gamma^{a})_{\alpha\beta}\partial_{a}X^{b'})\partial_{b'} \ . \tag{11}$$

Since the commutator of two vectors of the submanifold must lead to a third we have

$$2D_{(\alpha}\theta^{\gamma'}(\Gamma^{b'})_{\beta)\gamma'} = (\Gamma^{a})_{\alpha\beta}\partial_{a}X^{b'} \ . \tag{12}$$

In fact, this constraint is not independent; it follows directly from 10 by differentiation, as indeed it must since the algebra of the covariant derivatives on the brane is preserved.

Equation 10 above is the key equation for branes since it determines the structure of the worldsurface supermultiplet. It can be one of three types: on-shell, in which case it leads directly to the dynamics of the physical fields; off-shell Lagrangian, in which case it determines an off-shell multiplet which can be used in a Lagrangian to determine the dynamics; or off-shell non-Lagrangian, in which case the multiplet is off-shell but there is not a Lagrangian, at least of conventional type, which can be constructed which leads to the dynamics. In the third case further conditions are required, but most examples fall into the first two classes.

In eleven dimensions (with 32 odd dimensions), there are two possible branes, the two-brane and the five-brane[1]. In both cases equation 10 defines an on-shell supermultiplet, the $d = 3, N = 8$ scalar multiplet and the $d = 6, N = 2$ tensor multiplet, respectively. In both cases the leading components of $X^{a'}$ and $\theta^{\alpha'}$ can be interpreted as Goldstone fields corresponding to the breaking of supertranslational symmetry, but in the five-brane there is an extra component field which appears at leading order in $D_{\alpha}\theta^{\beta'}$. In general one has

$$D_{\alpha}\theta^{\beta'} = \frac{1}{2}(\Gamma^{ab'})_{\alpha}{}^{\beta'}\partial_{a}X_{b'} + h_{\alpha}{}^{\beta'} \ , \tag{13}$$

where

$$h_{(\alpha}{}^{\gamma'}(\Gamma^{a'})_{\beta)\gamma'} = 0 \ , \tag{14}$$

[1] The possibility that there might be a nine-brane has been raised (Howe, Sezgin (1997)); such an object, if it exists, would have to have additional worldsurface fermion fields.

but the latter equation only has non-trivial solutions for type II branes. For example, for the eleven-dimensional five-brane one finds

$$h_\alpha{}^{\beta'} = \frac{1}{6}(\Gamma^{abc})_\alpha{}^{\beta'} h_{abc} \,. \tag{15}$$

The field h_{abc} is totally antisymmetric and self-dual, and at the linearised level is closed. Its leading component is therefore the self-dual field strength tensor of a two-form gauge field. The quantity $D_\alpha\theta^{\beta'}$, evaluated at $\theta = 0$, is the analogue of u^α in equation 1, at least in the linearised theory. The term involving $\partial_a X^{b'}$ generalises the momentum which arises in the particle case, while the h-term is present only for type II branes.

3 D = 11 Supergeometry

In the rest of the paper we shall focus on superembeddings in eleven dimensions. We briefly recall the salient features of eleven-dimensional supergeometry. One has a real (11|32)-dimensional supermanifold $\underline{M}$ with a choice of odd tangent bundle $\underline{F} \subset \underline{T}$, the full tangent bundle, such that the associated Frobenius tensor $\underline{\phi}$, defined by

$$\underline{\phi}(X,Y) = [X,Y] \bmod \underline{F} \,, \tag{16}$$

where X and Y are odd vector fields, is invariant under the group $\mathrm{Spin}(1,10) \times \mathbb{R}^+$. This implies that there exist local bases $(E_{\underline{\alpha}})$ for $\underline{F}$ and $(E^{\underline{a}})$ for $\underline{B}^*$, where $\underline{B}$ is the quotient of $\underline{T}$ by $\underline{F}$ and the star denotes dual, in which the components of $\underline{\phi}$ are given by

$$\underline{\phi}_{\underline{\alpha}\underline{\beta}}{}^{\underline{c}} = \langle [E_{\underline{\alpha}}, E_{\underline{\beta}}], E^{\underline{c}} \rangle = \mathrm{i}(\Gamma^{\underline{c}})_{\underline{\alpha}\underline{\beta}} \,. \tag{17}$$

This set up defines what one might call a special superconformal structure; more generally one can allow for additional terms in $\underline{\phi}$ involving two-index and five-index Γ-matrices and we shall come back to this possibility shortly. If 17 holds it can be shown that it implies the equations of motion of eleven-dimensional supergravity must be satisfied modulo certain topological niceties which we shall ignore here (Howe (in preparation)). More precisely, one can show, given 17, that one can find a choice of $\underline{B}$ as a subbundle of $\underline{T}$ and a choice of $\mathrm{Spin}(1,10)$ connection such that the torsion and curvature tensors in superspace are related to those of on-shell supergravity by a super-Weyl transformation. One can therefore make such a transformation to eliminate the conformal factor thereby arriving at the standard geometry (Cremmer, Ferrara (1980), Brink, Howe (1980)). This geometry has structure group $\mathrm{Spin}(1,10)$ and therefore admits an invariant Lorentzian metric $\underline{g}_B$ on $\underline{B}$ and also an invariant fermionic metric $\underline{g}_F$ on $\underline{F}$ whose components in the standard basis are the components of the charge-conjugation matrix.

We note that when a connection is introduced it is natural to equate $\underline{\phi}$ with the dimension zero component of the torsion tensor (with a minus sign), although it is a perfectly well-defined tensor belonging to the space $\wedge^2 \underline{F}^* \otimes \underline{B}$ even if a connection is not introduced. Thus we have

$$\underline{\phi_{\alpha\beta}}^{\underline{c}} = -T_{\underline{\alpha\beta}}{}^{\underline{c}} . \tag{18}$$

4 Embeddings

We consider embeddings $M \xrightarrow{f} \underline{M}$ of the worldsurface M into the target space $\underline{M}$ which we shall take to have dimension $(11|32)$ for definiteness, although the discussion below is applicable more generally with appropriate modifications. It will be assumed that $\underline{M}$ has a superconformal structure but initially at least we shall not suppose that the Frobenius tensor $\underline{\phi}$ is invariant under the structure group $\mathrm{Spin}(1,10) \times \mathbb{R}^+$. Without loss of generality we can take it to be of the form

$$\underline{\phi_{\alpha\beta}}^{\underline{c}} = \mathrm{i}(\Gamma^{\underline{c}})_{\underline{\alpha\beta}} + (\Gamma^{\underline{bc}})_{\underline{\alpha\beta}} X_{\underline{bc}}{}^{\underline{a}} + (\Gamma^{\underline{bcdef}})_{\underline{\alpha\beta}} Y_{\underline{bcdef}}{}^{\underline{a}} , \tag{19}$$

where the antisymmetric components of X and Y vanish, as well as their traces. Both of these restrictions are compatible with the group structure. We shall call such a structure a general superconformal structure.

There is a natural choice of odd tangent bundle F on M given by

$$F = T \cap \underline{F} . \tag{20}$$

Dually, one has

$$B^* = T^* \cap \underline{B}^* . \tag{21}$$

The only other requirement we need to impose on the embedding is that the metric iduced on B^* (from any of the conformal class of Lorentzian metrics on $\underline{B}^*$) should be Lorentzian with signature $(p-1)$, $p = 2, 5$. Again this condition is automatically conformally invariant.

The Frobenius tensor ϕ of M is defined in the same way as above, namely as the commutator of two odd vector fields modulo the odd tangent bundle. The relation between the worldsurface and target space Frobenius tensors is given by

$$\underline{\phi}(X, Y, \underline{\omega}) = \phi(X, Y, f^*\underline{\omega}) , \tag{22}$$

where X and Y are odd vector fields on M (which may be considered as vector fields on $\underline{M}$) and $\underline{\omega}$ is a one-form on $\underline{M}$.

For any embedding one has three natural bundles (on M), the tangent bundle T, the tangent bundle, $\underline{T}$, of $\underline{M}$ restricted to M and the normal bundle T' which fit together in a short exact sequence

$$0 \to T \to \underline{T} \to T' \to 0 . \tag{23}$$

However, in the super case we have even and odd tangent bundles which themselves fit into an exact sequence

$$0 \to F \to T \to B \to 0\,, \tag{24}$$

and similarly for the target space as well as the corresponding normal bundles. In fact there are nine bundles in all and it can be shown that they fit together into the following diagram,

$$\begin{array}{ccccccccc} & & 0 & & 0 & & 0 & & \\ & & \uparrow & & \uparrow & & \uparrow & & \\ 0 & \to & F' & \to & T' & \to & B' & \to & 0 \\ & & \uparrow & & \uparrow & & \uparrow & & \\ 0 & \to & \underline{F} & \to & \underline{T} & \to & \underline{B} & \to & 0 \\ & & \uparrow & & \uparrow & & \uparrow & & \\ 0 & \to & F & \to & T & \to & B & \to & 0 \\ & & \uparrow & & \uparrow & & \uparrow & & \\ & & 0 & & 0 & & 0 & & \end{array} \tag{25}$$

where each of the rows and columns is exact and where each square is commutative. The proof of these assertions is straightforward. The dual bundles give rise to a similar diagram with the arrows reversed. In practice one wishes to split the sequences, so that the central bundle of each sequence becomes a direct sum of the other two. However, when this is carried out it is important to note that, although B as a quotient bundle is a subbundle of $\underline{B}$ the same is not true when B and $\underline{B}$ are regarded as subbundles of T and $\underline{T}$ respectively.

5 Brane Integrability

The basic embedding condition described above is extremely natural given the geometrical structures that arise in supergeometry. Moreover, it is also extremely restrictive. In fact, in the eleven-dimensional examples we are discussing, one has the following results: if an $(11|32)$-dimensional supermanifold $\underline{M}$ with a general superconformal structure admits embeddings of the type described in the previous section of either two-branes or five-branes through every point, then:

- any such brane is dynamical, that is the embedding implies that the worldsurface multiplet is on-shell,
- the superconformal structure on the target space must be special, which implies, as we have discussed above, that the equations of motion of eleven-dimensional supergravity must be satisfied,
- the worldsurface supergeometry is completely specified up to gauge freedoms.

In some respects this may not be too surprising as it has been known for some time that the requirement of κ-symmetry in the GS formulation of the membrane forces the target space supergeometry to be equivalent to eleven-dimensional supergravity (Bergshoeff, Sezgin, Townsend (1987)). However, in the present approach we have achieved this with the bare minimum of assumptions; everything, including κ-symmetry follows from the simple embedding condition 20.

The complete proof of the above assertions is extremely long and rather complicated in terms of details, but it is simple to understand how it comes about in principle. Consider the relation 22 between the Frobenius tensors of the two manifolds. We introduce local bases (E_α), (E^a) for F and B^* respectively and note that 20 implies that

$$E_\alpha = E_\alpha{}^{\underline{\alpha}} E_{\underline{\alpha}} \,, \tag{26}$$

for some 16×32 matrix $E_\alpha{}^{\underline{\alpha}}$, while the dual condition 21 implies

$$f^* E^{\underline{a}} = E^a E_a{}^{\underline{a}} \,. \tag{27}$$

Equation 22 then reads in components with respect to these bases,

$$E_\alpha{}^{\underline{\alpha}} E_\beta{}^{\underline{\beta}} \phi_{\underline{\alpha}\underline{\beta}}{}^{\underline{c}} = \phi_{\alpha\beta}{}^{c} E_c{}^{\underline{c}} \,. \tag{28}$$

If $(E_{\underline{\alpha}})$ is a spin basis for $\underline{F}$ any other such basis will be related to it by an element u of Spin(1,10) up to a conformal factor which we shall ignore for the moment. We write $u = (u_\alpha{}^{\underline{\alpha}}, u_{\alpha'}{}^{\underline{\alpha}})$, with $\alpha' = 1, \ldots 16$. Since $E_\alpha{}^{\underline{\alpha}}$ has maximal rank there will be a choice of u such that E_α is related to $u_\alpha{}^{\underline{\alpha}} E_{\underline{\alpha}}$ by a non-singular matrix. Hence, without loss of generality, we can write

$$E_\alpha{}^{\underline{\alpha}} = A_\alpha{}^{\beta} u_\beta{}^{\underline{\alpha}} + B_\alpha{}^{\beta'} u_{\beta'}{}^{\underline{\alpha}} \,, \tag{29}$$

where $\det A \neq 0$. Making a change of basis for F we arrive at

$$E_\alpha{}^{\underline{\alpha}} = u_\alpha{}^{\underline{\alpha}} + h_\alpha{}^{\beta'} u_{\beta'}{}^{\underline{\alpha}} \,. \tag{30}$$

On the bosonic space B^* the situation resembles more closely the case of a Lorentzian embedding and we may choose, again up to a conformal factor

$$E_a{}^{\underline{a}} = u_a{}^{\underline{a}} \,, \tag{31}$$

where $(u_a{}^{\underline{a}}, u_{a'}{}^{\underline{a}})$ is the element of SO(1,10) corresponding to $u = (u_\alpha{}^{\underline{\alpha}}, u_{\alpha'}{}^{\underline{\alpha}}) \in$ Spin(1,10). Thus, at any point $p \in M$, the embedding is specified by $u_a{}^{\underline{a}}, u_\alpha{}^{\underline{\alpha}}$ and $h_\alpha{}^{\beta'}$.

We can now decompose equation 28 into components tangent and normal to M. We then find

$$\phi_{\underline{\alpha\beta}}{}^{c'} + 2h_{(\alpha}{}^{\gamma'} \phi_{\underline{\beta)\gamma'}}{}^{c'} + h_\alpha{}^{\gamma'} h_\beta{}^{\delta'} \phi_{\underline{\gamma'\delta'}}{}^{c'} = 0 \,, \tag{32}$$

and

$$\phi_{\underline{\alpha\beta}}{}^{c} + 2h_{(\alpha}{}^{\gamma'} \phi_{\underline{\beta)\gamma'}}{}^{c} + h_\alpha{}^{\gamma'} h_\beta{}^{\delta'} \phi_{\underline{\gamma'\delta'}}{}^{c} = \phi_{\alpha\beta}{}^{c} \,, \tag{33}$$

where

$$\phi_{\underline{\alpha\beta}}{}^{c'} = u_\alpha{}^{\underline{\alpha}} = u_\beta{}^{\underline{\beta}} \phi_{\underline{\alpha}\underline{\beta}}{}^{\underline{c}} u_{\underline{c}}{}^{c'} \,, \tag{34}$$

and similarly for the other projections of $\phi_{\underline{\alpha\beta}}{}^{\underline{c}}$. Now in order for there to be embeddings of branes in general we require that these equations be satisfied for arbitrary embeddings passing through a given point $p \in \underline{M}$ and furthermore that this should be true for all points of $\underline{M}$. Since we may vary u and h independently this requires

$$\phi_{\underline{\alpha\beta}}{}^{c'} = 0 \,. \tag{35}$$

This can only be satisfied for arbitrary embeddings if the superconformal structure on $\underline{M}$ is special, i.e. if

$$\phi_{\underline{\alpha\beta}}{}^{\underline{c}} = \mathrm{i}(\Gamma^{\underline{c}})_{\underline{\alpha\beta}} \,. \tag{36}$$

Given this one finds that equations 32 and 33 are solved by

$$h_\alpha{}^{\beta'} = \begin{cases} 0 & \text{two-brane} \\ \frac{1}{6}(\Gamma^{abc})_\alpha{}^{\beta'} h_{abc} & \text{five-brane} \end{cases} \tag{37}$$

where h_{abc} is self-dual, and

$$\phi_{\alpha\beta}{}^c = \begin{cases} \mathrm{i}(\Gamma^c)_{\alpha\beta} & \text{two-brane} \\ \mathrm{i}(\Gamma^b)_{\alpha\beta} m_b{}^c & \text{five-brane} \end{cases} \tag{38}$$

where

$$m_a{}^b = \delta_a{}^b - 2h_{acd}h^{bcd} \,. \tag{39}$$

The above argument establishes brane-integrablity; to see that the embedding implies the dynamics it is sufficient to consider the linearised case, i.e. take the target space to be flat and assume that the embedded submanifold is also nearly flat. It is not too difficult to see that in this limit one recovers the equations describing the deformations of flat branes, and hence the worldsurface fields are indeed on-shell. By working to second order in the transverse fields one can quickly see that the worldsurface supergravity fields are also determined.

6 Some Geometrical Aspects of Superembeddings

We briefly recall some aspects of Riemannian embeddings (see, for example, (Kobayashi, Nomizu (1963))). Let M be a manifold embedded in a Riemannian manifold $(\underline{M}, \underline{g})$. The metric on the target space induces natural metrics on the embedded space and on the normal bundle T' as well as determining a natural orthogonal decomposition of $\underline{T}$ into tangential and normal components. Explicitly

$$\begin{aligned} \underline{g}(X,Y) &= g(X,Y) \,, \\ \underline{g}(X,Y') &= 0 \,, \\ \underline{g}(X',Y') &= g'(X',Y') \,, \end{aligned} \tag{40}$$

where X, Y are tangential vector fields, X', Y' normal vector fields, g is the induced metric on M and g' is the metric induced on the normal bundle. Metric connections ∇ and ∇' are determined in T and T' respectively from the metric connection $\underline{\nabla}$ on $\underline{M}$ by the Gauss-Weingarten equations

$$\underline{\nabla}_X Y = \nabla_X Y + K'(X,Y) ,$$
$$\underline{\nabla}_X Y' = \nabla'_X Y' + K(X,Y') , \tag{41}$$

where $K'(X,Y)$ is normal and $K(X,Y')$ tangential. K' is the second fundamental form, and K is related to K' by

$$\underline{g}(K'(X,Y),Z') + \underline{g}(X,K(Y,Z')) = 0 . \tag{42}$$

¿From 41 one can derive the torsion equations

$$[\underline{T}(X,Y)]^t = T(X,Y) ,$$
$$[\underline{T}(X,Y)]^n = K'(X,Y) - K'(Y,X) , \tag{43}$$

where the superscripts t and n denote tangential and normal respectively. Finally, we have the equations of Gauss and Codazzi relating the curvature tensors of T and T' to the Riemann curvature tensor of $\underline{M}$:

$$\underline{R}(X,Y,Z,\omega) = R(X,Y,Z,\omega) + (K(X,K'(Y,Z),\omega) - X \leftrightarrow Y) ,$$
$$\underline{R}(X,Y,Z',\omega') = R'(X,Y,Z',\omega') + (K'(X,K(Y,Z'),\omega') - X \leftrightarrow Y) , \tag{44}$$

where ω and ω' are respectively tangential and normal one-forms.

The above equations can be generalised to the supersymmetric case although the situation is more complicated due to the even-odd split. In view of the discussion of the preceding section we can assume that the target space supergeometry is the standard geometry describing on-shell supergravity. We begin with the membrane. The tensor $\underline{\phi}$ on $\underline{M}$ gives rise to the following tensors via embedding:

$$\underline{\phi}(X,Y,\omega) = \phi(X,Y,\omega) ,$$
$$\underline{\phi}(X,Y',\omega') = \tilde{\phi}(X,Y',\omega') ,$$
$$\underline{\phi}(X',Y',\omega) = \phi'(X',Y',\omega) , \tag{45}$$

while

$$\underline{\phi}(X,Y,\omega') = 0 ,$$
$$\underline{\phi}(X,Y',\omega) = 0 ,$$
$$\underline{\phi}(X',Y',\omega') = 0 , \tag{46}$$

where, in both equations, the vectors are all odd, the forms are even and normal vectors or forms are distinguished by a prime. From the bosonic metric we derive

$$\begin{aligned} \underline{g}_B(X,Y) &= g_B(X,Y) , \\ \underline{g}_B(X,Y') &= 0 , \\ \underline{g}_B(X',Y') &= g'_B(X,Y) , \end{aligned} \tag{47}$$

for even tangential and normal vectors X, Y and X', Y', thus defining induced metrics for B and B'. Starting from the fermionic metric we get

$$\begin{aligned} \underline{g}_F(X,Y) &= g_F(X,Y) , \\ \underline{g}_F(X,Y') &= 0 , \\ \underline{g}_F(X',Y') &= g'_F(X,Y) , \end{aligned} \tag{48}$$

for odd tangential and normal vectors X, Y and X', Y', thus defining induced fermionic metrics for F and F'. We also have

$$\begin{aligned} \underline{g}_F(X,Y) &= 0 , \\ \underline{g}_F(X,Y') &= 0 , \\ \underline{g}_F(X',Y) &= Ł(X',Y) , \\ \underline{g}_F(X',Y) &= 0 , \end{aligned} \tag{49}$$

for odd vectors X, X' and even vectors Y, Y', with the primes denoting normal vectors as usual. The above equations determine a decomposition of $\underline{T}$ with respect to the tangential and normal bundles. Explicitly, we have

$$\underline{F} \cong F \oplus F' , \tag{50}$$

while

$$\underline{B} \subset B \oplus B' \oplus F' . \tag{51}$$

The field Ł can be thought of as providing a gauge-invariant representation of the worldsurface multiplet. Indeed, in the linearised case it reduces to the even derivative of the transverse odd coordinate functions.

The generalisations of the Gauss-Weingarten equations are

$$\begin{aligned} \underline{\nabla}_X Y &= \nabla_X Y + K'(X,Y) + L(X,Y) , \\ \underline{\nabla}_X Y' &= \nabla'_X Y' + K(X,Y') + L'(X,Y') , \end{aligned} \tag{52}$$

where $K'(X,Y)$ and $L'(X,Y)$ are normal while $K(X,Y')$ and $L(X,Y)$ are tangential. The additional tangential terms are required because even vectors on M have non-vanishing projections on $\underline{F}$. For Y Y' odd $L(X,Y)$ and $L'(X,Y')$ both vanish while $K(X,Y)$ and $K'(X,Y')$ are odd. The torsion equations are

$$\begin{aligned} [\underline{T}(X,Y)]^t &= T(X,Y) + L(X,Y) - L(Y,X) , \\ [\underline{T}(X,Y)]^n &= K(X,Y) - K(Y,X) , \end{aligned} \tag{53}$$

while the Gauss-Codazzi equations have the same form as in the bosonic case despite the presence of the additional terms in the Gauss-Weingarten equations,

$$\underline{R}(X,Y,Z,\omega) = R(X,Y,Z,\omega) + (K'(X,K(Y,Z),\omega) - X \leftrightarrow Y) \ , \quad (54)$$

$$\underline{R}(X,Y,Z',\omega') = R'(X,Y,Z',\omega') + (K(X,K'(Y,Z'),\omega') - X \leftrightarrow Y) \ , (55)$$

where the last two arguments of the curvatures are either both even or both odd.

One can obtain many relations for the tensors defined above by differentiating the invariant tensors. It is straightforward to check that the connections defined on T and T' preserve the induced bosonic and fermionic metrics, and that the tensors constructed from $\underline{\phi}$ are also invariant. The structure groups for F and F' are both $\text{Spin}(1,2) \cdot \widetilde{\text{Spin}}(8)$, although different representations are involved, while the structure groups for B and B' are $SO_o(1,2)$ and SO(8), where the superscript "o" denotes the component connected to the identity.

Many of the above equations can be taken over in the case of the five-brane, but there are some differences. The equations for the bosonic metric remain the same but the fermionic ones change. One finds, instead of 48, the equations

$$\underline{g}_F(X,Y) = h(X,Y) \ , \quad (56)$$

$$\underline{g}_F(X,Y') = g_F(X,Y') \ , \quad (57)$$

$$\underline{g}_F(X',Y') = 0 \ , \quad (58)$$

for odd arguments. Note that in decomposing the eleven-dimensional charge conjugation matrix ($\underline{g}_F$) into $6+5$ one does not arrive at fermionic metrics on the tangential and normal subspaces but rather at an off-diagonal tensor which we have called g_F above although it is not a metric but rather determines an isomorphism between F^* and F'. The tensor h departs from this expected behaviour and is a signal of a type II brane. In index notation,

$$h_{\alpha\beta} = \frac{1}{3}(\Gamma^{abc})_{\alpha\beta} h_{abc} \ . \quad (59)$$

For mixed arguments ($X's$ odd Y's even) one has

$$\begin{aligned} \underline{g}_F(X,Y) &= \text{Ł}(X,Y) \ , \\ \underline{g}_F(X,Y') &= 0 \ , \\ \underline{g}_F(X',Y) &= 0 \ , \\ \underline{g}_F(X',Y) &= 0 \ . \end{aligned} \quad (60)$$

Again the field Ł is related to the worldsurface multiplet. For the Frobenius tensor one finds, as before,

$$\underline{\phi}(X,Y,\omega) = \phi(X,Y,\omega) ,$$
$$\underline{\phi}(X,Y',\omega') = \tilde{\phi}(X,Y',\omega') ,$$
$$\underline{\phi}(X',Y',\omega) = \phi'(X',Y',\omega) , \tag{61}$$

and

$$\underline{\phi}(X,Y,\omega') = 0 ,$$
$$\underline{\phi}(X',Y',\omega') = 0 , \tag{62}$$

where the vectors are odd and the forms even. However, one now has

$$\underline{\phi}(X,Y',\omega) \neq 0 . \tag{63}$$

In fact, this tensor is also linearly proportional to h.

One can take over the Gauss-Weingarten, torsion and curvature equations formally without change, but there are differences between the two and five-brane cases. For the five-brane the induced connections for the even tangent and normal bundles correspond to the groups $SO_o(1,5)$ and SO(5) respectively, but the connections for the odd tangent bundles do not give $\mathrm{Spin}(1,5) \cdot \mathrm{Spin}(5)$ connections. This is again due to the type II embedding structure. One would have had this result if h had been zero, but the intervention of this term complicates matters somewhat. An alternative procedure is to define connections which do preserve the natural groups in both the even and odd sectors, and this is the route that has been taken in the literature (Howe, Sezgin (1997), Howe, Sezgin, West (1997)).

We conclude with a few remarks on Wess-Zumino forms. So far we have made no mention of these, even though they play such a crucial rôle in the *GS* formalism. The reason for this is that supersymmetry implies that they are present, so that one does not have to introduce them separately by hand in the superspace formalism. In eleven dimensional superspace with the standard constraints it is easy to show that there exists a closed four-form $\underline{H}_4$ which has non-trivial components only at dimension zero and one, the dimension one component reflecting the presence of a non-trivial spacetime three-form potential. The pull-back of this form defines a four-form on M, obviously closed, and which is flat in the case of the membrane, i.e. its only non-vanishing component in a standard basis is a Γ-matrix contribution at dimension zero, and which obeys the equation

$$\mathrm{d}H_3 = -\frac{1}{4} f^* \underline{H}_4 , \tag{64}$$

in the case of the five-brane. The only non-vanishing component of H_3 is the purely vectorial component which is given by

$$H_{abc} = (m^{-1})_a{}^d h_{bcd} . \tag{65}$$

We emphasize that 64 is not a new equation; it is identically true provided that one defines H_3 as above and uses the results which follow from the torsion equations of the embedding. We refer the reader to the literature for more details on how one deduces the full equations of motion describing the brane dynamics from the basic embedding condition 20 using the superspace formalism (Howe, Sezgin (1997), Howe, Sezgin, West (1997), Howe, Sezgin, West (1997), Howe, Sezgin, West (in preparation)).

References

Aganagic, M., Popescu, C. and Schwarz, J.H. (1996): *Gauge-invariant and gauge-fixed D-brane actions*, hep-th/9612080.

Aganagic, M., Popescu, C. and Schwarz, J.H. (1997): *D-brane actions with local kappa symmetry*, Phys. Lett. **B393**, 311, hep-th/9610249.

Aharony, O. (1996): *String theory dualities from M theory*, Nucl. Phys. **B476**, 470, hep-th/9604103.

Bandos, I., Lechner, K., Nurmagambetov, A., Pasti, P., Sorokin, D. and Tonin, M. (1997): *Covariant action for the superfivebrane of M-theory*, hep-th/9701149.

Bandos, I.A., Sorokin, D., Tonin, M., Pasti, P. and Volkov, D. (1995): *Superstrings and supermembranes in the doubly supersymmetric geometrical approach*, Nucl. Phys. **B446**, 79, hep-th/9501113.

Bergshoeff, E., de Roo, M. and Ortin, T. (1996): *The eleven dimensional fivebrane*, Phys. Lett. **B386**, 85, hep-th/9606118.

Bergshoeff, E., Sezgin, E. and Townsend, P.K. (1987): *Supermembranes and eleven-dimensional supergravity*, Phys. Lett. **189B**, 75.

Bergshoeff, E. and Townsend, P.K. (1997): *Super D-branes*, Nucl. Phys. **B490**, 145, hep-th/9611173.

Brink, L. and Howe, P.S. (1980): *Eleven dimensional supergravity on the mass shell in superspace*, Phys. Lett. **91B**, 384.

Brink, L. and Schwarz, J.H. (1981): *Quantum superspace*, Phys. Lett. **100B**, 310.

Cederwall, M., von Gussich, A., Nilsson, B.E.W. and Westerberg, A. (1997): *The Dirichlet super-three-brane in ten-dimensional Type IIB supergravity*, Nucl. Phys. **B490**, 163, hep-th/9610148.

Cederwall, M., von Gussich, A., Nilsson, B.E.W., Sundell, P. and Westerberg, A. (1997): *The Dirichlet super p-branes in ten dimensional Type IIA and IIB supergravity*, Nucl. Phys. **B490**, 179, hep-th/9611159.

Cremmer, E. and Ferrara, S. (1980): *Formulation of 11-dimensional supergravity in superspace*, Phys. Lett. **91B**, 61.

Delduc, F., Galperin, A., Howe, P.S. and Sokatchev, E. (1993): *A twistor formulation of the heterotic D=10 superstring with manifest (8,0) worldsheet supersymmetry*, Phys. Rev. **D47**, 578, hep-th/9207050.

Duff, M.J. and Lu, J.X. (1993): *Type II p-branes: the brane scan revisited*, Nucl. Phys. **B390**, 276, hep-th/9207060.

Gliozzi, F., Scherk, J. and Olive, D. (1977): *Supersymmetry, supergravity theories and the dual spinor model*, Nucl. Phys. **B122**, 253.

Green, M.B. and Schwarz, J.H. (1984): *Covariant description of superstrings*, Phys. Lett. **136B**, 367.

Hughes, J., Liu, J. and Polchinski, J. (1986): *Supermembranes*, Phys. Lett. **B180**, 370.

Howe, P.S. (in preparation): *Weyl superspace*, in preparation.

Howe, P.S. and Sezgin, E. (1997): *Superbranes*, Phys. Lett. **B390**, 133, hep-th/9607227.

Howe, P.S. and Sezgin, E. (1997): *D=11,p=5*, Phys. Lett. **B394**, 62, hep-th/9611008.

Howe, P.S., Sezgin, E. and West, P.C. (1997): *Covariant field equations of the M-theory five-brane*, hep-th/9702008.

Howe, P.S., Sezgin, E. and West, P.C. (1997): *The six-dimensional self-dual tensor*, hep-th/9702111.

Howe, P.S., Sezgin, E. and West, P.C. (in preparation): *The geometry of super M-branes*, in preparation.

Kobayashi, S. and Nomizu, K. (1963): *Foundations of Differential Geometry, vol. II* (Wiley Interscience, 1963).

Neveu, A. and Schwarz, J.H. (1971): *Factorizable dual model of pions*, Nucl. Phys. **B31**, 86.

Pasti, P., Sorokin, D. and Tonin, M. (1997): *Covariant action for D=11 five-brane with the chiral field*, hep-th/9701037.

Perry, M. and Schwarz, J.H. (1997): *Interacting chiral gauge fields in six dimensions and Born-Infeld theory*, Nucl. Phys. **B489**, 47, hep-th/9611065.

Ramond, P. (1971): *Dual theory for free fermions*, Phys. Rev. **D3**, 2415.

Siegel, W. (1983): *Hidden local supersymmetry in the supersymmetric particle action*, Phys. Lett. **128B**, 397.

Schwarz, J.H. (1997): *Coupling of self-dual tensor to gravity in six dimensions*, Phys. Lett. **B395**, 191, hep-th/9701008.

Sorokin, D., Tkach, V. and Volkov, D.V. (1989): *Superparticles, twistors and Siegel symmetry*, Mod. Phys. Lett. **A4**, 901.

Sorokin, D., Tkach, V., Volkov, D.V. and Zheltukhin, A. (1989): *From superparticle Siegel supersymmetry to the spinning particle proper-time supersymmetry*, Phys. Lett. **B259**, 302.

Townsend, P.K. (1996): *D-branes from M-branes*, Phys. Lett. **B373**, 68, hep-th/9512062.

Witten, E. (1996): *Five-brane effective action in M-theory*, hep-th/9610234.

Superbrane Actions and Geometrical Approach

I. Bandos[1], P. Pasti[2], D. Sorokin[1] and M. Tonin[2]

[1] National Science Center, Kharkov Institute of Physics and Technology, Kharkov, 310108, Ukraine
[2] Università Degli Studi Di Padova, Dipartimento Di Fisica "Galileo Galilei" ed INFN, Sezione Di Padova, Via F. Marzolo, 8, 35131 Padova, Italia

Abstract. We review a generic structure of conventional (Nambu–Goto and Dirac–Born–Infeld–like) worldvolume actions for the superbranes and show how it is connected through a generalized action construction with a doubly supersymmetric geometrical approach to the description of super–p–brane dynamics as embedding world supersurfaces into target superspaces.

During last years Dmitrij Vasilievich Volkov actively studied geometrical and symmetry grounds underlying the theory of supersymmetric extended objects and we are happy to have been his collaborators in this work. One of the incentives for this research was to understand the nature of an important fermionic κ–symmetry of the target–superspace (or Green–Schwarz) formulation of the superparticles and superstrings with the aim to resolve the problem of its infinite reducibility, to relate the Green–Schwarz and Ramond–Neveu–Schwarz formulation of superstrings already at the classical level and to attack the problem of covariant quantization of superstrings. The κ–symmetry was conjectured to be a manifestation of local extended supersymmetry (irreducible by definition) on the world supersurface swept by a super–p–brane in a target superspace. This was firstly proved for $N = 1$ superparticles in three and four dimensions (Sorokin, Tkach and Volkov (1989)) and then for $N = 1$, $D = 6, 10$ superparticles (Delduc and Sokatchev (1992), Galperin A. and Sokatchev E.(1992)), $N = 1$ (Berkovits (1989),(1990); Tonin (1991),(1992), Delduc et al. (1992)), $N = 2$ (Galperin and Sokatchev (1993)) superstrings, $N = 1$ supermembranes (Pasti and Tonin (1994), Bergshoeff and Sezgin (1994)) and finally for all presently known super–p–branes (Bandos et al. (1995), Bandos, Sorokin and Volkov (1995), Howe and Sezgin (1997a), (1997b)) in all space–time dimensions where they exist. In (Volkov et. al. (1988),(1989), Aoyama et al. (1992), Berkovits (1994a,b), Polyakov (1997)) a twistor transform was applied to relate the Green–Schwarz and the Ramond–Neveu–Schwarz formulation.

The approach to describing the super–p–branes in this way is called the doubly supersymmetric geometrical approach, since it essentially exploits the theory of embedding world *super*surfaces into target *super*spaces. Apart from having clarified the geometrical nature of κ–symmetry and having made a substantial impact on the development of new methods of superstring covariant quantization (see Bandos and Zheltukhin (1992),(1994), Berkovits (1996) and references therein), the doubly supersymmetric approach has proved its power in studying new important class of super–p–branes (such as Dirichlet branes (Dai et al. (1989), Leigh (1989), Polchinski J. (1995),(1996)) and the M-theory five–brane (Duff and Stelle (1991), Güven (1992)) for which supersymmetric equations of motion were obtained in the geometrical approach (Howe and Sezgin (1997a), (1997b)) earlier than complete supersymmetric actions for them were constructed by standard methods (Cederwall et al. (1997a,b), Aganagic et al. (1997), Bergshoeff and Townsend (1997), Bandos et al. (1997b), Aganagic et al. (1997)). Thus, a problem arises to relate the super–p–brane equations obtained from the action with the field equations of the doubly supersymmetric geometrical approach, and to convince oneself that they really describe one and the same object. To accomplish this goal one should reformulate the action principle for the super–p–branes such that it would yield the embedding conditions of the geometrical approach in the most direct way. For ordinary super–p–branes such an action has been proposed in (Bandos, Sorokin and Volkov (1995)). The construction is based on generalized action principle of the group–manifold (or rheonomic) approach to superfield theories (Neeman and Regge (1978), D́ Auria et al. (1980) Castellani et al. (1991)). D. V. Volkov considered this approach as the most appropriate for implementing geometry of the supersymmetric extended objects into the description of their dynamics.

In this contribution we would like to review basic elements of the generalized action construction and to show that it is also applicable to the Dirichlet branes (Bandos et al. (1997a)) and, at least partially, to the M–theory five–brane (M–5–brane), thus allowing one to establish the relation between the formulations of (Cederwall et al. (1997a,b), Aganagic et al. (1997), Bergshoeff and Townsend (1997), Bandos et al. (1997b), Aganagic et al. (1997) and Howe and Sezgin (1997a), (1997b), Howe et al. (1997)).

On the way of reconstructing the super–p–brane actions we shall answer another question connected with their κ–symmetry transformations, namely, a puzzling fact that the κ–transformation of a "kinetic" part of the conventional super–p–brane actions is the integral of a $(p+1)$–form which compensates the κ–variations of a Wess–Zumino term of the actions. This puzzle is resolved in a formulation where the entire action of a super–p–brane is the integral of a differential $(p+1)$–form in the worldvolume of the brane (Bandos et al. (1995), Bandos, Sorokin and Volkov (1995)). To construct such an action one uses auxiliary harmonic (Galperin et al. (1984),(1985) or twistor–like variables which enable to get an irreducible realization of the κ–transformations (see Sorokin, Tkach and Volkov (1989), Bandos and Zheltukhin (1992),(1994), Bandos and Zheltukhin (1993),(1995) and references therein for superparticles, superstrings and type I super–p–branes). We shall also see that in the case of the D–branes and the M–5–brane this version of the action serves as a basis for the transition to a dual description of these objects.

Consider the general structure of the action for a super–p–brane propagating in a supergravity background of an appropriate space–time dimension (which is specified by a brane scan (Bergshoeff (1996)). We work with actions of a Nambu–Goto (or Dirac–Born–Infeld) type that do not involve auxiliary fields of intrinsic worldvolume geometry as in the Brink–Di Vecchia–Howe–Tucker form (Brink et al. (1976), Howe P. S. and Tucker R. W. (1977) of brane actions (see (Cederwall et al. (1997c), Abou Zeid and Hull (1997) for the BDHT approach to D–branes).

All known super–p–brane actions, except that of the M–5–brane which contains a third term (see below), generically consist of two terms:

$$S = I_1 + I_{WZ} = \int_{M_{p+1}} \mathrm{d}^{p+1}x \mathrm{e}^{-\frac{p-3}{4}\phi}\sqrt{-\det G_{mn}} + \int_{M_{p+2}} W_{p+2} \ . \quad (1)$$

The symmetric part g_{mn} of the matrix $G_{mn} \equiv g_{mn} + \mathcal{F}_{mn}$ in the first term of (1) describes a super–p–brane worldvolume metric induced by embedding into a target superspace which is parametrized by bosonic coordinates $X^{\underline{m}}(x)$ ($\underline{m} = 0, 1, ..., D-1$) and fermionic coordinates $\Theta^{\underline{\mu}}(x)$ ($\underline{\mu} = 1, ..., 2^{[\frac{D}{2}]}$) collectively defined as $Z^{\underline{M}} = (X^{\underline{m}}, \Theta^{\underline{\mu}})$. The worldvolume itself is parametrized by small x^m ($m = 0, ..., p$) with not underlined indices. $\phi(Z)$ is a background dilaton field. Note that there is no such a field in $D = 11$ supergravity.

The antisymmetric part $\mathcal{F}_{mn}$ of G_{mn}, which is absent from ordinary superbranes and nonzero for the D–branes and the M–5–brane, contains the field strength of a gauge field propagating in the brane worldvolume plus the worldvolume pullback of a Grassmann–antisymmetric field of target–space supergravity.

In the case of the D–branes in $D = 10$ the worldvolume field is a vector field $A_m(x)$ (Dai et al. (1989), Leigh (1989), Polchinski J. (1995),(1996), Cederwall et al. (1997a,b), Aganagic et al. (1997), Bergshoeff and Townsend (1997), the background field is a two–rank superfield $B_{\underline{MN}}(X, \Theta)$, and $\mathcal{F}_{mn}$ has the form

$$\mathcal{F}^{(D)}_{mn} = \mathrm{e}^{-\frac{\phi(Z)}{2}}(\partial_m A_n - \partial_n A_m + \partial_m Z^{\underline{N}} \partial_n Z^{\underline{M}} B_{\underline{MN}}) \ . \quad (2)$$

In the case of the M–5–brane the worldvolume gauge field is a self–dual (or chiral) tensor field $A_{mn}(x)$, and the background field is a three–rank superfield $C_{\underline{LMN}}(X, \Theta)$ of $D = 11$ supergravity (Duff and Stelle (1991), Güven (1992), Bandos et al. (1997b), Aganagic et al. (1997). The M–5–brane action also contains an auxiliary worldvolume scalar field $a(x)$ (Pasti et al. (1995),(1997a) whose presence ensures manifest $d = 6$ worldvolume covariance of the model (Pasti et al. (1997b), Bandos et al. (1997b), Aganagic et al. (1997). In this case the antisymmetric matrix takes the form

$$\mathcal{F}^{(M)}_{mn} = \frac{\mathrm{i}}{\sqrt{\partial_p a \partial^p a}} H^*_{mnl} \partial^l a(x) \ ,$$

$$H_{mnl} = 6\partial_{[l} A_{mn]} + \partial_l Z^{\underline{L}} \partial_m Z^{\underline{N}} \partial_n Z^{\underline{M}} C_{\underline{MNL}} \ , \tag{3}$$

where $*$ denotes Hodge operation, e.g. $H^*_{mnl} = \frac{\sqrt{-g}}{3!} \varepsilon_{mnlpqr} H^{pqr}$.

The second term in (1) is a Wess–Zumino (WZ) term. Generically it is more natural to define it as an integral of a closed differential $(p+2)$–form over a $(p+2)$–dimensional manifold whose boundary is the super–p–brane worldvolume. The structure of the WZ term depends on the p–brane considered and (in general) includes worldvolume pullbacks of antisymmetric gauge fields of target–space supergravity and their duals (see (Cederwall et al. (1997a,b), Aganagic et al. (1997), Bergshoeff and Townsend (1997), Bandos et al. (1997b), Aganagic et al. (1997) for details).

The third term which one must add to the action (1) to describe the M–5–brane dynamics is quadratic in H_{mnl} (Bandos et al. (1997b), Aganagic et al. (1997):

$$I_3 = \int \mathrm{d}^6 x \frac{\mathrm{i}}{\sqrt{-\partial_p a \partial^p a}} \mathcal{F}^{(M)}_{mn} H^{mnl} \partial_l a(x) \ . \tag{4}$$

In this case the action (1) plus (4) is invariant under the local symmetries (Pasti et al. (1995),(1997a), Pasti et al. (1997b))

$$\delta A_{mn} = \frac{\varphi(x)}{2(\partial a)^2}(H_{mnp}\partial^p a - \mathcal{V}_{mn}), \qquad \delta a(x) = \varphi(x) \tag{5}$$

and

$$\delta A_{mn} = \partial_{[m} a(x) \varphi_{n]}(x), \qquad \delta a(x) = 0 \tag{6}$$

where

$$\mathcal{V}^{mn} \equiv -2\sqrt{\frac{(\partial a)^2}{g}} \frac{\delta\sqrt{-\det(G_{pq})}}{\delta \mathcal{F}_{mn}},$$

and φ and φ_m are local gauge parameters. The local symmetry (5) allows one to gauge the field $a(x)$ away at the expense of manifest Lorentz invariance of the M–5–brane action and the local symmetry (6) is needed to ensure the self–duality condition for A_{mn}. These local symmetries are, in some sense, a bosonic analog of the fermionic κ–symmetry (see below) whose gauge fixing also results in the loss of Lorentz covariance.

The action (1) (plus (4) in the case of the M–5–brane) is invariant under the following κ–transformations of the worldvolume fields

$$i_\kappa E^{\underline{\alpha}} \equiv \delta_\kappa Z^{\underline{M}} E^{\underline{\alpha}}_{\underline{M}} = \kappa^{\underline{\alpha}}, \qquad i_\kappa E^{\underline{a}} = 0,$$

$$\delta_\kappa g_{mn} = -4\mathrm{i} E_{\{m} \Gamma_{n\}} i_\kappa E \ , \tag{7}$$

$$\delta_\kappa \mathcal{F}^{(D)} = i_\kappa \mathrm{d}B_{(2)}, \qquad \delta_\kappa H = i_\kappa \mathrm{d}C_{(3)}, \qquad \delta_\kappa a(x) = 0,$$

where

$$E^{\underline{A}} = \left(\mathrm{d}Z^{\underline{M}} E^{\underline{a}}_{\underline{M}}(X,\Theta), \mathrm{d}Z^{\underline{M}} E^{\underline{\alpha}}_{\underline{M}}(X,\Theta)\right) \tag{8}$$

are target–space supervielbeins pulled back into the worldvolume. They define the induced metric

$$g_{mn} = \partial_m Z^{\underline{M}} \partial_n Z^{\underline{N}} E^{\underline{a}}_{\underline{N}} \eta_{\underline{ab}} E^{\underline{b}}_{\underline{M}} \ , \tag{9}$$

and i_κ denotes the contraction of the forms with the κ–variation of $Z^{\underline{M}}$ as written above. The Grassmann parameter $\kappa^{\underline{\alpha}}(x)$ of the κ–transformations satisfies the condition

$$\kappa^{\underline{\alpha}} = \kappa^{\underline{\beta}} \bar{\Gamma}_{\underline{\beta}}{}^{\underline{\alpha}} \ , \tag{10}$$

where $\bar{\Gamma}$ is a traceless matrix composed of the worldvolume pullbacks of target–space Dirac matrices and the tensor $\mathcal{F}_{mn}$ such that $\bar{\Gamma}^2 = 1$. The form of $\bar{\Gamma}$ is specific for a p–brane considered and reflects the structure of the WZ term (Cederwall et al. (1997a,b), Aganagic et al. (1997), Bergshoeff and Townsend (1997), Bandos et al. (1997b), Aganagic et al. (1997)).

Eq. (10) reads that not all components (in fact only half) of $\kappa^{\underline{\alpha}}$ are independent, which causes the (infinite) reducibility of the κ–transformations. If one tries to get an irreducible set of κ–parameters in this standard formulation, one should break manifest Lorentz invariance of the models. The geometrical approach considered below provides us with a covariant way of describing independent κ–transformations.

For all super–p–branes the κ–variation (7) of the Wess–Zumino term is (up to a total derivative) the integral of a $(p+1)$–form

$$\delta_\kappa I_{WZ} = \int_{M_{(p+1)}} i_\kappa W_{(p+2)} \ . \tag{11}$$

For the complete action to be κ–invariant the WZ variation must be compensated by the variation of the NG or DBI-like term (and the term (4)). Thus, though these parts of the action are not the integrals of differential forms, their κ–variations are. To explain this puzzling fact it is natural to look for a formulation where the entire action is the integral of a $(p+1)$–form. When we deal with ordinary super–p–branes, for which $\mathcal{F}_{mn} = 0$, this can be easily done, since (apart from the presence of the dilaton field) the NG term in (1) is the integral volume of the world surface and can be written as the worldvolume differential form integral

$$I_1 = \int_{M_{(p+1)}} \frac{1}{(p+1)!} E^{a_0} \wedge E^{a_1} \wedge \ldots \wedge E^{a_p} \epsilon_{a_0 a_1 \ldots a_p} \ , \tag{12}$$

where $E^a = \mathrm{d}x^m E^a_m(x)$ is a worldvolume vielbein form. Since we consider induced geometry of the worldvolume, E^a is constructed as a linear combination of the target–space supervielbein vector components (8)

$$E^a = \mathrm{d}x^m \partial_m Z^{\underline{M}} E^{\underline{b}}_{\underline{M}} u_{\underline{b}}{}^a(x). \tag{13}$$

$u_{\underline{b}}{}^{a}(x)$ are components (vector Lorentz harmonics along the worldvolume) of an SO$(1, D-1)$–valued matrix

$$u_{\underline{b}}{}^{\tilde{a}} = (u_{\underline{b}}{}^{a},\ u_{\underline{b}}{}^{i}) \qquad a = 0, ..., p \quad i = p+1, ..., D-1 \ , \tag{14}$$

$$u_{\underline{a}}{}^{\tilde{c}} u_{\tilde{c}}{}^{\underline{b}} = \delta_{\underline{a}}{}^{\underline{b}}, \qquad u_{\tilde{a}}{}^{\underline{c}} u_{\underline{c}}{}^{\tilde{b}} = \delta_{\tilde{a}}{}^{\tilde{b}} = diag(\delta_{a}^{b}, \delta^{ij}) \ , \tag{15}$$

The orthogonality conditions (15) are invariant under the direct product of target–space local Lorentz rotations SO$(1, D-1) \times$ SO$(1, D-1)$ acting on u from the left and right, while the splitting (14) breaks one SO$(1, D-1)$ (tilded indices) down to its SO$(1, p) \times$ SO$(D-p-1)$ subgroup, which form a natural gauge symmetry of the p–brane embedded into D–dimensional space–time.

Surface theory tells us that (14) can always be chosen such that on the world surface

$$E^{i} = dZ^{\underline{M}} E_{\underline{M}}^{\underline{b}} u_{\underline{b}}{}^{i}(x)|_{M_{p+1}} = 0 \ , \tag{16}$$

i.e. orthogonal to the surface.

Dynamically one derives Eq. (16) from the action (12) by varying it with respect to the auxiliary variables $u_{\underline{b}}{}^{a}$ and taking into account the orthogonality condition (15).

In view of (13), (16) and (15) we see that the expression (9) for the induced metric reduces to $g_{mn} = E_m^a E_{an}$. Hence, we can replace the determinant of E_m^a written in (12) with $\sqrt{-\det g_{mn}}$ and return back to the NG form of the super-p-brane action. This demonstrates the equivalence of the two formulations.

Note that only vector components $E^{\underline{a}}$ of the target–space supervielbein (8) enter the action (12) through Eqs. (13). But in target superspace a supervielbein also has components along spinor directions (8) (i.e. $E^{\underline{\alpha}}$). When the supervielbein vector components undergo a local SO$(1, D-1)$ transformation with the matrix (14), the supervielbein spinor components are rotated by a corresponding matrix $v_{\underline{\beta}}{}^{\tilde{\alpha}}(x)$ of a spinor representation of the group SO$(1, D-1)$, the matrices $u_{\underline{b}}{}^{\tilde{a}}$ and $v_{\underline{\beta}}{}^{\tilde{\alpha}}$ being related to each other through the well–known formula (see for instance (Bandos and Zheltukhin (1992),(1994), Bandos and Zheltukhin (1993),(1995))

$$u_{\underline{b}}{}^{\tilde{a}} \Gamma^{\underline{b}}_{\underline{\alpha\beta}} = v_{\underline{\alpha}}{}^{\tilde{\delta}} \Gamma^{\tilde{a}}_{\tilde{\delta}\tilde{\gamma}} v_{\underline{\beta}}{}^{\tilde{\gamma}}, \qquad u_{\tilde{b}}{}^{\underline{a}} \Gamma^{\tilde{b}}_{\underline{\tilde{\alpha}\tilde{\beta}}} = u_{b}{}^{\underline{a}} \Gamma^{b}_{\underline{\tilde{\alpha}\tilde{\beta}}} - u^{i\underline{a}} \Gamma^{i}_{\underline{\tilde{\alpha}\tilde{\beta}}} = v_{\tilde{\alpha}}{}^{\underline{\delta}} \Gamma^{a}_{\underline{\delta\gamma}} v_{\tilde{\beta}}{}^{\underline{\gamma}} \ . \tag{17}$$

The matrix $v_{\underline{\beta}}{}^{\tilde{\alpha}}$ satisfies an orthogonality condition analogous to (15). Thus, it is natural to consider the spinor harmonic variables $v_{\underline{\beta}}{}^{\tilde{\alpha}}$ as independent and $u_{\underline{b}}{}^{\tilde{a}}$ composed of the former. The SO$(1, p) \times$ SO$(D-p-1)$ invariant splitting of v (analogous to (14)) is

$$v_{\underline{\beta}}{}^{\tilde{\alpha}} = (v_{\underline{\beta}}{}^{\alpha q}, v_{\underline{\beta}}{}^{\dot{\alpha}\dot{q}}) \ , \tag{18}$$

where $\alpha, \dot\alpha$ are the indices of (the same or different) spinor representations of $SO(1,p)$ and $q, \dot q$ correspond to representations of $SO(D-p-1)$. The choice of these representations depends on the dimension of the super–p–brane and the target superspace considered and is such that the dimension of the $SO(1,p)$ representations times the dimension of the $SO(D-p-1)$ representations is equal to the spinor representation of $SO(1,D-1)$.

To generalize Eq. (12) to the case of the D-branes one should take into account the presence of the antisymmetric tensor $\mathcal{F}_{mn}$ in (1) as follows:

$$I_1 = \int_{M_{(p+1)}} \Big(\frac{1}{(p+1)!} E^{a_0}\wedge E^{a_1}\wedge\ldots\wedge E^{a_p}\epsilon_{a_0a_1\ldots a_p} e^{-\frac{p-3}{2}\phi}\sqrt{-\det(\eta_{ab}+\mathcal{F}_{ab})} \qquad (19)$$

$$+Q_{p-1}\wedge[e^{-\frac{1}{2}\phi}(dA-B_{(2)}) - \frac{1}{2}E^b\wedge E^a\mathcal{F}_{ab}]\Big) + \int_{M_{p+2}} W_{p+2}$$

where we also included the WZ term, and the worldvolume scalar $\mathcal{F}_{ab}$ is an auxiliary antisymmetric tensor field with tangent space (Lorentz group) indices and Q_{p-1} is a Lagrange multiplier differential form which produces the algebraic equation

$$\mathcal{F}_2^{(D)} \equiv 1/2E^b\wedge E^a F_{ab} = e^{-\frac{\phi}{2}}(dA - B_2)\ . \qquad (20)$$

Eq. (20) relates $\mathcal{F}_{ab}$ to the 2–form $\mathcal{F}_2^{(D)}$ of the original action (1).

In the case of the M–5–brane its action (1) plus (4) is written as an integral of differential forms as follows

$$I_1 + I_3 = \int_{M_6} \Big(\frac{1}{6!}E^{a_0}\wedge E^{a_1}\wedge\ldots\wedge E^{a_p}\epsilon_{a_0a_1\ldots a_p}(\sqrt{-\det(\eta_{ab}+\mathcal{F}_{ab})} - \qquad (21)$$

$$\frac{i}{4\sqrt{-v_av^a}}\mathcal{F}^{ab}H_{abc}v^c) + Q_3\wedge[dA_2 - C_3 - \frac{1}{3}E^c\wedge E^b\wedge E^a H_{abc}]$$

$$+Q_5\wedge(da(x) - E^a v_a) + \int_{M_7} W_7.$$

In (21) $\mathcal{F}^{ab} = \frac{i}{\sqrt{-v_av^a}}H^{*abc}v_c$, and $H_{abc}(x)$ and $v_a(x)$ are auxiliary worldvolume scalar fields which are expressed in terms of original fields A_{mn} and $a(x)$ (3) upon solving the equations of motion for the Lagrange multiplier forms Q_3 and Q_5.

The variation of the actions (19) and (21) with respect to the auxiliary fields $\mathcal{F}_{ab}$, $H_{abc}(x)$ and $v_a(x)$ produces algebraic expressions for the Lagrange multipliers $Q_{(n)}$, which thus do not describe independent degrees of freedom of the models. Note also that, at least for the Dirichlet branes with $p\le 4$, one can invert the equations for Q_{p-1} in terms of $\mathcal{F}_{ab}$, express the latter in terms of the former and substitute them in the action. This gives a dual worldvolume description of the D–branes (Tseytlin A. A. (1996), Aganagic et al. (1997), Pasti et al. (1997b), Abou Zeid and Hull (1997).) Thus the actions

(19) and (21) have the form which provides one with a way to perform a dual transform of the superbrane models.

The worldvolume fields $Z^{\underline{M}}(x)$ and $A(x)$ (or $\mathcal{F}_{mn}$ and H_{mnl}) of the super-p-branes are transformed under the κ-transformations as above (see Eqs. (7)).

The κ-variation of the auxiliary fields and the Lagrange multipliers can be easily obtained from their expressions in terms of other fields whose κ-transformations are known.

To compute the κ-transformation of the actions (19) and (21) we should also know the κ-variations of the Lorentz-harmonic fields $u_{\underline{b}}{}^{\underline{a}}$, which are genuine worldvolume fields. However these variations are multiplied by algebraic field equations such as (16) and (20) and, therefore, they can be appropriately chosen to compensate possible terms proportional to the algebraic equations that arise from the variation of other terms. It means, in particular, that when computing $\delta_\kappa S$ we can freely use these algebraic equations and, at the same time, drop the κ-variations of these genuine worldvolume quantities if we are not interested in their specific form.

Thus by construction the actions (19) and (21) are integrals of $(p+1)$-forms $\mathcal{L}_{p+1}$ and one can show that their κ-variation has the following general structure (Bandos et al. (1997a)):

$$\delta S = \int i_\kappa d\mathcal{L}_{p+1} = \int i_\kappa \left(\mathrm{i} E^{(-)} \gamma^{(p)} E^{(-)} - 2E^{(-)} \gamma^{(p+1)} \Delta \phi \right) , \qquad (22)$$

where the second term is absent from the case of the M-5-brane,

$$E^{(-)\underline{\alpha}} = \frac{1}{2} E^{\underline{\beta}} (1 - \bar{\Gamma})_{\underline{\beta}}{}^{\underline{\alpha}}, \qquad \Delta_{\underline{\alpha}} \phi(X, \Theta) = E_{\underline{\alpha}}^{\underline{M}} \partial_{\underline{M}} \phi , \qquad (23)$$

$\gamma^{(p)}_{\underline{\alpha\beta}}$ and $\gamma^{(p+1)}_{\underline{\alpha\beta}}$ are, respectively, differential p-form and $p+1$-form constructed of worldvolume-projected target-space gamma-matrices and the tensor $\mathcal{F}_{mn}$ (see Cederwall et al. (1997a,b), Aganagic et al. (1997), Bergshoeff and Townsend (1997), Bandos et al. (1997a), Bandos et al. (1997c) for details).

The fact that $i_{\bar\kappa} \mathrm{d}\mathcal{L} = 0$ and (22) is κ-invariant follows from Eqs. (7) and (10) which imply $i_\kappa E^{(-)\hat{\underline{\alpha}}} = 0 = i_{\bar\kappa} E^{\underline{a}}$.

Note that, since the spinor parameter κ corresponds to a particular class of general variations of $\Theta(x)$, the knowledge of the κ-variation (22) of the super-p-brane action enables one to directly get equations of motion of $\Theta(x)$ as differential form equations

$$i\gamma^{(p)}_{\underline{\alpha\beta}} E^{(-)\underline{\beta}} - (\frac{1}{2}(1 - \bar{\Gamma})\gamma^{(p+1)})_{\underline{\alpha}}{}^{\underline{\beta}} \Delta_{\underline{\beta}} \phi = 0 . \qquad (24)$$

Let us now demonstrate how the presence of the Lorentz-harmonic fields allows one to extract in a covariant way the independent parameters of the κ-transformations (see Bandos and Zheltukhin (1992),(1994), Bandos and

Zheltukhin (1993),(1995) for ordinary super–p–branes)[1]. For this we use the $SO(1,p) \times SO(D-p-1)$ decomposition of the spinor harmonics (18). To be concrete, consider the example of a Dirichlet 3–brane (p=3) in a background of type IIB $D = 10$ supergravity (Bandos et al. (1997a)). In this case the decomposition (18) of a 16×16 matrix $v_{\underline{\beta}}^{\tilde{\underline{\alpha}}}$ takes the form

$$v_{\underline{\beta}}^{\tilde{\underline{\alpha}}} = (v_{\underline{\alpha} q}^{\ \alpha}, \bar{v}_{\underline{\alpha}}^{\ \dot{\alpha} q}) \ , \tag{25}$$

where $\alpha = 1,2$, $\dot{\alpha} = 1,2$ are Weyl spinor indices of $SO(1,3)$, $q = 1, ..., 4$ are $SO(6)$ spinor indices and bar denotes complex conjugation.

Using the exact form of the matrix $\bar{\Gamma}$ and the condition (10) one can show (Bandos et al. (1997a)) that the following 16 complex conjugate components of the complex κ–parameter are independent:

$$\kappa_q^{\alpha}(x) = \kappa^{\underline{\beta}} v_{\underline{\beta} q}^{\ \alpha}, \qquad \bar{\kappa}^{\dot{\alpha} q}(x) = \bar{\kappa}^{\underline{\beta}} \bar{v}_{\underline{\beta}}^{\ \dot{\alpha} q} \ . \tag{26}$$

By the use of independent parameters (such as (26)) the κ–transformations (7) can be rewritten in an irreducible form. This realization of κ–symmetry is target–space covariant since the parameters (26) are target–space scalars and carry the indices of the $SO(1,p) \times SO(D-p-1)$ group, the first factor of which is identified with the Lorentz rotations in the tangent space of the superbrane worldvolume and the second factor corresponds to an internal local symmetry of the super–p–brane. Because of the fermionic nature of these worldvolume κ–parameters it is tempting to treat them as the parameters of n–extended local supersymmetry[2] in the worldvolume of the super–p–brane, and this was just a basic idea of (Sorokin, Tkach and Volkov (1989)), which has been fruitfully developed in the framework of the doubly supersymmetric approach.

To make the local worldvolume supersymmetry manifest one should extend the worldvolume to a world supersurface parametrized by x^m and n $SO(1,p)$–spinor variables $\eta^{\alpha q}$ all fields of the super–p–brane models becoming worldvolume superfields.

Now, the differential form structure (19) of super–p–brane actions admits of an extension to worldvolume superspace by the use of generalized action principles of the group–manifold (or rheonomic) approach (Neeman and Regge (1978), D́ Auria et al. (1980) Castellani et al. (1991) to supersymmetric field theories. This has been carried out for the ordinary super–p–branes (Bandos, Sorokin and Volkov (1995) and the Dirichlet branes (Bandos et al. (1997a)). As to the M–5–brane, the presence of the term $Q_5 \wedge (da(x) - E^a v_a)$

---

[1] An alternative possibility of getting irreducible covariant κ–transformations and their covariant gauge fixing has been discussed in Bergshoeff et al. (1997), Kallosh (1997).

[2] The number n of worldvolume supersymmetries is such that $n \times \dim Spin(1,p) = \dim Spin(1, D-1)$.

in (21) causes problems to be solved yet to lift the M–5–brane action to worldvolume superspace, since (without some modification) this term would lead to rather strong (triviallizing) restrictions on worldvolume supergeometry. Thus for the time being further consideration is not applicable in full measure to the M–5–brane action (21), though final superfield equations for the superbranes which one gets as geometrical conditions of supersurface embedding are valid for the M–5–brane as well. The relation of the M–5–brane action (1) and (4) (Bandos et al. (1997b), Aganagic et al. (1997)) and component field equations of the M–5–brane obtained from the doubly supersymmetric geometrical approach (Howe and Sezgin (1997a), (1997b), Howe et al. (1997)) was established in (Bandos et al. (1997c)).

The rheonomic approach exhibits in a vivid fashion geometrical properties of supersymmetric theories, and when the construction of conventional superfield actions for them fails the generalized action principle allows one to get the superfield description of these models. As we shall sketch below, in the case of super–p–branes the generalized action serves for getting geometrical conditions of embedding world supersurfaces into target superspaces, which completely determine the on–shell dynamics of the superbranes (Bandos et al. (1995), Bandos, Sorokin and Volkov (1995), Howe and Sezgin (1997a), (1997b), Bandos et al. (1997a)). However an open problem is how to extend the generalized action approach to the quantum level.

Main points of this doubly supersymmetric construction are the following.

The generalized action for superbranes has the same form as (19) but with all fields and differential forms replaced with superforms in the worldvolume superspace $\Sigma=(x^m,\ \eta^{\alpha q})$. The integral is taken over an arbitrary $(p+1)$–dimensional bosonic surface $\mathcal{M}_{p+1} = (x^m, \eta^{\alpha q}(x))$ in the worldvolume superspace Σ. Thus, the surface $\mathcal{M}_{p+1}$ itself becomes a dynamical variable, i.e. one should vary (19) also with respect to $\eta(x)$, however it turns out that this variation does not produce new equations of motion in addition to the variation with respect to other fields, the equations of motion of the latter having the same form as that obtained from the component action we started with. But now the fact that the surface $\mathcal{M}_{p+1}$ is arbitrary and that the full set of such surfaces spans the whole worldvolume superspace makes it possible to consider these equations of motion as equations for the superforms and superfields defined in the whole worldvolume superspace Σ. The basic superfield equations thus obtained are Eqs. (16) and (24) (note that now the external differential also includes the η–derivative). Eqs. (16) and (24) tell us that induced worldvolume supervielbeins $(\mathrm{e}^a(x,\eta), \mathrm{e}^{\alpha q}(x,\eta))$ can be chosen as a linear combination of $E^{\underline{b}} u_{\underline{b}}{}^a \equiv E^a$ and $E^{\underline{\beta}} v_{\underline{\beta}}{}^{\alpha q} \equiv E^{\alpha q}$ (Bandos et al. (1995), Howe and Sezgin (1997a), (1997b), Bandos et al. (1997a)):

$$\mathrm{e}^a = E^b (m^{-1})_b{}^a \quad \Rightarrow \quad E^{\ a}_{\alpha q} = 0 \ , \tag{27}$$

$$\mathrm{e}^{\alpha q} = E^{\alpha q} + E^a \chi_a{}^{\alpha q}(x,\eta) \ , \tag{28}$$

as well as that

$$E^{(-)\underline{\alpha}}_{\alpha q} \equiv \left(E_{\alpha q}\frac{1}{2}(1-\bar{\Gamma})\right)^{\underline{\alpha}} = 0 \ , \qquad (29)$$

The choice of the matrix ${m_b}^a(x,\eta)$ is a matter of convenience and can be used to get the main spinor–spinor component of the worldvolume torsion constraints in the standard form $T^a_{\alpha q,\beta r} = i\delta_{qr}\gamma^a_{\alpha\beta}$. In this case ${m_b}^a$ is constructed out of worldvolume gauge fields (Howe and Sezgin (1997a), (1997b), Bandos et al. (1997a), Bandos et al. (1997c)).

Eq. (27) together with eq. (16) implies the basic geometrodynamical condition

$$E^{\underline{a}}_{\alpha q} = 0 \qquad (30)$$

which in the doubly supersymmetric approach to super–p–branes determines the embedding of the worldvolume superspace into the target superspace. In many interesting cases such as $D = 10$ type II superstrings (Galperin and Sokatchev (1993), Bandos et al. (1995)) and D–branes (Howe and Sezgin (1997a), (1997b), Bandos et al. (1997a)), and $D = 11$ branes (Bandos et al. (1995), Howe and Sezgin (1997a), (1997b)) the integrability conditions for (16), (27) and (28) reproduce all the equations of motion of these extended objects and also lead to torsion constraints on worldvolume supergravity (Bandos, Sorokin and Volkov (1995), Bandos et al. (1997a)).

Note that for the D–branes and the M–5–brane the embedding conditions analogous to (16), (27), (28) and (29) were initially not derived from an action, which was not known at that time, but postulated (Howe and Sezgin (1997a), (1997b)) on the base of the previous knowledge of analogous conditions for ordinary super–p–branes (Bandos et al. (1995), Bandos, Sorokin and Volkov (1995)).

To conclude, we have demonstrated how the super–p–brane action can be reconstructed as the worldvolume integral of a differential $(p + 1)$–form. The use of the Lorentz–harmonic variables in this formulation makes the κ–symmetry transformations to be performed with an irreducible set of fermionic parameters being worldvolume spinors. This indicates that the κ–symmetry originates from extended local supersymmetry in the worldvolume. We have shown how this worldvolume supersymmetry becomes manifest in a worldvolume superfield generalization of the super–p–brane action. The superfield equations derived from the latter are the geometrical conditions of embedding worldvolume supersurfaces swept by the superbranes in target superspaces. Thus, the approach reviewed in this article serves as a bridge between different formulations developed for describing superbrane dynamics.

Acknowledgements. We would like to thank our collaborators Kurt Lechner and Alexei Nurmagambetov with whom we obtained many results reported herein. Work of P.P. and M.T. was supported by the European Commission TMR programme ERBFMRX–CT96–0045 to which P.P. and M.T. are associated. I.B. thanks Prof. M. Virasoro for hospitality at the ICTP. I.B., and D.S. acknowledge partial support from grants of the Ministry of Science and Technology of Ukraine and the INTAS Grants N 93–127–ext, N 93–493–ext and N 93–0633–ext.

References

Abou Zeid, M. and Hull, C. M. (1997): Phys. Lett. **B404**, 264.

Bandos, I. A. and Zheltukhin, A. A. (1992): Phys. Lett. **B288**, 77
Bandos, I. A. and Zheltukhin, A. A. (1994): Sov. J. Elem. Part. Atom. Nucl. **25**, 453–477.

Bandos, I. A. and Zheltukhin, A. A. (1993): Int. J. Mod. Phys. **A8**, 1081
Bandos, I. A. and Zheltukhin, A. A. (1995): Class. Quantum Grav. **12**, 609.

Bandos, I., Sorokin, D. and Volkov, D. (1995): Phys. Lett **B352**, 269.

Bandos, I., Pasti, P., Sorokin, D., Tonin, M. and Volkov, D. (1995): Nucl.Phys. **B446**, 79.

Bandos, I., Sorokin, D. and Tonin, M. (1997a): Nucl. Phys. **B497**, 275.

Bandos, I., Lechner, K., Nurmagambetov, A., Pasti, P., Sorokin, D. and Tonin, M. (1997): Phys. Rev. Lett. **78**, 4332
Aganagic, M., Park, J., Popescu, C. and Schwarz, J. H. (1997): Nucl. Phys. **B496**, 191.

Bandos, I., Lechner, K., Nurmagambetov, A., Pasti, P., Sorokin, D. and Tonin, M. (1997): On the Equivalence of Different Formulations of the M Theory Five-Brane, hep-th/9703127.

Berkovits, N. (1989): **232B**, 184;
Berkovits, N. (1990): Phys. Lett. **241B**, 497.
Tonin, M. (1991): Phys. Lett. **B266**, 312;
Tonin, M. (1992): Int. J. Mod.Phys **7**, 613.
Delduc, F., Galperin, A., Howe, P. and Sokatchev, E. (1992): Phys. Rev. **D47**, 587.

Berkovits, N.(1996): Preprint IFUSP-P-1212, April 1996, hep-th/9604123.

Bergshoeff, E. (1996): P–brane and D–brane actions, hep–th/9607238.

Bergshoeff, E., Kallosh, R., Ortin, T. and Papadopulos, G. (1997): K-symmetry, Supersymmetry and Intersecting Branes, hep–th/9705040.
Kallosh, R. (1997): Volkov–Akulov theory and D–branes, hep–th/9705118 (a contribution to these Proceedings).

Brink, L., Di Vecchia, P. and Howe, P. S. (1976): Phys. Lett. **B65**, 471
Howe, P. S. and Tucker, R. W. (1977): J. Phys. **A10**, L155.

Cederwall, M., von Gussich, A., Nilsson, B. E. W. and Westerberg, A. (1997a): Nucl. Phys. **B490**, 163.
Aganagic, M., Popescu, C. and Schwarz, J. H. (1997): Phys. Lett. **393**, 311
Cederwall, M., von Gussich, A., Nilsson, B. E. W., Sundell, P. and Westerberg, A. (1997b): Nucl. Phys. **B490**, 179.
Bergshoeff, E. and Townsend, P. K. (1997): Nucl. Phys. **B490**, 145.

Cederwall, M., von Gussich, A., Micovic, A., Nilsson, B.E.W. and Westerberg, A. (1997c): Phys. Lett. **B390**, 148.

Dai, J., Leigh, R. G. and Polchinski, J. (1989): Mod. Phys. Lett. **A4**, 2073
Leigh, R. G. (1989): Mod. Phys. Lett. **A4**, 2767
Polchinski, J. (1995): Phys. Rev. Lett. **75**, 4724
Polchinski, J. (1996): Tasi lectures on D–branes, NSF-96-145, hep-th/9611050.

Delduc, F. and Sokatchev, E. (1992): Class. Quantum Grav. **9**, 361.
Galperin, A. and Sokatchev, E.(1992): Phys. Rev. **D46**, 714.

Duff, M. J. and Stelle, K. S. (1991): Phys. Lett. **B253**, 113 .
Güven R. (1992): Phys. Lett. **B276**, 49

C. G. Callan, J. A. Harvey and A. Strominger (1991): Nucl. Phys. **B367**, 60.
Gibbons, G. W. and Townsend, P. K. (1993): Phys. Rev. Lett. **71**, 3754
Kaplan, D. and Michelson, J. (1996): Phys. Rev. **D53**, 3474
Becker, K. and Becker, M.(1996): Nucl. Phys. **B472**, 221.
Galperin, A., Ivanov, E., Kalitzin, S., Ogievetsky, V. and Sokatchev, E. (1984): Class. Quantum Grav. **1**, 498
Galperin, A., Ivanov, E., Kalitzin, S., Ogievetsky, V. and Sokatchev, E. (1985): Class. Quantum Grav. **2**, 155.
Galperin, A. and Sokatchev, E. (1993): Phys. Rev. **D**, 48 4810.
Howe, P. and Sezgin, E. (1997a): Phys. Lett. **B390**, 133,
Howe, P. and Sezgin, E. (1997a): Phys. Lett. **B394**, 62.
Howe, P.S., Sezgin, E. and West, P. (1997): Phys. Lett. **B399**, 49.
Neeman, Y. and Regge, T. (1978): Phys. Lett **B74**, 31, Revista del Nuovo Cim. 1 1978 1
D́ Auria, R., Fré, P. and Regge, T. (1980): Revista del Nuovo Cim. **3**, 1;
Castellani, L., D́ Auria, R., Fré, P. (1991): *Supergravity and superstrings, a geometric perspective*, World Scientific, Singapore, (and references therein).
Pasti, P. and Tonin, M. (1994): Nucl.Phys. **B418**, 337
Bergshoeff, E. and Sezgin, E. (1994): Nucl.Phys. **B422**, 329.
Pasti, P., Sorokin, D. and Tonin, M. (1995): Phys. Rev. **D52**, R4277;
Pasti, P., Sorokin, D. and Tonin, M. (1997a): Phys. Rev. **D55**, 6292.
Pasti, P., Sorokin, D. and Tonin, M. (1997b): Phys. Lett. **398B**, 41.
Sorokin, D., Tkach, V. and Volkov, D. V.(1989): Mod. Phys. Lett. **A4**, 901.
Tseytlin, A. A. (1996): Nucl.Phys. **B469**, 51.
Aganagic, M., Park, J., Popescu, C. and Schwarz, J.H. (1997): Nucl. Phys. **B496**, 215.
Volkov, D. V. and Zheltukhin, A. (1988): Sov. Phys.JETP Lett **48**, 61
Volkov, D. V. and Zheltukhin, A. (1989): Lett. Math. Phys. **17**, 141.
Sorokin, D., Tkach, V., Volkov, D. and Zheltukhin, A. (1989): Phys. Lett. **B216**, 302.
Aoyama, S., Pasti, P. and Tonin, M. (1992): Phys. Lett. **B283**, 213.
Berkovits, N. (1994a): Nucl.Phys. **B420**, 332
Berkovits, N. (1994b): Nucl.Phys. **B431**, 258
Polyakov, D. (1997): Nucl.Phys. **B485**, 128.

The M Theory Five-Brane and the Heterotic String

J.H. Schwarz

California Institute of Technology
Pasadena, CA 91125, USA

Abstract. Brane actions with chiral bosons present special challenges. Recent progress in the description of the two main examples – the M theory five-brane and the heterotic string – is described. Also, double dimensional reduction of the M theory five-brane on K3 is shown to give the heterotic string.

1 The Bosonic Part of the Five-Brane Action

The M theory five-brane action contains a tensor gauge field, which (in linearized approximation) has a self-dual field strength. Ref. Schwarz (1997) analyzed the problem of coupling a 6d self-dual tensor gauge field to a metric field so as to achieve general coordinate invariance. It presented a formulation in which one direction is treated differently from the other five. At the time that work was done, the author knew of no straightforward way to make the general covariance manifest. However, shortly thereafter a paper appeared Pasti et al. (1997a) that presents equivalent results using a manifestly covariant formulation Pasti et al. (1995), (1997b), which we refer to as the PST formulation. In the following both approaches and their relationship are described. These results have been generalized to supersymmetric actions with local kappa symmetry (Bandos et al. (1997), Aganagic et al. (1997), Howe et al. (1997)), but here we will only consider the bosonic theories.

1.1 The Noncovariant Formulation

Let us denote the 6d (world volume) coordinates by $\sigma^{\hat{\mu}} = (\sigma^\mu, \sigma^5)$, where $\mu = 0,1,2,3,4$. The σ^5 direction is singled out as the one that will be treated differently from the other five. (This is a space-like direction, but one could also choose a time-like one.) The 6d metric $G_{\hat{\mu}\hat{\nu}}$ contains 5d pieces $G_{\mu\nu}, G_{\mu 5}$, and G_{55}. All formulas will be written with manifest 5d general coordinate invariance. As in refs. Perry and Schwarz (1996), Schwarz (1997), we represent the self-dual tensor gauge field by a 5×5 antisymmetric tensor $B_{\mu\nu}$, and its 5d curl by $H_{\mu\nu\rho} = 3\partial_{[\mu}B_{\nu\rho]}$. A useful quantity is the dual

$$\tilde{H}^{\mu\nu} = \frac{1}{6}\epsilon^{\mu\nu\rho\lambda\sigma} H_{\rho\lambda\sigma} \quad . \tag{1}$$

It was shown in ref. Schwarz (1997) that a class of generally covariant bosonic theories can be represented in the form $L = L_1 + L_2 + L_3$, where

$$\begin{aligned} L_1 &= -\frac{1}{2}\sqrt{-G} f(z_1, z_2) \ , \\ L_2 &= -\frac{1}{4}\tilde{H}^{\mu\nu}\partial_5 B_{\mu\nu} \ , \\ L_3 &= \frac{1}{8}\epsilon_{\mu\nu\rho\lambda\sigma}\frac{G^{5\rho}}{G^{55}}\tilde{H}^{\mu\nu}\tilde{H}^{\lambda\sigma} \ . \end{aligned} \tag{2}$$

The notation is as follows: G is the 6d determinant ($G = \det G_{\hat{\mu}\hat{\nu}}$) and G_5 is the 5d determinant ($G_5 = \det G_{\mu\nu}$), while G^{55} and $G^{5\rho}$ are components of the inverse 6d metric $G^{\hat{\mu}\hat{\nu}}$. The ϵ symbols are purely numerical with $\epsilon^{01234} = 1$ and $\epsilon^{\mu\nu\rho\lambda\sigma} = -\epsilon_{\mu\nu\rho\lambda\sigma}$. A useful relation is $G_5 = GG^{55}$. The z variables are defined to be

$$\begin{aligned} z_1 &= \frac{\mathrm{tr}(G\tilde{H}G\tilde{H})}{2(-G_5)} \\ z_2 &= \frac{\mathrm{tr}(G\tilde{H}G\tilde{H}G\tilde{H}G\tilde{H})}{4(-G_5)^2} \ . \end{aligned} \tag{3}$$

The trace only involves 5d indices:

$$\mathrm{tr}(G\tilde{H}G\tilde{H}) = G_{\mu\nu}\tilde{H}^{\nu\rho}G_{\rho\lambda}\tilde{H}^{\lambda\mu} \ . \tag{4}$$

The quantities z_1 and z_2 are scalars under 5d general coordinate transformations.

Infinitesimal parameters of general coordinate transformations are denoted $\xi^{\hat{\mu}} = (\xi^\mu, \xi)$. Since 5d general coordinate invariance is manifest, we focus on the ξ transformations only. The metric transforms in the standard way

$$\delta_\xi G_{\hat{\mu}\hat{\nu}} = \xi\partial_5 G_{\hat{\mu}\hat{\nu}} + \partial_{\hat{\mu}}\xi G_{5\hat{\nu}} + \partial_{\hat{\nu}}\xi G_{\hat{\mu}5} \ . \tag{5}$$

The variation of $B_{\mu\nu}$ is given by a more complicated rule, whose origin is explained in ref. Schwarz (1997):

$$\delta_\xi B_{\mu\nu} = \xi K_{\mu\nu} \ , \tag{6}$$

where

$$K_{\mu\nu} = 2\frac{\partial(L_1 + L_3)}{\partial\tilde{H}^{\mu\nu}} = K^{(1)}_{\mu\nu} f_1 + K^{(2)}_{\mu\nu} f_2 + K^{(\epsilon)}_{\mu\nu} \tag{7}$$

with

$$\begin{aligned} K^{(1)}_{\mu\nu} &= \frac{\sqrt{-G}}{(-G_5)}(G\tilde{H}G)_{\mu\nu} \\ K^{(2)}_{\mu\nu} &= \frac{\sqrt{-G}}{(-G_5)^2}(G\tilde{H}G\tilde{H}G\tilde{H}G)_{\mu\nu} \\ K^{(\epsilon)}_{\mu\nu} &= \epsilon_{\mu\nu\rho\lambda\sigma}\frac{G^{5\rho}}{2G^{55}}\tilde{H}^{\lambda\sigma} \ , \end{aligned} \tag{8}$$

and we have defined

$$f_i = \frac{\partial f}{\partial z_i} \ , \quad i = 1, 2 \ . \tag{9}$$

Assembling the results given above, ref. Schwarz (1997) showed that the required general coordinate transformation symmetry is achieved if, and only if, the function f satisfies the nonlinear partial differential equation Gibbons and Rasheed (1995)

$$f_1^2 + z_1 f_1 f_2 + \left(\frac{1}{2} z_1^2 - z_2\right) f_2^2 = 1 \ . \tag{10}$$

As discussed in Perry and Schwarz (1996), this equation has many solutions, but the one of relevance to the M theory five-brane is

$$f = 2\sqrt{1 + z_1 + \frac{1}{2} z_1^2 - z_2} \ . \tag{11}$$

For this choice L_1 can reexpressed in the Born–Infeld form

$$L_1 = -\sqrt{-\det\left(G_{\hat{\mu}\hat{\nu}} + i G_{\hat{\mu}\rho} G_{\hat{\nu}\lambda} \tilde{H}^{\rho\lambda} / \sqrt{-G_5}\right)} \ . \tag{12}$$

This expression is real, despite the factor of i, because it is an even function of $\tilde{H}$.

1.2 The PST Formulation

In ref. Pasti et al. (1997a) (using techniques developed in ref. Pasti et al. (1995), (1997b)) equivalent results are described in a manifestly covariant way. To do this, the field $B_{\mu\nu}$ is extended to $B_{\hat{\mu}\hat{\nu}}$ with field strength $H_{\hat{\mu}\hat{\nu}\hat{\rho}}$. In addition, an auxiliary scalar field a is introduced. The PST formulation has new gauge symmetries (described below) that allow one to choose the gauge $B_{\mu 5} = 0$, $a = \sigma^5$ (and hence $\partial_{\hat{\mu}} a = \delta^5_{\hat{\mu}}$). In this gauge, the covariant PST formulas reduce to the ones given above.

Equation (12) expresses L_1 in terms of the determinant of the 6×6 matrix

$$M_{\hat{\mu}\hat{\nu}} = G_{\hat{\mu}\hat{\nu}} + i \frac{G_{\hat{\mu}\rho} G_{\hat{\nu}\lambda}}{\sqrt{-G G^{55}}} \tilde{H}^{\rho\lambda} \ . \tag{13}$$

In the PST approach this is extended to the manifestly covariant form

$$M^{\text{cov.}}_{\hat{\mu}\hat{\nu}} = G_{\hat{\mu}\hat{\nu}} + i \frac{G_{\hat{\mu}\hat{\rho}} G_{\hat{\nu}\hat{\lambda}}}{\sqrt{-G (\partial a)^2}} \tilde{H}^{\hat{\rho}\hat{\lambda}}_{\text{cov.}} \ . \tag{14}$$

The quantity

$$(\partial a)^2 = G^{\hat{\mu}\hat{\nu}} \partial_{\hat{\mu}} a \partial_{\hat{\nu}} a \tag{15}$$

reduces to G^{55} upon setting $\partial_{\hat{\mu}} a = \delta^5_{\hat{\mu}}$, and

$$\tilde{H}^{\hat{\rho}\hat{\lambda}}_{\mathrm{cov.}} \equiv \frac{1}{6}\epsilon^{\hat{\rho}\hat{\lambda}\hat{\mu}\hat{\nu}\hat{\sigma}\hat{\tau}} H_{\hat{\mu}\hat{\nu}\hat{\sigma}}\partial_{\hat{\tau}} a \tag{16}$$

reduces to $\tilde{H}^{\rho\lambda}$. Thus $M^{\mathrm{cov.}}_{\hat{\mu}\hat{\nu}}$ replaces $M_{\hat{\mu}\hat{\nu}}$ in L_1. Furthermore, the expression

$$L' = -\frac{1}{4(\partial a)^2}\tilde{H}^{\hat{\mu}\hat{\nu}}_{\mathrm{cov.}} H_{\hat{\mu}\hat{\nu}\hat{\rho}} G^{\hat{\rho}\hat{\lambda}}\partial_{\hat{\lambda}} a \ , \tag{17}$$

which transforms under general coordinate transformations as a scalar density, reduces to $L_2 + L_3$ upon gauge fixing. It is interesting that L_2 and L_3 are unified in this formulation.

Let us now describe the new gauge symmetries of ref. Pasti et al. (1997a). Since degrees of freedom a and $B_{\mu 5}$ have been added, corresponding gauge symmetries are required. One of them is

$$\delta B_{\hat{\mu}\hat{\nu}} = 2\phi_{[\hat{\mu}}\partial_{\hat{\nu}]} a \ , \tag{18}$$

where $\phi_{\hat{\mu}}$ are infinitesimal parameters, and the other fields do not vary. In terms of differential forms, this implies $\delta H = \mathrm{d}\phi\wedge \mathrm{d}a$. $\tilde{H}^{\hat{\rho}\hat{\lambda}}_{\mathrm{cov.}}$ is invariant under this transformation, since it corresponds to the dual of $H\wedge \mathrm{d}a$, but $\mathrm{d}a\wedge\mathrm{d}a = 0$. Thus the covariant version of L_1 is invariant under this transformation. The variation of L', on the other hand, is a total derivative.

The second local symmetry involves an infinitesimal scalar parameter φ. The transformation rules are $\delta G_{\hat{\mu}\hat{\nu}} = 0, \delta a = \varphi$, and

$$\delta B_{\hat{\mu}\hat{\nu}} = \frac{1}{(\partial a)^2}\varphi H_{\hat{\mu}\hat{\nu}\hat{\rho}} G^{\hat{\rho}\hat{\lambda}}\partial_{\hat{\lambda}} a + \varphi V_{\hat{\mu}\hat{\nu}} \ , \tag{19}$$

where the quantity $V_{\hat{\mu}\hat{\nu}}$ is to be determined. Rather than derive it from scratch, let's see what is required to agree with the previous formulas after gauge fixing. In other words, we fix the gauge $\partial_{\hat{\mu}} a = \delta^5_{\hat{\mu}}$ and $B_{\mu 5} = 0$, and figure out what the resulting ξ transformations are. We need

$$\delta a = \varphi + \xi\partial_5 a = \varphi + \xi = 0 \ , \tag{20}$$

which tells us that $\varphi = -\xi$. Then

$$\begin{aligned} \delta_\xi B_{\mu\nu} &= \frac{1}{(\partial a)^2}\varphi H_{\mu\nu\hat{\rho}} G^{\hat{\rho}\hat{\lambda}}\partial_{\hat{\lambda}} a + \varphi V_{\mu\nu} + \xi H_{5\mu\nu} \\ &= -\xi\left(\frac{G^{\rho 5}}{G^{55}} H_{\mu\nu\rho} + V_{\mu\nu}\right) = \xi(K^{(\epsilon)}_{\mu\nu} - V_{\mu\nu}). \end{aligned} \tag{21}$$

Thus, comparing with eqs. (6) and (7), we need the covariant definition

$$V_{\hat{\mu}\hat{\nu}} = -2\frac{\partial L_1}{\partial \tilde{H}^{\hat{\mu}\hat{\nu}}_{\mathrm{cov.}}} \tag{22}$$

to achieve agreement with our previous results.

2 A New Heterotic String Action

There are two main approaches to constructing the world-sheet action of the heterotic string that have been used in the past Gross et al. (1985). In one of them, the internal torus is described in terms of bosonic coordinates. The fact that these bosons are chiral (*i.e.*, the left-movers and right-movers behave differently) is imposed through external constraints. In the second approach these bosonic coordinates are replaced by world-sheet fermions, which are Majorana–Weyl in the 2d sense. What will be most convenient for our purposes is a variant of the first approach. In this variant the coordinates of the Narain torus are still represented by bosonic fields, but the chirality of these fields is achieved through new gauge invariances rather than external constraints Cherkis and Schwarz (1997).

Consider the Narain compactified heterotic string in a Minkowski spacetime with $d = 10 - n$ dimensions Narain (1986), Narain et al. (1987). Let these coordinates be denoted by X^m with $m = 0, 1, \ldots, d-1 = 9-n$. To properly account for all the degrees of freedom, the Narain torus should be described by $16 + 2n$ bosonic coordinates Y^I, $I = 1, 2, \ldots, 16 + 2n$. These will be arranged to describe $26 - d = 16 + n$ left-movers and $10 - d = n$ right-movers. The Y^I are taken to be angular coordinates, with period 2π, so that $Y^I \sim Y^I + 2\pi$, and the conjugate momenta are integers. The actual size and shape of the torus is encoded in a matrix of moduli, denoted M_{IJ}, which will be described below.

The $(16 + 2n)$-dimensional lattice of allowed momenta should form an even self-dual lattice of signature $(n, 16 + n)$. Let us therefore introduce a matrix

$$\eta = \begin{pmatrix} I_n & 0 \\ 0 & -I_{16+n} \end{pmatrix} , \tag{23}$$

where I_n is the $n \times n$ unit matrix. An even self-dual lattice with this signature has a set of $16 + 2n$ basis vectors V_I, and the symmetric matrix

$$L_{IJ} = V_I^a \eta_{ab} V_J^b \tag{24}$$

characterizes the lattice. A convenient specific choice is

$$L = \Lambda_8 \oplus \Lambda_8 \oplus \sigma \oplus \ldots \oplus \sigma , \tag{25}$$

where Λ_8 is the negative of the E_8 Cartan matrix and $\sigma = \begin{pmatrix} 0 & 1 \\ 1 & 0 \end{pmatrix}$ appears n times.

The Narain moduli space is characterized, up to T duality equivalences that will be discussed below, by a symmetric matrix $M'_{ab} \in O(n, 16+n)$, which satisfies $M'\eta M' = \eta$. The fact that it is symmetric means that it actually parametrizes the coset space $\mathrm{O}(n, 16+n)/\mathrm{O}(n) \times O(16+n)$, which has $n(16+n)$ real dimensions. To describe the T duality equivalences, it is convenient to change to the basis defined by the basis vectors of the self-dual lattice. Accordingly, we define

$$M_{IJ} = V_I^a M'_{ab} V_J^b = (V^T M' V)_{IJ} \ . \tag{26}$$

This matrix is also symmetric and satisfies

$$M L^{-1} M = L \ , \tag{27}$$

from which it follows that $(L^{-1}M)^2 = 1$. This allows us to define projection operators

$$\mathcal{P}_\pm = \frac{1}{2}(1 \pm L^{-1}M) \ . \tag{28}$$

$\mathcal{P}_+$ projects onto an n-dimensional subspace, which will correspond to right-movers. Similarly, $\mathcal{P}_-$ projects onto the $(16+n)$-dimensional space of left-movers. The theory we are seeking should be invariant under an infinite discrete group of T duality transformations, denoted $\Gamma_{n,16+n}$,[1] so that the actual moduli space is the standard Narain space

$$\mathcal{M}_{n,16+n} = \Gamma_{n,16+n} \backslash \mathrm{O}(n, 16+n)/\mathrm{O}(n) \times O(16+n) \ . \tag{29}$$

The desired equations of motion for the Y coordinates are (Cecotti et al. (1988), Duff (1990), Tseytlin (1990),(1991a),(1991b), Maharana. and Schwarz (1993))

$$\mathcal{P}_- \partial_+ Y = 0 \quad \text{and} \quad \mathcal{P}_+ \partial_- Y = 0 \ , \tag{30}$$

where $\xi^\pm = \xi^1 \pm \xi^0$, so that $\partial_\pm = \frac{1}{2}(\partial_1 \pm \partial_0)$. ξ^0 and ξ^1 are the world-sheet time and space, respectively. The pair of equations in (30) can be combined in the form

$$M \partial_0 Y - L \partial_1 Y = 0 \ . \tag{31}$$

It is easy to write down a lagrangian that gives this equation (Floreanini and Jackiw (1987)):

$$\mathcal{L}_N = \frac{1}{2}(\partial_0 Y M \partial_0 Y - \partial_0 Y L \partial_1 Y) \ . \tag{32}$$

Two things are peculiar about this lagrangian. First, it does not have manifest Lorentz invariance. However, in ref. Schwarz and Sen (1994) it was shown that $\mathcal{L}_N$ has a global symmetry that can be interpreted as describing a non-manifest Lorentz invariance. Second, it gives the equation of motion

$$\partial_0 [M \partial_0 Y - L \partial_1 Y] = 0 \ , \tag{33}$$

which has a second, unwanted, solution $Y^I = f^I(\xi^1)$. The resolution of the second problem is quite simple. The transformation $\delta Y^I = f^I(\xi^1)$ is a gauge symmetry of $\mathcal{L}_N$, and therefore $f^I(\xi^1)$ represents unphysical gauge degrees of freedom.

[1] It is often called $\mathrm{O}(n, 16+n; \mathbb{Z})$.

The first problem, the noncovariance of $\mathcal{L}_N$, is more interesting. We will follow the PST approach (Pasti et al. (1995), (1997b)), and extend $\mathcal{L}_N$ to a manifestly Lorentz invariant action by introducing an auxiliary scalar field $a(\xi)$. The desired generalization of $\mathcal{L}_N$ is then

$$\mathcal{L}_{PST} = \frac{1}{2(\partial a)^2}(\tilde{Y} M \tilde{Y} - \tilde{Y} L \quad , \partial Y \cdot \partial a) \ , \tag{34}$$

where

$$\tilde{Y}^I = \epsilon^{\alpha\beta} \partial_\alpha Y^I \partial_\beta a \ . \tag{35}$$

Also, $(\partial a)^2$ and $\partial Y \cdot \partial a$ are formed using the 2d Lorentz metric, which is diagonal with $\eta^{00} = -1$ and $\eta^{11} = 1$.

The theory given by $\mathcal{L}_{PST}$ has two gauge invariances. The first is

$$\begin{aligned} \delta Y &= \varphi \left(\frac{1}{\partial_+ a} \mathcal{P}_- \partial_+ Y + \frac{1}{\partial_- a} \mathcal{P}_+ \partial_- Y \right) \ , \\ \delta a &= \varphi \ , \end{aligned} \tag{36}$$

where $\varphi(\xi^0, \xi^1)$ is an arbitrary infinitesimal scalar function. If this gauge freedom is used to set $a = \xi^1$, then $\mathcal{L}_{PST}$ reduces to $\mathcal{L}_N$. The second gauge invariance is

$$\delta Y^I = f^I(a) \ , \quad \delta a = 0 \ , \tag{37}$$

where f^I are arbitrary infinitesimal functions of one variable. This is the covariant version of the gauge symmetry of $\mathcal{L}_N$ that was used to argue that the undesired solution of the equations of motion is pure gauge.

2.1 Reparametrization Invariant Action

The formulas described above are not the whole story of the bosonic degrees of freedom of the toroidally compactified heterotic string, because they lack the Virasoro constraint conditions. The standard way to remedy this situation is to include an auxiliary world-sheet metric field $g_{\alpha\beta}(\xi)$, so that the world-sheet Lorentz invariance is replaced by world-sheet general coordinate invariance. Since we now want to include the coordinates X^m describing the uncompactified dimensions, as well, let us also introduce an induced world-sheet metric

$$G_{\alpha\beta} = g_{mn}(X) \partial_\alpha X^m \partial_\beta X^n \ , \tag{38}$$

where $g_{mn}(X)$ is the string frame target-space metric in d dimensions. It is related to the canonically normalized metric by a factor of the form $\exp(\alpha\phi)$, where ϕ is the dilaton and α is a numerical constant, which can be computed by requiring that the target-space lagrangian is proportional $\exp(-2\phi)$. We will mostly be interested in taking ϕ to be a constant and g_{mn} to be proportional to the flat Minkowski metric. Then the heterotic string coupling constant is $\lambda_H = \exp \phi$, and the desired world sheet lagrangian is

$$\mathcal{L}_g = -\frac{1}{2}\sqrt{-g}g^{\alpha\beta}G_{\alpha\beta} + \frac{\tilde{Y}M\tilde{Y}}{2\sqrt{-g}(\partial a)^2} - \frac{\tilde{Y}L\partial Y\cdot\partial a}{2(\partial a)^2} \ . \tag{39}$$

Now, of course, $(\partial a)^2 = g^{\alpha\beta}\partial_\alpha a\partial_\beta a$ and $\partial Y\cdot\partial a = g^{\alpha\beta}\partial_\alpha Y\partial_\beta a$. The placement of the $\sqrt{-g}$ factors reflects the fact that $\tilde{Y}/\sqrt{-g}$ transforms as a scalar.

There are a few points to be made about $\mathcal{L}_g$. First of all, the PST gauge symmetries continue to hold, so it describes the correct degrees of freedom. Second, just as for more conventional string actions, it has Weyl invariance: $g_{\alpha\beta} \to \lambda g_{\alpha\beta}$ is a local symmetry. This ensures that the stress tensor

$$T_{\alpha\beta} = -\frac{2}{\sqrt{-g}}\frac{\delta S_g}{\delta g^{\alpha\beta}} \ , \tag{40}$$

is traceless ($g^{\alpha\beta}T_{\alpha\beta} = 0$). Using the general coordinate invariance to choose $g_{\alpha\beta}$ conformally flat, and using the PST gauge invariance to set $a = \xi^1$, the Y equations of motion reduce to those described in the previous subsection. In addition, one obtains the classical Virasoro constraints $T_{++} = T_{--} = 0$.

The lagrangian $\mathcal{L}_g$ is written with an auxiliary world-volume metric, which is called the Howe–Tucker or Polyakov formulation. This is the most convenient description for many purposes. However, for the purpose of comparing to expressions derived from the M5-brane later in this paper, it will be useful to also know the version of the lagrangian in which the auxiliary metric is eliminated — the Nambu–Goto formulation. Note that $\mathcal{L}_g$ only involves the metric components in the combination $\sqrt{-g}g^{\alpha\beta}$, which has two independent components. It is a straightforward matter to solve their equations of motion and eliminate them from the action. This leaves the final form for the bosonic part of the heterotic string in $10-n$ dimensions

$$\mathcal{L} = -\sqrt{-G}\sqrt{1 + \frac{\tilde{Y}M\tilde{Y}}{G(\partial a)^2} + \left(\frac{\tilde{Y}L\tilde{Y}}{2G(\partial a)^2}\right)^2} - \frac{\tilde{Y}L\partial Y\cdot\partial a}{2(\partial a)^2} \ , \tag{41}$$

where $G = \det G_{\alpha\beta}$, and now

$$(\partial a)^2 = G^{\alpha\beta}\partial_\alpha a\partial_\beta a \ , \quad \partial Y\cdot\partial a = G^{\alpha\beta}\partial_\alpha Y\partial_\beta a \ . \tag{42}$$

2.2 Wrapping the M-Theory Five-Brane on K3

Let us now consider double dimensional reduction of the M5-brane on K3.[2] This is supposed to give the heterotic string in seven dimensions (Witten (1995), Harvey and Strominger (1995), Sen (1995), Townsend (1995)). Our starting point is the bosonic part of the M5-brane action (Perry and Schwarz (1996)) in the general coordinate invariant PST formulation. Since the other

[2] See ref. Aspinwall (1996) for a review of the mathematics of K3 and some of its appearances in string theory dualities.

11d fields are still assumed to vanish, $g_{MN}(X)$ must be Ricci flat. We will take it to be a product of a Ricci-flat K3 and a flat 7d Minkowski space-time.

Since the M5-brane is taken to wrap the spatial K3, the diffeomorphism invariances of the M5-brane action in these dimensions can be used to equate the four world-volume coordinates that describe the K3 with the four target-space coordinates that describe the K3. In other words, we set $\sigma^\mu = (\xi^\alpha, x^i)$ and $X^M = (X^m, x^i)$. Note that Latin indices i, j, k are used for the K3 dimensions (x^i) and early Greek letters for the directions (ξ^α), which are the world-sheet coordinates of the resulting string action. This wrapping by identification of coordinates, together with the extraction of the K3 zero modes, is what is meant by double dimensional reduction. With these choices, the 6d metric can be decomposed into blocks

$$(G_{\mu\nu}) = \begin{pmatrix} \tilde{G}_{\alpha\beta} & 0 \\ 0 & h_{ij} \end{pmatrix} , \tag{43}$$

with h_{ij} and $\tilde{G}_{\alpha\beta}$ being the K3 metric and the induced metric on the string world-sheet, respectively. The purpose of the tilde is to emphasize that $\tilde{G}_{\alpha\beta} = \tilde{g}_{mn}\partial_\alpha X^m \partial_\beta X^n$, where $\tilde{g}_{mn}$ is the 7d part of the canonical 11d metric. It differs from the metric introduced earlier by a scale factor, which will be determined below. It is convenient to take the PST scalar field a to depend on the ξ^α coordinates only. This amounts to partially fixing a gauge choice for the PST gauge invariance.

The two-form field B has the following contributions from K3 zero modes:

$$B_{ij} = \sum_{I=1}^{22} Y^I(\xi) b_{Iij}(x) \ , \quad B_{\alpha i} = 0 \ , \quad B_{\alpha\beta} = c_{\alpha\beta}(\xi) \ , \tag{44}$$

where b_{Iij} are the 22 harmonic representatives of $\mathrm{H}^2(\mathrm{K3}, \mathbf{Z})$, the integral second cohomology classes of K3. Any other terms are either massive or can be removed by gauge transformations. The nonzero components of $H_{\mu\nu\rho}$ and $\tilde{H}^{\mu\nu}$ are

$$H_{\alpha ij} = \sum_{I=1}^{22} \partial_\alpha Y^I b_{Iij} \tag{45}$$

$$\tilde{H}^{ij} = \sum_{I=1}^{22} \tilde{Y}^I \frac{1}{2} \epsilon^{ijkl} b_{Ikl} = \sum_{I=1}^{22} \sqrt{h} \tilde{Y}^I (*b_I)^{ij} \ , \tag{46}$$

where $\tilde{Y}^I = \epsilon^{\alpha\beta} \partial_\alpha Y^I \partial_\beta a$ as in eq. (35). Note that $c_{\alpha\beta}$ does not contribute.

Now we can compute the string action that arises from double dimensional reduction by substituting the decompositions (45) and (46) into the five-brane Lagrangian. To make the connection with the heterotic string action of the previous section, we make the identifications

$$L_{IJ} = \int_{K3} b_I \wedge b_J \ , \tag{47}$$

$$M_{IJ} = \int_{K3} b_I \wedge *b_J \ . \tag{48}$$

Note that $*b_I = b_J(L^{-1}M)^J{}_I$, and therefore $(L^{-1}M)^2 = 1$, as in sect. 2. Note also that $b_I \wedge b_J$ and $b_I \wedge *b_J$ are closed four-forms, and therefore they are cohomologous to the unique harmonic four-form of the K3, which is the volume form ω. It follows that

$$b_I \wedge b_J = *b_I \wedge *b_J = \frac{L_{IJ}}{\mathcal{V}}\omega + dT_{IJ} \ , \quad b_I \wedge *b_J = \frac{M_{IJ}}{\mathcal{V}}\omega + dU_{IJ} \ , \tag{49}$$

where $\mathcal{V} = \int_{K3}\omega$ is the volume of the K3 and $U_{IJ} = T_{IK}(L^{-1}M)^K{}_J$. The exact terms are absent when either two-form is self-dual, but there is no apparent reason why they should vanish when both of them are anti-self-dual. If we nevertheless ignore the exact pieces in these formulas, substitute into the Lagrangian, and integrate over the K3, we obtain

$$\mathcal{L}_1 = -\mathcal{V}\sqrt{-\tilde{G}}\sqrt{1 + \frac{\tilde{Y}^I M_{IJ}\tilde{Y}^J}{\tilde{G}(\partial a)^2\mathcal{V}} + \frac{1}{4}\left(\frac{\tilde{Y}^I L_{IJ}\tilde{Y}^J}{\tilde{G}(\partial a)^2\mathcal{V}}\right)^2} - \frac{\tilde{Y}^I L_{IJ}\partial_\alpha Y^J \tilde{G}^{\alpha\beta}\partial_\beta a}{2(\partial a)^2} \ . \tag{50}$$

This is precisely the heterotic string lagrangian (for $n = 3$) presented in eq. (41) of the previous section provided that the 7d metric g_{mn} in the string frame is related to the metric $\tilde{g}_{mn}$ derived from 11d by

$$g_{mn} = \mathcal{V}\tilde{g}_{mn} \tag{51}$$

so that $G_{\alpha\beta} = \mathcal{V}\tilde{G}_{\alpha\beta}$. This is the same scaling rule found by a different argument in ref. Witten (1995). Then, following ref. Witten (1995), the Einstein term in the 7d lagrangian is proportional to $\mathcal{V}\sqrt{-\tilde{g}}R(\tilde{g}) = \mathcal{V}^{-3/2}\sqrt{-g}R(g)$, from which one infers that $\mathcal{V} \sim \lambda_H^{4/3}$.

To complete the argument we must still explain why terms that have been dropped make negligible contributions. It is not at all obvious that the exact pieces in eq. (49) can be neglected, but it is what is required to obtain the desired answer. The other class of terms that have been dropped are the Kaluza–Klein excitations of the five-brane on the K3. By simple dimensional analysis, one can show that in the heterotic string metric these contributions to the mass-squared of excitations are of order λ_H^{-2}. Therefore they represent non-perturbative corrections from the heterotic viewpoint. Since our purpose is only to reproduce the perturbative heterotic theory, they can be dropped. Another class of contributions, which should not be dropped, correspond to simultaneously wrapping the M2-brane around a 2-cycle of the K3. These wrappings introduce charges for the 22 U(1)'s, according to how many times each cycle is wrapped. The contribution to the mass-squared of excitations depends on the shape of the K3, of course, but in the heterotic metric it is independent of its volume and hence of the heterotic string coupling constant.

Acknowledgment

I am grateful to M. Aganagic, S. Cherkis, J. Park, M. Perry, and C. Popescu for collaborating on portions of this work. This work is supported in part by the U.S. Dept. of Energy under Grant No. DE-FG03-92-ER40701.

References

Aganagic, M., Park, J., Popescu, C., and Schwarz, J. H. (1997): *World-Volume Action of the M Theory Five-Brane,* hep-th/9701166.
Aspinwall, P.S. (1996): *K3 Surfaces and String Duality,* hep-th/9611137.
Bandos, I., Lechner, K., Nurmagambetov, A., Pasti, P., Sorokin, D. and Tonin, M. (1997): *Covariant Action for the Super-Five-Brane of M-Theory,* hep-th/9701149.
Cecotti, S., Ferrara, S. and Girardello, L. (1988): Nucl. Phys. **B308**, 436.
Cherkis, S. and Schwarz, J.H. (1997): *Wrapping the M Theory Five-brane on K3,* hep-th/9703062.
Duff, M. (1990): Nucl. Phys. **B335**, 610.
Floreanini, R. and Jackiw, R. (1987): Phys. Rev. Lett. **59**, 1873.
Harvey, J.A. and Strominger, A. (1995): Nucl. Phys. **B449** 535, hep-th/9504047; Sen, A. (1995): Nucl. Phys. **B450**, 103, hep-th/9504027.
Howe, P.S., Sezgin, E., West, P.C. (1997): *Covariant Field Equations of the M Theory Five-Brane,* hep-th/9702008.
Gibbons, G.W. and Rasheed, D.A. (1995): Nucl. Phys. **B454** 185, hep-th/9506035.
Gross, D.J., Harvey, J.A., Martinec, E. and Rohm, R. (1985): Nucl. Phys. **B256**, 253.
Maharana, J. and Schwarz, J.H. (1993): Nucl. Phys. **B390** 3, hep-th/9207016.
Narain, K.S. (1986): Phys. Lett. **B169**, 41;
Narain, K.S., Sarmadi, M.H. and Witten, E. (1987): Nucl. Phys. **B279**, 369.
Pasti, P., Sorokin, D. and Tonin, M. (1997a): Phys. Lett. **398B** 41.
Pasti, P., Sorokin, D. and Tonin, M. (1995): Phys.Rev. **D52**, R4277;
Pasti, P., Sorokin, D. and Tonin, M. (1997b): Phys. Rev. **D55** 6292.
Perry, M. and Schwarz, J.H. (1996): *Interacting Chiral Gauge Fields in Six Dimensions and Born–Infeld Theory,* hep-th/9611065.
Schwarz, J.H. and Sen, A. (1994): Nucl. Phys. **B411**, 35, hep-th/9304154.
Schwarz, J.H. (1997): *Coupling a Self-Dual Tensor to Gravity in Six Dimensions,* hep-th/9701008.
Townsend, P.K. (1995): Phys. Lett. **B354**, 247, hep-th/9504095.
Tseytlin, A. (1990): Phys. Lett. **B242**, 163;
Tseytlin, A. (1991a): Nucl. Phys. **B350**, 395;
Tseytlin, A. (1991b): Phys. Rev. Lett. **66**, 545.
Witten, E. (1995): Nucl. Phys. **B443**, 85, hep-th/9503124.

A Linear Representation for the Topological Extensions of the Poincaré Superalgebra in $d = 11$

A.A. Deriglazov and A.V. Galajinsky

Department of Mathematics and Mathematical Physics,
Tomsk Polytechnical University, 634034 Tomsk, Russia

One of the promising approaches to construct a superfivebrane theory in eleven dimensions is to consider it as a model invariant under certain topological extension of the Poincaré supergroup (Sezgin (1986)). The structure of the new generators in the topologically extended superalgebra is connected (Sezgin (1986), Bergshoeff and Sezgin (1989), (1995); Sezgin (1995)) with the existence of supermembrane (Duff and Stelle (1991)) and superfivebrane (Güven (1992)) solitons of eleven dimensional supergravity. Recently, the supermembrane theory in d=11 was reformulated along these lines (Bergshoeff and Sezgin (1989), (1995); Sezgin (1995)) which generalizes the earlier results for the d=10 superstring (Siegel (1994), Green (1989)) and superparticle (Deriglazov and Galajinsky (1994)).

A conventional way to build a theory associated with a given Lie (super) algebra is to apply the group-theoretic construction (Ogievetsky (1974)). Given extended super Poincaré group G, with an element

$$\tilde{g} = \mathrm{e}^{-\mathrm{i}a^n P_n + \mathrm{i}\epsilon^\alpha Q_\alpha + \mathrm{i}\sigma_{\dots}\Sigma^{\dots} + \frac{\mathrm{i}}{2}\omega^{mn} M_{mn}} , \tag{1}$$

one has to construct the coset space G/H, with H the Lorentz subgroup. A point in the space

$$g(x, \theta, \psi) = \mathrm{e}^{-\mathrm{i}x^n P_n + \mathrm{i}\theta^\alpha Q_\alpha + \mathrm{i}\psi_{\dots}\Sigma^{\dots}} \times \mathrm{SO}(1, d-1) \tag{2}$$

is parametrized by the set of coordinates $(x^m, \theta^\alpha, \psi_{\dots})$. Left multiplication with a group element

$$\tilde{g} : \quad g(x, \theta, \psi) \to g(x', \theta', \psi') = \tilde{g} g(x, \theta, \psi) \tag{3}$$

defines an action of the group on the coset. Invariants of the group should be used to construct a theory.

The essential feature of the new superalgebras is that the translation generators do not commute with the supertranslations (Sezgin (1986), Bergshoeff and Sezgin (1989), (1995); Sezgin (1995), Green (1989))

$$[P, Q] \sim \Sigma . \tag{4}$$

In view of the construction just outlined it means highly nonlinear transformation laws for the coordinates parametrizing the coset.

The purpose of this talk is to demonstrate that the recent extension of the Poincaré superalgebra by the super two-form charge (Bergshoeff and Sezgin (1989), (1995); Sezgin (1995)) can be realized without spoiling the linear structure of the original super Poincaré transformations. The Γ-matrix identities that underlie the supermembrane theory in $d = 11$ turn out to be important for the linearization.

We use (anti)symmetrization "without strength", i.e., $A_{[m}B_{n]} \equiv A_m B_n - A_n B_m$, $A_{(m}B_{n)} \equiv A_m B_n + A_n B_m$. The conventions adopted for $d = 11$ are those of Ref. 9.

In trying to formulate a supermembrane theory with a manifest supersymmetry and inspired by the Γ-matrix identities, upon which the original formulation of super p-brane relies (Bergshoeff, Sezgin and Townsend (1987)), Bergshoeff and Sezgin suggested (Bergshoeff and Sezgin (1989), (1995); Sezgin (1995), Sezgin (1986)) new extensions of the Poincaré superalgebra in $d = 11$. The simplest of them reads

$$\begin{aligned} \{Q_\alpha, Q_\beta\} &= -2\Gamma^m{}_{\alpha\beta} P_m + (\Gamma_{mn}C)_{\alpha\beta}\Sigma^{mn} \ , \\ [Q_\alpha, P_m] &= -\mathrm{i}(\Gamma_{mn}C)_{\alpha\beta}\Sigma^{n\beta} \ , \\ [Q_\alpha, \Sigma^{mn}] &= \mathrm{i}\Gamma^{[m}{}_{\alpha\beta}\Sigma^{n]\beta} \ , \end{aligned} \tag{5}$$

where C is the charge conjugation matrix. The Jacobi identities for the algebra (5) hold due to the Γ-matrix identity

$$\Gamma^m{}_{(\alpha\beta}(\Gamma_{mn}C)_{\gamma\delta)} = 0 \ , \tag{6}$$

the latter satisfied in $d = 4, 5, 7$ and 11.

Application of the group-theoretic construction to the case at hand results in the transformation laws (Bergshoeff and Sezgin (1989), (1995); Sezgin (1995))

$$\begin{aligned} &\delta_\epsilon\theta^\alpha = \epsilon^\alpha, \qquad \delta_\epsilon x^m = \mathrm{i}\theta\Gamma^m\epsilon \ , \\ &\delta_\epsilon\Phi_{mn} = \mathrm{i}(\theta\Gamma_{mn}C\epsilon) \ , \\ &\delta_\epsilon\Phi_{n\alpha} = -\frac{1}{2}(\epsilon\Gamma_{mn}C)_\alpha x^m - \frac{1}{2}(\epsilon\Gamma^m)_\alpha\Phi_{mn} + \\ &\qquad + \frac{1}{6}\mathrm{i}(\theta\Gamma^m\epsilon)(\theta\Gamma_{mn}C)_\alpha + \frac{1}{6}\mathrm{i}(\theta\Gamma_{mn}C\epsilon)(\theta\Gamma^m)_\alpha \ ; \end{aligned} \tag{7a}$$

$$\delta_a x^m = a^m, \qquad \delta_a\Phi_{n\alpha} = \frac{1}{2}a^m(\theta\Gamma_{mn}C)_\alpha \ ; \tag{7b}$$

$$\delta_{\epsilon_{mn}}\Phi_{mn} = \epsilon_{mn}, \qquad \delta_{\epsilon_{mn}}\Phi_{n\alpha} = \frac{1}{2}(\theta\Gamma^m)_\alpha\epsilon_{mn} \ ; \tag{7c}$$

$$\delta_{\epsilon_{n\alpha}}\Phi_{n\alpha} = \epsilon_{n\alpha} \ , \tag{7d}$$

where $(x^m, \theta^\alpha, \Phi_{mn}, \Phi_{n\alpha})$ are the coordinates parametrizing the coset and $(a^m, \epsilon^\alpha, \epsilon_{mn}, \epsilon_{n\alpha})$ are the parameters associated to the generators $(P_m, Q_\alpha, \Sigma^{mn}, \Sigma^{n\alpha})$ respectively. Note that Eq.(7a) essentially involves nonlinear contributions.

In terms of the transformations (7) the algebra (5) has the form:

$$\begin{aligned}
&[\delta_{\epsilon_1},\delta_{\epsilon_2}] = \delta_a + \delta_{\epsilon_{mn}}, \; a^n = 2\mathrm{i}\epsilon_1\Gamma^n\epsilon_2, \quad \epsilon_{mn} = 2\mathrm{i}\epsilon_1\Gamma_{mn}C\epsilon_2 \;; \\
&[\delta_a,\delta_\epsilon] = \delta_{\epsilon_{n\alpha}}, \qquad \epsilon_{n\alpha} = -a^m(\epsilon\Gamma_{mn}C)_\alpha \;; \\
&[\delta_{\epsilon_{mn}},\delta_\epsilon] = \delta_{\epsilon_{n\alpha}}, \qquad \epsilon_{n\alpha} = -(\epsilon\Gamma^m)_\alpha\epsilon_{mn} \;,
\end{aligned} \tag{8}$$

Evaluation of this algebra turns out to be more instructive than it may seen at a glance. Actually, making use of Eq.(6) it is straightforward to check that the linear and nonlinear terms in the variation $\delta_\epsilon\Phi_{n\alpha}$ make the same contribution into the first line of Eq.(8) . This observation suggests that one can find another parametrization of the coset on which Bergshoeff–Sezgin superalgebra (5) would be linearly realized. The suitable parametrization looks like

$$g = \mathrm{e}^{-\mathrm{i}x^nP_n + \mathrm{i}\Phi_{n\alpha}\Sigma^{n\alpha} + \frac{\mathrm{i}}{2}\Phi_{mn}\Sigma^{mn} + \frac{2}{3}\mathrm{i}\theta^\alpha Q_\alpha}\mathrm{e}^{\frac{1}{3}\mathrm{i}\theta^\alpha Q_\alpha} \times \mathrm{SO}(1,d-1) \tag{9}$$

and the linear version for Eqs. (7a)-(7d) reads

$$\begin{aligned}
&\delta_\epsilon\theta^\alpha = \epsilon^\alpha, \qquad \delta_\epsilon x^m = \mathrm{i}\theta\Gamma^m\epsilon, \qquad \delta_\epsilon\Phi_{mn} = \mathrm{i}(\theta\Gamma_{mn}C\epsilon) \;, \\
&\delta_\epsilon\Phi_{n\alpha} = -\frac{2}{3}x^m(\epsilon\Gamma_{mn}C)_\alpha - \frac{2}{3}\Phi_{mn}(\epsilon\Gamma^m)_\alpha \;; && (10a) \\
&\delta_a x^m = a^m, \qquad \delta_a\Phi_{n\alpha} = \frac{1}{3}a^m(\theta\Gamma_{mn}C)_\alpha \;; && (10b) \\
&\delta_{\epsilon_{mn}}\Phi_{mn} = \epsilon_{mn}, \qquad \delta_{\epsilon_{mn}}\Phi_{n\alpha} = \frac{1}{3}\epsilon_{mn}(\theta\Gamma^m)_\alpha \;; && (10c) \\
&\delta_{\epsilon_{n\alpha}}\Phi_{n\alpha} = \epsilon_{n\alpha} \;; && (10d)
\end{aligned}$$

It is interesting to note that the generators of the nonlinear transformations (7)

$$\begin{aligned}
Q_\alpha &= \mathrm{i}\partial_\alpha + (\theta\Gamma^n)_\alpha\partial_n + (\theta\Gamma_{mn}C)_\alpha\partial^{mn} - \Big(\frac{\mathrm{i}}{2}x^m(\Gamma_{mn}C)_{\alpha\beta} + \\
&\quad + \frac{\mathrm{i}}{2}\Gamma^m{}_{\alpha\beta}\Phi_{mn} - \frac{1}{6}(\theta\Gamma^m)_\alpha(\theta\Gamma_{mn}C)_\beta - \frac{1}{6}(\theta\Gamma_{mn}C)_\alpha(\theta\Gamma^m)_\beta\Big)\partial^{n\beta} \;, \\
P_m &= -\mathrm{i}\partial_m - \frac{\mathrm{i}}{2}(\theta\Gamma_{mn}C)_\alpha\partial^{n\alpha} \;, \\
\Sigma^{mn} &= \mathrm{i}\partial^{[mn]} + \frac{\mathrm{i}}{2}(\theta\Gamma^{[m})_\alpha\partial^{n]\alpha} \\
\Sigma^{n\alpha} &= \mathrm{i}\partial^{n\alpha}
\end{aligned} \tag{11}$$

and ones of the linear transformations (10)

$$\begin{aligned}
Q_\alpha &= \mathrm{i}\partial_\alpha + (\theta\Gamma^n)_\alpha\partial_n + (\theta\Gamma_{mn}C)_\alpha\partial^{mn} - \\
&\quad - \Big(\frac{2}{3}\mathrm{i}x^m(\Gamma_{mn}C)_{\alpha\beta} + \frac{2}{3}\mathrm{i}\Phi_{mn}\Gamma^m{}_{\alpha\beta}\Big)\partial^{n\beta} \;, \\
P_m &= -\mathrm{i}\partial_m - \frac{\mathrm{i}}{3}(\theta\Gamma_{mn}C)_\alpha\partial^{n\alpha} \;, \\
\Sigma^{mn} &= \mathrm{i}\partial^{[mn]} + \frac{\mathrm{i}}{3}(\theta\Gamma^{[m})_\alpha\partial^{n]\alpha} \;, \\
\Sigma^{n\alpha} &= \mathrm{i}\partial^{n\alpha}
\end{aligned} \tag{12}$$

belong to equivalent representations[1] of the superalgebra (5)

$$\begin{aligned} &T_{i(\text{nonlinear})} = S T_{i(\text{linear})} S^{-1} \ , \\ &S = \mathrm{e}^{-\frac{1}{6}(x^m(\theta\Gamma_{mn}C)_\alpha + \Phi_{mn}(\theta\Gamma^m)_\alpha)\partial^{n\alpha}} \ , \end{aligned} \tag{13}$$

where we denoted $\partial/\partial\theta^\alpha = \partial_\alpha$, $\partial/\partial x^n = \partial_n$, $\partial/\partial\Phi_{n\alpha} = \partial^{n\alpha}$, $\partial/\partial\Phi_{mn} = \partial^{mn}$ and set $T_i \equiv (Q_\alpha, P^m, \Sigma^{mn}, \Sigma^{n\alpha})$. Eq.(13) can easily be checked by making use of the formula

$$\begin{aligned} &\mathrm{e}^{-B} A \mathrm{e}^{B} = \sum_{n=0}^{\infty} \frac{1}{n!}[A, B]_{(n)}, \\ &[A, B]_{(0)} \equiv A, \qquad [A, B]_{(n+1)} = [[A, B]_{(n)}, B] \ . \end{aligned} \tag{14}$$

Close examination of invariants of the group (7) (or (10)) shows that they are not sufficient to construct a supermembrane theory with local k-symmetry. Further extension by $\Sigma^{\alpha\beta}$-generator was proven to be necessary (Bergshoeff and Sezgin (1989), (1995); Sezgin (1995)). The commutation relations in the new algebra read (Bergshoeff and Sezgin (1989), (1995); Sezgin (1995))

$$\begin{aligned} &\{Q_\alpha, Q_\beta\} = -2\Gamma^m{}_{\alpha\beta} P_m + (\Gamma_{mn}C)_{\alpha\beta}\Sigma^{mn} \ , \\ &[Q_\alpha, P_m] = -\mathrm{i}(\Gamma_{mn}C)_{\alpha\beta}\Sigma^{n\beta} \ , \\ &[P_m, P_n] = \mathrm{i}(\Gamma_{mn}C)_{\alpha\beta}\Sigma^{\alpha\beta} \ , \\ &[P_m, \Sigma^{np}] = -\frac{1}{2} i\delta_m{}^{[n}\Gamma^{p]}{}_{\alpha\beta}\Sigma^{\alpha\beta} \ , \\ &[Q_\alpha, \Sigma^{mn}] = \mathrm{i}\Gamma^{[m}{}_{\alpha\beta}\Sigma^{n]\beta} \ , \\ &\{Q_\alpha, \Sigma^{n\beta}\} = \left(\frac{1}{2}\Gamma^n{}_{\gamma\delta}\delta_\alpha{}^\beta + 4\Gamma^n{}_{\gamma\alpha}\delta_\delta{}^\beta\right)\Sigma^{\gamma\delta} \ . \end{aligned} \tag{15}$$

The Jacobi identities for (15) hold due to Eq. (6). As compared to the usual Poincaré superalgebra, one finds the super two-form charge $\Sigma^{AB} = (\Sigma^{mn}, \Sigma^{m\alpha}, \Sigma^{\alpha\beta})$ in Eq. (15) which was connected (Bergshoeff and Sezgin (1989), (1995); Sezgin (1995), Sezgin (1986)) with the existence of a supermembrane solution (Duff and Stelle (1991)) of $d = 11$ supergravity.

Although in this case the commutation relations look rather complicated, the modified algebra (15) can be linearly realized like its contraction (5).

The transformations on the coset space (with the standard parametrization (2) adopted) are given by Eq. (7) with the following transformation laws for the coordinates $\Phi_{\alpha\beta}$ associated to the new generators $\Sigma^{\alpha\beta}$ added

---

[1] A comparison of the parametrizations (2) and (9) with the latter being rewritten in the form

$$\begin{aligned} g &= \mathrm{e}^{-\mathrm{i}x^n P_n + \mathrm{i}\Phi_{n\alpha}\Sigma^{n\alpha} + \frac{1}{2}\Phi_{mn}\Sigma^{mn} + \frac{2}{3}\mathrm{i}\theta^\alpha Q_\alpha} \mathrm{e}^{\frac{1}{3}\mathrm{i}\theta^\alpha Q_\alpha} \times \mathrm{SO}(1, d-1) = \\ &= \mathrm{e}^{-\mathrm{i}x^n P_n + \mathrm{i}\theta^\alpha Q_\alpha + \frac{1}{2}\Phi_{mn}\Sigma^{mn} + \mathrm{i}\left(\Phi_{n\alpha} + \frac{1}{6}x^m(\theta\Gamma_{mn}C)_\alpha + \frac{1}{6}\Phi_{mn}(\theta\Gamma^m)_\alpha\right)\Sigma^{n\alpha}} \ , \end{aligned}$$

suggests the explicit form for the operator S.

(compare with Ref. 2 where another parametrization of the coset has been chosen)

$$\begin{aligned}
\delta_\epsilon \Phi_{\alpha\beta} &= \frac{i}{2}(\Phi_{m\gamma}\epsilon^\gamma)\Gamma^m{}_{\alpha\beta} + 2i\Phi_{m(\alpha}(\epsilon\Gamma^m)_{\beta)} - \\
&- \frac{1}{4}\Phi_{nm}i(\theta\Gamma^n\epsilon)\Gamma^m{}_{\alpha\beta} + \frac{1}{3}x^n i(\epsilon\Gamma_{nm}C)_{(\alpha}(\theta\Gamma^m)_{\beta)} + \\
&+ \frac{1}{3}\Phi_{nm}i(\epsilon\Gamma^n)_{(\alpha}(\theta\Gamma^m)_{\beta)} - \frac{1}{3}x^n i(\theta\Gamma^m\epsilon)(\Gamma_{mn}C)_{\alpha\beta} + \\
&+ \frac{1}{12}x^n i(\theta\Gamma_{nm}C\epsilon)\Gamma^m{}_{\alpha\beta} \; ; \qquad (16a)\\
\delta_a \Phi_{\alpha\beta} &= -a^n x^m(\Gamma_{nm}C)_{\alpha\beta} - \frac{1}{2}a^m\Phi_{mn}\Gamma^n{}_{\alpha\beta} - \\
&- \frac{1}{3}a^n i(\theta\Gamma_{nm}C)_{(\alpha}(\theta\Gamma^m)_{\beta)} \; ; \qquad (16b)
\end{aligned}$$

$$\begin{aligned}
\delta_{\epsilon_{mn}}\Phi_{\alpha\beta} &= \frac{1}{2}x^n\epsilon_{nm}\Gamma^m{}_{\alpha\beta} - \frac{1}{3}\epsilon_{nm}i(\theta\Gamma^n)_{(\alpha}(\theta\Gamma^m)_{\beta)} \; ; \qquad (16c)\\
\delta_{\epsilon_{n\alpha}}\Phi_{\alpha\beta} &= \frac{1}{2}i(\theta^\gamma\epsilon_{n\gamma})\Gamma^n{}_{\alpha\beta} - 2i\epsilon_{n(\alpha}(\theta\Gamma^n)_{\beta)} \; ; \qquad (16d)\\
\delta_{\epsilon_{\alpha\beta}}\Phi_{\alpha\beta} &= \epsilon_{\alpha\beta} \; . \qquad (16e)
\end{aligned}$$

Taking the linearization for Eq.(7) from Eq.(10) and supplementing it with

$$\begin{aligned}
\delta_\epsilon\Phi_{\alpha\beta} &= 3i(\Phi_{n\gamma}\epsilon^\gamma)\Gamma^n{}_{\alpha\beta} + 3i\Phi_{n(\alpha}(\epsilon\Gamma^n)_{\beta)} \; ; \qquad (17a)\\
\delta_a\Phi_{\alpha\beta} &= -2a^m\Phi_{mn}\Gamma^n{}_{\alpha\beta} + a^m x^n(\Gamma_{nm}C)_{\alpha\beta} \; ; \qquad (17b)\\
\delta_{\epsilon_{mn}}\Phi_{\alpha\beta} &= -\epsilon_{mn}x^m\Gamma^n{}_{\alpha\beta} \; ; \qquad (17c)\\
\delta_{\epsilon_{n\alpha}}\Phi_{\alpha\beta} &= 2i(\epsilon_{n\gamma}\theta^\gamma)\Gamma^n{}_{\alpha\beta} - i\epsilon_{n(\alpha}(\theta\Gamma^n)_{\beta)} \; ; \qquad (17d)\\
\delta_{\epsilon_{\alpha\beta}}\Phi_{\alpha\beta} &= \epsilon_{\alpha\beta} \; , \qquad (17e)
\end{aligned}$$

one gets the linear version for Bergshoeff–Sezgin superalgebra (15). Making use of Eq. (6), it is straightforward to check that the transformations (10), (17) satisfy the algebra

$$\begin{aligned}
&[\delta_{\epsilon_1}, \delta_{\epsilon_2}] = \delta_a + \delta_{\epsilon_{mn}}, \; a^m = 2i\epsilon_1\Gamma^m\epsilon_2, \quad \epsilon_{mn} = 2i(\epsilon_1\Gamma_{mn}C\epsilon_2) \; ;\\
&[\delta_a, \delta_\epsilon] = \delta_{\epsilon_{n\alpha}}, \quad \epsilon_{n\alpha} = -a^m(\epsilon\Gamma_{mn}C)_\alpha \; ;\\
&[\delta_{\epsilon_{mn}}, \delta_\epsilon] = \delta_{\epsilon_{n\alpha}}, \quad \epsilon_{n\alpha} = -\epsilon_{mn}(\Gamma^m\epsilon)_\alpha \; ;\\
&[\delta_\epsilon, \delta_{\epsilon_{n\alpha}}] = \delta_{\epsilon_{\alpha\beta}}, \quad \epsilon_{\alpha\beta} = -i(\epsilon_{n\gamma}\epsilon^\gamma)\Gamma^n{}_{\alpha\beta} - 4i\epsilon_{n(\alpha}(\Gamma^n\epsilon)_{\beta)} \; ;\\
&[\delta_{a_1}, \delta_{a_2}] = \delta_{\epsilon_{\alpha\beta}}, \quad \epsilon_{\alpha\beta} = 2a_2{}^m a_1{}^n(\Gamma_{nm}C)_{\alpha\beta} \; ;\\
&[\delta_a, \delta_{\epsilon_{mn}}] = \delta_{\epsilon_{\alpha\beta}}, \quad \epsilon_{\alpha\beta} = a^m\epsilon_{mn}\Gamma^n{}_{\alpha\beta} \; ,
\end{aligned} \qquad (18)$$

which, being rewritten in terms of the generators, coincides with Eq. (15).

Thus, in this talk we have demonstrated that the super two-form charge topological extension of the $d = 11$ Poincaré superalgebra admits a linear realization. Recently, an extension of Eq. (15) by the super five-form charge has been proposed (Sezgin (1986)). We hope that analogous construction will work in the case of the M-algebra which may suggest considerable simplification in constructing a super five-brane theory in eleven dimensions.

References

Bergshoeff, E., Sezgin, E. and Townsend, P.K. (1987): Phys. Lett. **B189**, 75.
Bergshoeff, E. and Sezgin, E., (1989): Phys. Lett. **B232**, 96;
Bergshoeff, E. and Sezgin, E., (1995): Phys. Lett. **B354**, 256; hep-th/9504140;
Sezgin, E., (1995): hep-th/9512082.
Deriglazov, A.A. and Galajinsky, A.V.,(1994): Mod. Phys. Lett. **A9**, 3445.
Deriglazov, A.A. and Galajinsky, A.V., (1997): hep-th/9703104
Duff, M.J. and Stelle, K.S., (1991): Phys. Lett. **B253**, 113.
Green, M.B., (1989): Phys. Lett. **B223**, 157.
Güven, R., (1992): Phys. Lett. **B276**, 49.
Ogievetsky, V.I. (1974): Non-linear realizations of internal and spacetime symmetries, in: Proc. 10th Karpacz Winter School of Theoretical Physics.
Sezgin, E. (1986): hep-th/9609086.
Siegel, W., (1994): Phys. Rev. **D50**, 2799.

On the Construction of Global Duality Maps in Strings and Supermembrane Theories

I. Martin and A. Restuccia

Universidad Simon Bolivar, Departamento de Fisica, Caracas, Venezuela.
e-mail: isbeliam@usb.ve, arestu@usb.ve

Abstract. Global aspects of duality transformations between string and supermembrane theories are discussed. The global constraints arising from the requirement of the existence of duality maps are explicitly introduced into the action. The equivalence between the supermembrane and the large N limit of U(N) Yang-Mills theory in the presence of the global duality constraints is analyzed. Arguments are given pointing to the possibility of a non equivalence of the theories under such conditions.

In this presentation, we analyze the duality transformations between string and supermembrane theories emphasizing its global aspects. This approach requires in general the introduction of a geometrical structure on euclidean manifolds, introduced earlier in (Caicedo, Martin, Restuccia (1997)), which is described locally in terms of p-forms with nontrivial transitions. In the case considered here, however, we will be concerned only with duality in terms of 1-form connections over line bundles.

Duality maps for theories with p-forms have been discussed in (Witten (1995), Verlinde (1995), Lozano (1995), Barbon (1995), Kehagias (1995)) and appear naturally in the description of D-brane theories. The interesting result related to this global structure is that duality between theories of local p-forms and (d-p)-forms not only imply the quantization of couplings, the known generalized Dirac quantization condition, but also determine completely, the configuration space of these local p-forms with non trivial transitions. The geometrical structure introduces higher order co-cycles consistency conditions which may be viewed as gluing together non equivalent line bundles with associated connections (Brylinski (1992)).

For completion of the presentation we first consider the duality for bosonic strings emphasizing the need for a new topological term in the action in order to obtain the correct dual boundary conditions. We then study the dual map between d=11 supermembrane and d=10 IIA Dirichlet supermembrane and we give some arguments, based on the global condition for duality, pointing to the possibility of a non equivalence between the compactified d=11 supermembrane and the large N limit of the U(N) Yang Mills theory model proposed in (de Witt (1989)). The latter has been used to show that supermembranes, without the restrictions coming from a requirement of duality, have continuous spectra.

The string action is

$$S(\chi) = \frac{1}{2\alpha'} \int_{\sum} \mathrm{d}^2\xi \sqrt{g} g^{ij} \partial_i \chi^\mu \partial_j \chi_\mu \tag{1}$$

where g^{ij} is the world sheet metric and ξ^i , i=1,2 are the local coordinates of the Riemann surface $\sum$ of a fixed topology. We analyze first the closed string theory with one coordinate χ compactified over S^1. Associated to that coordinate we introduce a constrained 1-form $L = L_i \mathrm{d}\xi^i$ satisfying

$$\mathrm{d}L = 0 \tag{2}$$

$$\oint_{C^I} L = 2\pi n^I R \tag{3}$$

where C^I denotes a basis of the integer homology of dimension 1 over the worldsheet. Constraint (2) implies L is a closed 1-form, while (3) ensures the compactification over S^1, R is the compactification radius. The solution to (2) and (3) is the string map $\chi(\xi^1, \xi^2)$. We introduce Lagrange multipliers associated to constraints (2) and (3) and obtain the quantum equivalent action

$$S(L,V) = \frac{1}{2\alpha'} \int_{\sum} L \wedge^* L + \frac{\mathrm{i}}{\alpha'} \int_{\sum} L \wedge W(V) \tag{4}$$

where $V(\xi^1, \xi^2)$ is the dual map to $\chi(\xi^1, \xi^2)$:

$$W(V) = \mathrm{d}V \tag{5}$$

$$\oint_{C^J} W = 2\pi m^J R' \tag{6}$$

(6) is uniquely determined to obtain quantum equivalence between $S(L, V)$ and $S(\chi)$.

After summation on all n in the functional integral, we obtain

$$R' = \frac{\alpha'}{R} \tag{7}$$

That is the dual radius arises directly from the off-shell construction of the dual action. From (4) we obtain the standard on-shell duality relation

$$^*L + \mathrm{i}W = 0 \tag{8}$$

From (4) after functional integration on L we obtain

$$S(V) = \frac{1}{2\alpha'} \int_{\Sigma} W(V) \wedge^* W(V). \tag{9}$$

The duality between $L = \mathrm{d}\chi$ and $W = \mathrm{d}V$ is established in the global constraints (3), (6) and (8). Notice that (6) is uniquely determined from the off-shell construction while (3) implies the compactification of χ on S^1 of radius R. The quantum equivalence between (1) and (9) has been shown for any compact Riemann surface Σ, hence the T-duality is valid order by order in the perturbative expansion of closed string theories.

We now discuss the duality of open string theories.The standard open string boundary condition arises from (1) by considering the stationary points of $S(\chi)$. Its variation yields a boundary term

$$\left(\delta\chi\partial_i\chi.n^i\right)\big|_{\partial\Sigma} \tag{10}$$

It can be annihilated by assuming

$$\partial_i\chi^\mu.n^i = 0 \tag{11}$$

This boundary condition together with the usual string field equation gives a stationary point of (1) with respect to the space of variations $\delta\chi$ which are arbitrary even on the boundary. If instead we consider the space of maps χ restricted by a boundary condition and look for a stationary point of (1) restricted to that space, then

$$\chi^\mu|_{\partial\Sigma} = cte \tag{12}$$

would be also a solution, since then $\delta\chi = 0$. In this case one can have even a mixture of Dirichlet and Newmann conditions on the boundary as an acceptable solution. We will discuss the construction of the dual string action on the first case and show that a topological action term has to be added to (9) in order to have a dual action whose stationary points yields the dual boundary condition to (11). Notice that from the duality relation (8) one obtains

$$n.L = 0 \Rightarrow t.W = 0 \tag{13}$$

where t is tangent to the boundary. However from (9) if we consider arbitrary variations on the boundary we get

$$n.W = 0 \tag{14}$$

It turns out that the dual open string action is

$$\tilde{S}(V,Y) = \frac{1}{2\alpha'}\int_\Sigma W(V)\wedge^* W(V) + \frac{\mathrm{i}}{\alpha'}\int_\Sigma F(Y)\wedge W(V) \tag{15}$$

where $F = dY$ and Y is a map onto S^1 which can be determined to obtain the correct boundary condition for W. The new term in the action is a pure topological one. It does not modify the field equations, only contributes to the boundary terms. All the local dependence of $Y(\sigma)$ can be gauged away, only the boundary contribution remains.

We wish to discuss now the duality transformation between the covariant $d = 11$ supermembrane action with one coordinate X^{11} compactified on S^1, and the fully $d = 10$ Lorentz covariant worldvolumen action for the

$d = 10$ IIA Dirichlet supermembrane.The equivalence between the bosonic sectors was previously shown by Schmidhuber (Schmidhuber (1996)) using the Born-Infeld type action found by Leigh (Leigh (1989)). We will argue in a global way showing the equivalence between both theories, even when nontrivial line bundles are included in the construction of the D-brane action. In what follows we use as in (Howe, Tucker (1977, 1978)) the Howe-Tucker formulation of the $d = 11$ supermembrane over a target manifold with one coordinate compactified on S^1 (Townsend (1995)), that is we take X^{11} to be the angular coordinate φ on S^1. The action is then

$$\begin{aligned} S = -\frac{1}{2}\int_X \mathrm{d}^3\xi\sqrt{-\gamma}[\gamma^{ij}\pi_i^m\pi_j^n\eta_{mn} + \\ + \gamma^{ij}(\partial_i\varphi - \mathrm{i}\bar{\theta}\Gamma_{11}\partial_i\theta)(\partial_j\varphi - \mathrm{i}\bar{\theta}\Gamma_{11}\partial_j\theta) - 1] - \\ - \frac{1}{6}\int_X \mathrm{d}^3\xi\epsilon^{ijk}[b_{ijk} + 3b_{ij}\partial_k\varphi] \end{aligned} \tag{16}$$

where η is the Minkoswki metric in $d = 10$ spacetime, and

$$\begin{aligned} \pi^m &= \mathrm{d}x^m - \mathrm{i}\bar{\theta}\Gamma^m\mathrm{d}\theta \\ \epsilon^{ijk}b_{ijk} &= 3\epsilon^{ijk}\{\mathrm{i}\bar{\theta}\Gamma_{mn}\partial_i\theta[\pi_i^m\pi_j^n + i\pi_i^m(\bar{\theta}\Gamma^n\partial_j\theta) - \frac{1}{3}(\bar{\theta}\Gamma^m\partial_i\theta)(\bar{\theta}\Gamma^n\partial_j\theta)] \\ &\quad + (\bar{\theta}\Gamma_{11}\Gamma_m\partial_i\theta)(\bar{\theta}\Gamma_{11}\partial_j\theta)(\partial_k x^m - \frac{2\mathrm{i}}{3}\bar{\theta}\Gamma^m\partial_k\theta)\} \\ \epsilon^{ijk}b_{ij} &= -2\mathrm{i}\epsilon^{ijk}\bar{\theta}\Gamma_m\Gamma_{11}\partial_i\theta(\partial_j x^m - \frac{\mathrm{i}}{2}\bar{\theta}\Gamma^m\partial_j\theta) \ . \end{aligned} \tag{17}$$

We will now perform the same steps as we followed for the string. The constraints

$$\mathrm{d}L_1 = 0 \tag{18}$$

$$\oint_{\Sigma_1} L_1 = 2\pi n \ , \tag{19}$$

define a uniform map: $X \to S^1$

$$g = \exp \mathrm{i}\varphi \tag{20}$$

and

$$L_1 = -\mathrm{i}g^{-1}\mathrm{d}g = \mathrm{d}\varphi \ . \tag{21}$$

The converse being also valid. In this context Σ_1 is a basis of homology on the $d = 3$ worldsheet manifold.

The intermediate step in the construction of the duality map consists then in attaining an equivalent formulation to (16) in terms of the global 1-form L_1. The important point now is to realize that the Lagrange formulation of the constraints (18) and (19) may be obtained in terms of a connection 1-form over the space of all non trivial line bundles.

So we start with action

$$S_1 = -\frac{1}{2}\int_X \mathrm{d}^3\xi\sqrt{-\gamma}[\gamma^{ij}\pi_i^m\pi_j^n\eta_{mn} + \\ + \gamma^{ij}(L_{1i} - \mathrm{i}\bar\theta\Gamma_{11}\partial_i\theta)(L_{1j} - \mathrm{i}\bar\theta\Gamma_{11}\partial_j\theta) - 1] - \\ - \frac{1}{6}\int_X \mathrm{d}^3\xi\epsilon^{ijk}[b_{ijk} + 3b_{ij}L_{1k}] + \int_X F(A)L_1 \tag{22}$$

where L_1 is a globally defined 1-form over X and A is a connection on the space of all line bundles over X.

Functional integration on A yields

$$\delta(\mathrm{d}L_1)\delta(\oint_{\Sigma_1} L_1 - 2\pi n) \tag{23}$$

in the functional measure of the path integral.

We now use

$$\delta(\mathrm{d}L_1)\delta(\oint_{\Sigma_1} L_1 - 2\pi n) = \int [\mathrm{d}\varphi]\frac{\delta(L_1 - \mathrm{d}\varphi)}{\det d} \tag{24}$$

where φ defines a map from $X \to S^1$, that is $\mathrm{d}\varphi$ satisfies (19). We notice that the functional integral in (24) is over all maps from $X \to S^1$, it is not an integration over a cohomology class defined by an element of $H^1(X)$.

Using that zero modes in this case, are constants, we may directly integrate on L_1 and replace in (22) L_1 by $\mathrm{d}\varphi$. We thus arrive to the covariant $d = 11$ supermembrane action after elimination of L_1.

On the other hand, we may functionally integrate L_1 in (22) to arrive to the functional integral of the action

$$S_2 = -\frac{1}{2}\int_X \mathrm{d}^3\xi\sqrt{-\gamma}[\gamma^{ij}\pi_i^m\pi_j^n\eta_{mn} - \gamma^{ij}f_i(A)f_j(A) - 1] \\ - \frac{1}{6}\int_X \mathrm{d}^3\xi\epsilon^{ijk}b_{ijk} + \int_X \mathrm{d}^3\xi\gamma^{ij}f_i(A)\mathrm{i}\bar\theta\Gamma_{11}\partial_l\theta \tag{25}$$

Where

$$f_i(A) \equiv \epsilon_{imn}(F^{mn}(A) - \frac{1}{2}b^{mn}) \ . \tag{26}$$

The functional integral in A must now be performed over all line bundles over X. The result (25) was obtained by Townsend in (Townsend (1995)), for the case of a trivial line bundle. The equivalence between (25), the fully $d = 10$ Lorentz covariant worldvolume action for the $d = 10$ IIA Dirichlet supermembrane, and the $d = 11$ covariant supermembrane action (16) has then been established. In the functional integral for (16), integration over all maps between $X \to S^1$ must be performed while in the functional integral for (25) integration over the space of all connection 1-forms on all line bundles (modulo gauge transformations) must be performed.

It was shown in (de Witt (1989)) that the spectrum of the d=11 supermembrane is continuous. We will analyze the same problem when the configuration space of the supermembrane is restricted by the global constraint arising from the requirement of duality. The idea in (de Witt (1989)) is to regard the supermembrane (SM) as a limiting case of certain models in SUSY quantum mechanics. The Hamiltonian for the SM using light cone coordinates takes the form

$$H = \frac{\mathbf{P}_0^2}{2P_0^+} + \frac{\mathcal{M}^2}{2P_0^+} \tag{27}$$

Where $\mathbf{P}_0$ are the transverse momenta of the center of mass and $\mathcal{M}$ is the supermembrane mass operator. By expanding all coordinates and momenta into a complete orthonormal basis of functions $Y^A(\sigma)$, $A = 1, 2, ..., \infty$

$$\mathbf{X}(\sigma) = \mathbf{X}_0 + \sum_{\infty}^{A=1} \mathbf{X}_A Y^A(\sigma) \tag{28}$$

one may express the mass operator as

$$\mathcal{M}^{\in} = (\mathcal{P}_{\dashv}^{\mathcal{A}})^{\in} + \frac{\infty}{\in}\left(\{_{\mathcal{ABC}}\mathcal{X}_{\dashv}^{\mathcal{B}}\mathcal{X}_{\lfloor}^{\mathcal{C}}\right)^{\in} - \rangle\{_{\mathcal{ABC}}\theta^{\mathcal{A}}\gamma^{\dashv}\mathcal{X}_{\dashv}^{\mathcal{B}}\theta^{\mathcal{C}} \tag{29}$$

where f_C^{AB} are the structure constants of the invariant subgroup of the area preserving transformations,

$$\{Y^A, Y^B\} = f_C^{AB} Y^C \ . \tag{30}$$

An important property of the quantum mechanical models is that the Hamiltonian vanishes if the coordinates $\mathbf{X}_A$ take value in some abelian subalgebra. The classical theory is then unstable. However quantum mechanical effects may cure this classical problem. That is the case for example for the three dimensional Hamiltonian

$$H = p_x^2 + p_y^2 + p_z^2 + x^2y^2 + x^2z^2 + z^2y^2 \ . \tag{31}$$

It is classically unstable because of the zero energy valleys along the x,y and z axis. However it has a quantum mechanical discrete spectrum. This is so because the oscillators perpendicular to the valley direction give rise to a zero point energy, which induces an effective potential barrier which tends to confine the wave function. Supersymmetry change the situation because SUSY harmonic oscillators have no zero point energy so that the confining potential may vanish. Explicit calculations show indeed that the wave functions can go to infinity along the valleys of zero classical energy, so that the spectrum of the SM is continuous. The relevant point in the argument is the existence of potential valleys with zero energy extending to infinite and supersymmetry. Let us now consider the situation when the global restriction imposed by the requirement of duality is introduced.

Using the global condition (19) into (28) we obtain for the compactified coordinates

$$X_A n_I^A = m_I^A \tag{32}$$

where I denotes the element of the bases of the integer homology of dimension 1 on the world volume. If we replace this constraint into (29) we obtain a very different structure for the Hamiltonian. Together with the quartic terms we have now quadratic terms for each of the independent coordinates. We illustrate the point with the Hamiltonian (31). If we consider the constraint

$$P = nx + my + lz \tag{33}$$

where n,m,l and P are integers, and eliminate one of the coordinates $n \neq 0$,

$$X = \frac{P}{n} - \frac{m}{n}y - \frac{l}{n}z \tag{34}$$

we obtain after replacing (34) into (31) that the minima are located at isolated points. The argument of the zero energy valleys extending to infinity does not apply then in the presence of the constraint (32), at least for Hamiltonian (31).

This is then a straightforward example of how the global requirement of duality imposes restrictions on the configuration space with relevant consequences. We will soon report on the analysis of the spectrum for the SM under duality requirements.

References

Caicedo, M., Martin, I. and Restuccia, A. (1997):, hep-th/9701010.
Witten, E. (1995): hep-th/9505186;
Verlinde, D. (1995): Nucl.Phys. **B455** 211 ;
Lozano, Y. (1995): Phys. Lett. **B364** 19;
Barbon, J.L.F.(1995): Nucl. Phys. **B452** 313;
Kehagias, A. (1995): hep-th/9508159.
Brylinski, Jean-Luc (1992): Loop Spaces, Characteristic Classes and Geometric Quantization, Birkhäuser.
de Wit, B. (1989): M. Lüscher and H. Nicolai, Nucl. Phys. B320 135.
Schmidhuber, C. (1996): hep-th/9601003.
Leigh, R.G. (1989): Mod. Phys. Lett.**A4** 2767.
Howe, P. and Tucker, R.W. (1977): J. Phys.**A10** 155;
Howe, P. and Tucker, R.W. (1978): J. Math.Phys. **19** 981.
Townsend, P.K. (1995): hep-th/9512062.

Planckian Energy Scattering of D-Branes and M(atrix) Theory in Curved Space

I.V. Volovich

Steklov Mathematical Institute Gubkin St.8, GSP-1, 117966, Moscow, Russia

Abstract. We argue that black p-branes will occur in the collision of D0-branes at Planckian energies. This extents the Amati, Ciafaloni and Veneziano and 't Hooft conjecture that black holes occur in the collision of two light particles at Planckian energies. We discuss a possible scenario for such a process by using colliding plane gravitational waves. D-branes in the presence of black holes are discussed. M(atrix) theory and matrix string in curved space are considered. A violation of quantum coherence in M(atrix) theory is noticed.

1 Introduction

It is a great honour for me to submit an article in the volume dedicated to the memory of Dmitrii Vassilievich Volkov. His ideas on supersymmetry and on nonlinear theories are in the heart of modern fundamental physics.

It was suggested (Banks, Fischler, Shenker, Susskind (1996)) that the large N limit of the dimensional reduction of the ten dimensional U(N) supersymmetric Yang-Mills theory to one dimension might be interpreted as eleven dimensional M-theory in light cone gauge. The reduction of the supersymmetric Yang-Mills theory to two dimensions was interpreted as matrix string theory (Motl (1997)). This is based on the observation (Witten (1995)) that the effective low energy theory of N coincident parallel Dirichlet branes is described by the dimensional reduction of the U(N) supersummetric Yang-Mills theory.

M(atrix) theory (Banks, Fischler, Shenker, Susskind (1996)) is described by the following Lagrangian

$$L = \frac{1}{2}\mathrm{Tr}[\dot{Y}^i\dot{Y}^i + \frac{1}{2}[Y^i, Y^j]^2 + 2\theta^T\dot{\theta} + 2\theta^T\gamma_i[\theta, Y^i]] \tag{1}$$

where Y^i are Hermitian $N \times N$ matrices while θ is a 16-component fermionic spinor each component of which is an Hermitian $N \times N$ matrix and $i, j = 1, ..., 9$.

In (Danielsson, Ferretti, Sundborg (1996), Douglas, Kabat, Pouliot, Shenker (1996)) the interaction between D-branes in the non-relativistic approximation has been considered. It was found (Douglas, Kabat, Pouliot, Shenker (1996)) that supergravity is valid at distances greater than the string scale, while the description in terms of gauge theory in flat space is valid in the sub-stringy domain.

The application of the flat background is justified if the curvature is small. However if one has D-branes in the presence of a black hole we have to take into account the non-trivial background. The low energy bulk theory of D-branes is the supersymmetric Yang-Mills theory coupling with supergravity in ten dimensions. In this note we discuss the interpretation of the dimensional reduction of this theory as M-theory (if $p = 0$) or matrix string (if $p = 1$) in curved background. We will argue that (black) p-branes will occur in the collision of D0-branes at Planckian energies.

2 M(atrix) Theory in Curved Space

The low energy effective theory of D-branes in Minkowski spacetime is given by the dimensional reduction of the supersymmetric gauge theory in the ten dimensional Minkowski spacetime (Witten (1995)). If one has D-branes in Minkowski spacetime but in curved coordinates we have to start from the supersymmetric Yang-Mills theory in curved coordinates. Then we get a version of the M(atrix) theory Lagrangian (1) in the curved coordinates. If one has D-branes in a curved spacetime, for instance D-branes in the presence of black hole then it is natural to expect that the low energy effective theory will be given by the dimensional reduction of the supersymmetric gauge theory coupling with supergravity in the ten dimensional curved spacetime.

Let us consider the Yang-Mills theory in the D-dimensional space-time with the metric g_{MN}. The action is

$$I = -\frac{1}{4}\mathrm{Tr}\int \mathrm{d}^D x \, \sqrt{g} F_{MN} F_{PQ} g^{MP} g^{NQ} \tag{2}$$

where $F_{MN} = \mathrm{i}[D_M, D_N]$, $D_M = \nabla_M - \mathrm{i}A_M$. Let $\gamma : X^M = X^M(\sigma)$, $\sigma = (\sigma_0, ..., \sigma_p)$ be a $p+1$-dimensional submanifold (p-brane) and let us consider the dimensional reduction to γ. One has $A_M = (A_\alpha, Y_i), \alpha = 0, ..., p$; $i = p+1, ..., D-1$ and $F_{MN} = (F_{\alpha\beta}, F_{\alpha i}, F_{ij})$, $g^{MN} = (g^{\alpha\beta}, g^{\alpha i}, g^{ij})$. The Lagrangian is

$$L = -\frac{1}{2}\mathrm{Tr}[D_\alpha Y_i D_\beta Y_j g^{\alpha\beta} g^{ij} - \frac{1}{2}[Y_i, Y_j][Y_m, Y_n] g^{im} g^{jn} + ...] \tag{3}$$

Here $Y_i = Y_i(X^P(\sigma))$, $g_{MN} = g_{MN}(X^P(\sigma))$. If one takes $p = 1$ then the Lagrangian (3) describes the matrix string (Motl (1997), Banks, Seiberg (1997), Dijkgraaf, Verlinde, Verlinde (1997)) in curved background. For a 0-brane $X^M = X^M(\tau)$ in the gauge $A_0 = 0$, $g^{0i} = 0$ one has

$$L = -\frac{1}{2}\mathrm{Tr}[\dot{Y}_i \dot{Y}_j g^{00} g^{ij} - \frac{1}{2}[Y_i, Y_j][Y_m, Y_n] g^{im} g^{jn}] \tag{4}$$

The Lagrangian (4) describes the bosonic part of M(atrix) theory in curved space. It can be reduced to the bosonic part of (1) if we take $g^{00} = -1$, $g^{ij} = \delta^{ij}$ and $X^M(\tau) = \tau\delta_{M0}$.

Notice that in contrast to the picture with a noncommutative geometry in the short distance regime in (Banks, Fischler, Shenker, Susskind (1996)) here in fact we have the classical commutative spacetime coordinates X^M. The matrices Y_i describe the noncommutative dynamical system in the ordinary classical spacetime. The corresponding Hamiltonian is

$$H = \frac{1}{2}\mathrm{Tr}[P^i P^j g_{ij} - \frac{1}{2}[Y_i, Y_j][Y_m, Y_n]g^{im}g^{jn}] \tag{5}$$

One deals with quantum mechanics in the dependent on time background $g^{ij}(\tau) = g^{ij}(X(\tau))$. Now the properties of the matrix quantum mechanics depend on the choice of the curve $X(\tau)$. The one-loop effective action for the theory (4) with the p-brane metric g_{MN} can be computed using the background field method by the standard procedure. One gets corrections to the phase shift δ obtained in (Douglas, Kabat, Pouliot, Shenker (1996), Balasubramanian, Larsen (1997), Douglas, Polchinski, Strominger (1997)) which are now under consideration. If one takes a geodesic near the singularity then generically one gets the creation of particles (D0-branes) (Birrel, Davies (1982)).

If the metric g_{ij} in (4) describes a black hole then one can apply to M(atrix) theory the known Hawking arguments (Hawking (1976)) on the violation of quantum coherence. If one has M(atrix) theory in flat spacetime (1) then of course there exists the unitary evolution operator and there is no a violation of quantum coherence. But in this case simply there is no problem for discussion because there are no black holes in the flat spacetime.

To get the supersymmetric M(atrix) theory or matrix string theory in curved background one has to take the dimensional reduction of the supersymmetric Yang-Mills theory in ten dimensions coupling with supergravity,

$$L = \frac{1}{4g^2\Phi}\mathrm{Tr}(F_{MN}^2) - \frac{1}{2}\mathrm{Tr}(\overline{\chi}\Gamma^M D_M\chi) - \frac{1}{2\kappa^2}R - \frac{3\kappa^2}{8g^4\Phi^2}H_{MNP}^2 + \dots \tag{6}$$

For the p-brane

$$\mathrm{d}s^2 = f^{-\frac{1}{2}}(-\mathrm{d}t^2 + \mathrm{d}x_1^2 + \dots + \mathrm{d}x_p^2) + f^{\frac{1}{2}}(\mathrm{d}x_{p+1}^2 + \dots + \mathrm{d}x_9^2)\ , \tag{7}$$

$f = 1 + \frac{q}{r^{7-p}}$, which is the BPS-state we don't expect the Hawking radiation but there is the spacetime singularity here and there is back reaction of the gas of D-branes to the metric which is described by the equations:

$$R_{\mu\nu} - \frac{1}{2}Rg_{\mu\nu} = < T_{\mu\nu} > \tag{8}$$

where $T_{\mu\nu}$ is the energy-momentum tensor of D-branes.

It seems that M(atrix) theory being quantum mechanics in the curved spacetime suffers from the well known problems such as non-controllable spacetime singularities and the violation of quantum coherence. There is a hope (Volovich (1996), Douglas, Polchinski, Strominger (1997)) that large N limit gauge theory might be used to study these problems.

3 Scattering of D-Branes and Creation of Black Holes

Amati, Ciafaloni and Veneziano (Amati, Ciafaloni, Veneziano (1987, 1990)) and 't Hooft ('t Hooft (1988)) have argued that at extremely high energies interactions due to gravitational waves will dominate all other interactions. They conjectured that black holes will occur in the collision of two light particles at Planckian energies with small impact parameter. In (Amati, Ciafaloni, Veneziano (1987, 1990), Fabbrichesi, Pettorino, Veneziano, Vilkovisky (1994)) the elastic scattering amplitude in the eikonal approximation was found in the form

$$A(s,t) \propto s \int \mathrm{d}^2 b \, \mathrm{e}^{\mathrm{i}qb} \mathrm{e}^{\mathrm{i}I_{\mathrm{cl}}} \tag{9}$$

Here s and t are the Mandelstam variables and b is the impact parameter. I_{cl} was taken to be the value of the boundary term for the gravitational action calculated on the sum of two Aichelburg-Sexl shock waves,

$$I_{\mathrm{cl}} = Gs \log b^2 \tag{10}$$

The action I_{cl} is equal to the phase shift $\delta(b,v)$ for the process of the elastic scattering, see (Fabbrichesi, Pettorino, Veneziano, Vilkovisky (1994), Aref'eva, Viswanathan, Volovich (1994, 1995, 1996)).

One cannot see the creation of black holes in this approximation. In (Aref'eva, Viswanathan, Volovich (1994, 1995, 1996)) the following mechanism of the creation of black holes in the process of collisions of the Planckian energy particles has been suggested. Each of the two ultrarelativistic particles generates a plane gravitational wave. Then these plane waves collide and produce a singularity or black hole. The phase of the transition amplitude from plane waves to black holes was calculated as the value of the action on the corresponding classical solution.

It seems that scattering of D-branes at extremely high energies and small impact parameter will be similar to the described picture. In particular two colliding D0-branes at small impact parameter should produce p-branes.

Scattering of D-branes at large impact parameter has been considered in (Bachas (1995), Gubser, Klebanov, Hashimoto, Maldecena (1996), Lifschytz (1996), Douglas, Kabat, Pouliot, Shenker (1996), Balasubramanian, Larsen (1997)). The 0-brane metric lifted to 11 dimensions is

$$\mathrm{d}s^2 = \mathrm{d}u \, \mathrm{d}v + (1 + \frac{q}{r^7})\mathrm{d}u^2 + \mathrm{d}x_1^2 + \ldots + \mathrm{d}x_9^2 \tag{11}$$

It represents a plane-fronted wave moving in the x_{11} direction. At long distances the gravitational wave can be considered as a plane wave. Plane wave solutions in supergravity have been considered in (Gibbons (1982), Guven (1987), Bergshoeff, Kallosh, Ortin (1994), Horowitz, Tseytlin (1994), Russo, Tseytlin (1996)). Collision of two plane gravitational waves produce a spacetime that is locally isometric to an interior of black hole, see (Aref'eva,

Viswanathan, Volovich (1994, 1995, 1996)). To estimate the amplitude for the creation of black holes one can use the expression

$$A \propto \int d^9 b \, e^{iqb} e^{iI_{cl}} \tag{12}$$

where I_{cl} is the value of the boundary term for the gravitational action of the $D = 11$ supergravity (Horava, Witten (1996)) calculated on the solution describing colliding plane waves (Aref'eva, Viswanathan, Volovich (1994, 1995, 1996)).

Acknowledgment
The author is grateful to I.Ya. Aref'eva for the useful discussions. This work is partially supported by RFFI grant 96-01-00312.

References

Amati, D., Ciafaloni, M. and Veneziano, G. (1987, 1990): Phys. Let. **B197**, 81; Nucl. Phys. **B347**

Aref'eva, I.Ya., Viswanathan, K.S. and Volovich, I.V. (1994, 1995, 1996): Nucl. Phys. **B452**, 346; **B462**, 613; hep-th/9412157

Bachas, C. (1995): hep-th/9511043

Balasubramanian, V. and Larsen, F. (1997): hep-th/9703039

Banks, T., Fischler, W., Shenker, S. and Susskind, L. (1996): hep-th/9608086

Banks, T. and Seiberg, N. (1997): hep-th/9702187

Bergshoeff, E., Kallosh, R. and Ortin, T. (1994): hep-th/9406009

Birrell, N.D. and Davies, P.C.W. (1982): *Quantum fields in curved space*, Cambridge University Press, 1982

Danielsson, U.H., Ferretti, G. and Sundborg, B. (1996): hep-th/9603081

Dijkgraaf, R., Verlinde, E. and Verlinde, H. (1997): hep-th/9703030

Douglas, M., Kabat, D., Pouliot, P. and Shenker, S. (1996): hep-th/9608024

Douglas, M., Polchinski, J. and Strominger, A. (1997): hep-th/9703031

Fabbrichesi, M., Pettorino, R., Veneziano, G. and Vilkovisky, G.A. (1994): Nucl. Phys. **B419**, 147

Gibbons, G.W. (1982): Nucl.Phys. **B207**, 337

Gubser, S., Klebanov, I., Hashimoto, A. and Maldacena, J. (1996): hep-th/9601057

Guven, R. (1987): Phys. Let. 191 (1987)265

Hawking, S. (1976): Phys.Rev. **D14**, 2460

t'Hooft, G. (1988): Nucl.Phys. **B304**, 867

Horava, P. and Witten, E. (1996): hep-th/9603142

Horowitz, G.T. and Tseytlin, A.A. (1994): hep-th/9407099

Lifschytz, G. (1996): hep-th/9604156

Motl, L. (1997): hep-th/9701025

Russo, J.G. and Tseytlin, A.A. (1996): hep-th/9611047

Volovich, I.V. (1996): hep-th/9608137

Witten, E. (1995): hep-th/9510135

Self-Duality in Nonlinear Electromagnetism

M.K. Gaillard and B. Zumino

Physics Department, University of California, and
Theoretical Physics Group, Lawrence Berkeley National Laboratory

Abstract. We discuss duality invariant interactions between electromagnetic fields and matter. The case of scalar fields is treated in some detail.

1 Duality Rotations in Four Dimensions

The invariance of Maxwell's equations under "duality rotations" has been known for a long time. In relativistic notation these are rotations of the electromagnetic field strength $F_{\mu\nu}$ into its dual, which is defined by

$$\tilde{F}_{\mu\nu} = \frac{1}{2}\epsilon_{\mu\nu\lambda\sigma}F^{\lambda\sigma}, \quad \tilde{\tilde{F}}_{\mu\nu} = -F_{\mu\nu} \ . \tag{1}$$

This invariance can be extended to electromagnetic fields in interaction with the gravitational field, which does not transform under duality. It is present in ungauged extended supergravity theories, in which case it generalizes to a nonabelian group (Ferrara et al. (1977), Cremmer, Julia (1979)). In (Gaillard, Zumino (1981), Zumino (1982)) we studied the most general situation in which duality invariance of this type can occur. More recently (Gibbons, Rasheed (1995)) the duality invariance of the Born–Infeld theory, suitably coupled to the dilaton and axion (Gibbons, Rasheed (1996)), has been studied in considerable detail. In the present note we will show that most of the results of (Gibbons, Rasheed (1995), Gibbons, Rasheed (1996)) follow quite easily from our earlier general discussion. We shall also present some new results that were not made explicit in (Gaillard, Zumino (1981), Zumino (1982)), especially some properties of the scalar fields.

We begin by recalling and completing some basic results of our paper (Gaillard, Zumino (1981), Zumino (1982)). Consider a Lagrangian which is a function of n real field strengths $F^a_{\mu\nu}$ and of some other fields χ^i and their derivatives $\chi^i_\mu = \partial_\mu \chi^i$:

$$L = L\left(F^a, \chi^i, \chi^i_\mu\right) \ . \tag{2}$$

Since

$$F^a_{\mu\nu} = \partial_\mu A^a_\nu - \partial_\nu A^a_\mu \ , \tag{3}$$

we have the Bianchi identities

$$\partial^\mu \tilde{F}^a_{\mu\nu} = 0 \ . \tag{4}$$

On the other hand, if we define

$$\tilde{G}^a_{\mu\nu} = \frac{1}{2}\epsilon_{\mu\nu\lambda\sigma}G^{a\lambda\sigma} \equiv 2\frac{\partial L}{\partial F^{\mu\nu}_a} , \tag{5}$$

we have the equations of motion

$$\partial^\mu \tilde{G}^a_{\mu\nu} = 0 . \tag{6}$$

We consider an infinitesimal transformation of the form

$$\delta \begin{pmatrix} F \\ G \end{pmatrix} = \begin{pmatrix} A & B \\ C & D \end{pmatrix} \begin{pmatrix} F \\ G \end{pmatrix}, \tag{7}$$

$$\delta\chi^i = \xi^i(\chi), \tag{8}$$

where A, B, C, D are real $n \times n$ constant infinitesimal matrices and $\xi^i(\chi)$ functions of the fields χ^i (but not of their derivatives), and ask under what circumstances the system of the equations of motion (4) and (6), as well as the equation of motion for the fields χ^i are invariant. The analysis of (Gaillard, Zumino (1981)) shows that this is true if the matrices satisfy

$$A^T = -D, \quad B^T = B, \quad C^T = C , \tag{9}$$

(where the superscript T denotes the transposed matrix) and in addition the Lagrangian changes under (7) and (8) as

$$\delta L = \frac{1}{4}\left(FC\tilde{F} + GB\tilde{G}\right) . \tag{10}$$

The relations (9) show that (7) is an infinitesimal transformation of the real noncompact symplectic group $\mathrm{Sp}(2n, R)$ which has $\mathrm{U}(n)$ as maximal compact subgroup. The finite form is

$$\begin{pmatrix} F' \\ G' \end{pmatrix} = \begin{pmatrix} a & b \\ c & d \end{pmatrix} \begin{pmatrix} F \\ G \end{pmatrix} , \tag{11}$$

where the $n \times n$ real submatrices satisfy

$$c^T a = a^T c, \quad b^T d = d^T b, \quad d^T a - b^T c = 1 . \tag{12}$$

Notice that the Lagrangian is not invariant. In (Gaillard, Zumino (1981)) we showed, however, that the derivative of the Lagrangian with respect to an invariant parameter *is* invariant. The invariant parameter could be a coupling constant or an external background field, such as the gravitational field, which does not change under duality rotations. It follows that the energy-momentum tensor, which can be obtained as the variational derivative of the Lagrangian with respect to the gravitational field, is invariant under duality rotations. No explicit check of its invariance, as was done in (Gibbons, Rasheed (1995), Gibbons, Rasheed (1996), Born, Infeld (1934),Schrödinger (1935)), is necessary.

The symplectic transformation (11) can be written in a complex basis as

$$\begin{pmatrix} F' + \mathrm{i}G' \\ F' - \mathrm{i}G' \end{pmatrix} = \begin{pmatrix} \phi_0 & \phi_1^* \\ \phi_1 & \phi_0^* \end{pmatrix} \begin{pmatrix} F + \mathrm{i}G \\ F - \mathrm{i}G \end{pmatrix} , \tag{13}$$

where $*$ means complex conjugation and the submatrices satisfy

$$\phi_0^T \phi_1 = \phi_1^T \phi_0, \quad \phi_0^\dagger \phi_0 - \phi_1^\dagger \phi_1 = 1 . \tag{14}$$

The relation between the real and the complex basis is

$$\begin{aligned} 2a &= \phi_0 + \phi_0^* + \phi_1 + \phi_1^*, \quad & -2ib &= \phi_0 - \phi_0^* + \phi_1 - \phi_1^*, \\ 2ic &= \phi_0 - \phi_0^* - \phi_1 + \phi_1^*, \quad & 2d &= \phi_0 + \phi_0^* - \phi_1 - \phi_1^*. \end{aligned} \tag{15}$$

In (Gaillard, Zumino (1981), Zumino (1982)) we also described scalar fields valued in the quotient space $\mathrm{Sp}(2n, R)/\mathrm{U}(n)$. The quotient space can be parameterized by a complex symmetric $n \times n$ matrix $K = K^T$ whose real part has positive eigenvalues, or equivalently by a complex symmetric matrix $Z = Z^T$ such that $Z^\dagger Z$ has eigenvalues smaller than 1. They are related by

$$K = \frac{1 - Z^*}{1 + Z^*}, \quad Z = \frac{1 - K^*}{1 + K^*} . \tag{16}$$

These formulae are the generalization of the well-known map between the Lobachevskiĭ unit disk and the Poincaré upper half-plane: Z corresponds to the single complex variable parameterizing the unit disk; iK to the one parameterizing the upper half plane.

Under $\mathrm{Sp}(2n, R)$

$$\begin{aligned} K \to K' &= (-\mathrm{i}c + \mathrm{d}K)\,(a + \mathrm{i}bK)^{-1} , \\ Z \to Z' &= (\phi_1 + \phi_0^* Z)\,(\phi_0 + \phi_1^* Z)^{-1} , \end{aligned} \tag{17}$$

or, infinitesimally,

$$\delta K = -\mathrm{i}C + DK - KA - \mathrm{i}KBK, \quad \delta Z = V + T^* Z - ZT - \mathrm{i}ZV^* Z , \tag{18}$$

where

$$T = -T^\dagger, \quad V = V^T . \tag{19}$$

The invariant nonlinear kinetic term for the scalar fields can be obtained from the Kähler metric (Binétruy, Gaillard (1985))

$$\mathrm{Tr}\left(\mathrm{d}K^* \frac{1}{K + K^*} \mathrm{d}K \frac{1}{K + K^*}\right) = \mathrm{Tr}\left(\mathrm{d}Z \frac{1}{1 - Z^* Z} \mathrm{d}Z^* \frac{1}{1 - ZZ^*}\right) \tag{20}$$

which follows from the Kähler potential

$$\mathrm{Tr} \ln (1 - ZZ^*) \quad \text{or} \quad \mathrm{Tr} \ln (K + K^*) , \tag{21}$$

which are equivalent up to a Kähler transformation. It is not hard to show that the metric (20) is positive definite. Throughout this paper we assume a flat background space-time metric; the generalization to a nonvanishing gravitational field is straightforward (Gaillard, Zumino (1981), Zumino (1982), Gibbons, Rasheed (1995), Gibbons, Rasheed (1996)).

2 Born–Infeld Theory

As a particularly simple example we consider the case when there is only one tensor $F_{\mu\nu}$ and no additional fields. Our equations become

$$\tilde{G} = 2\frac{\partial L}{\partial F} , \tag{1}$$

$$\delta F = \lambda G, \quad \delta G = -\lambda F \tag{2}$$

and

$$\delta L = \frac{1}{4}\lambda\left(G\tilde{G} - F\tilde{F}\right) . \tag{3}$$

We have restricted the duality transformation to the compact subgroup U(1) ≅ SO(2), as appropriate when no additional fields are present. So $A = D = 0$, $B = -C = \lambda$.

Since L is a function of F alone, we can also write

$$\delta L = \delta F\frac{\partial L}{\partial F} = \lambda G\frac{1}{2}\tilde{G} . \tag{4}$$

Comparing (3) and (4), which must agree, we find

$$G\tilde{G} + F\tilde{F} = 0 . \tag{5}$$

Together with (1), this is a partial differential equation for $L(F)$, which is the condition for the theory to be duality invariant. If we introduce the complex field

$$M = F - \mathrm{i}G , \tag{6}$$

(5) can also be written as

$$M\widetilde{M}^* = 0 . \tag{7}$$

Clearly, Maxwell's theory in vacuum satisfies (5), or (7), as expected. A more interesting example is the Born–Infeld theory (Born, Infeld (1934)), given by the Lagrangian

$$L = \frac{1}{g^2}\left(-\Delta^{\frac{1}{2}} + 1\right) , \tag{8}$$

where

$$\Delta = -\det\left(\eta_{\mu\nu} + gF_{\mu\nu}\right) = 1 + \frac{1}{2}g^2F^2 - g^4\left(\frac{1}{4}F\tilde{F}\right)^2 . \tag{9}$$

For small values of the coupling constant g (or for weak fields) L approaches the Maxwell Lagrangian. We shall use the abbreviation

$$\beta = \frac{1}{4}F\tilde{F} . \tag{10}$$

Then

$$\frac{\partial \Delta}{\partial F} = g^2 F - \beta g^4 \tilde{F} \ , \tag{11}$$

$$\tilde{G} = 2\frac{\partial L}{\partial F} = -\Delta^{-\frac{1}{2}} \left(F - \beta g^2 \tilde{F} \right) \ , \tag{12}$$

and

$$G = \Delta^{-\frac{1}{2}} \left(\tilde{F} + \beta g^2 F \right) \ . \tag{13}$$

Using (12) and (13), it is very easy to check that $G\tilde{G} = -F\tilde{F}$: the Born–Infeld theory is duality invariant. It is also not too difficult to check that $\partial L/\partial g^2$ is actually *invariant* under (2) and the same applies to $L - \frac{1}{4}F\tilde{G}$ (which in this case turns out to be equal to $-g^2 \partial L/\partial g^2$). These invariances are expected from our general theory.

It is natural to ask oneself whether the Born–Infeld theory is the most general physically acceptable solution of (5). This was investigated in (Gibbons, Rasheed (1995)) where a negative result was reached: more general Lagrangians satisfy (5), the arbitrariness depending on a function of one variable.

3 Schrödinger's Formulation of Born's Theory

Schrödinger (Schrödinger (1935)) noticed that, for the Born–Infeld theory (8), F and G satisfy not only (5) [or (7)], but also the more restrictive relation

$$M \left(M\widetilde{M} \right) - \widetilde{M} M^2 = \frac{g^2}{8} \widetilde{M}^* \left(M\widetilde{M} \right)^2 \ . \tag{1}$$

We have verified this by an explicit, although lengthy, calculation using (6), (12), (13) and (9). Schrödinger did not give the details of the calculation, presenting instead convincing arguments based on particular choices of reference systems. One can write (1) as

$$\frac{\partial \mathcal{L}}{\partial M} = g^2 \widetilde{M}^* \ , \tag{2}$$

where

$$\mathcal{L} = 4 \frac{M^2}{\left(M\widetilde{M} \right)} \ , \tag{3}$$

and Schrödinger proposed $\mathcal{L}$ as the Lagrangian of the theory, instead of (8). Of course, $\mathcal{L}$ is a Lagrangian in a different sense than L, which is a field Lagrangian in the usual sense. Multiplying (1) by M and saturating the unwritten indices $\mu\nu$, the left hand side vanishes, so that (7) follows. Using (1) it is easy to see that $\mathcal{L}$ is pure imaginary: $\mathcal{L} = -\mathcal{L}^*$. Schrödinger also pointed out that, if we introduce a map

$$\frac{1}{g^2} \frac{\partial \mathcal{L}}{\partial M} = f(M) \ , \tag{4}$$

so that (1) or (2) can be written as

$$f(M) = \widetilde{M}^* \ , \tag{5}$$

the square of the map is the identity map

$$f\left(f(M)\right) = M \ . \tag{6}$$

This, together with the properties

$$f(\widetilde{M}) = -\tilde{f}(M), \quad f(M^*) = f(M)^* \ , \tag{7}$$

ensures the consistency of (1). Schrödinger used the Lagrangian (3) to construct a conserved, symmetric energy-momentum tensor. We have checked that, when suitably normalized, his energy-momentum tensor agrees with that of Born and Infeld up to an additive term proportional to $\eta_{\mu\nu}$.

Schrödinger's formulation is very clever and elegant and it has the advantage of being *manifestly* covariant under the duality rotation $M \to M\mathrm{e}^{i\lambda}$ which is the finite form of (2). It is also likely that, as he seems to imply, his formulation is fully equivalent to the Born–Infeld theory (8), which would mean that the more restrictive equation (1) eliminates the remaining ambiguity in the solutions of (7). This virtue could actually be a weakness if one is looking for more general duality invariant theories.

4 Axion, Dilaton and SL(2, R)

It is well known that, if there are additional scalar fields which transform nonlinearly, the compact group duality invariance can be enhanced to a duality invariance under a larger noncompact group (see, *e.g.*, (Gaillard, Zumino (1981)) and references therein). In the case of the Born–Infeld theory, just as for Maxwell's theory, one complex scalar field suffices to enhance the U(1) ≅ SO(2) invariance to the $SU(1,1) \cong \mathrm{SL}(2,R)$ noncompact duality invariance. This is pointed out in (Gibbons, Rasheed (1996)), but it also follows the considerations of our paper (Gaillard, Zumino (1981)). We shall use the letter S instead of K for the scalar field, which, in the example under consideration, is a single complex field, not an $n \times n$ matrix. In today's more standard notation

$$S = S_1 - \mathrm{i}S_2 = \mathrm{e}^{-\phi} - \mathrm{i}a, \quad S_1 > 0 \ , \tag{1}$$

where ϕ is the dilaton and a is the axion. For $\mathrm{SL}(2,R) \cong \mathrm{Sp}(2,R)$, the matrices A, B, C, D are real numbers and $A = -D$, B and C are independent. Then the infinitesimal $\mathrm{SL}(2,R)$ transformation is

$$\delta S = -2AS - \mathrm{i}BS^2 - \mathrm{i}C \ . \tag{2}$$

For the SO(2) ≅ U(1) subgroup, $A = 0$, $B = -C = \lambda$,

$$\delta S = -\mathrm{i}\lambda S^2 + \mathrm{i}\lambda \ . \tag{3}$$

The scalar kinetic term, proportional to

$$\frac{\partial_\mu S^* \partial^\mu S}{(S+S^*)^2} \ , \tag{4}$$

is invariant under the nonlinear transformation (2) which, in terms of S_1, S_2, takes the form

$$\delta S_1 = -2AS_1 - \mathrm{i}BS_1S_2, \quad \delta S_2 = -2AS_2 + B\left(S_1^2 - S_2^2\right) + C \ . \tag{5}$$

The full noncompact duality transformation on $F_{\mu\nu}$ is now

$$\delta F = AF + BG, \quad \delta G = DF + DG, \quad D = -A \ , \tag{6}$$

and we are seeking a Lagrangian $\hat{L}(F,S)$ which satisfies

$$\delta\hat{L} = \frac{1}{4}\left(FC\tilde{F} + GB\tilde{G}\right) \ , \tag{7}$$

where

$$\delta\hat{L} = \delta F\frac{\partial\hat{L}}{\partial F} + \delta S_1\frac{\partial\hat{L}}{\partial S_1} + \delta S_2\frac{\partial\hat{L}}{\partial S_2} \ , \tag{8}$$

and now

$$\tilde{G} = 2\frac{\partial\hat{L}}{\partial F} \ . \tag{9}$$

Equating (7) and (8) we see that $\hat{L}$ must satisfy

$$\frac{1}{4}\left(BG\tilde{G} - CF\tilde{F}\right) + \frac{1}{2}AF\tilde{G} + \delta S_1\frac{\partial\hat{L}}{\partial S_1} + \delta S_2\frac{\partial\hat{L}}{\partial S_2} = 0 \ . \tag{10}$$

This equation can be solved as follows. Assume that $L(\mathcal{F})$ satisfies (1) and (5), *i.e.*

$$\mathcal{G}\tilde{\mathcal{G}} + \mathcal{F}\tilde{\mathcal{F}} = 0 \ , \tag{11}$$

where

$$\tilde{\mathcal{G}} = 2\frac{\partial\mathcal{L}}{\partial\mathcal{F}} \ . \tag{12}$$

For instance, the Born–Infeld Lagrangian $L(\mathcal{F})$ does this. Then

$$\hat{L}(S,F) = L(S_1^{\frac{1}{2}}F) + \frac{1}{4}S_2F\tilde{F} \tag{13}$$

satisfies (10). Indeed

$$\frac{\partial\hat{L}(S,F)}{\partial F} = \frac{\partial L}{\partial\mathcal{F}}S_1^{\frac{1}{2}} + \frac{1}{2}S_2\tilde{F} \ . \tag{14}$$

So

$$\tilde{G} = \tilde{\mathcal{G}}S_1^{\frac{1}{2}} + S_2\tilde{F} \ , \tag{15}$$

$$G = \mathcal{G} S_1^{\frac{1}{2}} + S_2 F \ , \tag{16}$$

where we have defined

$$\mathcal{F} = S_1^{\frac{1}{2}} F \ , \tag{17}$$

and $\tilde{\mathcal{G}}$ is given by (12). Now

$$G\tilde{G} = \mathcal{G}\tilde{\mathcal{G}} S_1 + S_2^2 F\tilde{F} + 2S_2 \mathcal{F}\tilde{\mathcal{G}} \ . \tag{18}$$

Using (11) in this equation we find

$$G\tilde{G} = \left(S_2^2 - S_1^2\right) F\tilde{F} + 2S_2 \mathcal{F}\tilde{\mathcal{G}} \ . \tag{19}$$

We also have

$$F\tilde{G} = \mathcal{F}\tilde{\mathcal{G}} + S_2 F\tilde{F} \ . \tag{20}$$

On the other hand, since

$$\frac{\partial L}{\partial S_1^{\frac{1}{2}}} = \frac{\partial \mathcal{L}}{\partial \mathcal{F}} F = \frac{1}{2} \tilde{\mathcal{G}} F \ , \tag{21}$$

we obtain

$$\frac{\partial \hat{L}}{\partial S_1} = \frac{\partial L}{\partial S_1^{\frac{1}{2}}} \frac{1}{2} S_1^{-\frac{1}{2}} = \frac{1}{4} \tilde{\mathcal{G}} S_1^{-\frac{1}{2}} F = \frac{1}{4} \tilde{\mathcal{G}} \mathcal{F} S_1^{-1} \ . \tag{22}$$

In addition

$$\frac{\partial \hat{L}}{\partial S_2} = \frac{1}{4} F\tilde{F} \ . \tag{23}$$

Using (19), (20), (22) and (23), together with (5), we see that (10) is satisfied. It is easy to check that the scale invariant combinations $\mathcal{F}$ and $\mathcal{G}$, given by (17) and (12) have the very simple transformation law

$$\delta\mathcal{F} = S_1 B \mathcal{G}, \quad \delta\mathcal{G} = -S_1 B \mathcal{F} \ , \tag{24}$$

i.e., they transform according to the U(1) ≅ SO(2) compact subgroup just as F and G in (2), but with the parameter λ replaced by $S_1 B$. If $L(\mathcal{F})$ is the Born–Infeld Lagrangian, the theory with scalar fields given by $\hat{L}$ in (13) can also be reformulated à la Schrödinger. From (16) and (17) solve for $\mathcal{F}$ and $\mathcal{G}$ in terms of F, G, S_1 and S_2. Then $\mathcal{M} = \mathcal{F} - \mathrm{i}\mathcal{G}$ must satisfy the same equation (1) that M does when no scalar fields are present.

5 Connections to String Theory

The duality rotations considered here are relevant to effective field theories from superstrings. The supersymmetric extension (Binétruy, Gaillard (1996)) of the Lagrangian (13) with $L(\mathcal{F}) = -\frac{1}{4}\mathcal{F}^2$ describes the dilaton plus Yang-Mills sector of effective $N = 1$ supergravity theories obtained from superstrings in the weak coupling ($S_1 \to \infty$) limit. The SL$(2, Z)$ subgroup of SL$(2, R)$ that is generated by the elements $4\pi S \to 1/4\pi S$ and $S \to S - \mathrm{i}/4\pi$ relates different string theories (Schwarz, Sen (1993, 1994), Duff (1995), Witten (1995)) to one another. The generalization of (Gaillard, Zumino (1981)) to two dimensional theories (Cecotti, Ferrara, Girardello (1988)) has been used to derive the Kähler potential for moduli and matter fields in effective field theories from superstrings. In this case the scalars are valued on a coset space $\mathcal{K}/\mathcal{H}$, $\mathcal{K} \in \mathrm{SO}(n,n)$, $\mathcal{H} \in \mathrm{SO}(n) \times \mathrm{SO}(n)$. The kinetic energy is invariant under $\mathcal{K}$, and the full classical theory is invariant under a subgroup of $\mathcal{K}$. String loop corrections reduces the invariance to a discrete subgroup that contains the SL$(2, Z)$ group generated by $T \to 1/T$, $T \to T - \mathrm{i}$, where T is the squared radius of compactification in string units.
Acknowledgements. We are grateful for the hospitality provided by the Isaac Newton Institute where this work was initiated. We thank Gary Gibbons, David Olive, Harold Steinacker, Kelly Stelle and Peter West for inspiring conversations. This work was supported in part by the Director, Office of Energy Research, Office of High Energy and Nuclear Physics, Division of High Energy Physics of the U.S. Department of Energy under Contract DE-AC03-76SF00098 and in part by the National Science Foundation under grant PHY-95-14797.

References

Binétruy, P. and Gaillard, M.K. (1985): *Phys. Rev.* **D32**, 931.
Binétruy, P. and Gaillard, M.K. (1996): *Phys. Lett.* **B365**, 87.
Born, M. and Infeld, L. (1934): *Proc. Roy. Soc.* (London) **A144**, 425.
Cecotti, S., Ferrara, S. and Girardello, L. (1988): *Nucl. Phys.* **B308**, 436.
Ferrara, S., Scherk, J. and Zumino, B. (1977): *Nucl. Phys.* **B121**, 393;
Cremmer, E. and Julia, B. (1979): *Nucl. Phys.* **B159**, 141.
Gaillard, M.K. and Zumino, B. (1981): *Nucl. Phys.* **B193**, 221.
Gibbons, G.W. and Rasheed, D.A. (1995): *Nucl. Phys.* **B454**, 185.
Gibbons, G.W. and Rasheed, D.A. (1996): *Phys. Lett.* **B365**, 46.
Schwarz, J.H. and Sen, A. (1993, 1994): *Phys. Lett.* **B312**, 105 and *Nucl. Phys.* **B411**, 35;
Duff, M. (1995): *Nucl. Phys.* **B442**, 47;
Witten, E. (1995): *Nucl. Phys.*, **B443**, 85.
Schrödinger, E. (1935): *Proc. Roy. Soc.* (London) **A150**, 465.
Zumino, B. (1982): *Quantum Structure of Space and Time*, Eds. M.J. Duff and C.J. Isham (Cambridge University Press) p. 363.

Progress Toward A Classical $(\text{SUSY})^2$ 4D, $N = 1$ Green–Schwarz σ-Model Action

S.J. Gates, Jr.

Department of Physics
University of Maryland
College Park, MD 20742-4111 USA
e-mail: gates@umdhep.umd.edu**

Abstract. We investigate a (2,0) supergravity-matter action that is suggested as a starting point by the Berkovitz modification of the Green-Schwarz action.

1 Introduction

In 1986, H. Nishino and I authored a paper (Gates, Nishino (1986)) in which there appeared *for the first time in physics literature* the idea that the ultimate formulation of superstring actions must combine aspects of the Green-Schwarz (GS) and Neveu-Schwarz-Ramond (NSR) actions. We now call such models $(\text{SUSY})^2$ theories. This has been a topic to which I have periodically returned in subsequent investigations (Brooks, Gates, Muhammad (1986a), Gates, Majudar (1988, 1992), Gates (1989, 1990, 1997)). As well, numbers of other authors have independently investigated variations of this theme (Kowalski-Glikman (1986), Kowalski-Glikman, Kowalski-Glikman, van Holten, Aoyama, Lukierski (1988), van Holten (1987), Fisch (1992)). Among these there exist important works by Dr. Volkov (Sorokin, Tkach, Volkov (1989), Sorokin, Tkach, Volkov, Zheltukhin (1989)). This last work has led to the most complete yet realization of the idea that was proffered in ref. (Gates, Nishino (1986)). In particular motivated by this work, Berkovitz (Berkovits (1993, 1994)), has made great progress in constructing models that are the best to date at providing a complete *covariant* description of 4D, N = 1 superstrings. Much evidence of this progress can be seen in subsequent works (Berkovits, Siegel (1996), de Boer, Skenderis (1996)). Thus the contribution of Dr. Volkov are clearly visible in the current level of understanding.

However, even with the latest realization (Berkovits (1993, 1994)) of the idea expressed in ref. (Gates, Nishino (1986)), I am not completely satisfied. The model of Berkovitz realizes the $(\text{SUSY})^2$ symmetry *only* at the level of a quantized (i.e. conformal field theory) model. So it is my intuition that still we lack a realization of the $(\text{SUSY})^2$ principle at the purely *classical* level.

** Supported in part by National Science Foundation Grant PHY-91-19746 and by NATO Grant CRG-93-0789

The present paper is offered as one addition modest step toward this goal and dedicated to Dr. Volkov whose contributions to this and numerous other problems in theoretical physics are properly recognized by this volume.

2 Local (2,0) Superspace Geometry

The component field content of (2,0) world sheet supergravity consists of a graviton ($e_a{}^m$) a complex Weyl gravitino ($\chi_a{}^+$) and an SU(2) triplet gauge field (A_a). Some time ago, the geometry of (2,0) supergravity was initially developed (Brooks, Gates, Muhammad (1986)) and later modified (see the last work of ref. (Gates (1989, 1990, 1997))). The resulting (2,0) supergravity covariant derivative satisfies,

$$\begin{aligned} &[\nabla_+, \nabla_+\} = 0 \quad , \quad [\nabla_+, \overline{\nabla}_+\} = i\nabla_{\ddagger} \quad , \\ &[\nabla_+, \nabla_{\ddagger}\} = 0 \quad , \quad [\nabla_+, \nabla_{=}\} = i\,, \overline{\Sigma}^+ (\mathcal{M} + i\widehat{\mathcal{Y}}) \quad , \\ &[\nabla_{\ddagger}, \nabla_{=}\} = \Sigma^+\nabla_+ + \overline{\Sigma}^+\overline{\nabla}_+ + \mathcal{R}\mathcal{M} + \mathcal{F}\widehat{\mathcal{Y}} \quad , \end{aligned} \tag{2.1}$$

where Σ^+, $\mathcal{R}$ and $\mathcal{F}$ are superfields such that

$$\begin{aligned} \Sigma^+| &\equiv [\mathcal{D}_{\ddagger}\chi_{=} - \mathcal{D}_{=}\chi_{\ddagger} - c_{\ddagger}{}_{=}{}^c\chi_c{}^+] \quad , \\ \mathcal{R}| &\equiv r(e, \omega(e,\chi)) + i[\chi_{\ddagger}{}^+(\overline{\Sigma}^+|) + \overline{\chi}_{\ddagger}{}^+(\Sigma^+|)] \quad , \\ \mathcal{F}| &\equiv \mathcal{D}_{\ddagger}A_{=} - \mathcal{D}_{=}A_{\ddagger} - c_{\ddagger}{}_{=}{}^cA_c - [\chi_{\ddagger}{}^+(\overline{\Sigma}^+|) - \\ &\quad -\overline{\chi}_{\ddagger}{}^+(\Sigma^+|)] \quad , \end{aligned} \tag{2.2}$$

and $r(e, \omega(e,\chi))$ denotes the world sheet curvature. Above we use our long standing convention of denoting the limit as the Grassmann coordinates are set to zero by a | following the superfield. The field strength superfields satisfy

$$\overline{\nabla}_+\Sigma^+ = 0 \quad , \quad \nabla_+\Sigma^+ = \mathcal{R} + i\mathcal{F} \quad . \tag{2.3}$$

The (2,0) world sheet scale transformation laws of the covariant derivative take the forms

$$\begin{aligned} \delta_L\nabla_+ &= \tfrac{1}{4}(\Lambda + \overline{\Lambda})\nabla_+ + \tfrac{1}{2}(\nabla_+\Lambda)\mathcal{M} + i(\nabla_+\Lambda)\widehat{\mathcal{Y}} \quad , \\ \delta_L\nabla_{\ddagger} &= \tfrac{1}{2}(\Lambda + \overline{\Lambda})\nabla_{\ddagger} - i\tfrac{3}{4}[(\overline{\nabla}_+\overline{\Lambda})\nabla_+ + (\nabla_+\Lambda)\overline{\nabla}_+] \\ &\quad + \tfrac{1}{2}[\nabla_{\ddagger}(\Lambda + \overline{\Lambda})]\mathcal{M} + i[\nabla_{\ddagger}(\Lambda - \overline{\Lambda})]]\widehat{\mathcal{Y}} \quad , \\ \delta_L\nabla_{=} &= \tfrac{1}{2}(\Lambda + \overline{\Lambda})\nabla_{=} - \tfrac{1}{2}(\nabla_{=}(\Lambda + \overline{\Lambda}))\mathcal{M} \quad , \end{aligned} \tag{2.4}$$

in terms of a chiral superfield Λ. The transformation of the field strength Σ^+ is thus

$$\delta_L\,\Sigma^+ = \frac{3}{2}L\,\Sigma^+ + I(\nabla_{=}\nabla_+\Lambda) \quad , \tag{2.5}$$

and we note that the transformation laws of the other two field strength superfields follow from the application of δ_L to both sides of the latter result in (2.3).

A final result that is required to derive component results from (2,0) superspace results is to note the "density projection" formulae;

$$\begin{aligned}\int \mathrm{d}^2\sigma \mathrm{d}^2\zeta^{\ddagger}\, E^{-1}\,\mathcal{L} \equiv\ & \tfrac{1}{2}\int \mathrm{d}^2\sigma \mathrm{d}\zeta^{+}\, \mathcal{E}^{-1}\,((\nabla_{+} - \mathrm{i}\,2\overline{\chi}_{\ddagger}{}^{+})\overline{\nabla}_{+}\mathcal{L})| \\ & + \tfrac{1}{2}\int \mathrm{d}^2\sigma \mathrm{d}\overline{\zeta}^{+}\, \overline{\mathcal{E}}^{-1}\,((\overline{\nabla}_{+} - \mathrm{i}\,2\chi_{\ddagger}{}^{+})\nabla_{+}\mathcal{L})|\end{aligned} \tag{2.6}$$

valid for any (2,0) superfield $\mathcal{L}$ and as well

$$\int \mathrm{d}^2\sigma \mathrm{d}\zeta^{+}\, \mathcal{E}^{-1}\,\mathcal{L}_{-c} = \int \mathrm{d}^2\sigma\, e^{-1}\,((\nabla_{+} - \mathrm{i}\,2\overline{\chi}_{\ddagger}{}^{+})\,\mathcal{L}_{-c})| \quad , \tag{2.7}$$

valid for any chiral superfield $\mathcal{L}_{-c}$.

3 Manifest $(\mathrm{SUSY})^2$ in a New Green-Schwarz Type Action

In accord with the $(\mathrm{SUSY})^2$ principle, the object of primary interest is $\mathcal{Z}^{\underline{M}}$ that maps from (2,0) superspace into 4D, N = 1 superspace. We define $\mathcal{Z}^{\underline{M}} \equiv (\mathcal{Z}^{\mu},\, \mathcal{Z}^{\dot{\mu}},\, \mathcal{Z}^{\mu\dot{\mu}})$ where $\mathcal{Z}^{\underline{M}}$ is a (2,0) chiral superfield, i.e. $\overline{\nabla}_{+}\mathcal{Z}^{\underline{M}} = 0$. Since necessarily $\mathcal{Z}^{\underline{M}}$ is complex we may write

$$\mathcal{Z}^{\mu\dot{\mu}} \equiv X^{\mu\dot{\mu}} + \mathrm{i}Y^{\mu\dot{\mu}}, \quad \mathcal{Z}^{\mu} \equiv \frac{1}{2}[\Theta^{\mu} + \Delta^{\mu}], \quad \mathcal{Z}^{\dot{\mu}} \equiv \frac{1}{2}[\overline{\Theta}^{\dot{\mu}} - \overline{\Delta}^{\dot{\mu}}] \quad , \tag{3.1}$$

where $X^{\mu\dot{\mu}}$ and $Y^{\mu\dot{\mu}}$ are real. We identify the usual 4D, N = 1 space-time superstring coordinates as $X^{\mu\dot{\mu}}$, Θ^{μ} and $\overline{\Theta}^{\dot{\mu}}$ where

$$\begin{aligned}\tfrac{1}{2}[\mathcal{Z}^{\mu\dot{\mu}} + (\mathcal{Z}^{\mu\dot{\mu}})^{*}]| &\equiv \begin{pmatrix} X^0 + X^3 & X^1 - iX^2 \\ X^1 + iX^2 & X^0 - X^3 \end{pmatrix} \quad , \\ [\mathcal{Z}^{\mu} + (\mathcal{Z}^{\dot{\mu}})^{*}]| &= \Theta^{\mu}(\sigma, \tau) \quad .\end{aligned} \tag{3.2}$$

As is seen, manifest (2,0) supersymmetry forces the introduction of a sort of mirror superspace with coordinates $(\Delta^{\mu},\, Y^{\mu\dot{\mu}})$. Any successful (2,0) construction must ultimately eliminate these as dynamical entities.

To describe the 4D, N = 1 space-time supergeometry, we introduce a vielbein $\mathrm{E}_{\underline{M}}{}^{\underline{A}}$ that is, in general, a function of the coordinates $\mathcal{Z}^{\underline{M}}$ and $\overline{\mathcal{Z}}^{\underline{M}}$. However, we are also free to *restrict* the dependence so that $\mathrm{E}_{\underline{M}}{}^{\underline{A}}$ is *solely* a function of $X^{\mu\dot{\mu}}$, Θ^{μ} and $\overline{\Theta}^{\dot{\mu}}$. The quantities $\Pi_{A}{}^{\underline{A}} \equiv (\Pi_{+}{}^{\underline{A}},\, \Pi_{\ddagger}{}^{\underline{A}},\, \Pi_{=}{}^{\underline{A}})$ denote space-time supercovariant "normals" that are defined by,

$$\begin{aligned}\Pi_{+}{}^{\underline{A}} &= (\nabla_{+}\mathcal{Z}^{\underline{M}})\mathrm{E}_{\underline{M}}{}^{\underline{A}} \quad , \quad \Pi_{\ddagger}{}^{\underline{A}} = (\nabla_{\ddagger}\mathcal{Z}^{\underline{M}})\mathrm{E}_{\underline{M}}{}^{\underline{A}} \quad , \\ \Pi_{=}{}^{\underline{A}} &= (\nabla_{=}\mathcal{Z}^{\underline{M}})\mathrm{E}_{\underline{M}}{}^{\underline{A}} \quad ,\end{aligned} \tag{3.3}$$

and satisfy

$$F_{A\,B}{}^{\underline{C}} \equiv \nabla_A \Pi_B{}^{\underline{C}} - (-)^{AB} \nabla_B \Pi_A{}^{\underline{C}} \\ - T_{A\,B}{}^C \Pi_C{}^{\underline{C}} - (-)^{A\underline{B}} \Pi_A{}^{\underline{A}} \Pi_B{}^{\underline{B}} T_{\underline{A}\,\underline{B}}{}^{\underline{C}} = 0. \tag{3.4}$$

If $\Pi_A{}^{\underline{A}}$ is regarded as a world-sheet gauge field, it has a vanishing field strength where $T_{\underline{A}\,\underline{B}}{}^{\underline{C}}$ are the structure constants. Alternately, $\Pi_A{}^{\underline{A}}$ defines a linear mapping operator via $e_A = \Pi_A{}^{\underline{A}} \mathrm{E}_{\underline{A}}$ and $\mathrm{d}\omega^{\underline{A}} = \mathrm{d}\omega^A \Pi_A{}^{\underline{A}}$. Although (3.4) is a classical equation, it is interesting to conjecture that its expectation value in a quantized theory is related to anomalies and critical dimensions.

The only remaining component fields in $\mathcal{Z}^{\underline{M}}$ (complex bosonic twistor fields and complex NSR fermions) may be defined covariantly with respect to both world-sheet and space-time manifolds through the equation

$$\Pi_+{}^{\underline{A}}| \equiv (S_+{}^\alpha,\, \tilde{S}_+{}^{\dot\alpha},\, \psi_+{}^{\underline{a}}) \quad . \tag{3.5}$$

In the final work of ref. (Kowalski-Glikman (1986), Kowalski-Glikman, Kowalski-Glikman, van Holten, Aoyama, Lukierski (1988), van Holten (1987), Fisch (1992)), we suggested that a starting point for further study of this system is given by,

$$\begin{aligned} \mathcal{S} \;=\; & \Big\{ \int \mathrm{d}^2\sigma\, \mathrm{d}^2\zeta^{\ddagger}\; E^{-1} \,[\, \overline{\mathcal{Z}}^{\underline{M}} \mathrm{E}_{\underline{M}}{}^{\underline{A}} \Pi_{=}{}^{\underline{B}} t^{(0)}_{\underline{A}\,\underline{B}} \;+\; \Pi_+{}^{\underline{A}} \Pi_+{}^{\underline{B}} \Lambda_{==\underline{A}\,\underline{B}} \,] \;+\; \mathrm{h.c.} \Big\} \\ + & \Big\{ \int \mathrm{d}^2\sigma\, \mathrm{d}^2\zeta^{\ddagger}\; E^{-1} \,[\, \Pi_+{}^{\underline{A}} \Lambda_{=-}{}^{\underline{B}} t^{(1)}_{\underline{A}\,\underline{B}} \;+\; \Pi_{\ddagger}{}^{\underline{A}} \Lambda_{==}{}^{\underline{B}} t^{(2)}_{\underline{A}\,\underline{B}} \\ & \quad +\; \Pi_{=}{}^{\underline{A}} \Lambda^{\underline{B}} t^{(3)}_{\underline{A}\,\underline{B}} \,] \;+\; \mathrm{h.c.} \Big\} \\ + & \Big\{ \int \mathrm{d}^2\sigma\, \mathrm{d}\zeta^{+}\; \mathcal{E}^{-1} \,[\, \Sigma^+ \Phi(\mathcal{Z}) \,] \;+\; \mathrm{h.c.} \Big\} \quad , \end{aligned} \tag{3.6}$$

where the quantities $t^{(i)}_{\underline{A}\,\underline{B}}$ are a set of constant tensors. One parametrization of these is

$$t^{(i)}_{\underline{A}\,\underline{B}} \;=\; \begin{pmatrix} k_1^{(i)} C_{\alpha\beta} & 0 & 0 \\ 0 & k_2^{(i)} C_{\dot\alpha\dot\beta} & 0 \\ 0 & 0 & k_3^{(i)} C_{\alpha\beta} C_{\dot\alpha\dot\beta} \end{pmatrix} \quad . \tag{3.7}$$

For $k_1^{(0)} = k_2^{(0)} = 0$, $k_3^{(0)}$ nonvanishing, the first term produces the standard nonlinear σ-model with torsion for $\mathcal{Z}^{\underline{M}}|$ (Brooks, Gates, Muhammad (1986), Hull, Spence (1990), Hull (1990), Gates, Ketov (1990)). For $k_3^{(1)}$ nonvanishing, variation with respect to $\Lambda_{=-}{}^{\underline{b}}$ imposes the superfield equation $\Pi_+{}^{\underline{b}} = 0$. In the second work of reference (Sorokin, Tkach, Volkov (1989), Sorokin, Tkach, Volkov, Zheltukhin (1989)), the analog of this condition plays a critical role in eliminating would-be NSF fermions. We have introduced such terms in some of our previous work (see the third work of (Gates (1989, 1990, 1997))) and this seems crucial to realizing the condition of STVZ (Sorokin, Tkach, Volkov (1989), Sorokin, Tkach, Volkov, Zheltukhin (1989)). Finally, for $k_1^{(1)}$ and $k_2^{(3)}$ nonvanishing, a simple definition of propagators for the Grassmann coordinates (Θ^α) is possible. A explained previously (final work in ref. (Brooks, Gates, Muhammad (1986a), Gates, Majudar (1988, 1992))) no explicit factors of α' need appear in our action.

We note that additional restrictions arise from noting that the scale transformation in (2.4) also can be used. In particular, if we assume that that U(1) weights of the string coordinates vanishes then (2.4) implies,

$$\delta_L \Pi_+{}^{\underline{A}} = \frac{1}{4}(\Lambda + \overline{\Lambda})\,\Pi_+{}^{\underline{A}} \ ,$$

$$\delta_L \Pi_{\ddagger}{}^{\underline{A}} = \frac{1}{2}(\Lambda + \overline{\Lambda})\,\Pi_{\ddagger}{}^{\underline{A}} - \mathrm{i}\frac{3}{4}[\,(\overline{\nabla}_+\overline{\Lambda})\,\Pi_+{}^{\underline{A}} \ ,$$

$$\delta_L \nabla_{=} = \frac{1}{2}(\Lambda + \overline{\Lambda})\,\Pi_{=}{}^{\underline{A}} \ , \tag{3.8}$$

Imposing the classical condition of scale invariance demands that the coefficient $t^{(2)}_{\underline{A}\,\underline{B}}$ must vanish identically.

Although much more study is required before the action above can be considered as an acceptable model, it is encouraging that classical manifest (2,0) supersymmetry offers such possibilities. It is be hope that such construction will eventually lead to a consistent covariant quantization of the theory that evaded the Green-Schwarz model (Gates, Grisaru, Lindström, Roček, Siegel, van Nieuwenhuizen, van de Ven (1989), Henneaux, Fish (1989), Henneaux, Mezincescu (1985)). A clearly indicated next step is to investigate the component formulation via the projection operators discussed in this work. This seems like the most direct way to gain further insight.

Acknowledgment
The author wishes to thank the organizers of the meeting honoring Dr. Volkov for their invitation to make this contribution to this volume.

References

Berkovits, N. (1993, 1994): Nucl. Phys. **B395**, 77; idem. Nucl. Phys. **B420**, 332; Nucl. Phys. **B431**, 258.

Berkovits, N. and Siegel, W. (1996): Nucl. Phys. **B462**, 213;
de Boer, J. and Skenderis, K. (1996): Nucl. Phys. **B481**, 129.

Brooks, R., Gates, S. J., Jr. and Muhammad, F. (1986): Nucl. Phys. **B268**, 599.

Brooks R., Gates S. J., Jr. and Muhammad F. (1986a):, Class. Quant. Grav. **3**, 745;
Gates S. J., Jr. and Majumdar P. (1988, 1992):, Mod. Phys. Lett. **A4**, 339; idem. Phys. Lett. **248B**, 71.

Hull, C.M., Spence, B. (1990): Nucl. Phys. **B345**, 493
Hull, C.M. (1990): Mod. Phys. Lett. **A5**, 1793;
Gates, S.J., Jr. and Ketov, S.V. (1990): Phys. Lett. **271B**, 355.

Gates S. J., Jr. (1989, 1990, 1997): in the *Proceedings of the XXV Winter School of Theoretical Physics* in Karpacz, Poland, Birkhauser-Verlag Press (Feb., 1989) 169; idem. *Proceedings of the Superstrings and Particle Theory* meeting in Tuscaloosa, Alabama, World Scientific Press 1990, 57; idem., Phys. Lett. **390B**, 161.

Gates, S.J., Jr., Grisaru, M.T., Lindström, U., Roček, M., Siegel, W., van Nieuwenhuizen, P. and van de Ven, A.E. (1989): Phys. Lett. **225B**, 44;
Henneaux, M. and Fisch, J.M.L. (1989): Santiago, Centro Estudios Cientificos preprint, ULB-TH2/89-04, (June, 1989);
Henneaux, M. and Mezincescu, L. (1985): Phys. Lett. **152B**, 340.

Gates S. J., Jr. and Nishino H. (1986):, Class. Quant. Grav. **3**, 391.

Kowalski-Glikman, J. (1986): Phys. Lett. **180B**, 359;
Kowalski-Glikman, J. and van Holten, J.W (1986): Nucl. Phys. **B283**, 305;
Fisch, J.M.L. (1992): Phys. Lett. **219B**, 71;
Kowalski-Glikman, J., van Holten, J.W., Aoyama, S. and Lukierski, J. (1988): Phys. Lett. **201B**, (1988) 487;

Sorokin, D.P., Tkach, V.I. and Volkov, D. (1989): Mod. Phys. Lett. **A4**, 901;
Sorokin, D.P., Tkach, V.I., Volkov, D. and Zheltukin, A.A. (1989): Phys. Lett. **216B**, 302.

$N = 4$ Supersymmetric Integrable Systems

E.A. Ivanov

Bogoliubov Laboratory of Theoretical Physics, JINR
141 980, Dubna, Russian Federation

Abstract. I give an overview of recent progress in constructing the KdV, mKdV and NLS type hierarchies with extended $N = 4$ supersymmetry.

1 Introduction

It is widely believed nowadays that the ultimate theory of all fundamental forces is by no means a standard field theory, but rather that of extended objects like superstrings and supermembranes. Supersymmetry, the concept pioneered by D.V. Volkov (Volkov and Akulov (1973)), will surely be one of the key-stones of this future theory. Affine and W algebras and superalgebras are also expected to be necessary ingredients of the underlying symmetry structure of this theory as they naturally come out as the world-sheet (or world-volume) gauge symmetries of extended objects.

The WZW, Liouville - Toda and KdV type $2D$ integrable systems are intimately related to these symmetries. They are encountered and proved to be of high relevancy in a plenty of problems of modern mathematical physics (see, e.g., di Francesco et al. (1995)): in non-perturbative $2D$ gravity and related matrix models, in the geometric approaches to strings, superstrings and supermembranes, in Seiberg-Witten non-perturbative approach to supersymmetric Yang-Mills theory, etc. KdV, mKdV, NLS and KP type hierarchies of evolution equations exhibit a remarkable relationship with conformal, affine and W algebras: the latter provide a hamiltonian structure for the former (Gervais and Neveu (1982), Gervais (1985)). Supersymmetric extensions of these hierarchies and their interplay with superaffine and superconformal (W) algebras were under intense study for the last decade [4-10].

The subject of my talk is the KdV type integrable hierarchies with extended $N = 4$ supersymmetry. While a lot was known about $N = 1$ and $N = 2$ extensions, until recently it remained unclear whether consistent higher N hierarchies of this kind exist. Only $N = 4$ extensions of some exactly solvable Lorentz-covariant $2D$ systems, Liouville and WZW models, were known (Ivanov and Krivonos (1984), Ivanov et al. (1988)). Seeking higher N hierarchies is of extreme interest, in particular, because such systems can be relevant to the program of "grand-unification" of all known hierarchies: apparently unrelated lower supersymmetry (and purely bosonic) integrable systems can turn out to be various reductions of a single higher N supersymmetric system. This phenomenon can be already seen while passing from

$N = 1$ KdV hierarchy (Mathieu (1988), Mathieu (1988a), Mathieu (1988b) to the $N = 2$ ones (Laberge and Mathieu (1988), Labelle and Mathieu (1991)). The latter naturally incorporate both KdV and mKdV hierarchies in the bosonic sector.

In a series of papers [13-15] the first example of KdV system with higher supersymmetry, $N = 4$ KdV hierarchy, was constructed and analyzed, both in a manifestly supersymmetric $N = 4$ superfield form (Delduc and Ivanov (1993)) and via $N = 2$ superfields (Delduc et al. (1996), (1997)). It was found to be bi-hamiltonian, to have "small" $N = 4$ SCA (Ademollo et al. (1976)) as the second hamiltonian structure and to possess two different Lax formulations in terms of $N = 2$ super pseudo-differential operators (Delduc and Gallot (1996), Ivanov and Krivonos (1996)). A remarkable interplay between integrability of this system and breaking of the global automorphism SU(2) symmetry of $N = 4$ supersymmetry was revealed: it is integrable only provided this SU(2) is explicitly broken and the square of SU(2) breaking parameter is proportional to the inverse of the central charge of $N = 4$ SCA. It encompasses two different $N = 2$ KdV hierarchies, the $a = 4$ and $a = -2$ ones, as its two non-equivalent consistent reductions. Later it was found that "small" $N = 4$ SCA provides a hamiltonian structure for one more integrable hierarchy which is an extension of the $a = -2$, $N = 2$ KdV hierarchy and possesses only $N = 2$ global supersymmetry (Delduc et al. (1997)). Recently (Ivanov et al. (1997)), a simplest $N = 4$ supersymmetric affine hierarchy was constructed. It is defined on $N = 2$ extension of the bosonic affine algebra $\widehat{\mathrm{sl}(2) \oplus \mathrm{u}(1)}$ and underlies $N = 4$ KdV hierarchy much like ordinary mKdV hierarchy underlies (via a Miura map) the KdV hierarchy. It seems that any $N = 2$ affine algebra or superalgebra admitting a quaternionic structure exhibits hidden $N = 4$ supersymmetry and hence can give rise to $N = 4$ supersymmetric hierarchies. This provides a general clue to constructing and classifying such hierarchies.

2 KdV Example

To explain the basic idea of how to construct KdV type hierarchies via relating them to some infinite-dimensional algebras as second hamiltonian structure, let us start with the text-book KdV example.

As was shown in (Gervais and Neveu (1982), Gervais (1985)), the KdV equation

$$\dot{u} = -u''' + 6uu' \tag{1}$$

can be treated as a hamiltonian system,

$$\dot{u} = \{u, \mathcal{H}_3\} \ ,$$

with the hamiltonian and the Poisson brackets defined by

$$\mathcal{H}_3 = \frac{1}{2}\int \mathrm{d}x\, u^2(x) \quad , \quad \{u(x), u(y)\} = \left[-\partial^3 + 4u\partial + 2u'\right]\delta(x-y) \ . \quad (2)$$

For the Fourier modes of $u(x)$,

$$u(x) = \frac{6}{c}\sum_n \exp(-inx)L_n - \frac{1}{4}\,, \tag{3}$$

the Poisson brackets in (2) imply the structure relations of the Virasoro algebra

$$\mathrm{i}\,\{L_n, L_m\} = (n-m)L_{n+m} + \frac{c}{12}(n^3 - n)\delta_{n+m,0}\ . \tag{4}$$

So, the definition (2) means that the density of the KdV hamiltonian $\mathcal{H}_3$ is the square of a conformal stress-tensor $u(x)$ obeying the Virasoro algebra (2), (4). One says that the Virasoro algebra provides the second hamiltonian (or Poisson) structure for the KdV equation (historically, first hamiltonian formulation of KdV hierarchy was based upon a linear Poisson algebra, and the latter is referred to as the first hamiltonian structure). The higher order conserved quantities of the KdV equation can be regarded as the hamiltonians which generate, through the Poisson brackets (2), next equations from the KdV hierarchy.

3 N=1,2 KdVs

The same idea was applied for constructing $N = 1$ and $N = 2$ superextensions of the KdV equation [5-9]. They were related in an analogous way, via the second hamiltonian structure, to $N = 1$ and $N = 2$ SCAs. In the $N = 1$ case the basic object is $N = 1$ stress-tensor, the spin 3/2 fermionic $N = 1$ superfield

$$\Phi(t,x,\theta) = \xi(t,x) + \theta u(t,x). \tag{5}$$

It comprises the spin 3/2 fermionic current ξ and spin 2 stress-tensor u which generate, via appropriate PBs, the classical $N = 1$ SCA. The most general $N = 1$ supersymmetric dimension 3 hamiltonian reads

$$\mathcal{H}_3 = \int \mathrm{d}x\, \mathrm{d}\theta\, \Phi D\Phi\ ,\ D = \frac{\partial}{\partial\theta} + \theta\frac{\partial}{\partial x}\ ,\ D^2 = \partial_x\ , \tag{6}$$

and $N = 1$ KdV equation is the equation defining evolution of Φ with respect to $\mathcal{H}_3$

$$\dot{\Phi} = \{\Phi, \mathcal{H}_3\}\ , \tag{7}$$

with the superfield PB structure amounting to $N = 1$ SCA.

In the $N = 2$ case one deals with the $N = 2$ stress-tensor which is a spin 1 $N = 2$ superfield

$$J(t,x,\theta,\bar\theta) = j(t,x) + \theta\xi(t,x) + \bar\theta\bar\xi(t,x) + \theta\bar\theta u(t,x)\ , \tag{8}$$

with the components being currents of $N = 2$ SCA. Once again, the $N = 2$ KdV equation can be defined as an evolution equation

$$\dot{J} = \{J, \mathcal{H}_3\} \tag{9}$$

with respect to the most general $N = 2$ supersymmetric dimension 3 hamiltonian

$$\mathcal{H}_3 = \int \mathrm{d}x \, \mathrm{d}\theta \, \mathrm{d}\bar{\theta} \left(J[D, \bar{D}]J + \frac{a}{6} J^3 \right) \tag{10}$$

and the PB structure

$$\{J(1), J(2)\} = \left(J\partial + \partial J + DJ\bar{D} + \bar{D}JD + \partial[D, \bar{D}] \right) \delta(1, 2) \ , \tag{11}$$

which generates $N = 2$ SCA (we always choose the central charge equal to some number since on the classical level it can be fixed at will by proper rescalings of superfields and PBs). In these formulas

$$\{D, D\} = 0 \ , \ \{D, \bar{D}\} = -\partial_x \ , \quad \delta(1, 2) = \delta(x_1 - x_2)(\theta_1 - \theta_2)(\bar{\theta}_1 - \bar{\theta}_2) \ , \tag{12}$$

and the differential operator in the r.h.s. of 11 is evaluated at the first point of $N = 2$ superspace (all derivatives are assumed to act freely to the right).

We observe two differences of $N = 2$ case compared to the two previous cases. Firstly, $N = 2$ supersymmetry requires two fields in the bosonic sector, the spin 2 stress-tensor $u(t, x)$ and the spin 1 current $j(t, x)$ which generates the affine $\widehat{\mathrm{u}(1)}$ subalgebra of $N = 2$ SCA. The bosonic sector of $N = 2$ KdV equation is a coupled system of KdV and mKdV equations for these fields. Secondly, there is a free parameter a in the hamiltonian and, respectively, in $N = 2$ KdV equation. It was shown [8-10] that this equation is completely integrable, i.e. gives rise to an infinite hierarchy of conserved hamiltonians in involution and possesses a Lax formulation, only for the three special values of a

$$a = -2, \ 4, \ 1 \ . \tag{13}$$

These are just the values at which the coupled KdV-mKdV system in the bosonic sector of $N = 2$ KdV is integrable.

4 N=4 KdV Hierarchy

A natural generalization of $N = 2$ SCA in the list of Ademollo *et al.* (Ademollo et al. (1976)) is the "small" $N = 4$ SCA. Alongside with the conformal stress-tensor, it contains a triplet of the spin 1 currents of affine algebra $\widehat{\mathrm{su}(2)}$ and a complex su(2) doublet of the spin 3/2 fermionic currents. It can be formulated in a manifestly supersymmetric way as a set of $N = 4$ superfield PBs (Delduc and Ivanov (1993), Delduc et al. (1996), (1997)). We will prefer here a $N = 2$ superfield notation in which this SCA is represented by the $N = 2$ stress-tensor J and chiral and anti-chiral spin 1 supercurrents Φ, $\bar{\Phi}$, $D\Phi = \bar{D}\bar{\Phi} = 0$ (Delduc et al. (1996), (1997)). Together with 11, the PBs

$$\{J(1),\Phi(2)\} = -\left(\Phi\bar{D}D + \bar{D}\Phi D\right)\delta(1,2)\,,$$
$$\{J(1),\bar{\Phi}(2)\} = -\left(\bar{\Phi}D\bar{D} + D\bar{\Phi}\bar{D}\right)\delta(1,2)\,,$$
$$\{\Phi(1),\bar{\Phi}(2)\} = \left(\partial D\bar{D} + DJ\bar{D}\right)\delta(1,2)\,, \quad \{\Phi(1),\Phi(2)\} = 0 \qquad (14)$$

form the classical "small" $N = 4$ SCA.

In terms of these supercurrents the transformations promoting manifest $N = 2$ supersymmetry to $N = 4$ are given by

$$\delta J = -\epsilon\bar{D}\Phi - \bar{\epsilon}D\bar{\Phi}\,, \quad \delta\Phi = \bar{\epsilon}DJ\,, \quad \delta\bar{\Phi} = \epsilon\bar{D}J\,. \qquad (15)$$

It is straightforward to check covariance of 11, 14 under these transformations. Then the problem of constructing $N = 4$ KdV is reduced to constructing most general $N = 4$ supersymmetric dimension 3 hamiltonian out of $J, \Phi, \bar{\Phi}$. It is given by the following expression

$$\mathcal{H}_3 = \int \mathrm{d}x\, \mathrm{d}\theta\, \mathrm{d}\bar{\theta} \left\{ J[D,\bar{D}]J - 2\Phi'\bar{\Phi} + \frac{a}{6}J^3 - aJ\Phi\bar{\Phi} - \frac{1}{2}b\left(\Phi^2 + \bar{\Phi}^2\right) \right\} \qquad (16)$$

and contains two real parameters a and b, arbitrary for the moment. The evolution equations for $J, \Phi, \bar{\Phi}$ can be constructed in the standard way, their explicit form can be found in (Delduc et al. (1996), (1997)). They also include the parameters a and b.

In ref. (Delduc et al. (1996), (1997)) we investigated the issue of existence of higher-order non-trivial conserved hamiltonians for this $N = 4$ KdV system and found that they exist only for the following three options

$$(i).\ a = 4,\ b = 0; \quad (ii).\ a = -2,\ b = 6; \quad (iii).\ a = -2,\ b = -6\,. \qquad (17)$$

Just with these choices the $N = 4$ KdV system turns out bi-hamiltonian. Both the existence of non-trivial higher-order conserved quantities and the bi-hamiltonian property were strong indications that $N = 4$ KdV system is integrable and gives rise to the whole hierarchy for these values of the parameters.

These three choices are essentially different only at first sight. Actually, they are related to each other by hidden SU(2) symmetry transformations which form an automorphism group of $N = 4$, $1D$ supersymmetry. The realization of these transformations on $N = 2$ superfields J, Φ, $\bar{\Phi}$ looks not too illuminating; it can be found in ref. (Delduc et al. (1996), (1997)).

Both $N = 4$ supersymmetry and SU(2) covariance become transparent and manifest while formulating the $N = 4$ KdV system in $N = 4$, $1D$ harmonic superspace (Delduc and Ivanov (1993)). There, the "small" $N = 4$ SCA is represented by the analytic doubly-charged harmonic superfield $V^{++}(\zeta)$ subjected to the supplementary constraint [1]

[1] $\zeta \equiv (x, \theta^+, \bar{\theta}^+, u_i^+, u_k^-)$ are coordinates of an analytic subspace of harmonic $N =$ 4, $1D$ superspace (Galperin et al. (1984)), $u_i^\pm$, $u^{+\,i}u_i^- = 1$ being harmonic coordinates, $\theta^+, \bar{\theta}^+$ projections of the $N = 4$ grassmann coordinates $\theta^i, \bar{\theta}^k$ on the harmonics u_i^+.

$$D^{++}V^{++} = 0, \quad \left(D^{++} = u^{+i}\frac{\partial}{\partial u^{-i}} + \theta^+\bar{\theta}^+\partial_x\right) . \tag{18}$$

It restricts the harmonic dependence of V^{++} so that the irreducible set of component fields of the latter amounts to the currents contents of "small" $N = 4$ SCA. In the ordinary $N = 4$ superspace this "harmonic shortness" condition implies

$$V^{++} = V^{(ik)}(x, \theta, \bar{\theta})u_i^+ u_k^+ ,$$

while the analyticity is expressed as the following constraints on $V^{(ik)}$

$$D^{(i}V^{kl)} = \bar{D}^{(i}V^{kl)} = 0 ,$$

$D^i, \bar{D}^k$ being the appropriate spinor derivatives. The automorphism SU(2) symmetry in this manifestly $N = 4$ supersymmetric formulation is realized as rotations of the doublet indices i, k, l.

The $N = 2$ superfields $J, \Phi, \bar{\Phi}$ are first components (up to numerical coefficients) in the decomposition of such V^{12}, V^{11} and V^{22} with respect to the grassmann coordinates which enlarge $N = 2$ superspace to the $N = 4$ one.

The PB structure 11, 14 can be rewritten as a single PB for the superfields V^{++} (it is explicitly given in Delduc and Ivanov (1993)). The hamiltonian 16, being expressed through V^{++}, takes the following form (Delduc and Ivanov (1993))

$$\mathcal{H}_3 = \int \mathrm{d}Z[\mathrm{d}u](D^{--}V^{++})^2 + \int \mathrm{d}\zeta^{-4}[\mathrm{d}u](a^{--})^2(V^{++})^3 . \tag{19}$$

Here, D^{--} is the second harmonic derivative (not preserving the harmonic analyticity), $\mathrm{d}Z[\mathrm{d}u]$ and $\mathrm{d}\zeta^{-2}[\mathrm{d}u]$ are measures of integration over the whole harmonic superspace and its analytic subspace, the SU(2) breaking parameter $a^{--} = a^{(ik)}u_i^- u_k^-$ is needed for balance of the harmonic $U(1)$ charges in the second piece of $\mathcal{H}_3$. Now it is a matter of straightforward though tedious calculation to check that the three options 17 just correspond to the three (up to reflections) independent orientations of the SU(2) breaking constant vector a^{ik} $((a^{ik})^\dagger = -\epsilon_{il}\epsilon_{kt}a^{lt})$

$$\begin{aligned} &(i).\ a^{12} = \pm\sqrt{5}\,,\ a^{11} = a^{22} = 0\,; \quad (ii).\ a^{12} = 0\,,\ a^{11} = a^{22} = \pm i\sqrt{5}\,; \\ &(iii).\ a^{12} = 0\,,\ a^{11} = -a^{22} = \pm\sqrt{5}\,, \end{aligned} \tag{20}$$

this vector having in all cases the same fixed norm

$$|a|^2 = -a^{ik}a_{ik} = 2(a^{12}a^{12} - a^{11}a^{22}) = 10 . \tag{21}$$

The latter is the main condition for the $N = 4$ KdV system to possess an infinite number of higher-order hamiltonians in involution and to be bi-hamiltonian (Delduc and Ivanov (1993)). If from the beginning we would keep the central charge k of $N = 4$ SCA unfixed, in the r.h.s. of eq. 21 there appeared the factor $\frac{1}{k}$.

In refs. (Ivanov and Krivonos (1996), Delduc and Gallot (1996)) two different $N = 2$ superfield Lax operators for this $N = 4$ KdV hierarchy have been proposed. Both of them are pseudo-differential and are adapted to the first choice in eqs. 17, 20, taking account of the fact that all the three options are indeed equivalent by hidden SU(2) covariance. These Lax operators are given by

$$L_1 = \partial - J - \bar{D}\partial^{-1}(DJ) - F\bar{D}\partial^{-1}(D\bar{F}) + \bar{D}\partial^{-1}(D(F\bar{F})) \ ,$$
$$DF = \bar{D}\bar{F} = 0 \ , \tag{22}$$
$$(\Phi = D\bar{F}, \ \bar{\Phi} = \bar{D}F)$$
$$L_2 = D\bar{D} + D\bar{D}\partial^{-1}(J + \bar{\Phi}\partial^{-1}\Phi)\partial^{-1}D\bar{D} \ . \tag{23}$$

In both cases the flows and the corresponding conserved hamiltonians are given by

$$\frac{\partial L}{\partial t_k} = [L^k_{\geq 1}, L] \ , \ \mathcal{H}_n = \int \mathrm{d}x \, \mathrm{d}\theta \, \mathrm{d}\bar{\theta} \ \mathrm{res} L^n \ , \tag{24}$$

the suffix ≥ 1 meaning the pure differential part of pseudo-differential operator. Note different definitions of the residue of the pseudo-differential operators: in the first case it is defined as a coefficient before 1, while in the second case as that before $D\bar{D}\partial^{-1}$.

A natural reduction to $N = 2$ KdV systems is to put in 22 - 24

$$\Phi = \bar{\Phi} = 0 \ , \tag{25}$$

which leads to the $a = 4$, $N = 2$ kdV hierarchy as a consistent reduction of the $N = 4$ KdV one. All the conserved quantities, as well as the above Lax formulations, are reduced to those of this $N = 2$ KdV hierarchy. However, there exists another consistent reduction of $N = 4$ KdV. Namely, one can choose the SU(2) frame corresponding to the second or third options in 17 and also impose the conditions 25. Though before reductions these options are related to each other by the hidden SU(2) symmetry, the reductions break SU(2) down to $U(1)$ and so give rise to non-equivalent $N = 2$ KdV systems. One can show that under the second reduction all even-dimensional conserved $N = 4$ KdV hamiltonians vanish (their densities are proportional to Φ, or $\bar{\Phi}$) while the odd-dimensional ones go into those of the $a = -2$, $N = 2$ KdV hierarchy. For the flows corresponding to these hamiltonians this reduction is self-consistent in the sense that both the l.h.s and r.h.s. of the evolution equations for $\Phi, \bar{\Phi}$ vanish upon imposing 25. Thus two different $N = 2$ KdV hierarchy, the $a = 4$ and $a = -2$ ones, are encoded in the single $N = 4$ KdV hierarchy as its two non-equivalent reductions. The same property can be established in the $N = 1$ superfield formulation of $N = 4$ KdV system (Delduc et al. (1997)). It would be interesting to find another Lax formulation of $N = 4$ KdV, such that the existence of these two reductions and the equivalence of different options in 17 were manifest. Hopefully, such a formulation can be constructed in harmonic superspace.

5 "Quasi" N=4 KdV Hierarchy

In ref. (Delduc et al. (1997)) we have found one more integrable hierarchy with the small $N = 4$ SCA as the second hamiltonian structure. It was naturally assigned the name "quasi" $N = 4$ KdV hierarchy as the global $N = 4$ supersymmetry is explicitly broken down to $N = 2$ in this system. Also, it reveals no SU(2) covariance and goes over to the $a = -2$, $N = 2$ KdV upon imposing the conditions 25. So it can be treated as an integrable extension of this $N = 2$ KdV hierarchy by chiral and anti-chiral superfields Φ, $\bar{\Phi}$.

It is interesting that, at cost of introducing new parameter c (not confuse it with the central charge!), the dimension 3 hamiltonian of this system (and actually all higher-order hamiltonians) can be written uniformly with the $a = 4, b = 0$ hamiltonian of genuine $N = 4$ KdV system

$$\mathcal{H}_3^c = \int \mathrm{d}x \, \mathrm{d}\theta \, \mathrm{d}\bar{\theta} \left\{ J[D, \bar{D}]J - \frac{c-3}{3} J^3 - 4J\Phi\bar{\Phi} - 2c\Phi'\bar{\Phi} \right\} . \tag{26}$$

At $c = 1$ the hamiltonian 16 with $a = 4, b = 0$ is reproduced while at $c = 4$ one gets the quasi $N = 4$ KdV system which goes into the $a = -2$, $N = 2$ KdV hierarchy upon the reduction 25. Lacking $N = 4$ supersymmetry can be easily observed already at the level of linear pieces of the corresponding evolution equations

$$\dot{J} = -J''' + ..., \quad \dot{\Phi} = -c\Phi''' + ..., \quad \dot{\bar{\Phi}} = -c\bar{\Phi}''' + ... \tag{27}$$

Since $N = 4$ supersymmetry 15 linearly transforms J, Φ and $\bar{\Phi}$ through each other, $N = 4$ supercovariance strictly requires $c = 1$ in these equations. So, the case $c = 4$ clearly corresponds to the situation with broken $N = 4$ supersymmetry.

In (Delduc et al. (1997), Delduc and Gallot (1996)) a scalar Lax formulation for this hierarchy has been constructed

$$L = D\left(\partial + J - \Phi\partial^{-1}\bar{\Phi}\right)\bar{D} , \quad \frac{\partial L}{\partial t_k} = \left[L_{\geq 1}^{k/2}, L\right] . \tag{28}$$

Note that, like the $a = -2$, $N = 2$ KdV, this system admits also a matrix Lax formulation along the lines of ref. (Inami and. Kanno (1992)).

An interesting property of this new $N = 2$ hierarchy is that it gives rise, via consistent reductions, to two new lower-supersymmetry hierarchies with $N = 2$ SCA as the second hamiltonian structure. They were missed in the previous studies. One of them possesses only $N = 1$ global supersymmetry and no any kind of internal symmetry. The other possesses $U(1)$ symmetry but lacks supersymmetry. It is still different from the non-supersymmetric system constructed in (Labelle and Mathieu (1991)): it contains the mKdV hierarchy for the spin 1 current $j(t,x)$ in its bosonic sector, while in the system of ref. (Labelle and Mathieu (1991)) this field satisfies the trivial equation, $\dot{j} = 0$. These observations suggest the existence of a "horizontal" sequence of hierarchies associated with the given SCA. It is parametrized by an extra parameter c which takes, similarly to the parameters a, b, some special values for the integrable cases. These systems range from the maximally supersymmetric one to lower-supersymmetric and even non-supersymmetric hierarchies. This conjecture implies that $N = 4$ SCA can serve as the second hamiltonian structure for more hierarchies, e.g., respecting only $N = 1$ supersymmetry or having no supersymmetry at all.

6 N=4 NLS-mKdV Hierarchies

There exists a remarkable and well-known relation between (super)affine and (super)conformal algebras: the latter can be mapped on the former through various Sugawara-Feigin-Fuks (SFF) or coset constructions of (super)conformal stress-tensors in terms of the (super)affine currents. Being translated into the language of integrable hierarchies, this correspondence manifests itself as the relation between two types of hierarchies: the KdV type ones associated with (super)conformal algebras as the second hamiltonian structure and the mKdV type ones which are hierarchies of evolution equations for the (super)affine currents with the (super)affine algebra as the hamiltonian structure. In this setting, the SFF representations for the stress-tensors come out as Miura-type maps between these two sorts of hierarchies.

Let us again apply to the KdV example. Introduce a spin 1 current $v(x)$ generating $\widehat{\mathrm{u}(1)}$ affine algebra through the PB

$$\{v(x), v(y)\} = \partial_x \delta(x-y) \ . \tag{29}$$

Then, defining

$$u = v^2 + v', \tag{30}$$

one observes that, as a consequence of PB 29, the so defined u generates a classical Virasoro algebra

$$\{u(x), u(y)\} = [-\partial^3 + 4u\partial + 2u']\delta(x-y) \ , \tag{31}$$

which is the same as in eq. (2). Thus eq. 30 gives the simplest example of SFF construction relating Virasoro algebra to the affine algebra $\widehat{\mathrm{u}(1)}$. On the other hand, substituting 30 into the KdV hamiltonian in (2), one gets

$$\mathcal{H}_3 = \frac{1}{2} \int \mathrm{d}x \, (v^4 + v'v') \ . \tag{32}$$

Through the PB 29 this hamiltonian gives rise to the evolution equation for v

$$\dot{v} = \{v, \mathcal{H}_3\} = -v''' - 6v'v^2 \tag{33}$$

which is the familiar mKdV equation. One can directly check that eq. 33 yields the standard KdV equation for u defined by eq. 30. Thus the SFF representation 30 is at the same time the Miura map relating KdV and mKdV hierarchies.

This correspondence more or less directly extends to the case of supersymmetric hierarchies. E.g., it is easy to check that an $N = 2$ superextension of the algebra $\widehat{\mathrm{u}(1)}$

$$\{H(1), \bar{H}(2)\} = D\bar{D}\delta(1,2), \ \{H(1), H(2)\} = 0, \ DH = \bar{D}\bar{H} = 0 \ , \qquad (34)$$

$H, \bar{H}$ being spin 1/2 fermionic chiral and anti-chiral superfields (actually, it collects two algebras $\widehat{\mathrm{u}(1)}$ in its bosonic sector), yields just the $N = 2$ SCA (11) via the SFF construction

$$J = H\bar{H} + D\bar{H} + \bar{D}H \ . \qquad (35)$$

After substituting this expression into the hamiltonian 10, one gets a set of evolution equations for H, $\bar{H}$ which give rise to $N = 2$ mKdV hierarchies for the values of the parameter a listed in eq. 13. For J defined by eq. 35 one gets just the related $N = 2$ KdV hierarchies.

One can ask whether analogous underlying affine hierarchies can be found for $N = 4$ KdV and "quasi" KdV hierarchies. The answer is affirmative, though the proof is not straightforward.

First of all, it is clear that the relevant superaffine algebras should reveal some $N = 4$ structure. At present, explicit superfield constructions of superextensions of affine algebras and superalgebras exist up to $N \geq 1$ (N is as before the number of independent $1D$ supercharges). In particular, $N = 2$ extensions exist for any affine (super)algebra admitting a complex structure (Hull and Spence (1989)). It is natural to assume that a hidden $N = 4$ supersymmetry is inherent in those $N = 2$ affine (super)algebras which possess a quaternionic structure, namely those which contain as their local part the algebras listed in (Spindel et al. (1988)) (actually, this list can be readily extended to superalgebras). Then an $N = 2$ extension of two affine algebras $\widehat{\mathrm{u}(1)}$ could be the simplest algebra of this sort. It contains in the bosonic sector just four copies of the $\widehat{\mathrm{u}(1)}$ algebras, the set on which one can already define a quaternionic structure (Spindel et al. (1988)). This $N = 2$ algebra is generated by two pairs of chiral and anti-chiral superfields $H_\alpha, \bar{H}_\alpha$, $(\alpha = 1,2)$ with the following PBs

$$\{H_\alpha(1), H_\beta(2)\} = 0 \ , \quad \{H_\alpha(1), \bar{H}_\beta(2)\} = \delta_{\alpha\beta} D\bar{D}\delta(1,2) \ . \qquad (36)$$

Indeed, it is easy to see the covariance of these relations, as well as of the chirality conditions for $H_\alpha, \bar{H}_\alpha$, under the transformations

$$\delta H_1 = \epsilon D\bar{H}_2, \ \ \delta\bar{H}_1 = \bar{\epsilon}\bar{D}H_2, \ \ \delta H_2 = -\epsilon D\bar{H}_1, \ \ \delta\bar{H}_2 = -\bar{\epsilon}\bar{D}H_1 \ . \qquad (37)$$

They possess the same Lie bracket structure as 15 and so, together with the manifest $N = 2$ supersymmetry transformations, yield a representation of the same $N = 4$ supersymmetry. Hence, the above $N = 2$ affine algebra indeed supplies the simplest example of $N=4$ affine algebra (its supercurrents form an $N=4$ supermultiplet).

One may wonder whether it gives rise to "small" $N=4$ SCA via some SFF construction and has any relation to $N=4$ KdV hierarchy. However, it can be checked that one cannot construct, out of the superfields $H_\alpha, \bar{H}_\beta$, any $N=4$ multiplet of composite currents including $N=2$ conformal stress-tensor with a Feigin-Fuks term (the latter is absolutely necessary for producing a central term in $N=2$ SCA and thus generating at least an $N=2$ KdV hierarchy as a subsystem). The only possibility is the $N=4$ multiplet

$$\hat{J} = H_1\bar{H}_1 + H_2\bar{H}_2, \quad \hat{\Phi} = H_1 H_2, \quad \hat{\bar{\Phi}} = \bar{H}_2\bar{H}_1 \ , \tag{38}$$

which, via PBs 36, generates a topological (i.e. centreless) "small" $N=4$ SCA. So, possible $N=2$ affine hierarchies (even possessing rigid $N=4$ supersymmetry) constructed on the basis of this $N=2$ affine algebra seem to have no any direct relation to $N=4$ KdV hierarchy.

Next in complexity is $N=2$ extension of non-abelian affine algebra $\widehat{\mathrm{sl}(2)\oplus \mathrm{u}(1)}$ whose local bosonic part $\mathrm{sl}(2)\oplus\mathrm{u}(1)$ also contains four generators and is among the algebras given in (Spindel et al. (1988)). The PBs of these $N=2$ superalgebra read (Hull and Spence (1989))

$$\{H(1), \bar{H}(2)\} = D\bar{D}\delta(1,2) \tag{39}$$

$$\{H(1), F(2)\} = DF\delta(1,2)\ , \quad \{H(1), \bar{F}(2)\} = -D\bar{F}\delta(1,2)\ ,$$
$$\{\bar{H}(1), F(2)\} = -\bar{D}F\delta(1,2)\ , \quad \{\bar{H}(1), \bar{F}(2)\} = \bar{D}\bar{F}\delta(1,2)\ , \tag{40}$$

$$\{F(1), \bar{F}(2)\} = \left[(D+H)(\bar{D}+\bar{H}) + F\bar{F}\right]\delta(1,2)\ , \tag{41}$$

all other PBs vanishing. Here, H and $\bar{H}$ satisfy the standard chirality conditions while F and $\bar{F}$ are subject to the *nonlinear* version of chirality

$$(D+H)\,F = 0, \quad (\bar{D}-\bar{H})\,\bar{F} = 0\ . \tag{42}$$

These constraints are necessary for closure of Jacobi identities of the algebra (Hull and Spence (1989)).

Once again, it is a matter of direct calculation to check that these PBs together with the above linear and nonlinear chirality conditions are covariant under the following hidden *nonlinear* $N=2$ transformations (Ivanov et al. (1997))

$$\delta H = \epsilon D\bar{F} + \bar{\epsilon} HF, \quad \delta\bar{H} = \bar{\epsilon}\bar{D}F - \epsilon\bar{H}\bar{F}$$
$$\delta F = -\epsilon D\bar{H} - \epsilon(H\bar{H} + F\bar{F}), \quad \delta\bar{F} = -\bar{\epsilon}\bar{D}H - \bar{\epsilon}(H\bar{H} + F\bar{F})\ . \tag{43}$$

It can be easily checked that the above transformations, despite their nonlinearity, indeed realize the same extra $N=2$ supersymmetry as the transformations 15. Thus, combined with manifest $N=2$ supersymmetry, they again form the previously defined $N=4$ supersymmetry. The two pairs of affine $N=2$ supercurrents $F, \bar{F}$ and $H, \bar{H}$ are unified into an irreducible

$N = 4$ supermultiplet, so the $N = 2$ extension of $\widehat{\mathrm{sl}(2) \oplus \mathrm{u}(1)}$ algebra is in fact an $N = 4$ extension. This $N = 4$ structure might be made manifest by passing to $N = 4$ superfields, but we will not elaborate on this possibility here.

What is indeed important for our consideration is that this superaffine algebra allows for a transparent SFF construction of small $N = 4$ SCA on its basis. The explicit formulas expressing $N = 4$ supercurrents in terms of the affine supercurrents are as follows (Ivanov et al. (1997))

$$J = H\bar{H} + F\bar{F} + D\bar{H} + \bar{D}H\ , \quad \Phi = D\bar{F}\ , \quad \bar{\Phi} = \bar{D}F\ , \ (D\Phi = \bar{D}\bar{\Phi} = 0)\ . \tag{44}$$

These objects obey just the $N = 4$ SCA PB relations 11, 14 as a consequence of the affine PB relations 39 - 41. Due to the Feigin-Fuks term in J in 44 the resulting $N = 4$ SCA possesses a non-zero central charge, which, as was already mentioned, is crucial for getting $N = 4$ KdV system.

Now it is a standard routine to substitute these SFF expressions into the hamiltonians of $N = 4$ KdV hierarchy and to derive the flow equations for the affine currents $H, \bar{H}, F, \bar{F}$ using the PBs of $N = 2$ affine $\widehat{\mathrm{sl}(2) \oplus \mathrm{u}(1)}$ algebra. Note that the lowest flow equations were derived in (Ivanov et al. (1997)) also in another way, by the direct construction of the proper dimension $N = 4$ invariant hamiltonians and requiring them to be in involution. We do not present here these equations in view of their considerable complexity (see ref. Ivanov et al. (1997)). We only notice the existence of a few consistent reductions of them.

The first one is effected by putting

$$F = \bar{F} = 0 \rightarrow \Phi = \bar{\Phi} = 0,\ J = H\bar{H} + D\bar{H} + \bar{D}H\ . \tag{45}$$

This yields the $a = 4$, $N = 2$ mKdV [2].

The second reduction goes as

$$H = \bar{H} = 0 \rightarrow J = F\bar{F},\ \ DF = \bar{D}\bar{F} = 0\ . \tag{46}$$

The resulting system is the $N = 2$ NLS hierarchy of refs. Krivonos and Sorin (1995), Krivonos et al.(1995). The existence of such a reduction has been firstly noticed in (Ivanov and Krivonos (1996)) at the level of $N = 4$ KdV hierarchy, with F and $\bar{F}$ interpreted as the prepotentials of the spin 1 chiral supercurrents Φ and $\bar{\Phi}$ in a fixed gauge with respect to the prepotential gauge freedom.

This consideration shows that the $N = 4$ supersymmetric system constructed can be treated as $N = 4$ extension of at once two $N = 2$ supersymmetric hierarchies, $N = 2$ mKdV and NLS ones. This is why it has been named in (Ivanov et al. (1997)) the "$N = 4$ NLS-mKdV hierarchy".

[2] Using another choice of the frame with respect to the hidden automorphism SU(2), it is possible to perform a reduction to the $a = -2$, $N = 2$ mKdV hierarchy as well.

Needless to say, the "quasi" $N = 4$ KdV hierarchy can also be associated with some underlying $N = 2$ $\widehat{\mathrm{sl}(2) \oplus \mathrm{u}(1)}$ affine hierarchy. One simply substitutes the STT expressions 44 for the $N = 4$ supercurrents into the hamiltonians of the "quasi" $N = 4$ KdV hierarchy and Lax operator and derives the appropriate evolution equations for the affine supercurrents via the PB structure 39 - 41.

7 Concluding Remarks

Thus, now we are aware of general method of setting up $N = 4$ supersymmetric KdV type hierarchies: each $N = 2$ affine algebra or superalgebra admitting a hidden $N = 4$ supersymmetry (quaternionic structure) can be used to construct such hierarchies via appropriate superfield SFF maps. We hope to naturally come in this way to $N = 4$ extensions of W algebras. An example of $N = 2$ affine algebra with $N = 4$ structure, next in complexity to $N = 2$ $\widehat{\mathrm{sl}(2) \oplus \mathrm{u}(1)}$, is the algebra $N = 2$ $\widehat{\mathrm{sl}(3)}$. This case is under study.

There remain many conceptual and technical problems to be solved. In particular, it would be useful to work out convenient $N = 4$ superfield techniques of treating $N = 4$ hierarchies, based, e.g., on the harmonic superspace approach. Up to now, it has been successfully applied only to one example of $N = 4$ KdV type hierarchies, the $N = 4$ KdV system itself (Delduc and Ivanov (1993), Delduc et al. (1996), (1997)). An interesting unsolved problem is to construct super KdV hierarchy associated with the "large" $N = 4$ SCA as the second hamiltonian structure. There are indications that such a system could yield all the three known $N = 2$ supersymmetric KdV hierarchies as its different consistent reductions. One may also think about higher N hierarchies, say, with $N = 8$ supersymmetry. To my knowledge, no any "no-go" theorems are known which could forbid the existence of such systems. Also, it could happen that a number of the already known $N = 2$ hierarchies exhibit hidden higher N supersymmetries. For instance, in a recent preprint (Sorin (1997)) it was found that $N = 4$ KdV hierarchy allows a map on the so called $(1,1)$ GNLS (Generalized NLS) system (Bonora et al. (1996)) which involves one chiral fermionic and one chiral bosonic superfields.

Perhaps, the most urgent problem is to identify the place and to reveal possible implications of this novel wide class of supersymmetric integrable systems in the modern superstring and p- brane stuff. I believe this certainly can be done.

Acknowledgement

I am grateful to Organizers of the D.V. Volkov Memorial Seminar for inviting me to give this Talk. I thank Loriano Bonora for hospitality at S.I.S.S.A., Trieste, where this work was finalized. A partial support from RFBR grant

RFBR 96-02-17634, INTAS grant INTAS-94-2317 and a grant of the Dutch NWO organization is acknowledged.

References

Ademollo, M. et al. (1976): Nucl. Phys. **B111**, 77
Bonora, L., Krivonos, S. and Sorin, A. (1996): Nucl. Phys. **B477**, 835
Chaichian, M. and Kulish, P.(1987): Phys. Lett. **B183**, 169;
Chaichian, M. and Lukierski, J. (1988): Phys. Lett. **B212**, 461
Delduc, F. and Gallot, L. (1996): "N=2 KP ad KdV hierarchies in extended superspace", ENSLAPP-L-617/96, solv-int/9609008
Delduc, F., Gallot, L. and Ivanov, E. (1997): Phys. Lett, **B 396** 141
Delduc, F. and Ivanov, E. (1993): Phys. Lett. **B309**, 312
Delduc, F., Ivanov, E. and Krivonos, S. (1996), (1997): J. Math. Phys. **37**, 1356; **38**, 1224 E
Hull, C.M. and Spence, B. (1989): Phys. Lett. **B241**, 357
Galperin, A., Ivanov, E., Ogievetsky, V. and Sokatchev, E. (1984): JETP Lett. **40**, 912
Gervais, J.L. and Neveu, A. (1982): Nucl. Phys. **B209**, 125,
Gervais, J.L. (1985): Phys. Lett. **B160**, 277, 279
di Francesco, P., Ginsparg, P. and Zinn-Justin, J. (1995): Phys. Reports **254**, 1
Inami, T. and Kanno, H. (1992): Int. J. Mod. Phys. **A7**,, Suppl. **1A**, 418
Ivanov, E. A. and Krivonos, S. O. (1984): J. Phys. A: Math. and Gen. **17**, L671
Ivanov, E.A., Krivonos, S.O. and Leviant, V.M. (1988): Nucl. Phys. **B304**, 601
Ivanov, E. A. and Krivonos, S. O. (1996): "New integrable extensions of $N = 2$ KdV and Boussinesq hierarchies", JINR E2-96-344, hep-th/9609191 (Phys. Lett. A, in press)
Ivanov, E., Krivonos, S. and Toppan, F. (1997): "N=4 super NLS-mKdV hierarchies", JINR E2-97-108, DFPD 97-TH-11, hep-th/9703224
Krivonos, S. and Sorin, A.(1995): Phys. Lett. **B357**, 94,
Krivonos, S., Sorin, A. and Toppan, F. (1995): Phys. Lett. **A206**, 146
Labelle, P. and Mathieu, P. (1991): J. Math. Phys. **32**, 923
Laberge, C. and Mathieu, P. Phys. Lett. **B215**, 718
Manin, Yu. I. and Radul, A.O. (1985): Commun. Math. Phys. **98**, 65,
Kulish, P.P. (1985): Lett. Math. Phys. **10**, 87
Mathieu, P. (1988): J. Math. Phys. **29**, 2499;
Mathieu, P. (1988a): Phys. Lett. **B203**, 287
Mathieu, P.(1988b): J. Math. Phys. **29**, 2499
Popowicz, Z. (1993): Phys. Lett. **A174**, 411
Sorin, A. (1997): "The discrete symmetries of the $N = 2$ supersymmetric GNLS hierarchies", solv-int/9701020
Spindel, Ph., Sevrin, A., Troost, W. and Van Proeyen, A. (1988): Phys. Lett. **B206**, 71
Volkov, D.V. and Akulov, V.P. (1973): Phys. Lett. **B46**, 109

On Some Puzzles in $N = 2$ Supersymmetric Gauge Theory

L. O'Raifeartaigh

Dublin Institute for Advanced Studies

Abstract. Supersymmetry, of which D.V.Volkov was one of the earliest proponents, has had a number of spectacular successes. The most recent, which would have pleased him very much, is its use to obtain an exact expression for the low-energy effective Lagrangian in $N = 2$ supersymmetric gauge theory. The Seiberg-Witten Ansatz which provided this solution has been checked by direct computation in the 1 and 2 instanton approximations. In this note some puzzles presented by the instanton computations are pointed out and partially resolved.

1 Introduction

Recently it has been shown Seiberg and Witten (1994) (S-W) that the low energy (Cartan) sector of the $N = 2$ supersymmetric Yang-Mills theory is tractable in the sense that
(a) the effective Lagrangian for the Cartan fields can be expressed in terms of a single function $F(A)$ of the chiral scalar $N = 1$ superfield A and
(b) it is argued that $F(A)$ is a specific Fuchsian function, in fact is simply the ratio of the derivatives of two independent solutions of a specific hypergeometric equation.

The S-W result (a) is exact but (b) was obtained using an Ansatz based on electro-magnetic duality. Accordingly, in order to check the validity of the Ansatz, a number of direct computations (Finnell and Pouliot (1995), Dorey et al. (1996a), (1996b), (1996c), Ito and Sasakura (1996),(1997), Yung (1997), Aoyama et al. (1996), Harano and Sato (1997)) have been made. As the perturbative part of the theory reduces to a one-loop contribution and the non-perturbative part of $F(A)$ is assumed to be due to instantons, these direct computations have concentrated on the instanton contributions. So far only the charge 1 and 2 instanton contributions have proved tractable but the results for these are in agreement with the S-W Ansatz for $F(A)$.

Although the instanton computations, especially the $N = 2$ ones, are technically impressive, they raise a number of puzzling questions of principle, as follows:
(a) Given that the computational results are claimed to be valid only for low orders in an expansion in gv, where g is the coupling constant and v is the vacuum value of the Higgs field, why are the results in *exact* agreement with the S-W Ansatz?
(b) Given that the background configurations chosen for the computations

are neither stationary with respect to the Action nor supersymmetrically invariant how can the computations based on them be exact?
(c) Given that the residual Lagrangian cannot then be supersymmetric, why should the bosonic and fermionic quantum fluctuations cancel, as is assumed?

The purpose of this note is to propose a resolution of the some of the above puzzles. The main point is that by using a specific realization of the supersymmetry algebra the background configurations can be made exactly supersymmetric.

2 Statement of the Problems

To present the problems just mentioned in a quantitative manner we recall that the $N = 2$ supersymmetric Yang-Mills Action takes the form

$$A_g = \mathrm{Tr} \int \frac{1}{4}F^2 + \bar{\psi}D\!\!\!/\psi + \frac{1}{2}|D\phi|^2 + g[\bar{\psi},\psi]\phi - gD[\phi^\dagger,\phi] + \frac{1}{2}(D^2 + F^\dagger F) \quad (1)$$

where $F_{\mu\nu}$ is a gauge-field, ψ is a Dirac spinor, ϕ is a complex scalar field, the fields D and F are real and complex dummy-fields belonging the real-vector and chiral-scalar $N = 1$ sub-multiplets respectively, and all fields belong to the adjoint representation of a compact simple Lie group. The instanton computations are carried out in background in which the instanton number N is not zero and the scalar field ϕ satisfies the non-trivial boundary condition $\phi(x) \to v \neq 0$ as $x \to \infty$.

Because of the way in which the dummy fields occur in the Action, it is convenient to replace the dummy-fields F and D by a real triplet X_a defined by

$$F = X_1 + \mathrm{i}X_2 \qquad D = X_3 + g[\phi^\dagger,\phi] \quad (2)$$

In that case the Action (1) becomes

$$A_g = \mathrm{Tr} \int \frac{1}{4}F^2 + \bar{\psi}D\!\!\!/\psi + \frac{1}{2}|D\phi|^2 + g[\bar{\psi},\psi]\phi - \frac{g}{2}[\phi^\dagger,\phi]^2 + \frac{1}{2}X_aX_a \quad (3)$$

The important point is that the dummy-fields X_a now decouple.

First let us consider the puzzle (a) above. It is clear from the way that the coupling constant g occurs in the Action (1) that by rescaling the fields A_μ, ϕ, ψ and D by a factor g^{-1}, the coupling constant can be removed at the expense of an overall factor g^{-2}. Hence an expansion in g is not physically meaningful.

The Action (1.1) is also manifestly scale-invariant. Hence, since ϕ has scale-dimension -1, the boundary condition $|\phi| \to v \neq 0$ can be converted to $|\phi| \to 1$ without changing the Action. Thus for each instanton charge N the vacuum-value v can appear in the effective potential only in the factor $(\Lambda/v)^N$, where Λ is the renormalization parameter. It follows that an expansion in v-expansion within a given instanton sector is not physically meaningful.

It follows from these considerations that an expansion in powers of gv is not physically meaningful and thus, although the instanton results were obtained using such an expansion as a guide, they must actually be exact results. But this is a point that remains to be clarified.

3 Previous Background Configurations

We now turn to puzzles (b) and (c). The $N = 2$ supersymmetric transformations (West (1986)) are

$$[Q^i, \delta\phi_1] = \mathrm{i}\psi^i \qquad \{Q^i, \delta\phi_2\} = \gamma_5\psi^i \qquad \{Q^i, \delta A_\mu\} = \gamma_\mu\psi^i \tag{4}$$

$$\{Q_i, \psi^j\} = \delta^i_j \left(\frac{1}{2}\sigma.F + \not{D}\Phi + \gamma_5[\phi^\dagger, \phi]\right) + (\tau \cdot X)^i_j \tag{5}$$

and

$$[Q^i, X_a] = (\tau_a)^i_j \left(\not{D}\psi_j + g[\Phi, \psi_j]\right) \qquad \Phi = \phi_1 - \mathrm{i}\gamma_5\phi_2 \tag{6}$$

where the σ's and γ's are the usual Dirac matrices and the τ's are a set of Pauli matrices belonging to the $SU(2)_s$ group that connects chiral scalar and real vector $N = 1$ submultiplets.

The background configurations chosen in (Finnell and Pouliot (1995), Dorey et al. (1996a), (1996b), (1996c), Ito and Sasakura (1996),(1997), Yung (1997), Aoyama et al. (1996), Harano and Sato (1997)) are

$$F = F^* \qquad \not{D}\psi = 0 \qquad D^2\phi = g[\psi^\dagger, \psi] \tag{7}$$

Since the equations for ψ and ϕ in (7) are not the Euler-Lagrange field equations it is clear that the configurations are not stationary points of the Action. Furthermore, if we make a supersymmetric variation of (7) the first equation remains invariant but for the other two we obtain

$$[Q_i, \not{D}\psi_j] = \delta_{ij}\gamma_5 D[\phi^\dagger, \phi] + D(\tau.X)^i_j \tag{8}$$

and

$$[Q_i, D^2\phi - g[\bar{\psi}, \psi]] = g\left([(\tau \cdot X + \gamma_5[\phi^\dagger, \phi]), \psi_i]\right) \tag{9}$$

Thus the configurations (7) are manifestly not supersymmetric-invariant. Finally, for these background configurations we have from (3)

$$A \quad \rightarrow \quad \int \mathrm{d}\Omega(\phi, D\phi) + \int \mathrm{d}^4x \left(X.X - g^2[\phi^\dagger, \phi]^2\right) \tag{10}$$

where the 3-dimensional integral is over the surface Ω at space-time infinity. The surface term in (10) is the quantity that is actually computed in Finnell and Pouliot (1995), Dorey et al. (1996a), (1996b), (1996c), Ito and Sasakura (1996),(1997), Yung (1997), Aoyama et al. (1996), Harano and Sato (1997) and from (7) it would seem that there should be further classical contributions coming from the volume integral. Furthermore, since the background is not supersymmetric, one would expect that there would be still further contributions from the quantum fluctuations.

In Finnell and Pouliot (1995), Dorey et al. (1996a), (1996b), (1996c), Ito and Sasakura (1996),(1997), Yung (1997), Aoyama et al. (1996), Harano and Sato (1997) the dummy fields X_a are ignored and one possibility to get rid of the volume integral in (12) would be to choose the X_a so that the integrand vanishes. But this would lead to a further violation of supersymmetry, as can be seen either directly or by noting that it would correspond to a spontaneous breakdown of $SU(2)_s$ symmetry. Indeed so long as the dummy fields X_a are simple scalar fields there would appear to be no choice of their background values that would improve the situation. We turn now to the resolution of these puzzles.

4 Proposed Supersymmetric Background

We first note that since the fields X_a decouple from all other fields in (4) we have a certain freedom in deciding how they should be realized. We shall assume that they are actually matrix-valued, in particular that they are of the form

$$X_a = (\tau_a)Y \tag{11}$$

where Y is a single scalar field. *The crucial point is that with the Ansatz (11) the supersymmetric transformations (4)-(6) still close.* In fact they reduce to

$$[Q^i, \delta\phi_1] = i\psi^i \qquad [Q^i, \delta\phi_2] = \gamma_5\psi^i \qquad [Q^i, \delta A_\mu] = \gamma_\mu\psi^i \tag{12}$$

$$\{Q_i, \psi^j\} = \delta^i_j \left(\frac{1}{2}\sigma.F + \gamma.D\Phi + \gamma_5[\phi^\dagger, \phi] + Y\right) \tag{13}$$

and

$$[Q^i, Y] = \gamma.D\psi_j + g[\Phi, \psi_j] \tag{14}$$

Thus, they consitute a realization of the original supersymmetric system (2). An interesting feature of this realization is that, since the three bosonic dummy fields X_a are *represented* by a single field Y the usual supersymmetric rule that the number of bosonic fermionic fields be equal is violated. Thus the realization is not a *faithful* one. But since the fields X_a decouple this has no physical consequences. In fact it is obvious that the Lagrangian (3) is invariant with respect to the supersymmetry transformations (12)-(14).

We now claim that a background configuration can be chosen so that
(a) it is invariant with respect to the supersymmetry (12)-(14) and
(b) The Action (3) reduces to the surface term.

To show this we choose as background configuration

$$F = F^* \qquad \not{D}\psi = 0 \qquad D^2\phi = g[\psi^\dagger, \psi] \quad \text{and} \quad Y = -g\gamma_5[\phi^\dagger, \phi] \tag{15}$$

These background conditions differ from the previous ones by the last equality. Note that setting Y equal to the ϕ-commutator is *not* the same as setting

setting $D = g[\phi^\dagger, \phi]$ because the latter condition violates supersymmetry but (15) does not. In fact it is easy to verify that on the surface (15) we have

$$[Q_i, D^2\phi - g[\phi^\dagger, \phi]] = 0 \quad \text{and} \quad [Q_i, Y + g\gamma_5[\phi^\dagger, \phi]] = 0 \tag{16}$$

Furthermore, on this surface one sees by inspection that the four-dimensional integral in (10) vanishes identically so the Action reduces to the surface term $\int \mathrm{d}\Omega(\phi, D\phi)$ as required. The surface term is the quantity computed in Finnell and Pouliot (1995), Dorey et al. (1996a), (1996b), (1996c), Ito and Sasakura (1996),(1997), Yung (1997), Aoyama et al. (1996), Harano and Sato (1997). Since the background configuration is now supersymmetric it follows at once that the surface term must be supersymmetric. The supersymmetry of the surface term was verified in [3] for $N = 1$ and $N = 2$ using the explicit solutions of (15) but we see that in the present situation the supersymmetry of the surface integral is automatic and is valid for all N.

Thus we see puzzle (b) is solved by the observation that although the background was not supersymmetric with respect to the original supersymmetry (4)-(6) it is supersymmetric with repect to the realization (12)-(14).

5 Comparison of Instanton Computation and Ansatz

To consider puzzle of the background fluctuations (c) we need to recall in a little more detail how the S-W Ansatz is verified: According to S-W the effective low energy Action is

$$\int \mathrm{d}^4x \mathrm{d}^2\theta F(A) W_\alpha W_\alpha \tag{17}$$

where $A(x)$ and $W(x)$ are the chiral scalar and real vector $N = 1$ sub-superfields respectively, and $F(A)$ has the functional form

$$F(A) = \frac{\mathrm{i}}{\pi} A^2 \ln(A^2) + \sum_N F_N(A) \quad \text{where} \quad F_N(A) = c_N \left(\frac{\Lambda}{A}\right)^{4N-2} \tag{18}$$

N denoting the contribution from the N-th instanton sector. The form of $F(A)$ is determined on general grounds and what the S-W Ansatz determines is the actual values of the numerical coefficients c_N. Because of the integration over Grassman variables the effective Lagrangian (17) describes truncated n-point functions only of the form

$$< \Phi^n >^T = < \phi^{n-4}\psi^4 >^T \quad < \phi^{n-3}\psi^2 F_{\mu\nu} >^T \quad \text{and} \quad < \phi^{n-2} F_{\mu\nu}^2 > \tag{19}$$

Expanding $A(x)$ in the form $A(x) = v + \hat{A}(x)$ one sees that these are simply

$$< \Phi^n >^T = \frac{\partial^n}{(\partial v)^n} F(v) \quad \text{or} \quad < \Phi^n >^T_N = \frac{\partial^n}{(\partial v)^n} \frac{c_N}{v^{4N-2}} \quad n \geq 2 \tag{20}$$

The verification of the S-W Ansatz consists of comparing (20) with the truncated n-point functions obtained directly from

$$< \Phi^n >= \int D(\Phi)\Phi^n e^{-S_N(\Phi)} \qquad n \geq 2 \tag{21}$$

The trick used to compute (21) is to restrict oneself to the background configurations (15) and their supersymmetric transforms. The advantage is that the volume integral in the Action then vanishes. Furthermore, from (15), the configurations $\Phi(\alpha)$ and the value of the surface term, $\sigma(\alpha)$ say, are determined by the standard AHDMN parameters α which describe the instantons and their zero-modes. Hence, if one assumes that the quantum fluctuations do not contribute on account of supersymmetric cancellation, the truncated part of (21) reduces to the ordinary integral

$$< \Phi^n >_N^T= \int d\mu_N(\alpha)\Phi(\alpha)^n e^{-\sigma(\alpha)} \qquad n \geq 2 \tag{22}$$

where $d\mu_N(\alpha)$ is the measure in the space of the instanton parameters. The measure $d\mu_N(\alpha)$ is known explicitly only for the $N = 1$ and $N = 2$ sectors and this is why the S-W Ansatz has been checked only for these two sectors. From the properties of the background fields and their supersymmetrc vaiations it is possible to show that (22) can be written in the form

$$< \Phi^n >_N^T= \frac{\partial^n}{(\partial\lambda)^n} \int d\mu_N(\alpha)e^{-\lambda\sigma(\alpha)} \quad \text{at} \quad \lambda = 0 \qquad n \geq 2 \tag{23}$$

It is the quantity (23) that has been computed and compared successfully with (20) in references Finnell and Pouliot (1995), Dorey et al. (1996a), (1996b), (1996c), Ito and Sasakura (1996),(1997), Yung (1997), Aoyama et al. (1996), Harano and Sato (1997). An interesting consequence of the above results is that, although they are valid only for $n \geq 2$, they imply that

$$F_N(v) = \int d\mu_N(\alpha)e^{-\sigma(\alpha)} \tag{24}$$

Mathematically, (24) shows that (for $v = 1$) the coefficients c_N in the asymptotic expansion Fuchsian function $F(v)$ are equal to the volume of the instanton parameter-space compactified with the exponential factor shown. Physically, it shows that in each instanton sector c_N is a kind of partition function for the background fields. Finally equation (24) encapsulates all the information concerning the n-point functions.

We turn now to the question of the quantum fluctuations. The fact that the n-point functions are restricted to the background fields and their supersymmetric variations, would seem to justify the assumption the bosonic and fermionic contributions, *now that the background fields are truly supersymmetric* and this would appear to resolve the puzzle (c) above. But a number of important details remain to be cleared up. First, it should be demonstrated explicitly that the bosonic and fermionic quantum fluctuations do indeed cancel. Second, the role of the supersymmetric zero-modes i.e. the instanton modes which are *not* lifted by the exponential term in (22)-(24), needs to be clarified. Finally, in view of the simplicity and importance of equation (24) there ought to be a more direct way of obtaining it. These are problems to which we hope to consider in the future.

Acknowledgement

The author would like to thank R. Flume, M. Magro and I. Sachs for valuable discussions.

References

Aoyama, H., Harano, T., Sato, M. and Wada, S. (1996): Phys. Lett. **B388**, 331.
Dorey, N., Khoze, V.V. and Mattis, M.P. (1996a): Phys. Lett. **B388**, 324;
Dorey, N., Khoze, V.V. and Mattis, M.P. (1996b): Phys. Rev. **D54**, 2921;
Dorey, N., Khoze, V.V. and Mattis, M.P. (1996c): Phys. Rev. **D54**, 7832.
Finnell, D. and Pouliot, P. (1995): Nucl. Phys. **B453** 225.
Harano, T. and Sato, M. (1997): Nucl. Phys. **B484**, 167.
Ito, K. and Sasakura, N. (1996): Phys. Lett. **B382**, 95;
Ito, K. and Sasakura, N. (1997): Nucl. Phys. **B484**, 141;
Yung, A. (1997): Nucl. Phys. **B485**, 38.
Seiberg, N. and Witten, E. (1994): Nucl. Phys. **B426**, 19
West, P. (1986): *Introduction to supersymmetry and Supergravity*, Cambridge University Press

Alternative Formulations of $N = 2$ Supesymmetric Gauge Theory in Harmonic Superspace

B.M. Zupnik

Joint Institute for Nuclear Research, Dubna, Russia

Abstract. Analytical harmonic superfields are the basic variables of a standard harmonic formalism of SYM_4^2-theory. We consider superfield actions for alternative formulations of this theory using the unconstrained harmonic prepotentials. The corresponding equations of motion are equivalent to the component field equations of SYM_4^2-theory. The analyticity condition appears only on-shell as the zero-curvature equation in the alternative formulations.

Harmonic superspace has been introduced for the consistent off-shell description of the supersymmetric theories with $N = 2, D = 4$ supersymmetry (Galperin et al. (1984)). Analytical prepotentials of the SYM_4^2-theory V^{++} live in the analytic harmonic superspace with a restricted number of spinor coordinates. The action of this theory is a nonlinear functional of the analytic prepotentials (Galperin et al. (1985), Zupnik (1987)).

We shall use the basic notions and notation of Ref.(Galperin et al. (1985)). Let us consider the harmonic-zero-curvature equation for the harmonic connections V^{++}, A^{--} with a dimension $d = 0$

$$\partial_A^{++} A^{--} - \partial_A^{--} V^{++} + [V^{++}, A^{--}] = 0 \tag{1}$$

This equation has the following perturbative solution (Zupnik (1987)):

$$A^{--}(V) = \sum_{n=1}^{\infty} (-1)^n \int du_1 \dots du_n \frac{V^{++}(z, u_1) \dots V^{++}(z, u_n)}{(u^+ u_1^+) \dots (u_n^+ u^+)} \tag{2}$$

where the harmonic distributions $(u_1^+ u_2^+)^{-1}$ (Galperin et al. (1984)) are used.

The nonlinear equation of motion of SYM_4^2 has the following form (Zupnik (1987)):

$$F^{++} = (D^+)^2 (\bar{D}^+)^2 A^{--}(V) = 0 \tag{3}$$

where D_α^+ and $\bar{D}_{\dot\alpha}^+$ are the flat spinor derivatives in the harmonic superspace.

We use the analytic basis of the superfield gauge theory in the harmonic superspace with the analytic gauge parameters Λ. This basis is natural for the description of the covariantly analytic superfields.

Note that the analyticity conditions of the standard harmonic formalism are kinematic (off-shell):

$$D_\alpha^+ V^{++} = \bar{D}_{\dot\alpha}^+ V^{++} = 0 \tag{4}$$

One should use also the conventional constraint for the spinor connection $A^-_a = (A^-_\alpha, \bar{A}^-_{\dot\alpha})$ (Zupnik (1987))

$$[\nabla^{--}, \nabla^+_a] = \nabla^-_a = D^-_a + A^-_a = D^-_a - D^+_a A^{--} \tag{5}$$

Now we shall discuss the alternative (dual) formulation of the SYM^2_4-theory . Let us consider the following representation of the harmonic connection (Zupnik (1996)):

$$A^{--} = \hat{A}^{--} = D^+_\alpha A^{\alpha(-3)} + \bar{D}^+_{\dot\alpha} \bar{A}^{\dot\alpha(-3)} = D^+_a A^{a(-3)} \tag{6}$$

where $A^{a(-3)}(z,u)$ are the unconstrained 4-spinor prepotentials $a = (\alpha, \dot\alpha)$. This representation solves explicitly Eq(3) but does not quarantees the conservation of analyticity.

The equation (1) in the new *A-frame* is treated as an integrable equation for the function $V^{++}(A^{--})$ corresponding to a choice of the independent variables (6). Consider a dual in the $U(1)$-charge solution

$$\hat{V}^{++}(A^{a(-3)}) = \sum_{n=1}^{\infty}(-1)^n \int du_1 \ldots du_n \frac{\hat{A}^{--}(z,u_1)\ldots\hat{A}^{--}(z,u_n)}{(u^-u^-_1)\ldots(u^-_n u^-)} \tag{7}$$

where the new harmonic distributions $(u^-_1 u^-_2)^{-k}$ (Zupnik (1996)) are introduced. These distributions satisfy the following relations:

$$\partial^{--}_1 \frac{1}{(u^-_1 u^-_2)^k} = \frac{1}{(k-1)!}(\partial^{++}_1)^{k-1}\delta^{(-k,k)}(u_1,u_2) \tag{8}$$

which are completely analogous to the relations for the standard harmonic distributions with opposite $U(1)$-charges (Galperin et al. (1984)).

Let us write the alternative superfield action of the SYM^2_4-theory in the A-frame:

$$S(A^{a(-3)}) = \int d^{12}z \sum_{n=2}^{\infty} \frac{(-1)^n}{n} \int du_1 \ldots du_n \frac{\mathrm{Tr}[\hat{A}^{--}(z,u_1)\ldots\hat{A}^{--}(z,u_n)]}{(u^-u^-_1)\ldots(u^-_n u^-)} \tag{9}$$

where $\hat{A}^{--} = D^+_a A^{a(-3)}$.

Consider an arbitrary variation of this functional

$$\delta S = \int d^{12}z\, du \mathrm{Tr}\left[\delta\hat{A}^{--}\, \hat{V}^{++}\right] = \int d^{12}z\, du \mathrm{Tr}\left[\delta A^{a(-3)}\, D^+_a \hat{V}^{++}\right] \tag{10}$$

Thus, the action (9) generates a dynamical analyticity equation of SYM^2_4 (Zupnik (1996)):

$$D^+_a \hat{V}^{++}(A^{a(-3)}) = 0 \tag{11}$$

Thus, the nonlinear dynamical equation (3) and the kinematic analyticity of the standard harmonic formalism correspond to the linear constraint on the dual harmonic-superfield variable A^{--} and the dynamical zero-curvature equation. Note, that the chirality properties and the full set of Bianchi identities for the SYM_4^2 tensors $W = (\bar{D}^+)^2 A^{--}$ and $\overline{W} = (D^+)^2 A^{--}$ are the consequence of the dynamical analyticity and arise only on-shell in the dual formulation.

The A-frame prepotential possesses the following gauge transformations (Zupnik (1996)):

$$\delta A^{a(-3)} = R^{a(-3)} \Lambda + [\Lambda, A^{a(-3)}] + D_b^+ \Lambda^{ab(-4)} \tag{12}$$

where a general symmetrical spinor $\Lambda^{ab(-4)}$ and an analytic scalar Λ are the Lie-algebra valued superfield gauge parameters and $R^{a(-3)}$ is some spinor differential operator. The spinor derivative of $\delta A^{a(-3)}$ produces the standard gauge transformation of the harmonic connection

$$\delta \hat{A}^{--} = \partial^{--} \Lambda + \left[\hat{A}^{--}, \Lambda\right] = \nabla^{--} \Lambda \tag{13}$$

The A-frame action is invariant under the gauge transformations of new prepotentials

$$\delta_\Lambda S = \int \mathrm{d}^{12} z \; du \mathrm{Tr}[\nabla^{--} \Lambda \hat{V}^{++}] =$$
$$- \int \mathrm{d}^{12} z \; du \mathrm{Tr}\left[\Lambda \nabla^{--} \hat{V}^{++}\right] = - \int \mathrm{d}^{12} z \; du \mathrm{Tr}\left[\Lambda \partial^{++} D_a^+ \, A^{a(-3)}\right] = 0 \tag{14}$$

It should be stressed that prepotential $A^{a(-3)}$ of our harmonic formalism contains an infinite number of harmonic auxiliary fields , the physical component SYM_4^2 fields and pure gauge degrees of freedom in contrast with the analytic prepotential of the standard harmonic approach which has physical and pure gauge components only (Galperin et al. (1984)). The physical sector of the superfield $D_a^+ \, A^{a(-3)}$ contains the vector $A_{\alpha\dot{\beta}}$, the spinor ψ_i^a, the independent field-strengths $F^{\alpha\beta}$, $\bar{F}^{\dot{\alpha}\dot{\beta}}$, $f^{\alpha\dot{\beta}}$, $\bar{f}^{\dot{\alpha}\beta}$ and the scalars Φ, $\bar{\Phi}$. The dynamical analyticity condition (11) is equivalent to the component SYM_4^2 equations of motion. All auxiliary fields vanish on-shell and the standard first-order component SYM_4^2 equations for $F, \bar{F}, f, \bar{f}, A$ and ψ arise, too. It is evident that the different representations of SYM_4^2 theory are equivalent on-shell and have identical component solutions for the physical fields.

One can consider the intermediate version of the harmonic-superfield action of SYM_4^2 which interpolate between the action with the analytic prepotential and the dual action (9). Introduce the independent nonanalytic harmonic connection $\widetilde{V}^{++}$

$$\delta \widetilde{V}^{++} = \partial^{++} \Lambda + \left[\widetilde{V}^{++}, \Lambda\right] \tag{15}$$

The intermediate version of the SYM_4^2-action has the following form:

$$S(\widetilde{V}^{++}, A^{a(-3)}) = -\int \mathrm{d}^{12}z\, \mathrm{d}u \mathrm{Tr}\left[D^+_a\, A^{a(-3)}\widetilde{V}^{++}\right] +$$

$$+\sum_{n=2}^{\infty} \frac{(-1)^n}{n} \int \mathrm{d}^{12}z\, \mathrm{d}u_1 \ldots \mathrm{d}u_n \frac{\mathrm{Tr}\left[\widetilde{V}^{++}(z,u_1)\ldots\widetilde{V}^{++}(z,u_n)\right]}{(u_1^+u_2^+)\ldots(u_n^+u_1^+)} \quad (16)$$

where the dual superfield $A^{a(-3)}$ is treated as a Lagrange multiplier. This action is invariant under the analytical gauge transformations of $\widetilde{V}^{++}$ and $\hat{A}^{--}$ with the common analytical parameters.

The corresponding equations of motion are

$$\widetilde{A}^{--}(\widetilde{V}^{++}) = D^+_a\, A^{a(-3)} \quad (17)$$

where one should use the analogue of Eq(2) in the left-hand side and

$$D^+_a\widetilde{V}^{++} = 0 \quad (18)$$

One can discuss also the 2nd alternative form of the interpolating SYM_4^2-action. Let an unconstrained harmonic superfield $\mathcal{A}^{--}(z,u)$ and the analytic harmonic superfield V^{++} are the independent connections with the standard transformation laws. We suppose that Eq(1) is not valid off-shell in this formulation. The new interpolating action has the following form:

$$S(\mathcal{A}^{--}, V^{++}) = -\int \mathrm{d}^{12}z\, \mathrm{d}u \mathrm{Tr}\,[\mathcal{A}^{--}V^{++}] +$$

$$+\sum_{n=2}^{\infty} \frac{(-1)^n}{n} \int \mathrm{d}^{12}z\, \mathrm{d}u_1 \ldots \mathrm{d}u_n \frac{\mathrm{Tr}\,[\mathcal{A}^{--}(z,u_1)\ldots\mathcal{A}^{--}(z,u_n)]}{(u_1^-u_2^-)\ldots(u_n^-u_1^-)} \quad (19)$$

where $(u_1^-u_2^-)^{-1}$ is the harmonic distribution (8).

The corresponding harmonic-superfield equations of motion are equivalent to Eqs(1,3) of the standard harmonic formalism. It is easy to obtain the superfield Feynman rules for this action by the analogy with Ref.(Galperin et al. (1985)).

We hope that the dual formulations of SYM_4^2 in the harmonic superspace will be useful for the analysis of the remarkable quantum properties of this theory.

The author is grateful to A.S.Galperin and E.A.Ivanov for stimulating discussions. This work is partially supported by INTAS-grant 93-127, INTAS-grant 94-2317 and the grant RFBR 96-02-17634.

References

Galperin, A., Ivanov, E., Ogievetsky, V., and Sokatchev, E. (1984): Pis'ma ZhETF. **40**, 155

Galperin, A., Ivanov, E., Ogievetsky, V. and Sokatchev, E. (1985): Class. Quant. Grav. **2**, 601

Zupnik, B.M. (1987): Sov. J. Nucl. Phys. **44**, 512; Phys. Lett. **B183**, 175

Zupnik, B.M. (1996): Phys.Lett. **B375**, 170

Lie-Algebraic Characterization of 2D (Super-)Integrable Models

F. Toppan

Dipartimento di Fisica Università di Padova
Via Marzolo 8, I-35131 Padova

Abstract. It is pointed out that affine Lie algebras appear to be the natural mathematical structure underlying the notion of integrability for two-dimensional systems. Their role in the construction and classification of 2D integrable systems is discussed. The supersymmetric case will be particularly enphasized. The fundamental examples will be outlined.

The integrable hierarchies of differential equations in 1+1 dimensions have been widely studied in the last several years both in the physical and in the mathematical literature. In the time there has been an ever-growing evidence that their relevance should not be confined in the realm of pure mathematics, but rather their beautiful mathematical structures show themselves naturally when investigating physical problems.

It deserves being mentioned that 2-dimensional integrable systems appear in two different (neverthless related) ways: on one hand we have the non-relativistic integrable equations in 1 space and 1 time dimension. The basic example for these systems is the celebrated KdV equation and the systems of this kind will be referred as of KdV-type.

On the other hand the second big class of integrable systems is provided by the relativistic ones in two dimensions, that is the so-called Toda field theories, whose fundamental example is the even more celebrated Liouville equation.

As discussed later, both such classes are obtainable from one and the same mathematical construction, presenting the affine Lie-algebras as fundamental ingredient.

Before going ahead let us just mention some physical applications of both classes of theories, which motivated the interest of physicists in looking at them. At first the Liouville theory appeared in the Polyakov's geometrical attempt in quantizing the bosonic string off-criticality (Polyakov (1991)). More recently integrable systems of KdV-type were found associated to physical problems when it was realized they furnished the partition functions of the two-dimensional discretized gravity in the matrix-models approach (see Di Francescoet al. (1995) for a review). Quite new and rather unexpectedly integrable hierarchies appear even associated to 4-dimensional field theories as a sort of an underlying integrable structure of $N = 2$ SuperYang-Mills theories in the Seiberg-Witten framework Seiberg and Witten (1994).

Another way of associating Liouville theory to strings is a rather different one. It is based on the so-called geometrical approach, greatly developed by the Kharkov's group, to (classical) strings. In this approach the solution for the dynamical problem of a string moving on a $2+1$ flat-target is reduced to the solution of the Liouville equation (Omnés (1979)). It is worth to mention this point here because it is related to Prof. D.V. Volkov's last work. In fact in Bandos et al. (1996) it was constructed the geometrical approach for a Green-Schwarz superstring moving in a flat supersymmetric target of $2+1$ dimensions. Surprisingly, it was found that the equations of motions can now be reduced to a supersymmetrized version of the Liouville equation, but not the standard one. This result was the basic motivation for understanding the situation from purely Lie-algebraic data, work done in collaboration with D. Sorokin Sorokin and Toppan (1996a), (1996b) which will be reported later.

Let us come back now to the mathematical structure of integrable systems. The main difference between KdV-type hierarchies and Toda-type hierarchies is due to the fact that the former ones are recovered from a single copy of affine algebras, while the latter from two separated copies corresponding to the chiral and antichiral sectors respectively. Let us denote as J the currents valued on a given simple Lie algebra $\mathcal{G}$ and generating an affine Lie algebra $\hat{\mathcal{G}}$.

WZNZ models have dynamics expressed by group-manifold elements G, while the currents are defined as $J = \partial G \cdot G^{-1}$ and $\overline{J} = G^{-1}\overline{\partial} G$ (the antichiral one) respectively and satisfy free-equations. Neverthless in both cases, non-trivial equations are obtained by imposing constraints on the affine currents J (and $\overline{J}$ in the second case). Such constraints arise in two different ways, either as hamiltonian constraint, or as coset constraints. The first case correspond to the Dirac's theory of constraints, while the latter simply means that the dynamical quantities should have vanishing Poisson brackets with respect to some Kac-Moody subalgebra.

In the basic example of sl(2) (or A_1 algebra) these two constructions (denoted as a) and b)) lead respectively to (for non-relativistic, type-1 systems, and relativistic, type-2 systems) to:
1a) KdV equation;
1b) NLS equation;
2a) Liouville equation;
2b) Witten's $2D$'s black hole.

The arising of integrable hierarchies from constrained affine Lie algebras is particularly important because it provides the tools towards a classification of all hierarchies.

Let us now discuss the case of supersymmetric extensions of bosonic hierarchies. The interest in such extensions should not be thought being limited to the super-physics program (and more specifically to superstrings), instead a wider range of applications arises. For instance it is well-known that new bosonic hierarchies can arise when only the bosonic (or more generally the non-supersymmetric sector) of theories based on supergroups and superalgebras is considered.

One of the main sources of interest in investigating supersymmetries morever coincide with large N-supersymmetric extensions, because it correspond to a sort of "unification" or "grandunification" of known hierarchies. It happens indeed that seemingly unrelated bosonic or lower supersymmetric ($N = 1, 2$) hierarchies turn out to be a different manifestation of a single "unifying" large-N supersymmetric hierarchy.

Until recently it was commonly believed that supersymmetric hierarchies could be produced only from affinization of a particular kind of superalgebras. It must be explained that superalgebras, just as ordinary algebras, can be expressed through their Dynkin's diagrams which refer to their simple roots. However, since superalgebras admit two kinds of generators, even and odd, the simple roots could be either fermionic or bosonic. It was thought that only the special class of superalgebras admitting purely fermionic simple roots could provide supersymmetric integrable models. The reason for that was based on an argument related to the Drinfeld-Sokolov approach to integrable systems (based on simple roots). Since the basic derivative operator for supersymmetric theory is fermionic it was thought that the only consistent way to construct a fermionic matrix-Lax pair implied the use of the fermonic simple roots. It appeared at first as a surprise in Brunelli and Das (1992) that the supersymmetric version of the Non-Linear-Schrödinger equation admits a Lax pair based on sl(2) instead of osp(1|2) as expected. In Toppan (1995) it was shown that such supersymmetric equation admits a natural interpretation in terms of a coset construction based on the supersymmetric affinization of $\hat{sl(2)}$ (the tower of hamiltonian densities has vanishing Poisson brackets w.r.t. the supersymmetric $\hat{u}(1)$ subalgebra).

At this stage it was clear that, at least for coset construction, it was perfectly acceptable to recover supersymmetric integrable models from any bosonic or super Lie algebra.

More recently, in collaboration with D. Sorokin (Sorokin and Toppan (1996a), (1996b), we have shown how it is possible to bypass the requirement of purely fermionic Dynkin's diagram-type supealgebras even in the case of hamiltonian constraint (the case a) in the previous classification). For lack of space I cannot describe here our method and I refer to the cited papers for details. Let me just point out that the resulting hierarchies (in the Toda-type construction) are superconformally invariant, with the supersymmetry realized non-linearly and spontaneously broken. Such kind of systems give us automatically a non-standard Sugawara realization of the superconformal stress-energy tensor which involves fermionic $b-c$ systems of weight $(-\frac{1}{2}, \frac{3}{2})$. The simplest example of this kind is obtained in terms of the sl(2) algebra. It corresponds to the non-standard superLiouville equation found in Bandos et al. (1996) expressing the dynamics of a Green-Schwarz superstring on a $2+1$ Minkowski flat target.

In the last part of my talk I wish to introduce some new results, found in collaboration with E. Ivanov and S. Krivonos, concerning the large-N supersymmetric extension of integrable hierarchies (and their relation to affine algebras) (Ivanov et al. in preparation). A bosonic algebra such as $\mathrm{sl}(2)\oplus\mathrm{u}(1)$ turns out to be the basic structure underlying the $N=4$ KdV hierarchy.

Indeed the following features hold. Algebras admitting quaternionic structure have been classified (Sevrin et al. (1988)). They are thought to be related to $N=4$ theories. Indeed the simplest non-trivial case ($\mathrm{sl}(2)\oplus\mathrm{u}(1)$, the abelian $\mathrm{u}(1)^{\oplus 4}$ should be ruled out for our purposes) is such that their supersymmetric affinization admits a global $N=4$ structure (realized by non-linear transformations). Moreover an infinite number of $N=4$ hamiltonians in involution associated to the above superaffine algebra as Poisson bracket structure, can be found. The resulting integrable hierarchy can be denoted as $N=4$ NLS-mKdV hierarchy because different constraints compatible with the equations of motion lead respectively to the $N=2$ mKdV and to the $N=2$ NLS equation.

The $N=4$ hierarchy is not only supersymmetric but even $N=4$ superconformal because the Sugawara construction applied to the superaffine $\widehat{\mathrm{sl}(2)\oplus\mathrm{u}(1)}$ leads to a closed algebraic structure, where the $N=4$ transformations are linearized, which corresponds to the so-called minimal $N=4$ version of the SuperConformal Algebra (expressed in terms of three bosonic spin 1 $N=2$ superfields, one bosonic, one chiral and one antichiral). With respect to the generators of the $N=4$ SCA the equations of motions are closed and coincide with the equations of the $N=4$ KdV hierarchy (Delduc and Ivanov (1993)).

Therefore even large-N supersymmetric hierarchies find an interpretation in terms of (super-)affine Lie algebras. This result is particularly important because it paves the way towards an understanding and a Lie-algebraic classification of all $N=4$ hierarchies.

Acknowledgements

I wish to express my gratitude to the organizers of the conference in memory of Prof. D.V. Volkov for their kind invitation to give a talk.

References

Bandos, I., Sorokin, D. and Volkov, D. (1996): Phys. Lett. **B 372**, 77.

Brunelli, J.C. and Das, A. (1992): Jou. Math Phys. **33**, 63.

Delduc, F. and Ivanov, E.(1993): Phys. Lett. **B 309**, 312.

Di Francesco, P., Ginsparg, P. and Zinn-Justin, J. (1995): Phys. Reports **254**, 1.

Ivanov, E., Krivonos, S. anf Toppan, F. in preparation: "N=4 NLS-mKdV Hierarchies".

Omnés, R. (1979): Nucl. Phys. **B 149**, 269.

Polyakov, A.M. (1991): Phys. Lett. **B 103**, 210.
Seiberg, N. and Witten, E. (1994): Nucl. Phys. **B 426**, 19.
Sevrin, A., Troost, W. and Van Proeyen, A. (1988): Phys. Lett. **B 208**, 447.
Sorokin, D. and Toppan, F. (1996a): Nucl. Phys. **B 480**, 457,
Sorokin, D. and Toppan, F. (1996b): Preprint hep-th/9610038. To appear in Lett. Math. Phys.
Toppan, F. (1995): Int. Jou. Math. Phys. **A 10**, 895.

Universal Hidden Supersymmetry in Classical Mechanics and Its Local Extension

E. Gozzi

Dept. of Theor. Physics, University of Trieste
Strada Costiera 11, Miramare-Grignano 34014 Trieste, Italy

Abstract. We review here a path-integral approach to *classical* mechanics and explore the geometrical meaning of this construction. In particular we bring to light a universal hidden BRS invariance and its geometrical relevance for the Cartan calculus on symplectic manifolds. Together with this BRS invariance we also show the presence of a universal hidden genuine non-relativistic *supersymmetry*. In an attempt to understand its geometry we make this susy local following the analogous construction done for the supersymmetric quantum mechanics of Witten. [1]

1 Introduction

The discovery of supersymmetry (Gol'fand and Lichtman (1971); Gervais and Sakita (1971); Volkov and Akulov (1972) Wess J. and Zumino B. (1974)), first at the pure algebraic level then in two dimensions and finally in four dimensions both in its non-linear and linear versions, has been one of the most fascinating discovery of the last 30 years. Besides softening the ultraviolet behaviour of theories of fermions coupled with bosons, supersymmetry offered a way out, in its unbroken phase, to the cosmological constant problem.

Unfortunately the spectrum of particles known to-day indicates that supersymmetry must be broken. This brings back the cosmological constant problem unless the breaking of this *special* symmetry manages to avoid the usual theorems on symmetry breaking. In order to explore that possibility it is crucial to understand supersymmetry at a deeper level.

Supersymmetry, soon after its discovery, was made local (Volkov and Soroka (1974); Freedman et al. (1976); Deser and Zumino (1976)) producing theories of gravity that raised the hopes of not being afflicted by the problem of the non-removable infinities which plagued Einstein gravity. Despite the huge amount of work done on supergravity and later on superstrings, still the goal of a finite or renormalizable theory of gravity seems not to be near.

All of the above mentioned issues seem to indicate that supersymmetry is very important but its roots should be understood better. It is our opinion that these roots are deeply *geometrical*.

[1] This work is dedicated to the memory of Dmitrij Vasil'evich Volkov and to all those who are struggling to have their work recognized. I wish to warmly thank the organizers of the Volkov Memorial for the heroic effort of organizing it during a freezing winter in Ukraine.

In this respect there has been a 1-D model, proposed by Witten in 1981 (Witten (1981)), which has been a toy model for exploring several issues connected both with supersymmetry breaking and with geometry (Witten (1982)). The model has the limit of being just a model reproducing no aspect whatsoever of the natural world, except some phenomenological stochastic dynamics (Parisi and Sourlas (1982); Cecotti and Girardello (1983)). Quite independently of that model, the present author, together with M.Reuter and W.D.Thacker, has tried (Gozzi (1986); (1988); Gozzi E., Reuter M., Thacker W.D. (1989); Gozzi and Reuter (1990)) to give a path-integral representation to Classical Mechanics (CM) and the outcome has been a superlagrangian which has both a physical and geometrical meaning and *it is not just a toy model.* So we feel that some geometrical issues regarding supersymmetry could be tackled better using our Lagrangian. In fact it turns out that the 1-D superhamiltonian produced by this path-integral is nothing else than the *Lie-derivative* of the Hamiltonian flow and it has both a *universal* BRS-AntiBRS-like invariances and also a *universal* genuine non-relativistic supersymmetry (Gozzi and Reuter (1989)). The BRS-AntiBRS invariances have been interpreted geometrically as being the exterior derivative and coderivative on symplectic spaces, and all the standard Cartan calculus can be reproduced via our formalism. As we said before there is also a genuine supersymmetry whose physical meaning has been studied in (Gozzi and Reuter (1989)) as being related to the ergodicity-non-ergodicity of the Hamiltonian system under consideration. That supersymmetry (Gozzi and Reuter (1989)) had also other nice features. For example, due to the non-positivity of the Hilbert space, it could be broken without lifting the vacuum from zero and this can be very important for the cosmological constant problem. Being this susy universal it could be extracted not only from a point particle dynamics but also from any classical field theory once they are formulated via a Lie-brackets kind of formalism. Of course it would not be a relativistic supersymmetry even for relativistic field theories because it gives a privileged role to the forth component of space-time. It might be possible to made it relativistic if one uses the De Donder-Weyl (Kastrup (1983)) formulation of field-theory, but this has still to be explored in details (Gozzi, in progress). Of course we would obtain in that case a susy at the Lie-bracket level in which each boson and fermion present in the basic theory, whose Hamiltonian will act as a superpotential for our superhamiltonian, will get a partner and produce in that manner a huge pletora of fields which will live in a sort of shadow world degenerate with our world. The meaning of these partners would be purely geometrical as it is clear from (Gozzi (1986); (1988); Gozzi E., Reuter M., Thacker W.D. (1989); Gozzi and Reuter (1990)). It would be interesting if these shadow world, made only of auxiliary geometrical fields, had to be included in the counting needed for the cosmological constant (Gozzi, in progress).

To better understand the geometry behind this supersymmetry, in this brief note we take a first step in that direction by gauging it in the simple case of the point particle dynamics.

2 Path-Integral for Classical Mechanics

This section is meant to be a very quick and incomplete review of the path-integral approach to CM. For more details the reader should study (Gozzi (1986); (1988); Gozzi E., Reuter M., Thacker W.D. (1989); Gozzi and Reuter (1990)). The "propagator" which gives the *classical* probability for a particle to be at the point ϕ_2 at time t_2, if it it was at the point ϕ_1 at time t_1, is just a delta function $P(\phi_2, t_2|\phi_1, t_1) = \delta^{2n}(\phi_2 - \Phi_{\rm cl}(t_2, \phi_1))$ where $\Phi_{\rm cl}(t, \phi_0)$ is a solution of Hamilton's equation $\dot{\phi}^a(t) = \omega^{ab}\partial_b H(\phi(t))$ subject to the initial conditions $\phi^a(t_1) = \phi_1^a$. Here H is the conventional Hamiltonian of a dynamical system defined on some phase-space $\mathcal{M}_{2n}$ with local coordinates $\phi^a, a = 1 \cdots 2n$ and constant symplectic structure $\omega = \frac{1}{2}\omega_{ab}\mathrm{d}\phi^a \wedge \mathrm{d}\phi^b$. Slicing in infinitesimal parts the time interval $[t_2 - t_1]$ above and doing standard manipulations (Gozzi (1986); (1988); Gozzi E., Reuter M., Thacker W.D. (1989); Gozzi and Reuter (1990)), which for brevity we do not reproduce here, it is possible to give a path-integral representation to the transition probability above:

$$P(\phi_2, t_2|\phi_1, t_1) = \int_{\phi_1}^{\phi_2} \mathcal{D}\phi\, \mathcal{D}\lambda\, \mathcal{D}c\, \mathcal{D}\bar{c}\, \exp \mathrm{i} \int_{t_1}^{t_2} \widetilde{\mathcal{L}} \tag{1}$$

where $\widetilde{\mathcal{L}} \equiv \lambda_a[\dot{\phi}^a - \omega^{ab}\partial_b H(\phi)] + \mathrm{i}\bar{c}_a(\delta_b^a\partial_t - \partial_b[\omega^{ac}\partial_c H(\phi)])c^b$. The λ_a are c-number auxiliary variables while the c^a, $\bar{c}_a$ are Grassmannian auxiliary variables, so the new space is an enlargement of the phase-space from $2n$ dimensions to $8n$. The standard equations of motion for ϕ^a can be obtained from the variation with respect to λ_a of the above Lagrangian. They can also be obtained by the introduction of the the associated Hamiltonian $\widetilde{\mathcal{L}} = \lambda_a\dot{\phi}^a + \mathrm{i}\bar{c}_a\dot{c}^b - \widetilde{\mathcal{H}}$, with $\widetilde{\mathcal{H}}$ given by $\widetilde{\mathcal{H}} = \lambda_a h^a + \mathrm{i}\bar{c}_a\partial_b h^a c^b$ where $h^a(\phi) \equiv \omega^{ab}\partial_b H(\phi)$ are the components of what is called the the Hamiltonian vector field (Abraham and Marsden (1978)). To get the eqs. of motion via $\widetilde{\mathcal{H}}$ one needs also an extended Poisson bracket structure (*epb*) in the above mentioned enlarged space. It is easy to see that they are given by $\{\phi^a, \lambda_b\}_{\rm epb} = \delta_b^a$, $\{c^a, \bar{c}_b\}_{\rm epb} = -\mathrm{i}\delta_b^a$, *all others* $= 0$ Note that these are different from the normal Poisson brackets on $\mathcal{M}$ which were $\{\phi^a, \phi^b\}_{pb} = \omega^{ab}$. Via the extended Poisson brackets above we obtain from $\widetilde{\mathcal{H}}$ the same equations of motion as those which one would obtain (at least for ϕ^a) from H via the normal Poisson brackets: $\{\phi^a, H\}_{pb} = \{\phi^a, \widetilde{\mathcal{H}}\}_{\rm epb}$. Regarding the c^a one can easily see (Gozzi (1986); (1988); Gozzi E., Reuter M., Thacker W.D. (1989); Gozzi and Reuter (1990)) that they can be identified with the forms $c^a = \mathrm{d}\phi^a$, so the coordinates (ϕ^a, c^a) can be thought as labelling the cotangent bundle

$T^*\mathcal{M}$ to phase-space. It is also easy (Gozzi (1986); (1988); Gozzi E., Reuter M., Thacker W.D. (1989); Gozzi and Reuter (1990)), via the operatorial formulation of CM, to realize that the $\lambda_a, \bar{c}_a$ fields are instead a basis of the tangent bundle to the previous space. So the $8n$ variables $(\phi^a, c^a, \lambda_a . \bar{c}_a)$ are a set of coordinates for the tangent bundle to the cotangent bundle to phase-space $T(T^*\mathcal{M})$. This gives a complete *geometrical* description of all the various auxiliary variables which we had to introduce and makes this *not just a model* but something more fundamental. It is also possible (Gozzi (1986); (1988); Gozzi E., Reuter M., Thacker W.D. (1989); Gozzi and Reuter (1990)) to make a correspondence between higher forms and polynomials in c^a and between antisymmetric multivector fieldsAbraham and Marsden (1978)) and polynomials in $\bar{c}_a$. We will indicate this correspondence via a $\widehat{(\cdot)}$ symbol: $F^{(p)} = \frac{1}{p!} F_{a_1 \cdot a_p} d\phi^{a_1} \wedge \cdots \wedge d\phi^{a_p} \Longrightarrow \widehat{F}^{(p)} = \frac{1}{p!} F_{a_1 \cdots a_p} c^{a_1} \cdots \wedge c^{a_p}$, $v^{(p)} = \frac{1}{p!} V^{a_1 \cdots a_p} \partial_{a_1} \wedge \cdots \wedge \partial_{a_p} \Longrightarrow \widehat{V}^{(p)} = \frac{1}{p!} \bar{c}_{a_1} \cdots \bar{c}_{a_p}$. Before proceeding further we should also point out that the $\widetilde{\mathcal{H}}$ presents some universal invariance whose charges are the following (Gozzi (1986); (1988); Gozzi E., Reuter M., Thacker W.D. (1989); Gozzi and Reuter (1990): $Q \equiv ic^a\lambda_a$, $\bar{Q} \equiv i\bar{c}_a \omega^{ab} \lambda_b$, $Q_g \equiv c^a \bar{c}_a$, $K \equiv \frac{1}{2} \omega_{ab} c^a c^b$, $\bar{K} \equiv \frac{1}{2} \omega^{ab} \bar{c}_a \bar{c}_b$. Using the correspondence indicated above it is then possible to rewrite all the normal operations of the Cartan calculusAbraham and Marsden (1978), like doing an exterior derivative on forms dF , or doing an interior product between a vector field and a form $i_v F$, or building the Lie-derivative of a vector field l_h , by just using polynomials in c and $\bar{c}$ together with the extended Poisson brackets structure and the charges built above. These rules, which we called $\{\cdot,\cdot\}_{\text{epb}}$-rules, are summarized here: $dF^{(p)} \Longrightarrow i\{Q, \widehat{F}\}_{\text{epb}}$, $i_v F^{(p)} \Longrightarrow i\{\widehat{v}, \widehat{F}^{(p)}\}_{\text{epb}}$; $l_h F = di_h F + i_h dF \Longrightarrow -\{\widetilde{\mathcal{H}}, \widehat{F}\}_{\text{epb}}$, $pF^{(p)} \Longrightarrow i\{Q_g, \widehat{F}\}_{\text{epb}}$, $\omega(v, \cdot) \equiv v^\flat \Longrightarrow i\{\bar{K}, \widehat{V}\}_{\text{epb}}$, $(df)^\sharp \Longrightarrow i\{\bar{Q}, f\}_{\text{epb}}$ where the last three operations indicated above are, respectively, multiplying a form $F^{(p)}$ by its degree p, mapping a vector field V into its associated one form $V^\flat$ via the symplectic form, and building the associated Hamiltonian vector field $(df)^\sharp$ out of a function f. One sees from above that the various abstract derivations of the Cartan calculus are all implemented by some charges acting via the epb-brackets. From the third relation above one also can notice that the Lie-derivative of the Hamiltonian vector field of time evolution becomes nothing else than our $\widetilde{\mathcal{H}}$, thus confirming that the weight-function of our classical path-integral, generated by just a simple Dirac delta, is the right *geometrical* object associated to the time-evolution.

3 Universal Supersymmetry

The universal charges indicated in the previous section,Q,$\bar{Q}$,Q_g,K,$\bar{K}$, have all a geometrical meaning (Gozzi (1986); (1988); Gozzi E., Reuter M., Thacker W.D. (1989); Gozzi and Reuter (1990)) as it is clear from the previous section.

In particular Q plays the role of the exterior derivative while $\bar{Q}$ is the corresponding operation on vector fields. The reason they are conserved is because the exterior derivative anticommutes with the Lie-derivative. If we interpret $\widetilde{\mathcal{L}}$ as a 1-D field theory, then Q,$\bar{Q}$ are also analog of BRS-antiBRS chargesBecchi (1976) because their graded *epb*-brackets are: $\{Q,\bar{Q}\}_{\rm epb}=0$,$\{Q,Q\}_{\rm epb}=0$, $\{\bar{Q},\bar{Q}\}_{\rm epb}=0$. Besides these BRS-charges, there are also other charges conserved under $\widetilde{\mathcal{H}}$ which make up, once combined with the Q and $\bar{Q}$, a genuine supersymmetry. They are (Gozzi and Reuter (1989)) $N\equiv c^a\partial_a H$, $\bar{N}\equiv\bar{c}_a\omega^{ab}\partial_b H$. The brackets among themselves and with Q and $\bar{Q}$ are all zero except the following ones $\{Q,\bar{N}\}_{\rm epb}=\widetilde{\mathcal{H}}$, $\{{\rm i}\bar{Q},{\rm i}N\}_{\rm epb}=\widetilde{\mathcal{H}}$. This is a N=4 supersymmetry. In fact these charges can be combined in the following four independent ones: $Q_{(1)}\equiv\frac{{\rm i}(Q-\bar{N})}{\sqrt{2}}$, $Q_{(2)}\equiv\frac{(Q+\bar{N})}{\sqrt{2}}$, $Q_{(3)}\equiv\frac{{\rm i}(\bar{Q}+N)}{\sqrt{2}}$, $Q_{(4)}\equiv\frac{(\bar{Q}-N)}{\sqrt{2}}$, all of which give $\{Q_{(i)},Q_{(i)}\}_{\rm epb}=\widetilde{\mathcal{H}}$ with $(i)=(1)\cdots(4)$. This *universal* supersymmetry of CM has a nice *physical* interpretation deeply related to the fact that we are describing a *classical* system. It basically says (Gozzi and Reuter (1989)) that the spontaneous breaking of the invariance generated by the linear combinations of some of the $Q_{(i)}$ above occurs for non-ergodic systems while ergodic systems have that susy (at constant energy) always unbroken.

We would like here to explore instead the *geometrical* meaning of the susy charges . The reason is that all charges we found previously had a nice geometrical interpretation, so also the $Q_{(i)}$ must have one. To better understand that, a first question we asked ourselves is why God, in creating such a fundamental object as the Lie-derivative $\widetilde{\mathcal{H}}$, did not think of making the Susy local. To a mathematician this way of studying the geometrical meaning of a charge may sound strange, but not to a physicist. So the first thing we decided to do is to go from the Lie-derivative $\widetilde{\mathcal{H}}$ (or its lagrangian $\widetilde{\mathcal{L}}$), to a new one $\widetilde{\mathcal{H}}_g$ (or $\widetilde{\mathcal{L}}_g$) which has that susy local. First (using the standard Noether technique (Volkov and Soroka (1974); Freedman et al. (1976); Deser and Zumino (1976)) we do four *local* variations of $\widetilde{\mathcal{L}}$ generated by each of the four charges, Q,$\bar{Q}$,N,$\bar{N}$, which make up the four supersymmetric charges $Q_{(i)}$ and test which one of these variations just produces the associated charge itself multiplied by the time derivative of the symmetry parameter. The result (Gozzi, in progress) is that only the variation under Q and $\bar{N}$ do this. So these are the only two symmetry generators which can be made local in the sense that, by introducing auxiliary gauge-fields in $\widetilde{\mathcal{L}}$, the local variation of $\widetilde{\mathcal{L}}$ can be absorbed by the variation of the gauge-fields. As only two of the four charges can be made local, we can say that from global N=4 we go down to a local N=2 susy. In term of the susy charges $Q_{(i)}$ those which can be made local are only $Q_{(1)}$ and $Q_{(2)}$. The next step is to obtain the expression of the local lagrangian $\widetilde{\mathcal{L}}_g$. It is easy (Gozzi, in progress) to find that: calling the "gauge" fields e,ψ_1,ψ_2 where e is a c-number field while ψ_1,ψ_2 are Grassmannian ones, the $\widetilde{\mathcal{L}}_g$ is $\widetilde{\mathcal{L}}_g\equiv\lambda_a\dot{\phi}^a+{\rm i}\bar{c}_a\dot{c}^a-e\widetilde{\mathcal{H}}+\psi_1\bar{N}+\psi_2 Q$.

It is not the first time that a susy or a BRS has been made local in a 1-D system (Brink et al.(1976); Alvarez (1984); Urrutia and Zanelli (1990) Bastianelli and Consoli (1996)) but never for this action. The closest model for which it has been made local (Alvarez (1984)) is the supersymmetric quantum mechanics of Witten (Witten (1981))(susyqm). The question now to ask is if $\widetilde{\mathcal{L}}_g$ has any geometrical meaning at all. To do that we have to explore the last three pieces present in $\widetilde{\mathcal{L}}_g$ which were not present in $\widetilde{\mathcal{L}}$. The author of ref.Alvarez (1984) indicates that there is a gauge transformation which would turn the field e into the constant 1 while turning to zero both ψ_1 and ψ_2. In that manner the extra piece would disappear and $\widetilde{\mathcal{L}}_g$ would turn into $\widetilde{\mathcal{L}}$, at least for the susyqm. This fact does not happen for sure in our $\widetilde{\mathcal{L}}_g$ as we will show now and we have *strong doubts* it happens even in susyqm (Alvarez (1984)). The transformations of the gauge fields e,ψ_1, ψ_2 under the two local transformations associated to $\bar{N}$ and Q (with parameters ϵ_1, and ϵ_2) are respectively $\{\delta_1\psi_1 = \dot{\epsilon}_1,\ \delta_1\psi_2 = 0,\ \delta_1 e = \epsilon_1\psi_2\}$ and $\{\delta_2\psi_2 = \dot{\epsilon}_2,\ \delta_2\psi_1 = 0,\ \delta_2 e = -\epsilon_2\psi_1\}$. As we have to bring, following (Alvarez (1984)), ψ_1 and ψ_2 to zero, the infinitesimal parameters which do that are $\epsilon_1(t) = -\int_0^t \psi_1(t')\mathrm{d}t'$ and $\epsilon_2(t) = -\int_0^t \psi_1(t')\mathrm{d}t'$. Using these parameters we have now at the same time to bring e to zero. The combined transformation on e would be: $\delta e = [\int_o^t \psi_2\mathrm{d}t']\psi_2 + [\int_0^t \psi_1\mathrm{d}t']\psi_1$. This should be equal to $1-e$ to bring e to zero but this is impossible because it would impose a relation between e,ψ_1,ψ_2 which does not exist. So this fact gives us the information that it is not possible to turn $\widetilde{\mathcal{L}}_g$ into $\widetilde{\mathcal{L}}$. The fields e,ψ_1,ψ_2 remain, they have no dynamic but they remain. They are the analog of the gravitons and gravitinos in 1-D. They are basically Lagrangian multipliers and the fact that they cannot be eliminated indicates that they have a role. In fact if we insert $\widetilde{\mathcal{L}}_g$ into (1) we see that, once e,ψ_1,ψ_2 are integrated out, they impose some Dirac deltas constraints on the states between which we sandwich the propagator. This selection of states is related to the equivariant cohomology business (Ouvry et al. (1990)) and this should bring light to the geometrical meaning of $\widetilde{\mathcal{L}}_g$. More details will appear in (Gozzi, in progress).

References

Abraham, R.A. and Marsden, J., (1978): "*Foundations of Mechanics*" Benjamin, , New York.

Alvarez, E.(1984): Phys.Rev.**D29**, 320.

Becchi, C., Rouet, A. and Stora, R. (1976): Ann.Phys. **98**, 287.

Brink, L. et al.(1976): Phys.Lett. **B64**, 435;

Gol'fand, Yu. and Lichtman, E. (1971): JETP Lett **13**, 323;
Gervais, J.-L. and Sakita, B. (1971): Nucl. Phys. **B34**, 633;
Volkov, D. and Akulov, V. (1972): JETP Lett. **18**, 438;
Wess, J. and Zumino, B. (1974): Nucl.Phys. **B70**, 39.

Gozzi, E. (1986): **MPI-PAE/Pth 47/86**, Slac List PPF 8633;
Gozzi, E. (1988): Phys. Lett. **B201**, 525;

Gozzi, E., Reuter, M., Thacker, W.D. (1989): Phys. Rev. **D40**, 3363; ibidem **D46**, 757;
Gozzi, E. and Reuter, M. (1990): Phys. Lett. **240B**, 137.
Gozzi, E. and Reuter, M. (1989): Phys.Lett. **B233**, 383.
Gozzi, E. work in progress.
Kastrup, H. (1983): Phys. Rep. **101**, 1.
Ouvry, S., Tora, R.S, van Baal, P. (1990): Phys.Lett **B238**, 291.
Parisi, G. and Sourlas, N. (1982): Nucl.Phys. **B206**, 321;
Cecotti, S. and Girardello, L. (1983): Ann.Phys.N.Y. **145**, 81.
Urrutia, L.F. and Zanelli, J.(1990): Jour.Math.Phys. **31**, 2271;
de Carvalho, C.Aragao, Baulieu, L. (1992): Phys.Lett **B275**, 323l;
Bastianelli, F. and Consoli, L. (1996): it hep-th/9608065.
Volkov, D. and Soroka, V. (1974): Theor.Math.Phys. **18**, 28;
Freedman, D., Van Nieuwenhuizen, P. and Ferrara, S. (1976): Phys.Rev **D13**, 3214; Deser, S. and Zumino, B. (1976): Phys.Lett **62B**, 3214.
Witten, E.(1981): Nucl.Phys. **B188**, 513 .
Witten, E. (1982): Jour. Diff.Geom. **17**, 661.

The Hamiltonian Structure of the "Bosonic" and "Fermionic" Extensions of $N = 2$ KdV Hierarchy

L. Bonora[1] and S. Krivonos[2]

[1] International School for Advanced Studies (SISSA/ISAS)
Via Beirut 2, 34014 Trieste, Italy
INFN, Sezione di Trieste

[2] Bogoliubov Laboratory of Theoretical Physics, JINR,
141980 Dubna, Moscow Region, Russia

Abstract. We apply a manifestly $N = 2$ supersymmetric coset formalism to the "bosonic" and "fermionic" extensions of $N = 2$ KdV hierarchy. We construct the second Hamiltonian structure. It contains the same number of currents as the "small" $N = 4$ SCA but with reversed statistics for two of them, and it is non-local.

1 Introduction

The search for $N = 2$ and $N = 4$ extensions of well-known hierarchies of integrable equations, has stimulated new interest on integrable systems. This is motivated by the fact that such extensions are characterized by beautiful supersymmetric second hamiltonian structures, but, what is more important, by the discovery that they seem to provide an organization principle, in that different cases can be encompassed under a unique structure. The purpose of this contribution is to present a few examples of this fact and show some 'magic' of supersymmetric extensions of integrable hierarchies.

We start from $N = 4KdV$ (Delduc and Ivanov (1993)) and "bosonic" extension of $N = 2, a = -2$ KdV hierarchies (Delduc et al. (1997)) and make 'theme variations', i.e. we change statistics of some of the superfields and generate two "new" hierarchies (section 2). Their second hamiltonian structure is the same, but it is non-local (section 3). It turns out that the best way to unify these new hierarchies is via a coset construction which lift the second hamiltonian structure to a local algebra.

2 "Bosonic" and "Fermionic" Extensions of the $N = 2$ KdV

2.1 $N = 4$ KdV

The $N = 4$ supersymmetric KdV equation has been constructed in (Delduc and Ivanov (1993)). In terms of $N = 2$ superfields it can be written as the following coupled system of equations for the general bosonic superfield J and for the pair of chiral-anti-chiral fermionic superfields F and $\overline{F}$ (here we will explicitly write only second flow from $N = 4$ KdV hierarchy, while we recall that the $N = 4$ KdV equation itself corresponds to the third flow)

$$\frac{\partial J}{\partial t_2} = -\left[\mathcal{D},\overline{\mathcal{D}}\right] J' - 4JJ' - \left(\mathcal{D}F\overline{\mathcal{D}}\overline{F}\right)',$$
$$\frac{\partial F}{\partial t_2} = F'' + 4\mathcal{D}\left(J\overline{\mathcal{D}}F\right), \frac{\partial \overline{F}}{\partial t_2} = -\overline{F}'' + 4\overline{\mathcal{D}}\left(J\mathcal{D}\overline{F}\right), \tag{1}$$

where prime means differentiation over the space coordinate $\{z\}$, and $\mathcal{D}$ and $\overline{\mathcal{D}}$ are the $N = 2$ supersymmetric fermionic covariant derivatives

$$\mathcal{D} = \frac{\partial}{\partial\theta} - \frac{\bar{\theta}}{2}\frac{\partial}{\partial z}, \overline{\mathcal{D}} = \frac{\partial}{\partial\bar{\theta}} - \frac{\theta}{2}\frac{\partial}{\partial z}, \mathcal{D}^2 = \overline{\mathcal{D}}^2 = 0, \{\mathcal{D},\overline{\mathcal{D}}\} = -\partial_z. \tag{2}$$

The superfields F and $\overline{F}$ are constrained as follows

$$\mathcal{D}F = \overline{\mathcal{D}}\overline{F} = 0. \tag{3}$$

The integrability of the $N = 4$ KdV equation (1) has been proved in (Ivanov and Krivonos (1996), Delduc and Gallot (1996)) by explicit construction of the Lax operator and the Lax equation

$$L = \partial - 2J - 2\overline{\mathcal{D}}\partial^{-1}(\mathcal{D}J) + F\overline{\mathcal{D}}\partial^{-1}(\mathcal{D}\overline{F}) - \overline{\mathcal{D}}\partial^{-1}\left(\mathcal{D}(F\overline{F})\right),$$
$$\frac{\partial L}{\partial t_k} = \left[\left(L^k\right)_{\geq 1}, L\right], \tag{4}$$

where the subscript $\{\geq 1\}$ indicates the strictly differential part of the pseudodifferential operator. The parentheses in the above expression mean that the relevant operators act only on superfields within them. One can easily check that the second flow equations for the Lax operator (4) coincide with (1). The infinite set of conserved currents $\mathcal{H}_k$ for the $N = 4$ KdV equation (1) is given by the following expression (Ivanov and Krivonos (1996))

$$H_k \equiv \int \mathrm{d}z\mathrm{d}\theta\mathrm{d}\bar{\theta}\, \mathcal{H}_k = \int \mathrm{d}z\mathrm{d}\theta\mathrm{d}\bar{\theta}\, \left(L^k\right)_0, \tag{5}$$

where the residue is defined as the constant part of the operator.

Let us note that the $N = 4$ KdV hierarchy can be reformulated in terms of J and the bosonic chiral-anti-chiral superfields $\Phi, \overline{\Phi}$ (Delduc and Ivanov (1993)) defined by

$$\Phi \equiv \overline{\mathcal{D}}F, \overline{\Phi} \equiv \mathcal{\overline{D}}\overline{F}. \tag{6}$$

The second flow (1) now have the following form

$$\frac{\partial J}{\partial t_2} = -\left[\mathcal{D},\overline{\mathcal{D}}\right] J' - 4JJ' - \left(\Phi\overline{\Phi}\right)',$$
$$\frac{\partial \Phi}{\partial t_2} = -\Phi'' + 4\mathcal{D}\overline{\mathcal{D}}\left(J\Phi\right), \frac{\partial \overline{\Phi}}{\partial t_2} = \overline{\Phi}'' + 4\overline{\mathcal{D}}\mathcal{D}\left(J\overline{\Phi}\right). \tag{7}$$

The Lax operator and the Lax equation for the $N = 4$ KdV hierarchy in terms of the superfields J, Φ and $\overline{\Phi}$ (7) was constructed in (Delduc and Gallot (1996)) and reads

$$L = \mathcal{D}\overline{\mathcal{D}} + \mathcal{D}\overline{\mathcal{D}}\partial^{-1}\left(J + \overline{\Phi}\partial^{-1}\Phi\right)\partial^{-1}\mathcal{D}\overline{\mathcal{D}}\,, \quad \frac{\partial L}{\partial t_k} = -\left[L^k_{\geq 1}, L\right]. \qquad (8)$$

The main attractive feature of the $N = 4$ KdV hierarchy is that it possesses the classical "small" $N = 4$ SCA (with $su(2)$ affine subalgebra) as the second Hamiltonian structure (Delduc and Ivanov (1993)). Indeed, if we define the PB's for the "small" $N = 4$ SCA as

$$\begin{aligned}
\{J(1), J(2)\} &= -\left(J\partial + \partial J + \mathcal{D}J\overline{\mathcal{D}} + \overline{\mathcal{D}}J\mathcal{D} + \frac{1}{2}\partial\left[\mathcal{D}, \overline{\mathcal{D}}\right]\right)\delta(1,2)\,,\\
\{J(1), \Phi(2)\} &= \left(\Phi\overline{\mathcal{D}}\mathcal{D} + \overline{\mathcal{D}}\Phi\mathcal{D}\right)\delta(1,2)\,,\\
\left\{J(1), \overline{\Phi}(2)\right\} &= \left(\overline{\Phi}\mathcal{D}\overline{\mathcal{D}} + \mathcal{D}\overline{\Phi}\overline{\mathcal{D}}\right)\delta(1,2)\,,\\
\left\{\Phi(1), \overline{\Phi}(2)\right\} &= -2\left(\partial\mathcal{D}\overline{\mathcal{D}} + 2\mathcal{D}J\overline{\mathcal{D}}\right)\delta(1,2)\,,
\end{aligned} \qquad (9)$$

where $\delta(1,2)$ is the $N = 2$ superspace delta-function

$$\delta(1,2) = (\theta_1 - \theta_2)(\bar{\theta}_1 - \bar{\theta}_2)\delta(z_1 - z_2)\,, \qquad (10)$$

and the differential operators in the r.h.s are evaluated at the point Z_1 and the derivatives in the r.h.s are assumed to act freely to the right, then the flow equations for the $N = 4$ KdV hierarchy can be written in hamiltonian form as follows

$$\frac{\partial}{\partial t_k}\begin{pmatrix} J \\ \Phi \\ \overline{\Phi} \end{pmatrix} = \left\{\begin{pmatrix} J \\ \Phi \\ \overline{\Phi} \end{pmatrix}, H_k\right\}. \qquad (11)$$

The hamiltonian H_2 which gives rise to the second flow of $N = 4$ KdV hierarchy (7) has the following explicit form

$$H_2 = -\int \mathrm{d}z\mathrm{d}\theta\mathrm{d}\bar{\theta}\left(J^2 - \frac{1}{2}\overline{\Phi}\Phi\right). \qquad (12)$$

Finally we would like to recall that the second Hamiltonian structure for the $N = 4$ KdV hierarchy written in terms of the "pre-potentials" $F, \overline{F}$ (1) is non-local.

2.2 "Bosonic" Extension of N=2 KdV Hierarchy

In this subsection we are going to construct the "bosonic" extension of $N = 2$ KdV hierarchy and find its second Hamiltonian structure. The idea of such construction comes from a close inspection of the Lax operators for the $N = 4$ KdV hierarchy (4),(8). Indeed, in both cases the superfields $F, \overline{F}$ and $\Phi, \overline{\Phi}$ appear in the Lax operators only in pairs. So one may wonder whether the

Lax equations (4) and (8) will be consistent if we change the statistic of the superfields $F, \overline{F}$ and $\Phi, \overline{\Phi}$ to be bosonic and fermionic one, respectively, keeping the same form of the Lax operators. The answer is positive and the flow equations have the same form, e.g. the second flow equations (1) and (7) are

$$\frac{\partial J}{\partial t_2} = - \left[\mathcal{D}, \overline{\mathcal{D}}\right] J' - 4JJ' - \left(\mathcal{D}\mathcal{B}\overline{\mathcal{D}\mathcal{B}}\right)',$$
$$\frac{\partial \mathcal{B}}{\partial t_2} = \mathcal{B}'' + 4\mathcal{D}\left(J\overline{\mathcal{D}}\mathcal{B}\right), \quad \frac{\partial \overline{\mathcal{B}}}{\partial t_2} = -\overline{\mathcal{B}}'' + 4\overline{\mathcal{D}}\left(J\mathcal{D}\overline{\mathcal{B}}\right), \tag{13}$$

and

$$\frac{\partial J}{\partial t_2} = - \left[\mathcal{D}, \overline{\mathcal{D}}\right] J' - 4JJ' - \left(\Psi\overline{\Psi}\right)',$$
$$\frac{\partial \Psi}{\partial t_2} = -\Psi'' + 4\mathcal{D}\overline{\mathcal{D}}\left(J\Psi\right), \quad \frac{\partial \overline{\Psi}}{\partial t_2} = \overline{\Psi}'' + 4\overline{\mathcal{D}}\mathcal{D}\left(J\overline{\Psi}\right), \tag{14}$$

where $\mathcal{B}, \overline{\mathcal{B}}$ and $\Psi, \overline{\Psi}$ are bosonic and fermionic superfields, respectively, which obey the following constraints

$$\mathcal{D}\mathcal{B} = \overline{\mathcal{D}\mathcal{B}} = \mathcal{D}\Psi = \overline{\mathcal{D}\Psi} = 0 . \tag{15}$$

Thus, by simply changing of statistics of the superfields in the Lax operators we construct a "new" extension of the $N = 2$ KdV hierarchy. We would like to stress in fact that in passing to the superfields with opposite statistics we completely destroy the hidden $N = 2$ supersymmetry of the equations (1),(7) which together with the explicit $N = 2$ supersymmetry close on the $N = 4$ supersymmetry. So, the systems (13) and (14) possess only $N = 2$ supersymmetry and we can only speak of an extension of the $N = 2$ KdV hierarchy.

A natural question now arises: do the equations (14) possess a second Hamiltonian structure and what algebra, if any, corresponds to the latter? [1]. If such Hamiltonian structure exists it must contains at least all superfields that are present in equation (14), i.e. the bosonic J together with the fermionic superfields $\Psi, \overline{\Psi}$. But one can check that it is impossible to construct from these fields a *local* superalgebra that play the role of second Hamiltonian structure for the equations (14). This is in contrast with its "fermionic" counterpart, i.e. $N = 4$ KdV hierarchy, which has the local $N = 4$ SCA as the second Hamiltonian structure (9). However, it is possible to extend the number of superfields by adding one additional pair of chiral-anti-chiral fermionic superfields $H, \overline{H}$ to form a closed *local* although *non-linear* algebra

[1] We limit ourselves here to dealing with system (14), because the second Hamiltonian structure for its "pre-potential" form (13) is non-local even in the $N = 4$ KdV case.

$$
\begin{aligned}
\{J(1),J(2)\} &= -\left(J\partial+\partial J+\mathcal{D}J\overline{\mathcal{D}}+\overline{\mathcal{D}}J\mathcal{D}+\frac{1}{2}\partial\left[\mathcal{D},\overline{\mathcal{D}}\right]\right)\delta(1,2),\\
\{J(1),\Psi(2)\} &= \left(2\overline{\mathcal{D}}\mathcal{D}\Psi+\mathcal{D}\Psi\overline{\mathcal{D}}+\overline{\mathcal{D}}\Psi\mathcal{D}-\Psi\partial\right)\delta(1,2),\\
\{J(1),\overline{\Psi}(2)\} &= \left(2\mathcal{D}\overline{\mathcal{D}}\overline{\Psi}+\mathcal{D}\overline{\Psi}\overline{\mathcal{D}}+\overline{\mathcal{D}}\overline{\Psi}\mathcal{D}-\overline{\Psi}\partial\right)\delta(1,2),\\
\{J(1),H(2)\} &= H\overline{\mathcal{D}}\mathcal{D}\delta(1,2),\ \{J(1),\overline{H}(2)\}=\overline{H}\mathcal{D}\overline{\mathcal{D}}\delta(1,2),\\
\{H(1),\overline{H}(2)\} &= \frac{1}{2}\mathcal{D}\overline{\mathcal{D}}\delta(1,2),\ \{H(1),\Psi(2)\}=\mathcal{D}\Psi\delta(1,2)\ ,\\
\{H(1),\overline{\Psi}(2)\} &= -\mathcal{D}\overline{\Psi}\delta(1,2)\ ,\\
\{\overline{H}(1),\Psi(2)\} &= -\overline{\mathcal{D}}\Psi\delta(1,2),\ \{\overline{H}(1),\overline{\Psi}(2)\}=\overline{\mathcal{D}\overline{\Psi}}\ \delta(1,2),\\
\{\Psi(1),\overline{\Psi}(2)\} &= 2\left(-\partial\mathcal{D}\overline{\mathcal{D}}-2\mathcal{D}J\overline{\mathcal{D}}+\Psi\overline{\Psi}-4\mathcal{D}J\overline{H}-4JH\overline{\mathcal{D}}-8JH\overline{H}\right.\\
&\quad+2\mathcal{D}\overline{H}\mathcal{D}\overline{\mathcal{D}}-2\partial\mathcal{D}\overline{H}-2\partial H\overline{\mathcal{D}}-2H\partial\overline{\mathcal{D}}+2\overline{\mathcal{D}}H\mathcal{D}\overline{\mathcal{D}}\\
&\quad+4\mathcal{D}\overline{H}\mathcal{D}\overline{H}+4H\overline{\mathcal{D}}H\overline{\mathcal{D}}+8H\overline{H}\mathcal{D}\overline{H}+8H\overline{H}\overline{\mathcal{D}}H\\
&\quad\left.+4\mathcal{D}\overline{\mathcal{D}}H\overline{H}+4H\overline{\mathcal{D}}\mathcal{D}\overline{H}-8H\mathcal{D}\overline{\mathcal{D}}\overline{H}+4H\overline{H}\overline{\mathcal{D}}\mathcal{D}\right)\delta(1,2)\ ,
\end{aligned}
$$

which we claim is the right second Hamiltonian structure for the system (14) in the coset approach.

One can check that the Jacoby identities impose additional constraints on the superfields $\Psi,\overline{\Psi}$:

$$
\mathcal{D}\Psi=-2\,H\Psi\ ,\ \overline{\mathcal{D}\overline{\Psi}}=2\,\overline{H\overline{\Psi}}\ . \tag{16}
$$

The set of Poisson brackets (16) together with covariant chirality constraints (16) form the closed algebra we are looking for[2]. This algebra has the structure of non-linear W algebra due to the presence of non-linear terms in the right hand side.

The algebra (16) possesses manifest $N=2$ supersymmetry. The same is true also for the coset which is defined by modding it out by the subalgebra $\hat{U}(1)$ generated by $H,\overline{H}$. It is this coset that forms the second Hamiltonian structure for the "bosonic" $N=2$ KdV hierarchy (14). In other words, it is possible to construct an infinite number of Hamiltonians which have vanishing Poisson brackets with respect to $H,\overline{H}$. Thus, at the level of equations of motion, one can consistently put $H=\overline{H}=0$ and remain with the equations for the fields $J,\Psi,\overline{\Psi}$. Moreover, after putting H and $\overline{H}$ equal to zero, the covariant chirality conditions (16) will become the ordinary chiral-anti-chiral one

$$
\mathcal{D}\Psi=\overline{\mathcal{D}\overline{\Psi}}=0\ . \tag{17}
$$

The Hamiltonian densities $\mathcal{H}_k$ in the $N=2$ superspace for the integer $k=1,2,\ldots$ are bosonic and have conformal dimension k. The explicit form of the first three Hamiltonians is

[2] To our knowledge, this algebra first appeared in hidden form (as subalgebra of some extended algebra in a special basis) in the Ref.Ahn et al. (1997).

$$H_1 = \int dzd\theta d\bar{\theta}\, J\,,\; H_2 = \int dzd\theta d\bar{\theta} \left(-J^2 + \frac{1}{2}\overline{\Psi}\Psi \right)$$
$$H_3 = \int dzd\theta d\bar{\theta} \left(J \left[\mathcal{D},\overline{\mathcal{D}}\right] J + 2J\Psi\overline{\Psi} + \frac{4}{3}J^3 \right.$$
$$\left. - (\overline{\mathcal{D}}H + \mathcal{D}\overline{H})\Psi\overline{\Psi} + \frac{1}{2}\Psi'\overline{\Psi} \right) . \qquad (18)$$

The flows are defined as

$$\frac{\partial}{\partial t_k}\mathcal{F}(Z) = \{\mathcal{F}(Z), H_k\} \qquad (19)$$

for a given superfield $\mathcal{F}$. Since all the Hamiltonians (18) have vanishing Poisson brackets with $H, \overline{H}$, we can consistently put $H = \overline{H} = 0$ in the r.h.s. of (19) *after* evaluating of Poisson brackets.

The first flow just gives the following equations of motion for $J, \Psi, \overline{\Psi}$

$$\frac{\partial}{\partial t_1}J = J'\,,\; \frac{\partial}{\partial t_1}\Psi = \Psi'\,,\; \frac{\partial}{\partial t_1}\overline{\Psi} = \overline{\Psi}'\,, \qquad (20)$$

while the second Hamiltonian generates the equations (14).

2.3 "Fermionic" Extension of N=2 KdV Hierarchy

In a recent paper, Delduc et al. (1997), another integrable hierarchy with "small" $N = 4$ SCA as the second Hamiltonian structure was constructed. Despite the manifest $N = 4$ supersymmetric invariance of the second Hamiltonian structure the equations of motion possess only $N = 2$ supersymmetry and the whole system is a extension of $N = 2$ KdV hierarchy with $a = -2$.

This new hierarchy contains the same superfields as the $N = 4$ KdV – the usual bosonic J and a pair of bosonic chiral-anti-chiral superfields $\Phi, \overline{\Phi}$ – and can be described by the following Lax operator and Lax equation Delduc et al. (1997):

$$L = \mathcal{D}\overline{\mathcal{D}}\partial + 2\mathcal{D}J\overline{\mathcal{D}} + \mathcal{D}\Phi\partial^{-1}\overline{\Phi}\overline{\mathcal{D}}\,, \qquad \frac{\partial}{\partial t_k}L = -\left[\left(L^{\frac{k}{2}}\right)_{\geq 1}, L\right] . \qquad (21)$$

The first non-trivial (i.e. the second) flow equations of motion is

$$\frac{\partial J}{\partial t_2} = -\left(\overline{\Phi}\Phi\right)'\,,$$
$$\frac{\partial \Phi}{\partial t_2} = \Phi'' - 2\mathcal{D}\left(J\overline{\mathcal{D}}\Phi\right)\,,\; \frac{\partial \overline{\Phi}}{\partial t_2} = -\overline{\Phi}'' - 2\overline{\mathcal{D}}\left(J\mathcal{D}\overline{\Phi}\right) . \qquad (22)$$

Like the $N = 4$ KdV case one can check that it is possible to change the statistics of the superfields $\Phi, \overline{\Phi}$ in the Lax operator (21) by replacing them with fermionic ones $\Psi, \overline{\Psi}$. This Lax operator and Lax equation

$$L = \mathcal{D}\overline{\mathcal{D}}\partial + 2\mathcal{D}J\overline{\mathcal{D}} + \mathcal{D}\Psi\partial^{-1}\overline{\Psi}\overline{\mathcal{D}} , \qquad \frac{\partial}{\partial t_k}L = -\left[\left(L^{\frac{k}{2}}\right)_{\geq 1}, L\right] . \tag{23}$$

give rise to the same flow equations with $\Phi, \overline{\Phi}$ replaced by $\Psi, \overline{\Psi}$.

Again one can check that the coset approach works well in this case. More precisely, the algebra (16) provides the second Hamiltonian structure for the hierarchy (21) via coset construction. Explicitly, the first three Hamiltonian densities are

$$\mathcal{H}_1 = J \, , \; \mathcal{H}_2 = \frac{1}{2}\Psi\overline{\Psi} \, ,$$
$$\mathcal{H}_3 = J\left[\mathcal{D}, \overline{\mathcal{D}}\right] J + 2\Psi'\overline{\Psi} + 2J\Psi\overline{\Psi} - \frac{2}{3}J^3 - 4(\overline{\mathcal{D}}H + \mathcal{D}\overline{H})\Psi\overline{\Psi} \, , \tag{24}$$

where Ψ and $\overline{\Psi}$ are subject to the constraints (16) and the different flow equations are defined by (19).

Thus both, "bosonic" (18) and "fermionic" extensions of $N = 2$ KdV (24) can be treated within the coset approach as $N = 2$ quotients of the $N = 2$ nonlinear algebra (16).

3 Hamiltonian Reduction

In the previous section we used the coset formalism applied to the $N = 2$ algebra (16) to describe the "bosonic" (18) and the "fermionic" (24) extensions of the $N = 2$ KdV hierarchy. This approach allows for a simpler analysis than the usual Dirac's procedure, because the constraints $H = \overline{H} = 0$ can be imposed in the equations of motion. In this Section we will show how to apply the constraints $H = \overline{H} = 0$ to the algebra (16) and will demonstrate that the Dirac brackets for the remaining supercurrents $J, \Psi, \overline{\Psi}$ close on a non-local algebra.

Let us start with the obvious remark that the constraints

$$H = \overline{H} = 0 \tag{25}$$

we have to apply are incompatible with the Poisson brackets of our algebra

$$\left\{H(1), \overline{H}(2)\right\} = \frac{1}{2}\mathcal{D}\overline{\mathcal{D}}\delta(1,2) \, , \; \left\{\overline{H}(1), H(2)\right\} = -\frac{1}{2}\overline{\mathcal{D}}\mathcal{D}\delta(1,2) \, , \tag{26}$$

So we need to apply Dirac's procedure passing to the Dirac brackets

$$\{\mathcal{A},\mathcal{B}\}^* \equiv \{\mathcal{A},\mathcal{B}\} - \{\mathcal{A},H\}\left(-\frac{1}{2}\overline{\mathcal{D}}\mathcal{D}\right)^{-1}\{\overline{H},\mathcal{B}\} - \{\mathcal{A},\overline{H}\}\left(\frac{1}{2}\mathcal{D}\overline{\mathcal{D}}\right)^{-1}\{H,\mathcal{B}\} \tag{27}$$

for any two superfields $\mathcal{A}, \mathcal{B}$. One might expect some extra complications in the Dirac procedure because the operators $\mathcal{D}\overline{\mathcal{D}}$ and $\overline{\mathcal{D}}\mathcal{D}$ are not invertible on the whole $N = 2$ superspace. But fortunately this is not so for the case at hand. Indeed, due to the chiral-anti-chiral structure of H and $\overline{H}$, the Poisson brackets $\{\mathcal{A},H\}, \{\overline{H},\mathcal{B}\}, \{\mathcal{A},\overline{H}\}, \{H,\mathcal{B}\}$ can be always rewritten in the following form

$$\{\mathcal{A},H\} \sim (\ldots)\mathcal{D}\delta(1,2)\,,\ \{\mathcal{A},\overline{H}\} \sim (\ldots)\overline{\mathcal{D}}\delta(1,2)\,,$$
$$\{H,\mathcal{B}\} \sim \mathcal{D}(\ldots)\delta(1,2)\,,\ \{\overline{H},\mathcal{B}\} \sim \overline{\mathcal{D}}(\ldots)\delta(1,2)\,, \qquad (28)$$

where $(\ldots)$ is used to denote any combinations of operators and/or superfields. Thus, one can immediately conclude that we have to know the operators $(\mathcal{D}\overline{\mathcal{D}})^{-1}$ and $(\overline{\mathcal{D}}\mathcal{D})^{-1}$ not on the whole $N=2$ superspace but only on its chiral-anti-chiral subspaces, where they are well defined. Explicitly

$$\mathcal{D}\left(-\frac{1}{2}\overline{\mathcal{D}}\mathcal{D}\right)^{-1}\overline{\mathcal{D}} = \mathcal{D}\left(2\partial^{-1}\right)\overline{\mathcal{D}} \ \text{ because } \mathcal{D}\left(-\overline{\mathcal{D}}\mathcal{D}\partial^{-1}\right)\overline{\mathcal{D}} \equiv \mathcal{D}\overline{\mathcal{D}}\,,$$
$$\overline{\mathcal{D}}\left(\frac{1}{2}\mathcal{D}\overline{\mathcal{D}}\right)^{-1}\mathcal{D} = \overline{\mathcal{D}}\left(-2\partial^{-1}\right)\mathcal{D} \ \text{ because } \overline{\mathcal{D}}\left(-\mathcal{D}\overline{\mathcal{D}}\partial^{-1}\right)\mathcal{D} \equiv \overline{\mathcal{D}}\mathcal{D} \qquad (29)$$

With this in the mind all calculations can be done straightforwardly.

All Dirac brackets containing $H,\overline{H}$ are now equal to zero, so we can impose the constraints (25) in strong sense. The rest of the Dirac brackets between the superfields $J,\Psi,\overline{\Psi}$ form the following closed non-local algebra

$$\{J(1),J(2)\}^* = -\left(J\partial + \partial J + \mathcal{D}J\overline{\mathcal{D}} + \overline{\mathcal{D}}J\mathcal{D} + \frac{1}{2}\partial\left[\mathcal{D},\overline{\mathcal{D}}\right]\right)\delta(1,2)\,,$$
$$\{J(1),\Psi(2)\}^* = \left(-\overline{\mathcal{D}}\Psi\mathcal{D} + \Psi\left[\mathcal{D},\overline{\mathcal{D}}\right]\right)\delta(1,2)\,,$$
$$\{J(1),\overline{\Psi}(2)\}^* = \left(-\mathcal{D}\overline{\Psi}\overline{\mathcal{D}} + \overline{\Psi}\left[\mathcal{D},\overline{\mathcal{D}}\right]\right)\delta(1,2)\,,$$
$$\{\Psi(1),\Psi(2)\}^* = -4\Psi\overline{\mathcal{D}}\mathcal{D}\partial^{-1}\Psi\delta(1,2)\,,$$
$$\{\overline{\Psi}(1),\overline{\Psi}(2)\}^* = 4\overline{\Psi}\mathcal{D}\overline{\mathcal{D}}\partial^{-1}\overline{\Psi}\delta(1,2)\,,$$
$$\{\Psi(1),\overline{\Psi}(2)\}^* = -2\left(\mathcal{D}\overline{\mathcal{D}}\partial + 2\mathcal{D}J\overline{\mathcal{D}} - 2\Psi\mathcal{D}\overline{\mathcal{D}}\partial^{-1}\overline{\Psi}\right)\delta(1,2) \qquad (30)$$

with $\Psi,\overline{\Psi}$ obeying the chirality constraints

$$\mathcal{D}\Psi = \overline{\mathcal{D}\Psi} = 0\,. \qquad (31)$$

The algebra (30) is a particular case of the non-local algebras proposed in Bonora and Sorin (1997). Let us note that validity of the Jacoby identities for the non-local algebra (30) is guaranteed since we constructed it by applying the standard Dirac's procedure applied to the local algebra (16).

Now one can straightforwardly check that the Hamiltonians

$$H_1 = \int \mathrm{d}z\mathrm{d}\theta\mathrm{d}\bar{\theta}\, J\,,\ H_2 = \int \mathrm{d}z\mathrm{d}\theta\mathrm{d}\bar{\theta}\left(-J^2 + \frac{1}{2}\overline{\Psi}\Psi\right)$$
$$H_3 = \int \mathrm{d}z\mathrm{d}\theta\mathrm{d}\bar{\theta}\left(J\left[\mathcal{D},\overline{\mathcal{D}}\right]J + 2J\Psi\overline{\Psi} + \frac{4}{3}J^3 + \frac{1}{2}\Psi'\overline{\Psi}\right) \qquad (32)$$

and

$$H_1 = \int \mathrm{d}z\mathrm{d}\theta\mathrm{d}\bar{\theta}\, J\,,\ H_2 = \int \mathrm{d}z\mathrm{d}\theta\mathrm{d}\bar{\theta} \left(\frac{1}{2}\Psi\overline{\Psi}\right)\,,$$

$$H_3 = \int \mathrm{d}z\mathrm{d}\theta\mathrm{d}\bar{\theta} \left(J\left[\mathcal{D},\overline{\mathcal{D}}\right]J + 2\Psi'\overline{\Psi} + 2J\Psi\overline{\Psi} - \frac{2}{3}J^3\right) \tag{33}$$

produce the correct flow equations for the "bosonic" and "fermionic" extensions of the $N = 2$ KdV hierarchy; they thus provide the second Hamiltonian structure for both hierarchies.

Acknowledgments. S.K. is grateful to Organizers of D.V. Volkov Memorial Seminar inviting him to give this Talk and to E. Ivanov for many useful discussions. He also thanks SISSA for the hospitality and financial support extended to him during the course of this work.

This investigation has been supported in part by the Russian Foundation of Fundamental Research, grant No. 96-02-17634, RFBR-DFG grant No. 96-02-00180, INTAS grant 94-2317, and by a grant from the Dutch NWO organization.

References

Ahn, C., Ivanov, E. and Sorin, A. (1997): Commun. Math. Phys. **183**, 205.

Bonora, L. and Sorin, A. (1997): " *The Hamiltonian structure of the N=2 supersymmetric GNLS hierarchy*", SISSA 52/97/EP, hep-th/9704130.

Delduc, F. and Ivanov, E. (1993): Phys. Lett. **B 309**, 312.

Delduc, F. and Gallot, L. (1996): " *N=2 KP and KdV hierarchies in extended superspace*", ENSLAPP-L-617/96, solv-int/9609008.

Delduc, F., Gallot, L. and Ivanov, E. (1997): Phys. Lett. **B 396**, 141.

Ivanov, E. and Krivonos, S. (1996): "*New integrable extensions of N=2 KdV and Boussinesq hierarchies*",JINR E2-96-344,hep-th/9609191, (Phys. Lett. A in press).

Mass Generation in the Supersymmetric Nambu–Jona–Lasinio Model in an External Magnetic Field

I.A. Shovkovy

Bogolyubov Institute for Theoretical Physics
252143 Kiev, Ukraine

Abstract. The mass generation in the $(3+1)$–dimensional supersymmetric Nambu–Jona–Lasinio model in a constant magnetic field is studied. It is shown that the external magnetic field catalyzes chiral symmetry breaking.

It was shown in Gusynin et al. (1994), (1995a), Gusynin et al. (1995b), (1995c), (1996) and later confirmed in Leung et al. (1996), (1997), Hong et al. (1996) that a constant magnetic field is a strong catalyst of dynamical chiral symmetry breaking, leading to the generation of a fermion dynamical mass even at the weakest attraction between fermions (the prehistory of the question includes Klevansky and Lemmer (1988), Klimenko (1991), Suganuma and Tatsumi (1991), Schramm et al. (1992) among others).

The effect is accounted for the effective dimensional reduction $D \to D-2$ of the infrared dynamics responsible for the fermion pairing in a magnetic field. This reduction is a reflection of simple physics: the motion of charged particles is partly restricted in the plane perpendicular to the magnetic field. The latter is also related to the fact that the chiral condensate mainly appears due to the lowest Landau level whose dynamics is $(D-2)$–dimensional.

In this talk I shall briefly present the results for the supersymmetric Nambu–Jona–Lasinio (SNJL) model in a magnetic field.

The motivation for the problem is the following. As was heuristically proved in (Gusynin et al. (1994), (1995a), Gusynin et al. (1995b), (1995c), (1996)), the catalysis by an external magnetic field is a rather universal (model–independent) phenomenon. In non–supersymmetric models, chiral symmetry breaking is usually realized if the coupling constant is large enough. As for the influence of the magnetic field, it reduces the critical coupling to zero. On the other hand, there is no spontaneous chiral symmetry breaking in the SNJL model at all (Buchmüller and Love (1982)).

Below it will be shown that an external magnetic field changes the situation in the SNJL model dramatically: chiral symmetry breaking, in agreement with the universality of the effect (Gusynin et al. (1995b), (1995c), (1996)), occurs for any value of the coupling constant.

The action of the SNJL model with the $\mathrm{U}_L(1) \times \mathrm{U}_R(1)$ chiral symmetry in a magnetic field (in notations of Ref. Wess and Bagger (1992) except the metric $g^{\mu\nu} = \mathrm{diag}(1,-1,-1,-1)$) is:

$$\Gamma = \int \mathrm{d}^8 z \left[\bar{Q} e^V Q + \bar{Q}^c e^{-V} Q^c + G(\bar{Q}^c \bar{Q})(Q Q^c) \right] \ . \tag{1}$$

Here $\mathrm{d}^8 z = \mathrm{d}^4 x \mathrm{d}^2 \theta \mathrm{d}^2 \bar{\theta}$, Q^α and Q^c_α are chiral superfields carrying the color index $\alpha = 1, 2, \ldots, N_c$, i.e. Q^α and Q^c_α are assigned to the fundamental and antifundamental representations of the $\mathrm{SU}(N_c)$, respectively:

$$Q^\alpha = \varphi^\alpha + \sqrt{2}\theta \psi^\alpha + \theta^2 F^\alpha, \quad Q^c_\alpha = \varphi^c_\alpha + \sqrt{2}\theta \psi^c_\alpha + \theta^2 F^c_\alpha \tag{2}$$

(henceforth I shall omit color indices). The vector superfield $V(x, \theta, \bar{\theta}) = -\theta \sigma^\mu \bar{\theta} A^{\mathrm{ext}}_\mu$, with $A^{\mathrm{ext}}_\mu = B x^2 \delta^3_\mu$, describes an external magnetic field in the $+x_1$ direction.

The action (1) is equivalent to the following one:

$$\Gamma_A = \int \mathrm{d}^8 z \left[\bar{Q} e^V Q + \bar{Q}^c e^{-V} Q^c + \frac{1}{G} \bar{H} H \right] - \\ - \int \mathrm{d}^6 z \left[\frac{1}{G} H S - Q Q^c S \right] - \int \mathrm{d}^6 \bar{z} \left[\frac{1}{G} \bar{H} \bar{S} - \bar{Q} \bar{Q}^c \bar{S} \right] \ . \tag{3}$$

Here $\mathrm{d}^6 z = \mathrm{d}^4 x \mathrm{d}^2 \theta$, $\mathrm{d}^6 \bar{z} = \mathrm{d}^4 x \mathrm{d}^2 \bar{\theta}$, and H and S are two auxiliary chiral fields:

$$H = h + \sqrt{2}\theta \chi_h + \theta^2 f_h, \quad S = s + \sqrt{2}\theta \chi_s + \theta^2 f_s \ . \tag{4}$$

The Euler–Lagrange equations for these auxiliary fields take the form of constraints:

$$H = G Q Q^c, \quad S = -\frac{1}{4} \bar{D}^2 (\bar{H}) = -\frac{G}{4} \bar{D}^2 (\bar{Q} \bar{Q}^c) \ . \tag{5}$$

Here $\bar{D}$ is a SUSY covariant derivative (Wess and Bagger (1992)). The action (3) reproduces Eq.(1) upon application of the constraints (5).

In terms of the component fields, the action (3) is

$$\Gamma_A = \int \mathrm{d}^4 x \Big[- \varphi^\dagger (\partial_\mu - \mathrm{i} e A^{\mathrm{ext}}_\mu)^2 \varphi - \varphi^{c\dagger} (\partial_\mu + \mathrm{i} e A^{\mathrm{ext}}_\mu)^2 \varphi^c \\ + \mathrm{i} \bar{\psi} \bar{\sigma}^\mu (\partial_\mu - \mathrm{i} e A^{\mathrm{ext}}_\mu) \psi + \mathrm{i} \bar{\psi}^c \bar{\sigma}^\mu (\partial_\mu + \mathrm{i} e A^{\mathrm{ext}}_\mu) \psi^c + F^\dagger F + F^{c\dagger} F^c \\ + \frac{1}{G} \left(-h^\dagger \Box h + \mathrm{i} \bar{\chi}_h \bar{\sigma}^\mu \partial_\mu \chi_h + f_h^\dagger f_h \right) + \frac{1}{G} (\chi_h \chi_s - h f_s - s f_h + \mathrm{h.c.}) \\ - \Big(s \psi \psi^c + (\varphi \psi^c + \varphi^c \psi) \chi_s - s (\varphi F^c + \varphi^c F) - \varphi \varphi^c f_s + \mathrm{h.c.} \Big) \Big] \ . \tag{6}$$

To obtain the effective potential, all the auxiliary scalar fields are treated as (independent of x) constants and all the auxiliary fermion fields equal zero. Then, the Euler–Lagrange equations for the fields F, F^c, f_h, h and their conjugates leads to $F^\dagger = -s\varphi^c$, $F^{c\dagger} = -s\varphi$, $f_h^\dagger = s$, $f_s^\dagger = 0$, plus h.c. equations. After taking these into account, the action reads

$$\Gamma_A = \int d^4x \Bigg[-\varphi^\dagger \left[(\partial_\mu - ieA_\mu^{\rm ext})^2 + \rho^2\right]\varphi - \varphi^{c\dagger}\left[(\partial_\mu + ieA_\mu^{\rm ext})^2 + \rho^2\right]\varphi^c$$

$$+ i\bar\psi_D\gamma^\mu(\partial_\mu - ieA_\mu^{\rm ext})\psi_D - \sigma\bar\psi_D\psi_D - \pi\bar\psi_D i\gamma^5\psi_D - \frac{\rho^2}{G}\Bigg] , \qquad (7)$$

where $s = \sigma + i\pi$, $\rho^2 = |s|^2 = \sigma^2 + \pi^2$, and the Dirac fermion field ψ_D is introduced.

In leading order in $1/N_c$, the effective potential $V(\rho)$ can now be derived in the same way as in the ordinary NJL model. The difference is that, besides fermions, the two scalar fields φ^c and φ give a contribution to $V(\rho)$:

$$V(\rho) = \frac{\rho^2}{G} + V_{\rm fer}(\rho) + 2V_{\rm bos}(\rho) , \qquad (8)$$

where

$$V_{\rm fer}(\rho) = \frac{N_c}{8\pi^2 l^4} \int\limits_{1/(l\Lambda)^2}^{\infty} \frac{ds}{s^2} \exp\left(-s(l\rho)^2\right) \coth s , \qquad (9)$$

$$V_{\rm bos}(\rho) = -\frac{N_c}{16\pi^2 l^4} \int\limits_{1/(l\Lambda)^2}^{\infty} \frac{ds}{s^2} \exp\left(-s(l\rho)^2\right) \frac{1}{\sinh s} . \qquad (10)$$

This can be rewritten as (Elias et al. (1996)):

$$V(\rho) = \frac{N_c}{8\pi^2 l^4}\left[\frac{(l\rho)^2}{g} + (l\rho)^2\left(1 - \ln\frac{(l\rho)^2}{2}\right) + 4\cdot \int\limits_{(l\rho)^2/2}^{[(l\rho)^2+1]/2} dx \ln\Gamma(x)\right] +$$

$$+ \frac{N_c}{16\pi^2 l^4}\left[\ln(\Lambda l)^2 - \gamma - \ln(8\pi^2)\right] + O\left(\frac{1}{\Lambda}\right) , \qquad (11)$$

where the dimensionless coupling constant is $g = GN_c/8\pi^2 l^2$.

As the magnetic field B goes to zero ($l \to \infty$), one obtains:

$$V(\rho) = \frac{\rho^2}{G} . \qquad (12)$$

This potential is positive–definite, as has to be in a supersymmetric theory. The only minimum of this potential is $\rho = 0$ what corresponds to the chiral symmetric vacuum (Buchmüller and Love (1982)).

The presence of a magnetic field changes this situation dramatically: at $B \neq 0$, a non–trivial global minimum, corresponding to chiral symmetry breaking, exists for all $g > 0$.

The gap equation $dV/d\rho = 0$, following from Eq.(11), is

$$\frac{N\rho}{4\pi^2 l^2}\left[\frac{1}{g} - \ln\frac{(l\rho)^2}{2} + 2\ln\Gamma\left(\frac{(l\rho)^2+1}{2}\right) - 2\ln\Gamma\left(\frac{(l\rho)^2}{2}\right)\right] = 0 \; . \quad (13)$$

It is easy to check that, at $B \neq 0$, the trivial solution $\rho = 0$ corresponds to a maximum of V, since $\mathrm{d}^2V/\mathrm{d}\rho^2|_{\rho\to 0} \to -\infty$. Numerical analysis of equation (13) for $g > 0$ and $B \neq 0$ shows that there is a nontrivial solution $\rho = m_{\mathrm{dyn}}$ which is the global minimum of the potential. The analytic expression for m_{dyn} can be obtained for small g (when $m_{\mathrm{dyn}}l \ll 1$) and for very large g (when $m_{\mathrm{dyn}}l \gg 1$). In those two cases, the results are:

$$\frac{1}{g} \simeq -\ln\frac{\pi(\rho l)^2}{2}, \quad g \ll 1 \; ; \quad (14)$$

$$\frac{1}{g} \simeq \frac{1}{2(\rho l)^2}, \quad g \gg 1 \; , \quad (15)$$

i.e.

$$m_{\mathrm{dyn}} \simeq \sqrt{\frac{2|eB|}{\pi}}\exp\left[-\frac{4\pi^2}{|eB|N_c G}\right], \quad g \ll 1 \; ; \quad (16)$$

$$m_{\mathrm{dyn}} \simeq \frac{|eB|}{4\pi}\sqrt{GN_c}, \quad g \gg 1 \; . \quad (17)$$

At this point, it seems appropriate to note that the infrared dynamics in the SNJL model in a magnetic field is actually equivalent to that in the ordinary NJL model in a magnetic field as soon as the coupling is weak. This follows from direct comparison of the effective potentials and the kinetic terms of the models (Elias et al. (1996)). The physical picture underlying this equivalence is also clear. The spectra of charged free fermions and bosons in a magnetic field are essentially different:

$$E_n(k_1) = \pm\sqrt{m^2 + 2|eB|n + k_1^2}, \qquad n = 0, 1, 2, \ldots \; , \quad (18)$$

and

$$E_n(k_1) = \pm\sqrt{m^2 + |eB|(2n+1) + k_1^2}, \qquad n = 0, 1, 2, \ldots \; . \quad (19)$$

for fermions and bosons, respectively. The crucial difference between them is the existence of the gap $\Delta E = \sqrt{|eB|}$ in the spectrum of massless bosons and the absence of any gap in the spectrum of massless fermions. Thus at weak coupling, the infrared dynamics, responsible for chiral symmetry breaking, is dominated by fermions while bosonic degrees of freedom are irrelevant. So, it is not a surprise that the infrared dynamics in the SNJL and NJL models are equivalent.

In conclusion, I note that the results obtained here are in agreement with the general conclusion of Refs.(Gusynin et al. (1994), (1995a), Gusynin et al. (1995b), (1995c), (1996)), saying that the catalysis of chiral symmetry breaking by a magnetic field is a universal, model independent effect.

Acknowledgments

I would like to thank the organizers of the Seminar for financial support and for the opportunity to give this talk. I thank V. Elias, D.G.C. McKeon, and V.A. Miransky for enjoyable collaboration.

References

Buchmüller, W. and Love, S.T. (1982): Nucl. Phys. B**204,** 213.
Elias, V., McKeon, D.G.C., Miransky, V.A. and Shovkovy, I.A. (1996): Phys. Rev. D**54,** 7884.
Hong, D.K., Kim, Y. and Sin, S.-J. (1996): Phys. Rev. D**54,** 7879.
Gusynin, V.P., Miransky, V.A. and Shovkovy, I.A. (1994): Phys. Rev. Lett. **73,** 3499;
Gusynin, V.P., Miransky, V.A. and Shovkovy, I.A. (1995a): Phys. Rev. D**52,** 4718.
Gusynin, V.P., Miransky, V.A. and Shovkovy, I.A. (1995b): Phys. Lett. B**349,** 477;
Gusynin, V.P., Miransky, V.A. and Shovkovy, I.A. (1995c): Phys. Rev. D**52,** 4747;
Gusynin, V.P., Miransky, V.A. and Shovkovy, I.A. (1996): Nucl. Phys. B**462,** 249.
Klevansky, S.P. and Lemmer, R.H. (1988): Phys. Rev. D**38,** 3559;
Klimenko, K.G. (1991): Teor. Mat. Fiz. **89,** 211;
Suganuma, H. and Tatsumi, T. (1991): Ann. Phys. **208,** 470;
Schramm, S., Müller, B. and Schramm, A.J. (1992): Mod. Phys. Lett. A**7,** 973.
Leung, L.N., Ng, Y.J. and Ackley, A.W. (1996): Phys. Rev. D**54,** 4181;
Leung, L.N., Ng, Y.J. and Ackley, A.W. (1997): Chiral Symmetry Breaking in a Uniform External Magnetic Field, hep-th/9701172.
Wess, J. and Bagger, J. (1992): *Supersymmetry and Supergravity*, Princeton.

On Extension of Minimality Principle in Supersymmetric Electrodynamics

V.V. Tugai[1] and A.A. Zheltukhin[2]

[1] Scientific Physicotechnological Center, 310145 Kharkov, Ukraine
[2] Kharkov Physicotechnical Institute, 310108 Kharkov, Ukraine

Deep geometric ideas of Eli Cartan (Cartan (1923, 1924)) inspired Dmitrij Vasiljevitch Volkov in his investigations of the phenomenological lagrangians method (Volkov (1969)) and in his work at the construction of supersymmetry and supergravity (Volkov and Akulov (1972)), (Volkov and Soroka (1973)). The Cartan idea on the extension of the connection conception and the introduction of torsion was applied by Dmitrij Vasiljevitch Volkov in the formulation of minimality principle for the interactions of Goldstone particles with other fields (Volkov, Zheltukhin and Tkach (1972)). His profound intuition at once allowed to assume that the Pauli matricies $\sigma^{\mu}_{\alpha\dot{\alpha}}$ play a fundamental role of the torsion tensor components in $z^M = (x^\mu, \theta^\alpha_i, \bar{\theta}_{\dot{\alpha}i})$ superspace.

The Cartan's geometry naturally emerged in early papers by Scherk and Schwarz (Sherk and Schwarz (1974)) on the string theory of gravitation where the strength $H_{\mu\nu}{}^{\rho}$ of the antisymmetric Kalb-Ramond field plays the role of torsion (Kalb and Ramond (1974)). These and some other impressive results show ties between superspace torsion and spin.

Here we wish to single out a possibility of an extension of the minimality principle for electromagnetic interactions of charged and neutral particles having spin 1/2. This possibility is also based on the Cartan's idea of an extension of the connection conception.

1 Superspace Connections and Electromagnetic Interactions

An extention of the minimality principle can be achieved by the addition of new terms (Tugai and Zheltukhin (1996), Tugai and Zheltukhin (1995)) to the standard gauge covariant and supesymmetric derivatives $\mathcal{D}_M$ of supersymmetric electrodynamics (Wess and Bagger (1983))

$$\mathcal{D}_M = D_M + eA_M \quad \rightarrow \quad \nabla_M = \mathcal{D}_M + \mathrm{i}\mu\tilde{W}_M \equiv D_M + e_{(q)}A^{(q)}_M \ , \qquad (1)$$

where the additional superfield $\tilde{W}_M(z)$ is an invariant of the gauge group $U(1)$, $e_{(q)}A^{(q)}_M \equiv eA_M + \mathrm{i}\mu\tilde{W}_M$ is a new generalized connection, and the constant μ with the dimension of length ($\hbar = c = 1$) has the physical meaning of anomalous magnetic moment (AMM).

The algebra of the doubly lengthened covariant derivatives $\nabla_M = (\nabla_\mu, \nabla_\alpha{}^i, \bar{\nabla}^{\dot{\alpha}i})$ differs from the algebra of the standard N-extended supersymmetric derivatives (West (1986))

$$
\begin{aligned}
&[D_M, D_N\} = T_{MN}{}^L D_L , \\
&T_{MN}{}^L = -2\mathrm{i}\sigma^\mu_{\alpha\dot{\beta}} , \qquad \text{for } N = 1 \text{ SUSY} \\
&T_{MN}{}^L = 2\mathrm{i}\sigma^\mu_{\alpha\dot{\beta}}\delta^j_i , \qquad \text{for } N > 1 \text{ SUSY}
\end{aligned} \tag{2}
$$

by the presence of the electromagnetic superfield strength $\mathbb{F}^{(q)}_{MN}$

$$
[\nabla_M, \nabla_N\} = T_{MN}{}^L \nabla_L + e_{(q)} \mathbb{F}^{(q)}_{MN} . \tag{3}
$$

which takes into account AMM of superparticle and is defined as

$$
\begin{aligned}
&e_{(q)} \mathbb{F}^{(q)}_{MN} \equiv eF_{MN} + \mathrm{i}\mu F^{(\mu)}_{MN} , \\
&F_{MN} \equiv D_M A_N - (-1)^{MN} D_N A_M - T_{MN}{}^R A_R , \\
&F^{(\mu)}_{MN} \equiv D_M \tilde{W}_N - (-1)^{MN} D_N \tilde{W}_M - T_{MN}{}^R \tilde{W}_R .
\end{aligned} \tag{4}
$$

Conservation of supersymmetry and $U(1)$ gauge symmetry demands that the superfield $\tilde{W}_M(z)$ should be an invariant of these symmetries. Let us look for these invariants in the set of functions which are a linear combination of the strength components $F_{MN}(z)$ with constant coefficients. The dimensionality reasons, i.e. $[\tilde{W}_\mu] = L^{-2}$, $[\tilde{W}^i_\alpha] = L^{-3/2}$, together with the constraints imposed on F_{MN} (Wess and Bagger (1983), West (1986)) sharply restrict the form of the invariant $\tilde{W}_M$. As a result, the desired spinorial components of $\tilde{W}_M$ are taken in the form of the product of $F_{\mu\alpha}{}^i$ by the torsion of the flat superspace $\tilde{W}_{\dot{\alpha}i} \sim F_\mu{}^\alpha{}_j \sigma^\mu_{\alpha\dot{\alpha}} \delta^j_i$. In particular, for $N = 2$ case the expression $\tilde{W}_\mu \sim \partial_\mu F_\alpha{}^i{}_\beta{}^j \varepsilon^{\alpha\beta} \varepsilon_{ij}$ may be taken for the vector component of $\tilde{W}_\mu$. Due to the symmetry of $F_{\alpha\beta}$ this expression equals zero for $N = 1$ case. The desired representation for the superfield $\tilde{W}_M$ in $N = 1$ case has the following form (Tugai and Zheltukhin (1996), Tugai and Zheltukhin (1995))

$$
\tilde{W}_M = W_M \equiv \frac{\mathrm{i}}{4}(0, -\sigma_{\mu\alpha\dot{\alpha}} F^{\mu\dot{\alpha}}, \tilde{\sigma}^{\mu\dot{\alpha}\alpha} F_{\mu\alpha}) . \tag{5}
$$

The uniqueness of the representation (5) arises when taking account of the standard F_{MN} constraints (Wess and Bagger (1983))

$$
F_{\alpha\beta} = F_{\dot{\alpha}\dot{\beta}} = 0 , \tag{6}
$$

$$
F_{\alpha\dot{\beta}} = 0 . \tag{7}
$$

and the Bianchi identities

$$
\mathbb{C}\mathrm{ycl}_{MNR} (-1)^{(MNR)} \left(D_M F_{NR} - T_{MN}{}^L A_R \right) = 0 \tag{8}
$$

equivalent to the Jacobi identities for the covariant derivatives $\mathcal{D}_M$ of supersymmetric electrodynamics (SUSY ED)

$$\mathbb{C}\mathrm{ycl}_{MNR}(-1)^{MNR}[\mathcal{D}_M,[\mathcal{D}_N\ ,\mathcal{D}_R\}\}=0\ . \tag{9}$$

Eqs. (6,7) require that the superfield F_{MN} of $N=1$ SUSY ED should be expressed in terms of the spinor chiral superfields W_α and $\bar{W}_{\dot\alpha}$ (Wess and Bagger (1983))

$$W^\alpha=\frac{\mathrm{i}}{4}F_{\mu\dot\alpha}\tilde\sigma^{\mu\,\dot\alpha\alpha}\ ,\qquad \bar{W}^{\dot\alpha}=(W^\alpha)^*\ . \tag{10}$$

Therefore $\tilde{W}_\mu$ must be constructed of W_α and $\bar{W}_{\dot\alpha}$. Using these superfields one cannot construct the vector superfield $\tilde{W}_\mu$ having the dimension L^{-2}. As a result, the desired extension of the minimality principle for $N=1$ SUSY ED is defined by the superfield $\tilde{W}_M$ (10) having the form

$$\tilde{W}_M(z)=(0,W_\alpha,\bar{W}^{\dot\alpha}) \tag{11}$$

with W_α and $\bar{W}_{\dot\alpha}$ restricted by the chirality and reality conditions (Wess and Bagger (1983))

$$D_\alpha\bar{W}_{\dot\alpha}=\bar{D}_{\dot\alpha}W_\alpha=D^\alpha W_\alpha-\bar{D}_{\dot\alpha}\bar{W}^{\dot\alpha}=0\ . \tag{12}$$

The proof of the uniqueness is completed by the analysis of the Jacobi identities

$$\mathbb{C}\mathrm{ycl}_{MNR}(-1)^{(MNR)}[\nabla_M,[\nabla_N\ ,\nabla_R\}\}=0\ , \tag{13}$$

which take the form

$$e\mathbb{C}\mathrm{ycl}_{MNR}(-1)^{(MNR)}\left(D_MF_{NR}-T_{MN}{}^LF_{LR}\right)+\mathrm{i}\mu\mathbb{C}\mathrm{ycl}_{MNR}(-1)^{(MNR)}$$
$$\left(D_MF^{(\mu)}_{NR}-T_{MN}{}^LF^(\mu)_{LR}\right)=0 \tag{14}$$

after the substitution of Eqs.(1), (3) and (4) into Eq. (13). The first and the second term in Eq.(14) equal zero by construction and due to sufficient arbitrariness in the definition of W_α and $\bar{W}_{\dot\alpha}$, respectively

Therefore the extended derivatives (1, 11) satisfy the Bianchi identities and give a solution of the desired problem of selfconsistent extension of the minimality principle for $N=1$ SUSY ED particles with spin 1/2.

The superalgebra (3) of the generalized derivatives ∇_M takes the form

$$\begin{aligned}
[\nabla_\mu,\nabla_\nu]&=-\frac{e}{2}(\bar{D}\tilde\sigma_{\mu\nu}\bar{W}-D\sigma_{\mu\nu}W)\ ,\\
[\nabla_\mu,\nabla_\alpha]&=\mathrm{i}eW^\beta\sigma_{\mu\beta\dot\alpha}+\mathrm{i}\mu\partial_\mu W_\alpha\ ,\\
\{\nabla_\alpha,\nabla_\beta\}&=\mathrm{i}\mu(D_\alpha W_\beta+D_\beta W_\alpha)\ ,\\
\{\nabla_\alpha,\bar\nabla_{\dot\beta}\}&=-2\mathrm{i}\sigma^\mu_{\alpha\dot\beta}\nabla_\mu+\mathrm{i}\mu(D_\alpha\bar{W}_{\dot\beta}+\bar{D}_{\dot\beta}W_\alpha)\ ,
\end{aligned} \tag{15}$$

and we observe the noncommutativity effect for the generalized spinor derivatives ∇_α.

Due to the considered extension procedure, the action for charged or neutral superparticle in an external electromagnetic field gets an additional term which describes the electromagnetic interaction of superparticle taking into account its AMM

$$\begin{aligned} S_{\text{int}} &= \mathrm{i}\int \mathrm{d}z^M e_{(q)} A_M^{(q)} = S^{(e)} + S^{(\mu)} , \\ S^{(e)} &= \mathrm{i}e \int \mathrm{d}z^M A_M , \\ S^{(\mu)} &= -\mu \int \mathrm{d}z^M \tilde{W}_M = -\mu \int \mathrm{d}\theta^\alpha W_\alpha - \mu \int \mathrm{d}\bar{\theta}_{\dot{\alpha}} \bar{W}^{\dot{\alpha}} . \end{aligned} \tag{16}$$

The component analysis of the action (16) carried out in (Tugai and Zheltukhin (1996), Tugai and Zheltukhin (1995)) shows that $S^{(\mu)}$ term in Eq.(16) generates the nonminimal Pauli term with the constant μ which may be interpreted as the AMM of Dirac particle measured in values of Bohr magneton.

In the next section we consider a generalization of the extended minimality principle for the case $N = 2$.

2 Generalized Connections in $N = 2$ Extended Electrodynamics

In the standard $N = 2$ SUSY ED (Lusanna and Milevski (1984), Sohnius, Stelle and West (1980)) the strength components F_{MN} obey the constraints

$$F_{(\alpha}{}^{i}{}_{\beta)}{}^{j} = F_{(\dot{\alpha}|i|\dot{\beta})j} = 0 , \tag{17}$$

$$F_{\alpha}{}^{i}{}_{\dot{\beta}j} = 0 . \tag{18}$$

Together with the Bianchi identities, these constraints permit to express F_{MN} in terms of the scalar chiral superfield $W(z)$ and $\bar{W}(z)$ having the dimension L^{-1} and restricted by the conditions

$$\bar{D}_{\dot{\alpha}i} W = D_\alpha{}^i \bar{W} = 0 . \tag{19}$$

The explicit form of the superfield F_{MN} components is given by the following expressions

$$\begin{aligned} F_{\alpha}{}^{i}{}_{\beta}{}^{j} &= -\varepsilon_{\alpha\beta}\varepsilon^{ij}\bar{W} , \quad F_{\dot{\alpha}i\dot{\beta}j} = \varepsilon_{\dot{\alpha}\dot{\beta}}\varepsilon_{ij} W , \\ F_{\mu\alpha}{}^{i} &= \frac{\mathrm{i}}{4}(\sigma_\mu \bar{D}^i)_\alpha \bar{W} , \quad F_{\mu\dot{\alpha}i} = -\frac{\mathrm{i}}{4}(D_i \sigma_\mu)_{\dot{\alpha}} W , \\ F_{\mu\nu} &= \frac{1}{16}(\bar{D}_i \tilde{\sigma}_{\mu\nu} \bar{D}^i)\bar{W} + \frac{1}{16}(D^i \sigma_{\mu\nu} D_i) W . \end{aligned} \tag{20}$$

In order to construct an extended connection for the $N = 2$ SUSY ED, we must again use the dimensionality reasones and look for the required extension in terms of a linear combination of W, $\bar{W}$ and their covariant derivatives. Then the required representation for the additional superfield

T_M (which produces additional term to the connection A_M) may be written in the form

$$\tilde{W}_M(z) = T_M = (\partial_\mu(bW + \bar{b}\bar{W}) \ , aD_\alpha{}^i W, -\bar{a}\bar{D}^{\dot{\alpha}i}\bar{W}) \ . \tag{21}$$

In a way analogous to that used in $N = 1$ case (Tugai and Zheltukhin (1996), Tugai and Zheltukhin (1995)) it is convenient to introduce the two-component "charge" $e_{(q)} = (e, \mathrm{i}\mu)$ and the superconnection $A_M^{(q)} = \binom{A_M}{T_M}$ in the extended "charged" space $e_{(q)}$. Then the doubly lengthened covariant derivative (3) can be more compactly presented as

$$\nabla_M = D_M + e_{(q)} A_M^{(q)} \ . \tag{22}$$

Then the $U(1)$ gauge invariant superfield action for charged or neutral particles with AMM in an external superfield of the $N = 2$ Maxwell supermultiplet takes the form

$$\mathrm{S_{int}} = \mathrm{S_{int}^{(e)}} + \mathrm{S_{int}^{(\mu)}} = \mathrm{i}e_{(q)} \int\limits_{\Gamma^*} \omega^M(\mathrm{d}z) A_M^{(q)} = \mathrm{i} \int\limits_{\Gamma^*} (e\omega^M A_M + \mathrm{i}\mu\omega^M T_M) \tag{23}$$

where $N = 2$ Cartan's differential forms $\omega^M(\mathrm{d}z) = (\omega^\mu, \omega^\alpha{}_i, \omega^{\dot{\alpha}i})$ in $N = 2$ superspace z_M are given by (Volkov and Akulov (1972), West (1986)):

$$\begin{aligned} &\omega^\mu = \mathrm{d}x^\mu - \mathrm{i}(\mathrm{d}\theta_i \sigma^\mu \bar{\theta}^i) + \mathrm{i}(\theta_i \sigma^\mu \mathrm{d}\bar{\theta}^i) \ , \\ &\omega^\alpha{}_i = \mathrm{d}\theta^\alpha{}_i, \ \bar{\omega}_{\dot{\alpha}i} = \mathrm{d}\bar{\theta}_{\dot{\alpha}i} \ . \end{aligned} \tag{24}$$

The use of the relation

$$\frac{\mathrm{d}}{\mathrm{d}\tau} W = \omega_\tau^\mu \partial_\mu W + \dot{\theta}^\alpha{}_i D_\alpha{}^i W + \dot{\bar{\theta}}_{\dot{\alpha}i} \bar{D}^{\dot{\alpha}i} W \tag{25}$$

allows to present $\mathrm{S_{int}}$ (23) in the equivalent form

$$\mathrm{S_{int}} = \mathrm{i}e \int\limits_{\Gamma^*} \mathrm{d}\tau \omega_\tau^M A_M + \mathrm{i}\mu \int\limits_{\Gamma^*} \mathrm{d}\tau \left(\lambda_2 \dot{\theta}^\alpha{}_i D_\alpha{}^i W + \bar{\lambda}_2 \dot{\bar{\theta}}_{\dot{\alpha}i} \bar{D}^{\dot{\alpha}i} \bar{W} \right) \ . \tag{26}$$

Following from Eq.(26) is the existence of two physically equivalent ways for the representation of the required superfield T_M:

$$T_M = (\partial_\mu(\lambda_1 W + \bar{\lambda}_1 \bar{W}), 0, 0) \ , \tag{27}$$

and

$$T_M = (0, -\mathrm{i}\lambda_2 D_\alpha{}^i W, \mathrm{i}\bar{\lambda}_2 \bar{D}^{\dot{\alpha}i} \bar{W}) \ . \tag{28}$$

Without restricting generality of considerations, we shall use the T_M representation in the form

$$T_M = (0, -\frac{1}{4} D_\alpha{}^i W, -\frac{1}{4} \bar{D}^{\dot{\alpha}i} \bar{W}) \ . \tag{29}$$

Note that the equivalent representation for T_M (29) is the following: $T_M = (\frac{1}{4}\partial_\mu(W+\bar{W}),\, 0,\, 0)$. Eq.(29) creates the expressions for the components of $N=2$ generalized superfield strength of the Maxwell supermultiplet

$$
\begin{aligned}
e_{(q)}\mathbb{F}^{(q)}{}_{\alpha}{}^{i}{}_{\beta}{}^{j} &= eF_{\alpha}{}^{i}{}_{\beta}{}^{j} = -e\varepsilon_{\alpha\beta}\varepsilon^{ij}\bar{W}\ ,\\
e_{(q)}\mathbb{F}^{(q)}{}_{\dot{\alpha}i\dot{\beta}j} &= eF_{\dot{\alpha}i\dot{\beta}j} = e\varepsilon_{\dot{\alpha}\dot{\beta}}\varepsilon_{ij}W\ ,\\
e_{(q)}\mathbb{F}^{(q)}{}_{\alpha}{}^{i}{}_{\dot{\beta}j} &= eF_{\alpha}{}^{i}{}_{\dot{\beta}j} + \mathrm{i}\mu D_{\alpha}{}^{i}T_{\dot{\beta}j} + \mathrm{i}\mu\bar{D}_{\dot{\beta}j}T_{\alpha}{}^{i} = \frac{1}{2}\mu\delta^i_j\hat{\partial}_{\alpha\dot{\beta}}(W+\bar{W})\ ,\\
e_{(q)}\mathbb{F}^{(q)}{}_{\mu\alpha}{}^{i} &= eF_{\mu\alpha}{}^{i} + \mathrm{i}\mu\partial_\mu T_{\alpha}{}^{i} = e\frac{\mathrm{i}}{4}(\sigma_\mu\bar{D}^i)_\alpha\bar{W} - \frac{\mathrm{i}}{4}\mu\partial_\mu D_{\alpha}{}^{i}W\ ,\\
e_{(q)}\mathbb{F}^{(q)}{}_{\mu\dot{\alpha}i} &= eF_{\mu\dot{\alpha}i} + \mathrm{i}\mu\partial_\mu T_{\dot{\alpha}i} = -e\frac{\mathrm{i}}{4}(D_i\sigma_\mu)_{\dot{\alpha}}W - \frac{\mathrm{i}}{4}\mu\partial_\mu\bar{D}_{\dot{\alpha}i}\bar{W}\ ,\\
e_{(q)}\mathbb{F}^{(q)}{}_{\mu\nu} &= eF_{\mu\nu} = \frac{e}{16}(\bar{D}_i\tilde{\sigma}_{\mu\nu}\bar{D}^i)\bar{W} + \frac{e}{16}(D^i\sigma_{\mu\nu}D_i)W\ .
\end{aligned}
\tag{30}
$$

As a consequence of Eqs.(30), the algebra of $N=2$ doubly lengthened covariant derivatives ∇_M takes the form

$$
\begin{aligned}
\{\nabla_{\alpha}{}^{i}, \nabla_{\beta}{}^{j}\} &= -e\varepsilon_{\alpha\beta}\varepsilon^{ij}\bar{W}\ ,\\
\{\bar{\nabla}_{\dot{\alpha}i}, \bar{\nabla}_{\dot{\beta}j}\} &= e\varepsilon_{\dot{\alpha}\dot{\beta}}\varepsilon_{ij}W\ ,\\
\{\nabla_{\alpha}{}^{i}, \bar{\nabla}_{\dot{\beta}j}\} &= 2\mathrm{i}\sigma^{\mu}_{\alpha\dot{\beta}}\delta^i_j\nabla_\mu + \frac{1}{2}\mu\delta^i_j\hat{\partial}_{\alpha\dot{\beta}}(W+\bar{W})\ ,\\
[\nabla_\mu, \nabla_{\alpha}{}^{i}] &= e\frac{\mathrm{i}}{4}(\sigma_\mu\bar{D}^i)_\alpha\bar{W} - \frac{\mathrm{i}}{4}\mu\partial_\mu D_{\alpha}{}^{i}W\ ,\\
[\nabla_\mu, \bar{\nabla}_{\dot{\alpha}i}] &= = -e\frac{\mathrm{i}}{4}(D_i\sigma_\mu)_{\dot{\alpha}}W - \frac{\mathrm{i}}{4}\mu\partial_\mu\bar{D}_{\dot{\alpha}i}\bar{W}\ ,\\
[\nabla_\mu, \nabla_\nu] &= \frac{e}{16}(\bar{D}_i\tilde{\sigma}_{\mu\nu}\bar{D}^i)\bar{W} + \frac{e}{16}(D^i\sigma_{\mu\nu}D_i)W\ .
\end{aligned}
\tag{31}
$$

The motin equations of the charged $N=2$ superparticle with AMM generated by the action (26) and S_0

$$
\mathrm{S} = \mathrm{S}_0 + \mathrm{S}_{\mathrm{int}} = \frac{1}{2}\int_{\Gamma^*} \mathrm{d}\tau \left[\frac{\omega^{\mu}_{\tau}\omega_{\tau\mu}}{g_\tau} + g_\tau m^2\right] + \mathrm{i}\int_{\Gamma^*} \mathrm{d}\tau\, \omega^M_\tau e^{(q)} A^{(q)}_M \tag{32}
$$

take the form of the generalized Lorentz equations

$$
\begin{aligned}
(g_\tau^{-1}\omega_{\tau\mu})^{\cdot} &= -\mathrm{i}\omega^M_\tau e_{(q)}\mathbb{F}^{(q)}_{M\mu}\ ,\\
g_\tau^{-1}\omega_{\tau\mu}(\sigma^\mu\dot{\bar{\theta}}^i)_\alpha &= \frac{1}{2}\omega^M_\tau e_{(q)}\mathbb{F}^{(q)}_{M\alpha}{}^{i}\ ,\\
g_\tau^{-1}\omega_{\tau\mu}(\dot{\theta}_i\sigma^\mu)_{\dot{\alpha}} &= -\frac{1}{2}\omega^M_\tau e_{(q)}\mathbb{F}^{(q)}_{M\dot{\alpha}i}\ .
\end{aligned}
\tag{33}
$$

where $\mathbb{F}^{(q)}_{M\mu}$ is defined by Eqs.(33).

3 Anomalous Magnetic Moment and Extended Connection

The component expansion of the superfield $T_\alpha{}^i$ depending on the chiral variables has the form analogous to (West (1986))

$$\begin{aligned}T_\alpha{}^i(z_L) &= \frac{\mathrm{i}}{4}\lambda_\alpha{}^i(x_L) + \frac{1}{8}\theta_{\alpha k}C^{ki}(x_L) - \frac{1}{8}\theta^{\beta i}f_{\beta\alpha}(x_L) - \frac{\mathrm{i}}{2}(\hat{\partial}\bar{\theta}^i)_\alpha w(x_L)\\ &-\frac{1}{2}(\hat{\partial}\bar{\theta}^i)_\alpha(\theta_j\lambda^j(x_L)) - \frac{1}{4}\Xi^{(2,0)ij}(\hat{\partial}\bar{\lambda}_j(x_L))_\alpha - \frac{1}{4}\Xi^{(2,0)\beta}{}_\alpha(\hat{\partial}\bar{\lambda}^i(x_L))_\beta\\ &+\frac{\mathrm{i}}{8}\Xi^{(2,0)jk}(\hat{\partial}\bar{\theta}^i)_\alpha C_{jk}(x_L) - \frac{\mathrm{i}}{8}\Xi^{(2,0)\beta\gamma}(\hat{\partial}\bar{\theta}^i)_\alpha f_{\beta\gamma}(x_L)\\ &+\frac{1}{3}\theta_\alpha{}^i\Xi^{(2,0)}{}_j{}^i\Box\bar{w}(x_L) + \frac{\mathrm{i}}{3}\Xi^{(2,0)\gamma}{}_\beta\theta^\beta{}_j(\hat{\partial}\bar{\theta}^i)_\alpha(\hat{\partial}\bar{\lambda}^j(x_L))_\gamma\\ &+\frac{\mathrm{i}}{6}\Xi^{(4,0)}(\hat{\partial}\bar{\theta}^i)_\alpha\Box\bar{w}(x_L)\ .\end{aligned} \tag{34}$$

The explicit form for the monomials $\Xi^{(m,n)}$ is given in (Tugai and Zheltukhin (1997)).

The physical content of the theory is convenient to analyse in the central base coordinates. In this base the part of the action (26) which describes the contribution of AMM particle has the form

$$\begin{aligned}\mathrm{S}^{(\mu)}_{\mathrm{int}}\Big|_{\mathrm{photon}} &= -\mu\int_{\Gamma^*} \mathrm{d}\tau\left(\omega_\tau^M T_M(x^\mu,\theta^\alpha{}_i,\bar{\theta}_{\dot{\alpha}i})\Big|_{\mathrm{photon}}\right) =\\ &= \mu\int_{\Gamma^*}\mathrm{d}\tau\Big[\mathrm{i}(\dot{\theta}\sigma^{\mu\nu}\theta)v_{\mu\nu} + \frac{1}{2}(\dot{\theta}_i\sigma^\nu\bar{\theta}^j)\Xi^{(2,0)i}{}_j\partial^\mu v_{\mu\nu}\\ &+\frac{1}{2}(\dot{\theta}_i\sigma^{\mu\nu}\theta_k)(\theta^k\sigma^\rho\bar{\theta}^i)\partial_\rho v_{\mu\nu} - (\dot{\theta}\sigma^\rho\bar{\theta})(\theta\sigma^{\mu\nu}\theta)\partial_\rho v_{\mu\nu}\\ &+\frac{\mathrm{i}}{2}(\dot{\theta}_i\theta_j)\Xi^{(0,2)ij}(\theta\sigma^{\mu\nu}\theta)\Box v_{\mu\nu} + \frac{\mathrm{i}}{2}(\dot{\theta}\sigma^{\mu\nu}\theta)\Xi^{(2,2)}\Box v_{\mu\nu}\\ &+\mathrm{i}(\dot{\theta}\sigma^{\mu\nu}\theta)\Xi^{(2,2)\rho\sigma}\partial_\rho\partial_\sigma v_{\mu\nu} - \frac{\mathrm{i}}{2}(\dot{\theta}_i\sigma^\rho\bar{\theta}_k)(\theta^i\sigma^\sigma\bar{\theta}^k)(\theta\sigma^{\mu\nu}\theta)\partial_\rho\partial_\sigma v_{\mu\nu}\\ &-\frac{1}{2}(\dot{\theta}\sigma^\rho\bar{\theta})(\theta\sigma^{\mu\nu}\theta)\Xi^{(2,2)}\partial_\rho\Box v_{\mu\nu} - (\dot{\theta}\sigma^\rho\bar{\theta})(\theta\sigma^{\mu\nu}\theta)\Xi^{(2,2)\sigma\eta}\partial_\rho\partial_\sigma\partial_\eta v_{\mu\nu}\\ &+\frac{1}{4}(\dot{\theta}_i\sigma^\nu\bar{\theta}^j)\Xi^{(2,2)}\Xi^{(2,0)i}{}_j\Box\partial^\mu v_{\mu\nu} + \frac{1}{4}(\dot{\theta}_i\sigma^{\mu\nu}\theta_k)(\theta^k\sigma^\rho\bar{\theta}^i)\Xi^{(2,2)}\Box\partial_\rho v_{\mu\nu}\\ &+\frac{1}{2}(\dot{\theta}_i\sigma^\nu\bar{\theta}^j)\Xi^{(2,2)\rho\sigma}\Xi^{(2,0)i}{}_j\partial_\rho\partial_\sigma\partial^\mu v_{\mu\nu}\\ &+\frac{1}{2}(\dot{\theta}_i\sigma^{\mu\nu}\theta_k)(\theta^k\sigma^\rho\bar{\theta}^i)\Xi^{(2,2)\sigma\eta}\partial_\rho\partial_\sigma\partial_\eta v_{\mu\nu} + \mathrm{c.c.}\Big]\ .\end{aligned} \tag{35}$$

To elucidate the physical meaning of Eq.(34), note that it is a pseudoclassical limit $\hbar \to 0$ of field theory. Under such a consideration the grassmannian spinors $\theta_\alpha{}^i$ are treated as a limit of the fermionic Fock operators (Casalbuoni (1976), Berezin and Marinov (1975), Volkov and Akulov (1973)) $b_\alpha{}^i$ which describe spin degrees of freedom. Since ${\theta_\alpha}^i$ have dimensionality $L^{-1/2}$ and ${b_\alpha}^i$ are choosen dimensionless, they are connected by the relation

$$\sqrt{\frac{\hbar}{2mc}}\, {b_\alpha}^i \to {\theta_\alpha}^i \text{ when } \hbar \to 0 \ , \tag{36}$$

where $\hbar/mc$ is the Compton wavelength corresponding to the massive Dirac field quants. Further it is convenient to pass from ${\theta_\alpha}^i$ and ${\dot\theta_\alpha}^i$ having the dimension $L^{\frac{1}{2}}$ to a Dirac bispinor Θ^i

$$\Theta^i = \begin{pmatrix} \frac{1}{g_\tau}{\dot\theta_\alpha}^i \\ m^2\bar\theta^{\dot\alpha i} \end{pmatrix} , \qquad \Sigma_{\mu\nu} = \frac{\mathrm{i}}{4}[\gamma_\mu, \gamma_\nu] \ , \tag{37}$$

where γ_μ are the Dirac matrices. The bispinor Θ^i has the dimension $L^{-\frac{3}{2}}$ and is an invariant under the proper time reparametrization due to the presence of the einbein g_τ (32), which has the dimension L^{-2} . In the terms of Θ^i (37) the action (35) is presented as

$$\mathrm{S}^{(\mu)}_{\mathrm{int}}\Big|_{\mathrm{photon}} = -\mu \int_{\Gamma^*} \left(\frac{g_\tau \mathrm{d}\tau}{m^2}\right) \left[(\bar\Theta_i \Sigma_{\mu\nu}\Theta^i)v^{\mu\nu} + \text{h. order rel. corrections}\right] \tag{38}$$

where $\left(\frac{g\mathrm{d}\tau}{m^2}\right)$ is a 1-dimensional reparametrization invariant "volume" having the dimension L^4. Due to the fact that the grassmannian bispinor Θ^i are treated as a pseudoclassical limit of fermionic field operators Ψ^i (Casalbuoni (1976), Berezin and Marinov (1975), Volkov and Akulov (1973)), the first term in (38) has a sense of a pseudoclassical limit of the Pauli term. The latter describes the electromagnetic interaction of neutral particles possessing the AMM equal μ. This observation explains the physical meaning of the constant μ.

A possible field theory image of the action (38) can be restored by the substitution which conserves the reparametrization symmetry and all the dimensions

$$\Theta^i(\tau) \to \Psi^i(x), \quad \frac{g_\tau \mathrm{d}\tau}{m^2} \to d^4x \ , \tag{39}$$

where $\Psi^i(x)$ has the sense of the field operator of a neutral particle with spin 1/2 possesing the canonical dimensionality $L^{-\frac{3}{2}}$.

Passing to the photino part of the action and preserving the leading term of the expansion in powers of c^{-1}, we find

$$\mathrm{S}^{(\mu)}_{\mathrm{int}}\Big|_{\mathrm{photino}} = -i\frac{\mu}{4}\int_{\Gamma^*} g_\tau \mathrm{d}\tau \left[\bar\Lambda_{Ri}{\Theta_L}^i - \bar\Theta_{Li}{\Lambda_R}^i\right] , \tag{40}$$

where ${\Lambda^i}_{\binom{L}{R}} = \frac{1 \pm \mathrm{i}\gamma_5}{2}\Lambda^i$, $\Lambda^i = \begin{pmatrix} {\lambda_\alpha}^i \\ \bar\lambda^{\dot\alpha i} \end{pmatrix}$ is the photino field. The expression (40) is a pseudoclassical image of the field action

$$S_{\text{int}}^{(\mu)}\Big|_{\text{photino}} = -i\,\frac{\mu m^2}{4}\int dx^4\left[\bar{\Lambda}_{Ri}\Psi_L{}^i - \bar{\Psi}_{Li}\Lambda_R{}^i\right], \tag{41}$$

which can be interpreted as a hint for a possible conversion of the massive neutralino Ψ_L^i in to the photino Λ_{Ri} with the coupling constant proportional to the AMM of neutralino.

Note also that supersymmetry, together with the proposed extension of the minimality principle, reproduces rather complicated, but controlable structure of high-order terms in the expansion with respect to $1/c$. This structure can be restored by the study of the component structure of the extended superfield and the interpretation of Θ^i as pseudoclassical images of the spinor fields (Tugai and Zheltukhin (1996)).

Thus, we conclude that the proposed here possibility of a generalization of the minimality principle can be found helpful for studying possible spin effects in the electromagnetic interaction of charged and neutral fermions predicted by supersymmetry (Seiberg and Witten (1994)).

We are grateful to N. Berkovitz, E.A. Ivanov, A.P. Rekalo, J.P. Stepanovskij and B.M. Zupnik for fruitful discussions and critical remarks.

References

Cartan, A.H. (1923, 1924): Ann. Ec. Norm. Sup. **40**, 324; **41**, 1.

Volkov, D.V. (1969): Phenomenological lagrangian of the Goldstone particles. Preprint ITP-69-75. Kiev, 1969.

Volkov, D.V., Akulov, V.P. (1972): Pis'ma Zh. Eksp. Teor. Fiz. **16**, 621.

Volkov, D.V. Akulov, V.P. (1973): Zh. Eksp. Teor. Fiz. **17**, 367.

Volkov, D.V., Soroka, V.A. (1973): Pis'ma Zh. Eksp. Teor. Fiz. **18**, 529.

Volkov, D.V., Zheltukhin, A.A. and Tkach, V.I. (1972): Teor. Mat. Fiz. **10**, 329.

Sherk, J., Schwarz, J.H. (1974): Phys. Lett. **B52**, 347.

Kalb, M., Ramond, P. (1974): Phys. Rev. **D9**, 2273.

Tugai, V.V., Zheltukhin, A.A. (1995): Pis'ma Zh. Eksp. Teor. Fiz. **61**, 532.

Tugai, V.V., Zheltukhin, A.A. (1996): Phys. Rev. **D54**, 4160.

Tugai, V.V., Zheltukhin, A.A. (1997): Supersymmetry and minimality principle of electromagnetic interactions. Preprint KPTI-97-6. Kharkov, 1997.

Wess, J., Bagger, J. (1983): Supersymmetry and Supergravity. Prin. Univ. Press. Princeton. New Jersey, 1983.

West, P. (1986): Introduction to supersymmetry and supergravity. World Scientific, 1986.

Lusanna, L., Milevski, B. (1984): Nucl. Phys. **B247**, 396.

Sohnius, M., Stelle, K., West, P. (1980): Superspace and Supergravity. Cambr. Univ. Press. Cambridge, 1980.

Casalbuoni, R. (1976): Nuov. Cim. **33A**, 389.

Berezin, F.A., Marinov, M.S. (1975): Pis'ma Zh. Eksp. Teor. Fiz. **21**, 678.

Seiberg, N., Witten, E. (1994): Nucl. Phys. **B246**, 19.

Stochastic Wess–Zumino–Witten Models

R. Léandre

Département de Mathématiques Institut Elie Cartan
Faculté des Sciences Université de Nancy I
54000 Vandoeuvre-les-Nancy France

1 Introduction

The Wess-Zumino model in quantum field theory consists to introduce a tensorial perturbation to a free field operator in a supersymmetric theory. This tensorial perturbation is related to a functional, the Wess–Zumino–Witten functional. Witten (Witten (1982)) had shown that this model is related to the Morse theory. We look 1+1 dimensional theories, if we look at the propagation of the model (Guilarte (1990), Guilarte).

The purpose of this talk is to look at the one dimensional aspects of the Wess–Zumino–Witten model. We consider a set of paths in a symplectic compact manifold. We introduce the Brownian bridge measure, instead of using the difficult measures of physicists. We construct the Wess–Zumino–Witten functional (or the symplectic functional), when the configuration space is simply connected. This allows us to construct the perturbated exterior derivative, to compute its adjoint, because there is a measure, the associated supercharge, and the associated Wess–Zumino–Witten laplacian. We deform the measure and the considered Hilbert space. The model concentrates near the critical points of the functional, and we perform a short time asymptotic of the operator (Léandre (1988)). In particular, the Wess–Zumino–Witten laplacian tends in law to a supersymmetric brownian harmonic oscillator, whose kernel is known (Léandre).

If the configuration space is not simply connected, we introduce the stochastic universal cover of the brownian bridge. This allows us to define the Wess–Zumino–Witten functional (or the area functional) in whole generality. We show that the diffusion over the universal cover is the lift of the diffusion over the based loop space constructed in Driver, Rockner (1992).

The measures above considered are measures over the space of continuous loops. We introduce a measure over the space of finite energy loops (Léandre) The energy functional is in all the Sobolev spaces. This allows us to construct a stochastic Wess–Zumino–Witten model associated to the energy functional.

2 Stochastic Wess–Zumino–Witten Model of a Symplectic Manifold

Let M be a compact symplectic manifold associated to a symplectic form ω. We suppose that M is Riemannian. Let L and L' to compact Lagrangian submanifolds in transversal position. We introduce the set of paths going from L to L' denoted by $P(L, L')$ and we endow it with the measure

$$\mathrm{d}P_{L,L'} = \frac{p_1(x,y)\mathrm{d}x\mathrm{d}y \otimes \mathrm{d}P_{1,x,y}}{\int_{L\times L'} p_1(x,y)\mathrm{d}x\mathrm{d}y} \tag{1}$$

x describes L, y describes L', $P_{1,x,y}$ is the law of the Brownian bridge starting from x and arriving in time 1 in x. $p_t(x,y)$ is the heat kernel associated to the Laplace-Beltrami operator over M.

Let $\gamma_.$ be a path and τ_s be the parallel transport over the path. A tangent vector over a path is written as $X_s = \tau_s H_s$ where H_s is of finite energy, $X_0 \in T_L$, $X_1 \in T_{L'}$. We use Fourier modes in order to splitt the tangent space into a countable sum of tangent vectors spaces in $\gamma_{1/2}$, and of the tangent space of L at the starting point and the tangent space of L' and the end point. This decomposition allows us to introduce over the tangent space of γ denoted by T_γ a connection ∇^∞ which is compatible with the operation of time reversal and with the metric.

We have for enough simple vector fields the integration by parts formula:

$$E[< \mathrm{d}F, X >] = E[F\mathrm{div}X] \tag{2}$$

for a functional F and for a vector field X. $\mathrm{div}X$ belongs to all the L^p.

We intoduce the Fermionic Fock space associated to the Hilbert tangent space. Let us denote it by $\Lambda(T_\gamma)$. We get a bundle over $P(L, L')$.

We consider the one form, which is close over $P(L, L')$:

$$\sigma = \int_0^1 \omega(\mathrm{d}\gamma_s, .) \tag{3}$$

By looking a system of distinguished paths joining a path of reference and the generic path γ, we can show that there exists a functional F_s (called the Wess–Zumino–Witten functional or the symplectic action) such that $\mathrm{d}F_s = \sigma$: we integrate σ over these distinguished paths by using the theory of Stratonovitch integral. Moreover, we have for a small enough λ:

$$E[\exp[\lambda|F_s|] < \infty \ . \tag{4}$$

We put

$$\mathrm{d}_r = \sum \nabla^\infty_{X_n} \wedge X_n \ , \tag{5}$$

where X_n is an orthonormal basis of the tangent space of γ.

Definition 1.1: the regularized Wess–Zumino–Witten supercharge is

$$Q_{W.Z.W} = \mathrm{d}_r + \mathrm{d}F_s \wedge + \mathrm{d}_r^* + \mathrm{i}_{\mathrm{d}F_s} . \tag{6}$$

It is densely defined and symmetric. The regularized Wess–Zumino–Witten operator is $\Delta_{W.Z.W} = Q^2_{W.Z.W}$, which is densly defined, symmtric and positive.

In order to analyze $Q_{W.Z.W}$ and $\Delta_{W.Z.W}$, we introduce the measure $\mathrm{d}P_{L,L',\epsilon^2}$ by replacing in (1) the time 1 by ϵ^2. This measure concentrates over the finite set of intersection points M_i of L and L'. This suggests that it is suitable to consider the gaussian fluctuations around these intersection points. There are given by the following formula:

$$\mathrm{d}P_{l,M_i} = C\exp[-\frac{\|x-y\|^2}{2}]\mathrm{d}x\mathrm{d}y \otimes \mathrm{d}P_{l,x,y} \tag{7}$$

x is a generic element of $T_L(M_i)$, y is a generic element of $T_{L'}(M_i)$, and $\mathrm{d}P_{l,x,y}$ is the law of the flat brownian bridge starting from x and arriving in y in $T_M(M_i)$. The limit Wess–Zumino–Witten functional is closely related to the Paul Levy area. Namely it is nothing else but

$$F_l = 1/2\int_0^1 \omega(\mathrm{d}\gamma_{s,\mathrm{flat}}, \gamma_{s,\mathrm{flat}}) \tag{8}$$

with some clear notations. We can compute the limit Wess–Zumino–Witten supercharge and the limit Wess–Zumino–Witten laplacian. We get for the second one an infinite dimensional gaussian supersymmetric harmonic oscillator. The only difference with the curved case is that we do not have to introduce connections in order to determine these operators. The kernel of the limit Wess–Zumino–Witten operator is reduced to $\exp[-F_l]$.

We can compute the operators $Q_{\epsilon,W.Z.W}$ and $\Delta_{\epsilon,W.Z.W}$ for the measure P_{L,L',ϵ^2} and the functional $\frac{F_s}{\epsilon^2}$, which tends in law to the limit functional. Modulo some rescaling, which are the mirror of the fact that the limit measure analyzes the fluctuations of the measure near the critical points of the Wess–Zumino–Witten functional, Léandre shows the following theorem:

Theorem 1.2: $Q_{\epsilon,W.Z.W}$ tends in law to the limit Wess–Zumino–Witten supercharge. $\Delta_{\epsilon,W.Z.W}$ tends in law to the limit Wess–Zumino–Witten laplacian.

The reader can see Léandre for more details.

3 Cover of the Brownian Bridge

Let $L_x(M)$ be the based loop space of the symplectic manifold M. Let γ_i a dense se of finite energy loops in $L_x(M)$. Let $B(\gamma_i;\delta)$ the ball of center γ_i and radius δ for the uniform distance. If $\gamma \in B(\gamma_i;\delta)$, we schrink the loop in the constant loop γ_{ref} by the following procedure: in order to go from γ to γ_i, we choose the path:

$$l_{i,t}(s) = \exp_{\gamma_i(s)}[t(\gamma(s) - \gamma_i(s))] \tag{9}$$

and to go from γ_i to γ_{ref}, we choose any path l_i (We select in fact a dense countable set of paths). We say that two distinguished paths l and l' are equivalents if they have the same end loops and if the loop in the loop space constituted of l and l' circled in the opposite sense is the boundary of a stochastic surface (See Léandre for more details. This allows us to define an universal stochastic cover of $L_x(M)$, denoted by $\tilde{L}_x(M)$, and to introduce a measure over it $\tilde{P}$ which lifts the Brownian bridge measure. The closed form σ can be lifted to a form $\tilde{\sigma}$, and by constuction, by using the theory of Stratonovitch integrals, there exists a functional $\tilde{F}$ such that $\mathrm{d}\tilde{F} = \tilde{\sigma}$. There is an Ornstein-Uhlenbeck operator Δ over $L_x(M)$. It can be lifted into an Ornstein-Uhlenbeck operator $\tilde{\Delta}$ over $\tilde{L}_x(M)$. To Δ is associated a Dirichlet form A:

$$A(F,F) = E[F^2] + E[< \mathrm{d}F, \mathrm{d}F >] \tag{10}$$

A is lifted into a Dirichlet form $\tilde{A}$. Let $X_t(\gamma)$ the process of Driver, Rockner (1992) associated to A, defined outside a set of capacity 0:

$$\exp[-t\Delta]F(\gamma) = E[F(X_t(\gamma))] \ . \tag{11}$$

There is a unique lift $\tilde{X}_t(\gamma)$ of $X_t(\gamma)$, which represents the heat semi-group associated to it.

The reader can see Léandre for more details.

4 Energy Functional

We consider the following stochastic differential equation:

$$\begin{aligned} \mathrm{d}\gamma_s &= \tau_s(C + B_s)\mathrm{d}s \\ \gamma_0 &= x \end{aligned} \tag{12}$$

where τ_s is the parallel transport over the curve γ, B_s is a Brownian motion over $T_x(M)$ and C an independent finite dimensional non degenerate gaussian variable over $T_x(M)$.

The law of γ_1 has a density which is strictly positive in x. We can desintegrate the law of γ over $L_x(M)$. In order to show that, we introduce the tangent space of a finite energy curve γ which are constituted of path of the shape $\tau_s H_s$. H_s is now a path in $T_x(M)$ with two derivatives. We take as Hilbert norm:

$$||X||^2 = ||H_0'||^2 + \int_0^1 ||H"_s||^2 \mathrm{d}s \ . < \infty \ . \tag{13}$$

There is for enough simple vector fields an integration by parts formula similar to (2).

The energy functional $I(\gamma) = \int_0^1 ||\gamma_s'||^2 \mathrm{d}s$ belongs to all the Sobolev spaces. We can repeat in this context the considerations of the first part. We refer to Léandre for more details. But in Léandre, we did not do the limit theorem (B_s should be replaced by ϵB_s for small ϵ), where the closed geodesics of the manifold should play an important role, because the limit model has no simple geometrical structure in this model.

References

Arai, A. (1992): A general class of infinite dimensional operators and path representation of their index. J.F.A. 105, 342-408.

Atiyah, M. (1988): New invariants of 3- and 4- dimensional manifolds. Proceedings of symposia in pure mathematics. 48, 285-299.

Bismut, J.M. (1984): Large deviations and the Malliavin Calculus. Progress in Math. 45. Birkhauser..

Driver, B., Rockner, M. (1992): Construction of diffusions on path and loop spaces of compact Riemannian manifolds. C.R.A.S. t 315., Série I. 603-608.

Guilarte, J.M. (1990): The supersymmetric sigma model, topological quantum mechanics and knot invariants. Journ. Geometry and Physics 7, 255-302.

Guilarte, J.M.: *Sphalerons and instantons in two dimensional field theory.*

Hino, M. : Spectral properties of Laplacians on an abstract Wiener space with a weighted Wiener measure. Preprint.

Jaffe, A., Lesniewski, A., Weitsman, J. (1987): Index of a family of Dirac operators on loop spaces. C.M.P. 112, 75-88.

Léandre, R. (1988): Applications quantitatives et qualitatives du Calcul de Malliavin. Col. Franco-Japonais. Métivier M. Watanabe S. edit. L.N.M. 1322 Springer 109-133. English translation: Geometry of Random motion. Durrett R. Pinsky M. edit. Contemporary Maths. 73, 173-197.

Léandre, R. (1993): Integration by parts formulas and rotationally invariant Sobolev Calculus on the free loop space. XXVII Winter School of theoretical physic. Gielerak R. Borowiec A. edit. J. of Geometry and Physics. 11, 517-528.

Léandre, R.: Invariant Sobolev Calculus on the free loop space. To be published in Acta Applicandae Mathematicae.

Léandre, R.: Stochastic Wess–Zumino–Witten model over a symplectic manifold. To be published in Journal of Geometry and Physics.

Léandre, R: Cover of the brownian bridge ans stochastic symplectic action. To be published in Reviews in Mathematical Physics.

Léandre, R.: Stochastic Wess–Zumino–Witten model for the measure of Kontsevitch. Preprint.

Witten, E. (1982): Supersymmetry and Morse theory. J. diff. Geometry. 17, 661-692.

On $(k \oplus l|q)$-Dimensional Supermanifolds

A. Konechny and A. Schwarz

Department of Mathematics, University of California, Davis, CA 95616
konechny@math.ucdavis.edu, schwarz@math.ucdavis.edu

The present paper stems from an attempt to understand the precise mathematical meaning of some constructions used by physicists. We have in mind in particular the consideration of independent spin structures in holomorphic and antiholomorphic sectors of N=1 superconformal field theory and the notion of chiral (heterotic) supermanifold (for example see (Moore, Nelson, Polchinsky (1986))). In this paper we introduce a notion of $(k \oplus l|q)$-dimensional supermanifold and show that in many cases rigorous definitions can be based on this notion.

Let $\Lambda = \Lambda_0 + \Lambda_1$ be a Grassmann algebra with an even subspace Λ_0 and an odd subspace Λ_1. The space $\mathbf{R}_\Lambda^{p|q}$ can be defined as a space consisting of rows $(x^1, \dots, x^p, \xi^1, \dots, \xi^q)$ where $x^1, \dots, x^p \in \Lambda_0$ are even elements of Grassmann algebra Λ and $\xi^1, \dots, \xi^q \in \Lambda_1$ are odd elements from Λ. Physicists usually say that one can take as Λ any Grassmann algebra provided it is large enough. From the viewpoint of a mathematician it is better to consider a family of sets $\mathbf{R}_\Lambda^{p|q}$ corresponding to all Grassmann algebras Λ. It is easy to see that a parity preserving homomorphism $\alpha : \Lambda \to \Lambda'$ generates naturally a map $\tilde{\alpha} : \mathbf{R}_\Lambda^{p|q} \to \mathbf{R}_{\Lambda'}^{p|q}$ and that $\widetilde{\beta\alpha} = \tilde{\beta}\tilde{\alpha}$ for any parity preserving homomorphisms $\alpha : \Lambda \to \Lambda'$, $\beta : \Lambda' \to \Lambda''$. In the language of mathematics this means that the correspondence $\Lambda \mapsto \mathbf{R}_\Lambda^{p|q}$ determines a functor acting from the category of Grassmann algebras into the category of sets (or of vector spaces) (see (Schwarz (1984)) for details).

An (even) superfield on $\mathbf{R}^{p|q}$ can be defined as an expression of the form

$$\sum_{k=2l} \sum_{1 \le i_1 < \dots < i_k} f_{i_1,\dots,i_k}(x^1, \dots, x^p)\xi^{i_1} \dots \xi^{i_k} \tag{1}$$

Here $f_{i_1,\dots,i_k}$ are smooth functions on $\mathbf{R}^p$. It is important to notice that such an expression determines a map $F_\Lambda : \mathbf{R}_\Lambda^{p|q} \to \Lambda_0 = \mathbf{R}_\Lambda^{1|0}$. This follows from the fact that we can substitute an even element of a Grassmann algebra into any smooth function of real variable. To verify this statement we notice that every even element x of Grassmann algebra can be represented in the form $x = m + n$ where $m \in \mathbf{R}$ and n is nilpotent. We define $f(x)$ using the Taylor expansion $f(x) = \sum_{k=0}^\infty \frac{f^{(k)}(m)}{k!} n^k$ (the series terminates because n is nilpotent). If $\alpha : \Lambda \to \Lambda'$ is a parity preserving homomorphism then $\tilde{\alpha} \circ F_\Lambda = F_\Lambda \circ \tilde{\alpha}$. This means in mathematical terminology that a superfield (1) specifies a natural map of the functor $\mathbf{R}^{p|q}$ into the functor $\mathbf{R}^{1|0}$. Analogously, an odd superfield can be considered as a natural transformation of functors $\mathbf{R}^{p|q} \to$

$\mathbf{R}^{0|1}$. An arbitrary superfield can be viewed as a natural transformation of functors $\mathbf{R}^{p|q} \to \mathbf{R}^{1|1}$ (recall that $\mathbf{R}^{1|1}_{\Lambda} = \Lambda_0 \oplus \Lambda_1$). Note that in the above considerations the Grassmann algebra can be replaced by any algebra Λ every element of which can be represented as a sum of a real number and nilpotent element (we assume that Λ is associative, Z_2-graded , supercommutative algebra having unit element). Algebras of this kind will be called almost nilpotent algebras or AN algebras. In other words we can say that $\mathbf{R}^{p|q}$ can be considered as a functor on the category of AN algebras and a superfield determines a natural transformation of the functor $\mathbf{R}^{p|q}$ into the functor $\mathbf{R}^{1|0}$ or $\mathbf{R}^{0|1}$.

The main notions of superalgebra and of supergeometry can be formulated very easily in the language of functors. A superspace can be defined as an arbitrary functor on the category **AN** of AN algebras taking values in the category of sets. The body of a superspace can be defined as the set corresponding to the AN algebra $\Lambda = \mathbf{R}$. We introduce a $(p|q)$-dimensional supermanifold as a superspace that is locally equivalent to $\mathbf{R}^{p|q}$. In other words $(p|q)$-dimensional supermanifold can be pasted together from domains in $\mathbf{R}^{p|q}$ by means of smooth transformations. A body of a $(p|q)$-dimensional supermanifold is a p-dimensional smooth manifold. Replacing in the definition of a superspace the category of sets by the category of groups or by the category of Lie algebras we obtain the definitions of supergroup and super Lie algebra respectively.

It is convenient to generalize the notion of the superspace $\mathbf{R}^{p|q}$ as follows. Denote by $\mathbf{R}^{k\oplus l|q}_{\Lambda}$ a set of rows $(x^1, \ldots, x^k, y^1, \ldots, y^l, \xi^1, \ldots, \xi^q)$ where $x^1, \ldots, x^k$ are arbitrary even elements of AN algebra Λ , $y^1, \ldots, y^l$ are nilpotent even elements of Λ and $\xi^1, \ldots, \xi^q$ are odd elements of Λ. A superspace $\mathbf{R}^{k\oplus l|q}$ can be defined as a functor $\Lambda \to \mathbf{R}^{k\oplus l|q}_{\Lambda}$. Strictly speaking to define a functor we should also construct a homomorphism $\tilde{\alpha} : \mathbf{R}^{k\oplus l|q}_{\Lambda} \to \mathbf{R}^{k\oplus l|q}_{\Lambda'}$ for every parity preserving homomorphism $\alpha : \Lambda \to \Lambda'$ of AN algebras. We omit this obvious construction.

Let us define a superfield on $\mathbf{R}^{k+l|q}$ as an expression of the form

$$\sum_{s} \sum_{1 \le i_1 < \ldots < i_s} f_{i_1, \ldots, i_s}(x^1, \ldots, x^k, y^1, \ldots, y^l) \xi^{i_1} \ldots \xi^{i_s} \tag{2}$$

where $f_{i_1,\ldots,i_s}$ are smooth functions of variables $(x^1, \ldots, x^k) \in \mathbf{R}^k$ and formal power series with respect to $y^1, \ldots, y^l$. Such an expression determines a map of $\mathbf{R}^{k\oplus l|q}$ into $\mathbf{R}^{1|0}$ if expression (2) is even and into $\mathbf{R}^{0|1}$ if it is odd.

We define a $(k \oplus l|q)$-dimensional supermanifold as a superspace that is locally equivalent to $\mathbf{R}^{k\oplus l|q}$. Almost all notions of (super)geometry can be generalized to the case of $(k \oplus l|q)$-dimensional supermanifolds. In particular one can define the supergroup of transformations of such a manifold and corresponding Lie (super) algebra of vector fields. As usual, a vector field on a $(k \oplus l|q)$-dimensional superdomain with coordinates $(x^1, \ldots, x^k,\ y^1, \ldots, y^l,$ $\xi^1, \ldots, \xi^q)$ can be identified with a first order differential operator

$$A^i \frac{\partial}{\partial x^i} + B^j \frac{\partial}{\partial y^j} + C^s \frac{\partial}{\partial \xi^s} \tag{3}$$

where A^i, B^j, C^s are smooth functions of variables $x^1, \ldots, x^k$, formal power series with respect to $y^1, \ldots, y^l$ and polynomials in $\xi^1, \ldots, \xi^q$. If A^i, B^j are odd and C^s are even then the operator (3) is parity reversing ; we say that the corresponding vector field is odd. The definition of an even vector field is similar. A $(k \oplus l|q)$-dimensional supermanifold is pasted together from $(k \oplus l|q)$-dimensional superdomains by means of smooth transformations. If we have a $(p|q)$-dimensional supermanifold $\mathcal{A}$ and a k-dimensional submanifold $\mathcal{B}^{(0)}$ of its body $\mathcal{A}^{(0)}$ we can construct easily a $(k \oplus (p-k)|q)$-dimensional supermanifold in the following way. For each Λ there is a natural mapping $\tilde{m}_\Lambda : \mathcal{A}_\Lambda \to \mathcal{A}^{(0)}$ corresponding to the homomorphism $m_\Lambda : \Lambda \to \mathbf{R}$ that evaluates a numerical part of an element from Λ. Then, $\mathcal{B}_\Lambda$ consists of those points of $\mathcal{A}_\Lambda$ that project onto $\mathcal{B}^{(0)}$ under m_Λ. The mapping $\tilde{\alpha} : \mathcal{B}_\Lambda \to \mathcal{B}_{\Lambda'}$ assigned to a homomorphisms $\alpha : \Lambda \to \Lambda'$ is a restriction of the corresponding map defined for the supermanifold $\mathcal{A}$. Let us call the supermanifold $\mathcal{B}$ obtained this way as a restriction of supermanifold $\mathcal{A}$ to the subset $\mathcal{B}_0$ of its body.

One can define classes of supermanifolds with interesting geometric properties restricting the allowed class of transformations. For example one can define a complex analytic transformation of $(2k \oplus 2l|2q)$-dimensional superdomain generalizing the usual requirement that the Jacobian matrix of transformation commutes with a standard matrix J obeying $J^2 = -1$. Then, $(k \oplus l|q)$-dimensional complex manifold can be defined as a manifold glued together by complex analytic transformations. One can also use complex coordinates $(Z^A) = (z^1, \ldots, z^k, w^1, \ldots, w^l, \vartheta^1, \ldots, \vartheta^q)$ on a complex manifold. They take values in complex AN algebras with antilinear involution. Of course together with these coordinates we should consider complex conjugate coordinates $\bar{Z}^A$. Analytic transformations do not mix Z^A and $\bar{Z}^A$. Therefore usually we will not mention $\bar{Z}^A$.

Let U be a $(1|N)$-dimensional complex superdomain with complex coordinates $(z, \theta^1, \ldots, \theta^N)$. We define N-superconformal transformations as complex analytic transformations preserving up to a factor the one-form $\omega_N = dz + \theta^1 d\theta^1 + \ldots + \theta^N d\theta^N$. For example in case $N = 1$ superconformal transformations can be written explicitly as follows

$$\begin{aligned} \tilde{z} &= u(z) - u'(z)\epsilon(z)\theta \\ \tilde{\theta} &= \sqrt{u'(z)} \left(\theta + \epsilon(z) + \frac{1}{2}\epsilon(z)\epsilon'(z)\theta \right) \end{aligned} \tag{4}$$

where $u(z)$ and $\epsilon(z)$ are even and odd analytic functions of z respectively. By a N-superconformal manifold we mean a manifold pasted together from $(1|N)$-dimensional complex superdomains by N-superconformal transformations. Note that strictly speaking we are looking here not at a single supermanifold but rather at a family of supermanifolds parameterized by gluing functions similar to $u(z)$ and $\epsilon(z)$ in $N = 1$ case. The superspace M_N of equivalence classes of N-superconformal manifolds is called a moduli space of N-superconformal structures (supermoduli space).

A N-superconformal vector field defined in a complex superdomain $U^{1|N}$ is a vector field X such that the Lie derivative of the form ω_N restricted to $U^{1|N}$ with respect to X is proportional to ω_N, i.e. $L_X\omega_N = f\omega_N$ for some superfield f. Such vector fields correspond to infinitesimal N-superconformal transformations. Given a N-superconformal manifold $\mathcal{A}$ one can define a N-superconformal vector field on it as a vector field such that its restriction to each elementary coordinate patch $U^{1|N}$ is N-superconformal. It follows from standard results of deformation theory that a formal tangent space to the moduli space of N-superconformal structures at a "point" $\mathcal{A}$ (being a N-superconformal manifold) is isomorphic to $H^1(\mathcal{A},\gamma_N)$. Here $H^1(\mathcal{A},\gamma_N)$ stands for the first cohomology group of $\mathcal{A}$ with coefficients in the sheaf of N-superconformal vector fields γ_N.

Moduli spaces play the central role in the Segal's axiomatics of conformal field theory (CFT) (Segal (1989)). In this approach one considers conformal 2d surfaces (complex curves) with parametrized boundary components. Each boundary component is homeomorphic to a circle and it is assumed that the parametrization can be extended to a complex coordinate in a small neighborhood. We can think of a standard annulus in the complex plane $\{z \in \mathbf{C} | \frac{1}{2} \leq |z| \leq 1\}$ mapped by a biholomorphic mapping into a neighborhood of the boundary component. The neighborhoods of different boundary components are assumed to be nonoverlapping. The annuli are divided into two classes: "incoming" and "outgoing". The moduli space of such objects with m incoming and n outgoing annuli is denoted by $P_{m,n}$. Let us stress here that we allow disconnected surfaces as well. There arise naturally two operations on the sets $P_{m,n}$:

$$P_{m_1,n_1} \times P_{m_2,n_2} \to P_{m_1+n_1,m_2+n_2} \tag{5}$$

$$P_{m,n} \to P_{m-1,n-1} \tag{6}$$

Here the first operation corresponds to the disjoint union of surfaces and the second one corresponds to the identification of m-th incoming annulus with the n-th outgoing one by the rule: $z' = \frac{1}{4}z^{-1}$. In Segal's axiomatics CFT is specified by maps

$$\alpha_{m,n} : P_{m,n} \to \mathrm{Hom}_{\mathbf{C}}\{H^m, H^n\}$$

assigning to each point in $P_{m,n}$ a linear mapping $H^m \to H^n$ belonging to the trace class (here H is a fixed Hilbert space). The collection of mappings $\alpha_{m,n}$ should satisfy some set of axioms ensuring the compatibility of mappings $\alpha_{m,n}$ with mappings (5), (6) and with permutations of boundary components. One can generalize the Segal's axiomatics to the case of N-superconformal field theories replacing 2d conformal surfaces by N-superconformal manifolds with boundaries and complex annuli by N-superconformal annuli (see (Schwarz (1996)) for this type of axiomatics stated for N=2 SCFT).

Now let us describe supermanifolds that appear in 2D superconformal field theories (SCFT) having different number of supersymmetries for left movers and right movers. Consider a $(2|p+q)$-dimensional complex domain U with even coordinates z_L, z_R and odd coordinates $\theta_L^1, \ldots, \theta_L^p, \theta_R^1, \ldots, \theta_R^q$. The superdomain U can be considered as a superdomain in real superspace $\mathbf{R}^{4|2p+2q}$. We will single out a subspace V of U that has real dimension $(2 \oplus 2|2p+2q)$ by imposing the condition that $z_R - \bar{z}_L$ is nilpotent. By definition a transformation of V is called (p,q)-superconformal if it does not mix the left coordinates z_L, θ_L^i with the right coordinates z_R, θ_R^i and preserves up to a factor one-forms

$$\omega_L = \mathrm{d}z_L + \theta_L^1 d\theta_L^1 + \ldots + \theta_L^p \mathrm{d}\theta_L^p$$
$$\omega_R = \mathrm{d}z_R + \theta_R^1 d\theta_R^1 + \ldots + \theta_R^q \mathrm{d}\theta_R^q$$

We define a (p,q)-superconformal manifold as a superspace pasted together from several copies of V by means of (p,q)-superconformal transformations. Again we assume here that the odd gluing parameters such as $\epsilon(z)$ in (4) are allowed.

Let $\mathcal{A}^{(0)}$ be a compact connected oriented 2-dimensional surface of genus $g > 1$. It is easy to define a moduli space $\mathcal{M}_{p,q,g}$ of (p,q)-superconformal manifolds having $\mathcal{A}^{(0)}$ as a body. Denote by $\mathcal{M}_{p,q,g}^{(0)}$ the body of moduli space $\mathcal{M}_{p,q,g}$. One can prove that the superspace $\mathcal{M}_{p,q,g}$ can be constructed out of supermoduli spaces $\mathcal{M}_{p,g}$ and $\mathcal{M}_{q,g}$ of p and q-superconformal manifolds with body $\mathcal{A}^{(0)}$. The construction goes as follows. Like in the case of moduli space of conformal structures $\mathcal{M}_{p,g}$ is equipped with a canonical complex structure. Thus we may consider the superspace $\mathcal{M}_{p,g} \times \bar{\mathcal{M}}_{q,g}$ where $\bar{\mathcal{M}}_{q,g}$ denotes the space $\mathcal{M}_{q,g}$ with conjugate complex structure. Note that there is a natural projection $\pi_p : \mathcal{M}_{p,g} \to \mathcal{M}_g$ to the moduli space of complex structures on $\mathcal{A}^{(0)}$. This projection simply corresponds to the fact that each p-superconformal manifold has a complex structure on its body. If z is a complex coordinate on $\mathcal{M}_g$ and $\bar{z}$ is its counterpart on $\bar{\mathcal{M}}_g$ then we define the subset $\mathcal{M}_{p,q,g}^{(0)} \subset \mathcal{M}_{p,g}^{(0)} \times \bar{\mathcal{M}}_{q,g}^{(0)}$ as the set of points (a,b) such that $\pi_p(a) = (\pi_q(b))^*$. The superspace $\mathcal{M}_{p,q,g}$ has the body $\mathcal{M}_{p,q,g}^{(0)}$ and can be obtained as a restriction of $\mathcal{M}_{p,g} \times \bar{\mathcal{M}}_{q,g}$ to the corresponding subset of its body. Choosing a local coordinate system on $\mathcal{M}_{p,g} \times \bar{\mathcal{M}}_{q,g}$ in such a way that $z, \bar{z}$ are part of the coordinates we see that the condition above implies that $z - (\bar{z})^*$ can take only nilpotent values (as this is zero on the body). This means that if $\mathcal{M}_{p,g}$ is of dimension $(k|l)$ and $\mathcal{M}_{q,g}$ is of dimension $(\tilde{k}|\tilde{l})$ the superspace $\mathcal{M}_{p,q,g}$ has the dimension $((k+\tilde{k}-3g+3) \oplus (3g-3)|l+\tilde{l})$. The construction of $\mathcal{M}_{p,q,g}$ presented above is simply a more formal way to say that left and right superconformal structures on a (p,q)-superconformal manifold are independent up to a complex structure on the body that they share. It is easy to generalize this construction to the case when $\mathcal{A}_0$ is disconnected.

Now we are in a position to generalize the Segal's axiomatics to include (p,q)-superconformal field theories. To define a proper analog of the space $P_{m,n}$ one should consider a larger moduli space of (not necessarily connected) (p,q)-superconformal surfaces having parameterized boundary components of different type (determined by the boundary conditions for odd coordinates). Furthermore, the corresponding analogs of the mappings $\alpha_{m,n}$ should be holomorphic. The last requirement sets a connection between p-superconformal and (p,p)-superconformal theories as outlined below. Firstly, it is possible to embedd the supermoduli space $\mathcal{M}_{p,g}$ into $\mathcal{M}_{p,p,g}$. Then , given a real analytic function on $\mathcal{M}_{p,g} \subset \mathcal{M}_{p,p,g}$ one can extend it to a holomorphic function defined in some neighborhood of $\mathcal{M}_{p,g}$ in $\mathcal{M}_{p,p,g}$. Assuming that a continuation to the whole $\mathcal{M}_{p,p,g}$ is possible we see that a p-superconformal field theory determines a (p,p)-superconformal one.

Now let us make some remarks about the integration over $\mathcal{M}_{p,q,g}$. This question is important in particular in heterotic string theory. To calculate a string amplitude one defines a holomorphic volume element on $\mathcal{M}_{p,q,g} \times \bar{\mathcal{M}}_{p,q,g}$ and chooses a real cycle of integration on the body of this space. We would like to stress that possible reality conditions for odd and nilpotent variables do not affect the integration result (see Konechny, Schwarz (1997) or Deligne (1997) for discussion). Simply an integration over odd variables is essentially an algebraic operation. As for the choice of real cycle on the body of $\mathcal{M}_{p,q,g} \times \bar{\mathcal{M}}_{p,q,g}$ one can take the (real) diagonal of the body of $\mathcal{M}_{p,q,g} \times \bar{\mathcal{M}}_{p,q,g}$. The change of nilpotent variable roughly speaking corresponds to the infinitesimal change of integration cycle and therefore does not change the value of integral.

References

Deligne, P. (1997): *Real versus complex* , IAS QFT school notes (1997).

Konechny, A., Schwarz, A. (1997): *Geometry of N=1 super Yang-Mills theory in curved superspace* Jr. of Geom. and Phys. (1997); hep-th/9609081

Moore, G., Nelson, P., Polchinsky, J. (1986): *Strings and supermoduli* Phys. Lett. **B 169**, pp. 47-53

Segal, G. (1989): *The definition of conformal field theory*, preprint; *Two dimensional conformal field theories and modular functors*, in IXth International Conference on Mathematical Physics, B. Simon, A. Truman and I.M. Davies Eds. (Adam Hilger, Bristol, 1989).

Schwarz, A. (1984): *On the definition of superspace* Teor. Mat. Fiz. **60**, p. 37; English transl. in Theoret. and Math. Phys. **60**,

Schwarz, A. (1996): *Superanalogs of symplectic and contact geometry and their applications to quantum field theory* , Amer. Math. Soc. Transl. (2), **177**,pp. 203-218

Let the Spin and the Charges Unify

N.M. Borštnik

Department of Physics, University of Ljubljana, Jadranska 19
J. Stefan Institute, Jamova 39, Ljubljana , 1001, Slovenia
E-mail: norma.mankoc@ijs.si, tel.: 386 61 176 65 25

Abstract. In space of d ordinary and d Grassmann coordinates, with $d \geq 15$, the charges unify with the spin: the Lorentz group $SO(1, d-1)$ in Grassmann space manifests under certain conditions as $SO(1,3)$ (in $d = 4$ subspace) times $SO(10) \supset SU(3) \times SU(2) \times U(1)$ (in the rest of the space). The multiplets of SO(1,14) are discussed from the point of view of the Electroweak Standard Model. [1]

The fact that ordinary space-time is not sufficient to describe the dynamics of our world was first recognized more than 70 years ago, when, in addition to infinite dimensional vector space spanned over ordinary coordinate space, a space of two vectors - the internal space of the fermionic spin - was introduced. Since then the internal space of fermions has been enlarged to enable the description of the particle - antiparticle degrees of freedom or their handedness, the weak charge and the colour charge. The unification of electromagnetic and weak interactions makes it clear that the electromagnetic charge originates in internal space, as well. The internal space of bosons grew more or less parallel to the internal space of fermions.

Theories connect all the symmetries or the corresponding properties appearing in physics with appropriate groups, and define, accordingly, the quantum numbers: The spin is connected with the Lorentz group $SO(1,3)$. Charges are connected with the group $U(1)$, $SU(2)$ and $SU(3)$.

While in *ordinary* space time, only the *vectorial* types of representations for the Lorentz group are possible [the generators of the infinitesimal transformations $L^{mn} = (x^m p^n - x^n p^m)$ define spherical harmonics, with integer angular momenta], for the *internal* spaces two types of representations for either the Lorentz group or the groups describing charges are required: the *fundamental* and the *adjoint*. The former are used to describe the internal space of fermions, the later to describe the internal space of bosons - the gauge vector fields. Singlets are added to each type of representations in order to describe fermions and bosons which don't manifest the colour or the weak charge.

Generators of the infinitesimal transformations of the Lorentz group M^{mn} and the groups defining charges τ^{Ai}

[1] This talk is devoted to the memory of Volkov, with whom I have very constructive discussions in Palesau in 1995 about the work presented here.

$$M^{mn} = L^{mn} + \left\{ \begin{matrix} \mathcal{S}^{mn} \\ S^{mn} \end{matrix} \right\}, \; m, n \in \{0,1,2,3\}, \tag{1}$$

$$\tau^{Ai} = \quad \left\{ \begin{matrix} \mathcal{T}^{Ai} \\ T^{Ai} \end{matrix} \right\}, \; A \in \{1,2,3\}, i \in \{1, n_A\} \; . \tag{2}$$

define representations, with $n_A = N_A^2 - 1$ for $\mathrm{SU}(N_A)$ and $n_A = N_A^2$ for $\mathrm{U}(N_A)$.

The generators of the Lorentz group $\mathrm{SO}(1,3)$, M^{mn}, and those of the groups SU(3), SU(2) and U(1), τ^{Ai}, fulfil the commutation relations [2]

$$\begin{aligned} \{M^{ab}, M^{cd}\} &= \mathrm{i}(M^{ad}\eta^{bc} + M^{bc}\eta^{ad} - M^{ac}\eta^{bd} - M^{bd}\eta^{ac}), \\ \{\tau^{Ai}, \tau^{Bj}\} &= \mathrm{i} f^{Aijk} \delta^{AB} \, \tau^{Ak} \; . \end{aligned} \tag{3}$$

Here $\mathcal{S}^{mn}$ and $\mathcal{T}^{Ai}$ stand for the operators, defining the fundamental representations of the Lorentz group [3] and the group $\mathrm{SU}(N_A)$, respectively, and let S^{mn} and T^{Ai} stand for operators defining the adjoint representations of the corresponding groups. Both types of operators fulfil the algebra of Eq.(3), respectively. Operators of the adjoint representations are determined by the structure constants of the groups[4].

Modern theories try to *unify* the *internal spaces of charges*. This talk elaborates the idea of *unifying spins with charges*. It is not only the beauty of describing all the internal degrees of freedom in a unique way, which suggests this kind of unification. Motivation also comes from the experimental fact, which in the Electroweak Standard Model manifests in connection between the handedness of the fermions and their weak charge representations and

[2] Since besides the operators of an even Grassmann character, like M^{mn} and τ^{Ai}, operators of an odd Grassmann character shall also appear, we introduce the generalized commutation relations(Mankoč Borštnik (1992a), (1992b), (1993), (1994), (1995a))

$$\{A, B\} := AB - (-1)^{n_{AB}} BA, \qquad n_{AB} = \left\{ \begin{matrix} +1, \text{ if A and B have an odd Grass. char.} \\ 0, \quad \text{otherwise.} \end{matrix} \right\} \; .$$

[3] If $M_i^{\pm} = \frac{1}{2}(\frac{1}{2}\varepsilon_{ijk}M^{jk} \pm \mathrm{i}M^{0i})$, $i, j, k \in \{1,2,3\}$ is defined, we find $\{M_i^{\pm}, M_j^{\pm}\} = \mathrm{i}\varepsilon_{ijk}M_k^{\pm}$, $\{M_i^{\pm}, M_j^{\mp}\} = 0$, which demonstrates the $\mathrm{SU}(2) \times \mathrm{SU}(2)$ structure of the group $\mathrm{SO}(1,3)$, it follows that $M^{ij} = \varepsilon_{ijk}(M_{k+} + M_{k-})$, $M^{0k} = -\mathrm{i}(M_{k+} - M_{k-})$. Operators M_{k+} and M_{k-} define the left handed and the right handed representations of the Lorentz group of either fundamental or adjoint types. Here $\varepsilon_{a_1 a_2 \ldots a_n} = \varepsilon^{a_1 a_2 \ldots a_n}$ is the totally antisymmetrical tensor with $\varepsilon_{123..n} = 1$.

[4] To see this for the Lorentz group $\mathrm{SO}(1,3)$, we take into account the $\mathrm{SU}(2) \times \mathrm{SU}(2)$ structure of this group, presented above. We find $(S^{ij})_{lm} = \mathrm{i}\varepsilon^{ijk}\varepsilon_{klm}$, $(S^{0i})_{lm} = \mathrm{i}\varepsilon_{ilm}$. The operators, defining the adjoint representations of the group $\mathrm{SU}(N_A)$, are: $(T^{Ai})_{jk} = -\mathrm{i}f^{Aijk}$. The operators $\mathcal{S}^{mn}$ (as well as $\mathcal{S}_{\pm}^{k}$) are the Pauli 2×2 matrices, while S^{mn} (as well as $S_{\pm}^{k}$) are 3×3 matrices. Similarly, operators $\mathcal{T}^{Ai}$ are $N_A \times N_A$ matrices, while $(T^{Ai})_{jk} = -\mathrm{i}f^{Aijk}$ are $n_A \times n_A$ matrices.

in the breaking of the left-right symmetry. I see the unification of spins and charges as a way of the possibly connecting the handedness of the fields with their charge representations.

Unification of spins and charges was proposed in ref (Mankoč Borštnik (1992a), (1992b), (1993), (1994), (1995a) , Mankoč Borštnik (1995b), (1996), Mankoč Borštnik, Fajfer (1992)) within the approach that space time has d ordinary commuting ($x^a x^b - x^b x^a = 0,\ a,b \in \{0,1,2,3,5,..,d\}$) and d Grassmann anticommuting ($\theta^a\theta^b + \theta^b\theta^a = 0,\ a,b \in \{0,1,2,3,5,..,d\}$) coordinates[3], with $d \geq 15$, and that consequently all symmetries of a system are connected with only coordinate transformations in ordinary and Grassmann space.

In *Grassmann space there exist two kinds of operators of the Lorentz transformations* (Mankoč Borštnik (1992a), (1992b), (1993), (1994), (1995a)), one defining *spinorial kind of representations,*, the other defining the *vectorial kind of representations.* They are appropriate for describing spins and charges for fermions and bosons, respectively, *unifying spins with charges for each kind of representations separately.*

The talk comments on fermionic and bosonic representations, defined by these two kinds of operators, from the point of view of the unification of spins and charges and the Electroweak Standard Model, assuming that the dimension $d = 15$.

Let us first briefly present the approach.

A linear vector space spanned over a Grassmann coordinate space of d coordinates has the dimension 2^d. If monomials $\theta^{\alpha_1}\theta^{\alpha_2}....\theta^{\alpha_m}$ are taken as a set of basic vectors with $\alpha_j \neq \alpha_k$, half of the vectors have an odd (those with an odd m) and half of the vectors an even (those with an even m) Grassmann character. Any vector in this space may be represented as a linear superposition of monomials $f(\theta) = \alpha_0 + \sum_{i=1}^{d} \alpha_{a_1 a_2..a_i}\theta^{a_1}\theta^{a_2}....\theta^{a_i},\quad a_k < a_{k+1}$, where constants $\alpha_0, \alpha_{a_1 a_2..a_i}$ are complex numbers.

On this linear space the following linear operators (Mankoč Borštnik (1992a), (1992b), (1993), (1994), (1995a)) can be defined:

$$p^\theta{}_a := i\overrightarrow{\partial^\theta}_a,\ \ \tilde{a}^a := i(p^{\theta a} - i\theta^a),\ \ \tilde{\tilde{a}}^a := -(p^{\theta a} + i\theta^a)\ , \tag{4}$$

with $\overrightarrow{\partial^\theta}_a = \frac{\overrightarrow{\partial}}{\partial\theta^a}$. According to Eq.(4) we find

$$\begin{aligned}&\{p^{\theta a}, p^{\theta b}\} = 0 = \{\theta^a, \theta^b\},\ \ \{p^{\theta a}, \theta^b\} = -i\eta^{ab},\\ &\{\tilde{a}^a, \tilde{a}^b\} = 2\eta^{ab} = \{\tilde{\tilde{a}}^a, \tilde{\tilde{a}}^b\},\ \ \{\tilde{a}^a, \tilde{\tilde{a}}^b\} = 0\ .\end{aligned} \tag{5}$$

Any of the two bilinear forms(Mankoč Borštnik (1992a), (1992b), (1993), (1994), (1995a))

[3] The metric tensor $\eta_{ab} = \mathrm{diag}(1,-1,-1,-1,...,-1)$ lowers the indices of a vector $\{\theta^a\} = \{\theta^0,\theta^1,...,\theta^d\}, \theta_a = \eta_{ab}\theta^b$. Linear transformation actions on vectors $(\alpha\acute{\theta}^a + \beta\acute{x}^a) = L^a{}_b(\alpha\theta^b + \beta x^b)$, which leave forms $(\alpha\theta^a + \beta x^a)(\alpha\theta^b + \beta x^b)\eta_{ab}$ invariant, are called the Lorentz transformations.

$$\tilde{\mathcal{S}}^{ab} := -\frac{\mathrm{i}}{4}[\tilde{a}^a, \tilde{a}^b], \;\; \tilde{\tilde{\mathcal{S}}}^{ab} := -\frac{\mathrm{i}}{4}[\tilde{\tilde{a}}^a, \tilde{\tilde{a}}^b] \;, \tag{6a}$$

with $[A, B] := AB - BA$, closes the algebra of the Lorentz group $\mathrm{SO}(1, d-1)$ (Eq.(3)) and defines what will be called ***the spinorial representations*** of the Lorentz group and of subgroups of the Lorentz group (Mankoč Borštnik (1992a), (1992b), (1993), (1994), (1995a) , Mankoč Borštnik (1995b), (1996), Mankoč Borštnik, Fajfer (1992)). We shall use these representations to describe spins and charges of fermionic fields. Making use of only $\tilde{\mathcal{S}}^{ab}$, the sign $\tilde{()}$ will be omitted and a^a and $\mathcal{S}^{ab}$ will be written for the corresponding operators, using the symbol which was introduced for operators describing the fundamental representations.

The operators

$$S^{ab} := (\theta^a p^{\theta b} - \theta^b p^{\theta a}) \tag{6b}$$

which also close the algebra of the Lorentz group (Eq.(3)), define what will be called *the vectorial* representations of the Lorentz group $\mathrm{SO}(1, d-1)$ and of subgroups of this group (Mankoč Borštnik (1992a), (1992b), (1993), (1994), (1995a)). These representations will be used to describe spins and charges of bosonic fields.

Both kinds of operators are, according to Eqs.(4), bilinear forms of the differential operators in Grassmann space.

In d dimensional ordinary and d dimensional Grassmann space we may write:

$M^{ab} = L^{ab} + \left\{ \begin{matrix} \mathcal{S}^{ab} \\ S^{ab} \end{matrix} \right\}, \; a, b \in \{0, 1, 2, 3, 5, .., d\}.$

Assuming that $d - 4$ out of d coordinates stay compactified and that for energies $\ll \frac{1}{<x^h>}$, where $< x^h >$, $h \in \{5, 6, .., d\}$, is the radius of the $h - th$ coordinate, the contribution of L^{hk} to M^{hk}, $h, k \in \{5, 6, ..., d\}$ is nonnoticeable, it follows that at low enough energies then M^{ab} manifests approximately as in Eqs.(1,2), with

$\tau^{Ai} = c^{Ai}{}_{hk} M^{hk} = c^{Ai}{}_{hk} \left\{ \begin{matrix} \mathcal{S}^{hk} \\ S^{hk} \end{matrix} \right\}, \; h, k \in \{5, .., d\}, \; A \in \{1, 2, 3\}, \;\; i =\in \{1, .., n_A\},$

$c^{Ai}{}_{hk} = -c^{Ai}{}_{kh}, \;\; -4c^{Ai}{}_{hk} c^{Bjk}{}_{l} - \delta^{AB} f^{Aijk} c^{Ak}{}_{hl} = 0,$

so that for the operators τ^{Ai} the commutation relations of Eq.(3) are valid.

It can be proved for $d = 2n$, where n is an integer, that Γ, $\Gamma = \frac{-\mathrm{i}(-2\mathrm{i})^n}{(2n)!}$ $\varepsilon_{a_1 a_2 \ldots a_{2n}} M^{a_1 a_2} \ldots . M^{a_{2n-1} a_{2n}}$, are among invariants of the Lorentz group: $\{\Gamma, M^{cd}\} = 0$. This invariant will be used to define the basis. $\Gamma^{(4)}$, defined in $\mathrm{SO}(1,3)$ subspace, determine handedness of either fermions or bosons.

According to the above equations,the algebra of the group $\mathrm{SO}(1, 14)$ contains as subalgebras the algebras of subgroups $\mathrm{SO}(1, 4)$ and $\mathrm{SO}(10)$. The group $\mathrm{SO}(1, 4)$ rather than the group $\mathrm{SO}(1, 3)$ has to be used to describe the spin of the fermionic and bosonic fields. The generators of $\mathrm{SO}(1, 3)$, which is the subgroup of $\mathrm{SO}(1, 4)$, determine the spins of the fermionic

($\mathcal{S}^{mn}$, $m, n \in \{0,1,2,3\}$) and bosonic (S^{mn}, $m, n \in \{0,1,2,3\}$) fields. The remaining generators of group SO(1, 4), that is $-2iM^{5m}$, $m \in \{0,..,3\}$, may in the case of generators of a spinorial character be recognized as the Dirac γ^m matrices, with all the desired properties [5]. The group SO(10), containing subgroups SU(3), SU(2), U(1), is used to describe the charges of fermionic and bosonic fields.

Looking for the SU(3) × SU(2) × U(1) structure of the group SO(10), in accordance with the above equations, operators $\tau^{1i}, \tau^{2i}, \tau^{3i}$ close the subalgebras according to Eq.(3) and the coefficients f^{1ijk}, ε^{ijk} are the structure constants of the groups SU(3), SU(2), respectively, one finds:
$\tau^{1\,1} := \frac{1}{2}\,(M_{6\,9} - M_{7\,8})$, $\tau^{1\,2} := \frac{1}{2}\,(M_{6\,8} + M_{7\,9})$, $\tau^{1\,3} := \frac{1}{2}\,(M_{6\,7} - M_{8\,9})$, $\tau^{1\,4} := \frac{1}{2}\,(M_{6\,11} - M_{7\,10})$, $\tau^{1\,5} := \frac{1}{2}\,(M_{6\,10} + M_{7\,11})$, $\tau^{1\,6} := \frac{1}{2}\,(M_{8\,11} - M_{9\,10})$, $\tau^{1\,7} := \frac{1}{2}\,(M_{8\,10} + M_{9\,11})$, $\tau^{1\,8} := \frac{1}{2\sqrt{3}}\,(M_{6\,7} + M_{8\,9} - 2M_{10\,11})$,
$\tau^{2\,1} := \frac{1}{2}\,(M_{12\,15} - M_{13\,14})$, $\tau^{2\,2} := \frac{1}{2}\,(M_{12\,14} + M_{13\,15})$, $\tau^{2\,3} := \frac{1}{2}\,(M_{12\,13} - M_{14\,15})$,
$\tau^{3\,1} := \sqrt{\frac{3}{5}}\,\,[-\frac{1}{3}\,(M_{6\,7} + M_{8\,9} + M_{10\,11}) + \frac{1}{2}\,(M_{12\,13} + M_{14\,15})]$.
Operators τ^{Ai}, $A \in \{1,..,3\}$, $i \in \{1, n_A\}$, define either spinorial ($M^{hk} = \mathcal{S}^{hk}$, $\tau^{Ai} = \mathcal{T}^{Ai}$) or vectorial ($M^{hk} = S^{hk}$, $\tau^{Ai} = T^{Ai}$) representations.

To find the irreducible representations of the group SO(1, 14) in terms of the subgroups SO(1, 4) × SU(3) × SU(2) × U(1), the eigenvalue problem for the Casimir operators and all the commuting operators for each of subgroups has to be solved:

$< \theta|\mathcal{A}_i|\varphi >= a^f{}_i < \theta|\varphi >$, $< \theta|A_i|\phi >= a^b{}_i < \theta|\phi >$, $i = \{1, r\}$,

where $\mathcal{A}_i$ and A_i stand for r commuting operators of the spinorial and the vectorial character, respectively and $a^f{}_i$ and $a^b{}_i$ for the corresponding eigenvalues.

To solve the eigenvalue problem, one has to express the operators in the coordinate representation and write the eigenvectors as polynomials of θ^a, $a = 1, 15$. We assume that spinorial representations have an odd and vectorial an even Grassmann character[7], respectively.

One can first solve the eigenvalue problem separately in each of the subspaces in which generators of the groups SO(1, 4), SU(3) and SU(2) operate, respectively, and then get the representations in the whole space as the direct product of the representations in different subspaces. We then look for the multiplets of group SO(1, 14) by applying the operators S^{ab}, $a, b \in \{0,1,2,3,5,..15\}$ at any starting representation. I shall present only those

[5] Operators a^a are Grassmann odd operators. Operating on spinors they change fermions to bosons, changing the Grassmann character from odd to even, and therefore a^m can not be recognized as Dirac γ^m matrices (Mankoč Borštnik (1992a), (1992b), (1993), (1994), (1995a)). One also finds $\Gamma^{(4)} = ia^0a^1a^2a^3 = i\gamma^0\gamma^1\gamma^2\gamma^3$, where exponent (4) of $\Gamma^{(4)}$ denotes the four dimensional subspace and $\{a^m, a^n\} = 2\eta^{mn} = \{\gamma^m, \gamma^n\}$, $m, n \in \{0,1,2,3\}$.

[7] In the cannonical quantization of fields, spinorial representations should quantize to fermions, vectorial to bosons.

representations which are of interest on the level of the Electroweak Standard Model and comment on them.

In Table I, quantum numbers of fermions belonging to four multiplets are presented. The four multiplets occur due to four different representations in the subspace of SO(1,3) and differ among themselves with respect to the discrete symmetries of the Lorentz group in this part of the Grassmann space. It is important to notice that in each multiplet the left handed SU(3) triplets and SU(2) doublets with a U(1) charge $\frac{1}{6}$, together with right handed SU(3) triplets and SU(2) singlets with a U(1) charge equal to $\frac{2}{3}$ and $-\frac{1}{3}$, and also the left handed SU(3) singlets and SU(2) doublets with a U(1) charge equal to $-\frac{1}{2}$ and the right handed SU(3) singlets and SU(2) singlets with a U(1) charge equal to -1 and 0 appear. One also finds the corresponding antiparticles in a multiplet.

The four multiplets speak, and the *approach therefore predicts*, if quarks and leptons are elementary particles or if the discrete symmetries in the SO(1,3) part of the Grassmann space manifest on the level of quarks and leptons, *for four rather than three families.*

All gauge fields and the Higgs's field, being interesting for the Electroweak Standard model, appear among the representations of the SO(1,14) operators of the vectorial character, defined by Eq.6a): the representation forming the SO(1,3) triplet the SU(3) octet and the SU(2) singlet, left and right handed (the gluons), the one forming the SO(1,3) triplet, the SU(3) singlet and the SU(2) singlet, left and right handed (the photon), the one which is the SO(1,3) triplet, the SU(3) singlet and the SU(2) triplet, left handed (the weak bosons), as well as the one which is the $S(1,3)$ singlet, the SU(3) singlet and the SU(2) doublet (the Higgs scalar).

Table I: Quantum numbers of fermions for four chiral multiplets of the group SO(1,14), which may represent four families of massless quarks and leptons. Quantum numbers of the corresponding antifermions have the opposite values. We introduce $Q = \tau^{23} + \tau^{31}$. To the column of SU(2) doublets with $\Gamma = 1$, the column of SU(2) singlets with $\Gamma = -1$, correspond.

family	I II III IV	SU(2) doublets τ^{23}	τ^{31}	Q	$\Gamma^{(4)}$	SU(2) singlets τ^{23}	τ^{31}	Q	$\Gamma^{(4)}$
SU(3) triplets									
$\tau^{13} = (\frac{1}{2}, -\frac{1}{2}, 0)$	$u_1\ u_2\ u_3\ u_4$	1/2	1/6	2/3	± 1	0	2/3	2/3	∓ 1
$\tau^{18} = (\frac{1}{2\sqrt{3}}, \frac{1}{2\sqrt{3}}, -\frac{1}{\sqrt{3}})$	$d_1\ d_2\ d_3\ d_4$	-1/2	1/6	-1/3	± 1	0	-1/3	-1/3	∓ 1
SU(3) singlets									
$\tau^{13} = 0$	$\nu_1\ \nu_2\ \nu_3\ \nu_4$	1/2	-1/2	0	± 1	0	0	0	∓ 1
$\tau^{18} = 0$	$e_1\ e_2\ e_3\ e_4$	-1/2	-1/2	1	± 1	0	-1	-1	∓ 1

Although, in addition to the spinorial representations, presented in Tables I, and the vectorial representations, discussed above, there are also other representations, since the vector space, spanned over the Grassmann coordinate space has 2^{d-1} vectors of an even and 2^{d-1} of an odd Grassmann character, fermions which would be SU(3) octets or SU(2) triplets can not be found among them. However, we haven't yet studied the corresponding coset spaces.

To assure the reader that there is no contradiction between the proposed unification and the "no go" theorem of Colemann and Mandula (Coleman, Mandula (1967)), it may seem sufficient to notice that the dynamics of fields in our approach is defined in d dimensional ordinary (and d dimensional Grassmann) space-time rather than in four dimensional ordinary subspace-time, so that the scattering matrix has to be defined as a unitary matrix in d rather than in four ordinary dimensions, as it is assumed in the "no go" theorem. If all the coordinates but four are compactified, as we have supposed, then one expects that at energies, low compared to the inverse radii of the subspace of d-4 ordinary dimensions, the S - matrix, according to what was presented in this talk, manifests approximately as a unitary matrix in a four dimensional (sub)space-time, and as a analytic function of only the four momenta p^m, $m \in \{0,1,2,3\}$, with the connected symmetry group isomorfic to the direct product of the Poincaré group in four dimensions and the groups defining charges in d-4 Grassmann dimensions.

With the growing energy, however, one expects that not only the S matrix would start to manifest the unitarity in the d dimensional space-time, but charges and spins would start to manifest as a part of the Lorentz group in the d dimensional ordinary and Grassmann space, with gravity as the only interaction, which causes scattering among particles.

Acknowledgement

This work was supported by Ministry of Science and Technology of Slovenia. The author appreciates fruitful discussions with H.B. Nielsen.

References

Coleman, S., Mandula, J. (1967): Phys. Rev. **159**, 1251.

Mankoč Borštnik, N. (1992a): Phys.Lett. **B 292**, 22,
Mankoč Borštnik, N. (1992b): Il Nuovo Cimento **A 105**, 1461,
Mankoč Borštnik, N. (1993): J. of Math. Phys. **34**, 3731,
Mankoč Borštnik, N. (1994): Int. J. of Mod. Phys. **A 9**, 1731,
Mankoč Borštnik, N. (1995a): J. of Math. Phys. **36(4)**, 1593.

Mankoč Borštnik, N. (1995b): Modern Phys. Lett. **A 10**, 587,
Mankoč Borštnik, N. (1996): Unification of spins and charges in Grassmann space enables unification of all interactions, to appear in the Proceedings of ICSMP, Dubna, July 95, hep-th/9512050, hep-th/ 9610004,
Mankoč Borštnik, N., Fajfer, S. (1992): hep-th/9506175.

Kerr Spinning Particle and Superparticle Models

A. Burinskii

Gravity Research Group, IBRAE Academy of Sciences, B.Tulskaya 52, Moscow 113191 Russia, e-mail: grg@ibrae.msk.su

1 Introduction

It is known that the Kerr geometry displays some remarkable features suggesting certain relationships with the spinning elementary particles. In particular, the gyromagnetic ratio of the Kerr-Newman solution is the same as that of the Dirac electron. This fact stimulated consideration of various models of spinning particles based on the Kerr-Newman geometry (Carter (1968), Debney, Kerr, Schild(1969), Israel (1970), Burinskii (1974), Lòpez (1984), Ivanenko, Burinskii (1978)). There were also obtained some string-like structures in the Kerr geometry (Ivanenko, Burinskii (1978), Burinskii (1995), Lind, Newman (1974), Burinskii (1994)).

In 1992 Witten pointed out on the role of black holes in string theory, and a generalization of the Kerr solution to low energy string theory was obtained by Sen (Sen (1992)). It was shown (Burinskii (1995)) that the Kerr-Sen solution acquires near the Kerr singular ring a metric similar to the field around a heterotic string. A role of black holes as fundamental string states was obtained (Sen (1992), (Dabholkar et al. (1996)) that led to a conclusion, suggested from diverse points of view, that black holes should be treated as elementary particles (Sen (1995, 1996)).

On the another hand, the models of spinning particles basing on the Grassmann anticommuting parameters are well known (Volkov, Akulov (1972), Casalbuoni (1976), Brink, Schwarz (1981),and Misc (1975, 1980, 1989), Tugai, Zheltukhin (1995)).

In this paper a combined model of the Kerr spinning particle and superparticle is considered. The Kerr geometry may be represented in a complex form as being created by a mysterious complex "source" propagating along a complex world line. We consider a supergeneralization of the Kerr geometry caused by similar superparticle "source" propagating along a complex superworldline that gives rise to a metric of super-black-hole with nonlinear realization of supersymmetry (Volkov, Akulov (1972), Deser, Zumino (1977), Wess, Bagger (1983)). [1]

---

[1] Nontrivial supergeneralization of Reissner-Nordström solution was given by Aichelburg and Güven (Aichelburg, Güven (1981)).

2 Complex Structure of the Kerr Solution

We use formalism of Debney, Kerr and Schild (Debney, Kerr, Schild(1969)) adopted to complex representation of the Kerr solution. Starting from the Kerr-Schild form of metric $g_{ik} = \eta_{ik} + hk_i k_k$; where η_{ik}=diag(-1,1,1,1) is the auxiliary Minkowski metric in Cartesian coordinates (t,x,y,z), one can see that main peculiarities of the Kerr solution are connected with a form of harmonic scalar function h and vector field $k(x)$ of principal null directions (P.N. congruence). Function h is the Appel potential $h = 2mRe(1/\tilde{r})$, which may be expressed in the oblate spheroidal coordinates r, θ as $\tilde{r} = r + ia\cos\theta$. It has a ring-like singularity $r = \cos\theta = 0$ which is a branch line of the Kerr geometry. The space is covering by two sheets corresponding the positive and negative values of r. In the rest frame the function $\tilde{r}$ may be represented also as a complex radial distance

$$\tilde{r} = \sqrt{(x_a - x_{0a})(x^a - x_0^a)}, \quad a = 1,2,3 \ , \tag{1}$$

from the complex point $x_0 = (0,0,ia)$. It involves a complex interpretation of the Kerr solution (Burinskii (1995), Burinskii (1994)), initiated by Lind and Newman(Lind, Newman (1974)), in which the Kerr geometry is presented as a retarded-time field generated by a "complex point source" which propagates in complex Minkowski space CM^4 along a complex "world line" $x_o^i(\tau)$, $(i = 0,1,2,3)$, parametrized by a complex time parameter $\tau = t + \mathrm{i}\sigma = x_o^o(\tau)$.

An important role in this construction is played by the complex light cones, whose apexes lie on the complex "world line" $x_o(\tau)$. The complex light cone

$$K = \{x : x = x_o^i(\tau) + \psi_R^\alpha \sigma_{\alpha\dot{\alpha}}^i \bar{\psi}_L^{\dot{\alpha}}\} \tag{2}$$

may be split into two families of null planes: "right" (ψ_R =const; $\bar{\psi}_L$ -var.) and "left" ($\bar{\psi}_L$ =const; ψ_R -var.). The rays of the P.N. congruence $k(x)$ of the Kerr geometry are the tracks of these complex null planes (right or left) on the real slice of Minkowski space (Ivanenko, Burinskii (1978), Burinskii (1994), Burinskii, Kerr, Perjes (1995)). P.N. congruence propagates from "negative" sheet of 3-space onto "positive" one crossing the disk spanned by singular ring. In the null coordinates $u = (z+t)/\sqrt{2}; \quad v = (z-t)/\sqrt{2}; \quad \xi = (x + iy)/\sqrt{2}; \quad \bar{\xi} = (x - iy)/\sqrt{2}$ we have

$$k = k_i \mathrm{d}x^i = P^{-1}(\mathrm{d}u + \bar{Y}d\xi + Yd\bar{\xi} - Y\bar{Y}\mathrm{d}v) \ , \tag{3}$$

where $Y(x)$ is a complex projective spinor field $Y = \bar{\psi}^{\dot{2}}, \quad \bar{\psi}^{\dot{1}} = 1$. [2]

The condition for a complex light cone to have a real slice is

$$(x - x_0(\tau))^2 = 0 \ , \tag{4}$$

where x is a real point. In the rest frame and by gauge $x_0^0 = \tau$ this equation may be split as a complex retarded-time equation

[2] We use here the spinor notations of the book (Wess, Bagger (1983)).

$$t - \tau = \tilde{r} = -(x_i - x_{0i})\dot{x}_0^i \ . \tag{5}$$

It fixes the relation $\mathrm{Im}\tau = \sigma = a\cos(\theta)$ between imaginary part of the complex time and a family of the null rays with polar direction θ, ϕ.

The Kerr theorem (Burinskii (1995)) allows to describe the geodesic and shear-free P.N.-congruences in twistor terms via function $Y(x)$ which is a solution of the equation[3] $F(Y, \lambda_1, \lambda_2) = 0$, F being an analytical function. The complex radial distance $\tilde{r}$ may be expressed as $\tilde{r} = -\mathrm{d}F/\mathrm{d}Y$. Singular regions are defined as caustics of the congruence satisfying the system of equations $F = 0; \quad \mathrm{d}F/\mathrm{d}Y = 0$.

For the Kerr congruence function F can be expressed via parameters of complex world line $x_o(\tau)$ (Ivanenko, Burinskii (1978), Burinskii (1994), Aichelburg, Güven (1981))

$$F \equiv (\lambda_1 - \lambda_1^0)K\lambda_2^0 - (\lambda_2 - \lambda_2^0)K\lambda_1^0 \ , \tag{6}$$

where $K = [\partial_\tau x_0^i(\tau)]\partial_i$, and λ_1^0, λ_2^0 are values of the twistor coordinates on the world line $x_0(\tau)$. The resulting function F is quadratic in Y and the solution $Y(x)$ may be given in explicit form.

3 Geometry Generated by Superworldline

Now we would like to generalize this complex retarded-time construction to the case of complex "supersource" propagating along a superworldline

$$X_0^i(\tau) = x_0^i(\tau) - \mathrm{i}\theta\sigma^i\bar{\zeta} + \mathrm{i}\zeta\sigma^i\bar{\theta}; \quad \zeta^\alpha(\tau), \quad \bar{\zeta}^{\dot{\alpha}}(\tau) \ . \tag{7}$$

Similarly to the above "real slice" we introduce a "B-slice" as a "body" of superspace (Dabholkar et al. (1996)), where the nilpotent part of x^i is equal to zero. The "real slice" is a real subset of the "B-slice". The real slice condition (4) takes now the form $s^2 = (x_i - x_{0i}(\tau))(x^i - x_0^i(\tau)) = 0$. Selecting the nilpotent parts of this equation we obtain the above real slice condition (4) and the B-slice conditions

$$[x^i - x_0^i(\tau)](\theta\sigma_i\bar{\zeta} - \zeta\sigma_i\bar{\theta}) = 0 \ ; \tag{8}$$

$$(\theta\sigma\bar{\zeta} - \zeta\sigma\bar{\theta})^2 = 0 \ . \tag{9}$$

The equation (8) may be rewritten using (2) in the form

$$(\theta^\alpha\sigma_{i\alpha\dot{\alpha}}\bar{\zeta}^{\dot{\alpha}} - \zeta^\alpha\sigma_{i\alpha\dot{\alpha}}\bar{\theta}^{\dot{\alpha}})\psi^\beta\sigma^i_{\beta\dot{\beta}}\bar{\psi}^{\dot{\beta}} = 0 \tag{10}$$

which yields

$$\bar{\psi}\bar{\theta} = 0, \qquad \bar{\psi}\bar{\zeta} = 0 \tag{11}$$

---

[3] The three parameters $Y, \quad \lambda_1 = u + Y\bar{\xi}, \quad \lambda_2 = \xi - Yv$ are projective twistor coordinates.

which in turn is a condition of proportionality of the commuting spinors $\bar{\psi}(x)$ and anticommuting spinors $\bar{\theta}$ and $\bar{\zeta}$ providing the left null superplanes to reach B-slice. Taking into account that $\bar{\psi}^{\dot{2}} = Y(x), \quad \bar{\psi}^{\dot{1}} = 1$ we obtain

$$\bar{\theta}^{\dot{2}} = Y(x)\bar{\theta}^{\dot{1}}, \quad \bar{\theta}^{\dot{\alpha}} = \bar{\theta}^{\dot{1}}\bar{\psi}^{\dot{\alpha}}, \quad \bar{\zeta}^{\dot{2}} = Y(x)\bar{\zeta}^{\dot{1}}, \quad \bar{\zeta}^{\dot{\alpha}} = \bar{\zeta}^{\dot{1}}\bar{\psi}^{\dot{\alpha}} . \tag{12}$$

It also gives that $\bar{\theta}\bar{\theta} = \bar{\zeta}\bar{\zeta} = 0$, and equation (9) is satisfied automatically. Therefore, the B-slice condition fixes a correspondence between the coordinates $\bar{\theta}, \quad \bar{\zeta}$ and twistor null planes forming the Kerr congruence.

Conjugate sector also gives $\theta^{\alpha} = \bar{Y}(x)\theta^1$; however, coordinate of superworldline $\zeta(\tau)$ remains independent as well as θ^1, $\bar{\theta}^{\dot{1}}$, and $\bar{\zeta}^{\dot{1}}$.

The retarded time equation (5) takes the form $t - T = \tilde{R} = \tilde{r} + \eta$, where $\tilde{R} = -(x_i - X_{0i})\dot{X}_0^i$ is a superdistance. The "body-part" of this equation satisfies the above relation (5), $T = \tau - \eta$ is a supertime containing the nilpotent term

$$\eta = \mathrm{i}\theta\sigma^0\bar{\zeta} - \mathrm{i}\zeta(\tau)\sigma^0\bar{\theta} . \tag{13}$$

In stationary case $\dot{x}_0^i = (1,0,0,0)$, $\dot{\zeta} = 0$ we have on the B-slice

$$\tilde{R} = r + \mathrm{i}a\cos\theta + \mathrm{i}\theta\sigma^0\bar{\zeta} - \mathrm{i}\zeta(\tau)\sigma^0\bar{\theta} . \tag{14}$$

Supergeneralization of the Kerr theorem may be achieved by substitution superworldline $X_0(\tau)$ instead of $x_0(\tau)$ in the function F. As a result one can obtain a superfield $Y(x)$ which on the B-slice takes the usual form since all the nilpotent terms disappear. From the Kerr theorem one obtains the general expression for superdistance out of B-slice

$$\tilde{R} = -\mathrm{d}\hat{F}/\mathrm{d}Y = \tilde{r} - \mathrm{i}[x^i - x_0^i(\tau)]\dot{\zeta}(\tau)\sigma_i\bar{\theta} - \mathrm{i}[\dot{x}_0^i(\tau) + \mathrm{i}\dot{\zeta}(\tau)\sigma^i\bar{\theta}](\theta\sigma_i\bar{\zeta} - \zeta\sigma_i\bar{\theta})$$

which may be necessary by application the (anti)chiral differential operators $D_{\alpha}, \quad \bar{D}_{\dot{\alpha}}$ (Tugai, Zheltukhin (1995), Wess, Bagger (1983)).

4 Supershift of the Kerr Solution

One can note that the Kerr solution is a particular solution of supergravity with vanishing spin-3/2 field, and that in the stationary case $\dot{X}_0 = (1,0,0,0), \quad \dot{\zeta} = 0$ solution with supersource (7) can be obtained from the Kerr solution by supershift

$$x'^i = x^i + \mathrm{i}\theta\sigma^i\bar{\zeta} - \mathrm{i}\zeta\sigma^i\bar{\theta}; \qquad \theta' = \theta + \zeta, \quad \bar{\theta}' = \bar{\theta} + \bar{\zeta} , \tag{15}$$

which is a "trivial" supergauge transformation (Burinskii (1994), Dabholkar et al. (1996)) in supergravity. However, the subsequent B-slice constraint is a nonlinear operation breaking four-dimensional supersymmetry (Lòpez (1984), Burinskii (1994), Sen (1992)). As a result the arising spin-3/2 field can not be gauged away.

Starting from tetrad form of the Kerr solution $\mathrm{d}s^2 = e^1e^2 + e^3e^4$, where

$$e^1 = \mathrm{d}\xi - Y\mathrm{d}v; \quad e^2 = \mathrm{d}\bar{\xi} - \bar{Y}\mathrm{d}v \; ; \tag{16}$$

$$e^3 = \mathrm{d}u + \bar{Y}\mathrm{d}\xi + Y\mathrm{d}\bar{\xi} - Y\bar{Y}\mathrm{d}v \; ; \tag{17}$$

$$e^4 = \mathrm{d}v - he^3 \; , \tag{18}$$

and using the coordinate transformations (15) under constraints (12), and also substitution $\tilde{R} \to \tilde{r}$, one obtains the following tetrad

$$e'^1 = e^1 + (A - C^1\bar{\theta}^{\bar{1}})\mathrm{d}Y, \qquad e'^2 = e^2 + A\mathrm{d}\bar{Y} \; , \tag{19}$$

$$e'^3 = e^3 - C^3\bar{\theta}^{\bar{1}}\mathrm{d}Y, \qquad e'^4 = \mathrm{d}v + \tilde{h}e'^3 \; , \tag{20}$$

where $\mathrm{d}Y = \tilde{R}^{-1}(Pe^1 - P_{\bar{Y}}e^3)$, and

$$A = \mathrm{i}\sqrt{2}(\theta^1\bar{\zeta}^{\bar{1}}), \qquad C^a = \mathrm{i}e_i^a(\zeta\sigma^i\partial_Y\bar{\psi}), \qquad \tilde{h} = m(Re\tilde{R}^{-1})/P^3 \; . \tag{21}$$

This is a metric of superblackhole with broken four-dimensional supersymmetry. For parameters of spinning particles it corresponds to a specific state of "black hole" without horizons and very far from extreme.

Our derivation of super-Kerr metric is similar to the first derivation of the Kerr-Newman solution by complex shift from the Reissner-Nordström metric given by Newman and collaborators (1965). The first use of complex shift in scalar electrodynamics is traced back to Appel (1887) who found out the potential $eRe(1/\tilde{r})$ characterized by typical Kerr's singularity and twofoldedness of space. The first use of supershift in electrodynamics was considered in the recent work by Tugai, Zheltukhin (1995), where the B-slice was selected by δ-function. As a result a supermultiplet of Maxwell fields was generated from the Coulomb solution.

Therefore, at the moment there are known several applications of the method in consideration. For example, the simplest interesting new solutions can be obtained by a simultaneous performing the complex- and super-shift to the Coulomb solution in flat space. Similarly, a supergeneralization of the Kerr-Newman solution leading to supermultiplet of Maxwell fields on the Kerr background may be obtained, as well as a supergeneralization of the Kerr-Sen solution.

References

Aichelburg, P.C. and Güven, R. (1981): Phys.Rev. **D24**, 2066

Brink, I. and Schwarz, J. (1981): Phys.Lett. 100B (1981) 310

Burinskii, A. (1974): Sov. Phys. JETP **39**, 193.

Burinskii, A. (1994): String - like Structures in Complex Kerr Geometry, Proceedings of the Fourth Hungarian Relativity Workshop Edited. by R.P. Kerr and Z. Perjes, Academiai Kiado,Budapest 1994, gr-qc/9303003, Phys.Let A185, 441;

Burinskii, A. (1995): Phys.Rev. D52, 5826;

Burinskii, A., Kerr R.P. and Perjes Z. (1995): Nonstationary Kerr Congruences, Preprint gr-qc/9501012
Carter, B. (1968): Phys. Rev. **174**, 1559.
Casalbuoni, R. (1976): Nuovo Cim., 33A, 389,
Dabholkar, A. et al. (1996): Nucl. Phys.**B 474**, 85.
Debney, G.C., Kerr, R.P., Schild A. (1969): J.Math.Phys., **10**, 1842.
Deser, S. and Zumino, B. (1977): Phys. Rev. Lett., **38**, 1433
Ivanenko, D. and Burinskii, A. (1978): Izv. VUZ Fiz. (Sov. Phys. J. (USA)), nr.**7**, 113
Israel, W. (1970): Phys rev **D2**, 641.
Lind, R.W., Newman, E.T. (1974): J. Math. Phys.,**15**, 1103.
Lòpez, C.A. (1984): Phys. Rev. **D30**, 313.
Misc (1975, 1980, 1989):
Pashnev, A. and Volkov, D. (1980): Teor.Mat. Fiz. 44, 310.
Berezin, F.A and Marinov, M.S. (1975): Pis'ma ZhETP, 21, 678
Sorokin, D.P., Tkach, V.I., Volkov, D.V. and Zheltukhin, A.A. (1989): Phys. Lett.**B 216**, 302
Sen, A. (1992): Phys.Rev.Lett., 69, 1006,
Sen, A. (1995, 1996): Modern Phys. Lett. **A 10**, 2081, Nucl.Phys **B46** (Proc.Suppl), 198.
Tugai, V. and Zheltikhin, A. (1995): Phys.Rev. **D51**, R3997
Volkov, D.V., Akulov, V.P. (1972): Pis'ma Zh. Eksp.Teor.Fiz. **16**, 621
Wess, J. and Bagger, J. (1983): Supersymmetry and Supergravity, Princeton, New Jersey 1983

Spinons and Parafermions in Fermion Cosets

D.C. Cabra

Investigador CONICET, Argentina.
On leave from Universidad Nacional de La Plata, Argentina.
International Centre for Theoretical Physics, Strada Costiera 11, 34100 Trieste, Italy

Abstract. We introduce a set of gauge invariant fermion fields in fermionic coset models and show that they play a very central role in the description of several Conformal Field Theories (CFT's). In particular we discuss the explicit realization of primaries and their OPE in unitary minimal models, parafermion fields in $\mathbb{Z}_k$ CFT's and that of spinon fields in $\mathrm{SU}(N)_k$, $k = 1$ Wess-Zumino-Witten models (WZW) theories. The higher level case ($k > 1$) will be briefly discussed. Possible applications to QHE systems and spin-ladder systems are addressed.

1 Introduction

The purpose of this talk is to briefly review the fermionic coset description of some particular CFT's, whose relevance has shown up in particular in the construction of the primary operators in the minimal unitary models and the $\mathbb{Z}_k$ parafermion models, as well as in the identification of quasiparticle operators in $\mathrm{SU}(N)_k$ WZW models (Witten (1984),Knizhnik, Zamolodchikov (1984)). We will particularly emphasize the realization of order-disorder (OD) algebras as well as the realization of quasi-particle operators with generalized statistics, where the gauge invariant fermion fields play a central role.

We want to point out that we will only briefly comment on some of the basics of CFT which are relevant for the discussion we want to pursue. For more details see for example ref. (Ginsparg (1990)).

The so called coset construction was introduced by Goddard, Kent and Olive (Goddard, Kent, Olive (1986)) as a way to obtain CFT's with Virasoro central extension less than unity, and can be used to build up many interesting conformal field theories (Ginsparg (1990)). Lagrangian realizations of these coset theories have been presented, both in the bosonic formulation (Gawedzki et al. (1989, 1990)), in terms of gauged WZW models and in the fermionic versions (Bardakci et al. (1988), Cabra et al. (1990)), in terms of constrained (gauged) fermions.

Here we will use the fermionic description of coset models, and will discuss essentially three cases: the models in the so called minimal unitary or FQS series (Friedan et al. (1984)), the $\mathbb{Z}_k$ parafermion models (Fradkin and Kadanoff (1980), Zamolodchikov and Fateev (1985)) and the $\mathrm{SU}(N)_k$ WZW models (Knizhnik, Zamolodchikov (1984), Naculich, Schnitzer (1990)) (from the point of view of fermion cosets). However, the present approach can be applied to the study of arbitrary coset models. (Details on the computations that lead to the results reviewed here are contained in refs. Cabra and Rothe (1995, 1996), Cabra and Moreno (1996), Cabra and Rossini (to app.), Cabra, Moreno, Rossini (in prep.)).

In the case of minimal models we have shown that all the primaries can be constructed as certain fermion bilinears, and in the case of the Ising model, (the first model in the minimal unitary series), we have shown that the role of the gauge invariant fermions is crucial in the identification of the OD fields. This is the first case in which the relevance of gauge invariant fermions has shown up (Cabra and Rothe (1995, 1996)).

In ref. (Cabra and Moreno (1996)) we have shown how the construction of the OD operators in (Cabra and Rothe (1995, 1996)) can be generalized to construct all the primaries in $\mathbb{Z}_k$ parafermion models, and we used this construction to study the "thermal" perturbation of the system.

More recently (Cabra and Rossini (to app.)), we have shown that the gauge invariant fermions realize in a natural way the so called "spinon" fields, which were shown to play a crucial role in the quasiparticle description of the Hilbert space of the $SU(N)_1$ WZW theories, as motivated by the underlying Yangian symmetry (Bouwknegt et al. (1995, 1996), Bernard et al. (1994)). (The extension to higher level cases is in progress (Cabra, Moreno, Rossini (in prep.))). As a byproduct, we have also shown how to factorize the WZW primaries in holomorphic and anti-holomorphic parts, a problem that has received some attention recently (Bardakci et al. (1990, 1991), Halperin and Obers (1996)).

All these results clearly suggest that the gauge invariant fermions (gif's) play an important role in the description of CFT's: in the case of $SU(N)_k$ WZW models, these gif's are precisely the quasiparticle operators, and both in the minimal unitary models and in the $\mathbb{Z}_k$ parafermion models, all the primaries are composites (bound states) of these gif's. Hence, there is a good chance that the construction of quasi-particle operators in terms of gif's pursued in refs. (Cabra and Rossini (to app.), Cabra, Moreno, Rossini (in prep.)) could be extended to other CFT's. This construction could be of some interest in connection with the so called quasiparticle (or fermionic) representation of the characters for CFT's (Kedem et al. (1993)).

Another interesting point is that the quasiparticle operators constructed in ref. (Cabra, Moreno, Rossini (in prep.)) for the higher level cases satisfy non-abelian braiding relations, and hence could play a role in the description of some QHE systems, such as the so called Haldane-Rezayi and Pfaffian states (Moore and Read (1991), Block and Wen (1992)), where the elementary quasiparticles have non-abelian statistics.

Finally we want to mention the so called spin-ladder systems (see ref. (Dagotto and Rice (1996), Sierra (1996)), and references therein), which in the low energy regime can be studied as certain WZW models with interactions.

Our approach could be also useful in the study of these systems, where many interesting effects have a non-perturbative origin.

2 An Example: SU(N)$_1$ WZW Theory as a Fermion Coset

Let us first set up our conventions and describe the approach in a simple example, such as the fermionic coset representation of the SU(N)$_1$ WZW theory (Naculich, Schnitzer (1990)).

The idea is to describe this model as a fermionic coset U(N)/U(1), which is constructed starting with N free massless Dirac fermions and freezing the U(1) charge by imposing

$$j_\mu|\text{phys}>= 0 \ , \tag{1}$$

where j_μ is the U(1) fermionic current.

This is achieved in the path-integral by introducing a δ-functional as

$$\delta[j_\mu] = \int Da_\mu \exp\left(-\int \mathrm{d}^2x\psi^\dagger\gamma_\mu\psi a_\mu\right) \ . \tag{2}$$

The partition function of the U(N)/U(1) model is then given by

$$\mathcal{Z}_{\mathrm{U}(N)/\mathrm{U}(1)} = \int D\psi^\dagger D\psi Da_\mu \exp\left(-\int \mathrm{d}^2x\psi^\dagger(\mathrm{i}\partial\!\!\!/ + \mathrm{i}\,a\!\!\!/)\psi\right) \ , \tag{3}$$

and is equivalent to the partition function of the SU(N)$_1$ WZW model in the sense that correlators of corresponding fields in the two theories coincide (Naculich, Schnitzer (1990)).

In order to write the partition function in a more manageable form, we perform the following change of variables (Roskies and Schaposnik (1981), Polyakov and Wiegmann (1983), Saraví et al. (1981))

$$\begin{aligned} a &= \mathrm{i}(\bar\partial u)u^{-1} & \bar a &= \mathrm{i}(\partial\bar u)\bar u^{-1}, \\ \psi_1 &= u\chi_1 & \psi_2^\dagger &= \chi_2^\dagger u^{-1}, \\ \psi_2 &= \bar u\chi_2 & \psi_1^\dagger &= \chi_1^\dagger \bar u^{-1}. \end{aligned} \tag{4}$$

where $\partial \equiv \frac{\partial}{\partial z}$, $\bar\partial \equiv \frac{\partial}{\partial \bar z}$ and $\psi = \begin{pmatrix}\psi_1\\ \psi_2\end{pmatrix}$. The fields u and $\bar u$ are parametrized in terms of real scalar fields as $h = \exp(-\phi - \mathrm{i}\eta)$ and $\bar h = \exp(\phi - \mathrm{i}\eta)$.

Taking into account the gauge fixing procedure and the Jacobians associated to (4) (Roskies and Schaposnik (1981), Polyakov and Wiegmann (1983), Saraví et al. (1981)) one arrives at the desired decoupled form for the partition function:

$$\mathcal{Z} = \mathcal{Z}_{ff}\mathcal{Z}_{fb}\mathcal{Z}_{gh} \ , \tag{5}$$

where

$$\mathcal{Z}_{ff} = \int \mathcal{D}\chi^\dagger \mathcal{D}\chi \exp(-\frac{1}{\pi}\int (\chi_2^\dagger \bar{\partial}\chi_1 + \chi_1^\dagger \partial \chi_2) d^2x),$$
$$\mathcal{Z}_{fb} = \int \mathcal{D}\phi \exp(\frac{N}{2\pi}\int \phi \Delta \phi d^2x) \ . \tag{6}$$

(The explicit form of the ghost partition function is inessential in what follows).

Notice that, although the partition function of the theory is completely decoupled, BRST quantization conditions connect the different sectors in order to ensure unitarity (Gawedzki et al. (1989, 1990)).

The central charge is now easily evaluated as the sum of three independent contributions coming from the different sectors, $c_{ff} = N$, $c_{fb} = 1$ and $c_{gh} = -2$, thus giving $c = N - 1$ which coincides with the central charge of the $SU(N)_1$ WZW model. Similarly, conformal dimensions of primaries can be evaluated using this decoupled picture.

More general models, as for example the $SU(2)_k$ WZW models, can be also represented as fermionic cosets by making use of the general equivalence (Naculich, Schnitzer (1990))

$$U(2k)/\left(U(1) \times SU(k)_2\right) \ . \tag{7}$$

In this case, in addition to the constraint implemented by the abelian gauge field a_μ, we have to introduce another constraint associated with the $SU(k)$ currents. This constraint will be implemented by a non-abelian gauge field in the Lie algebra of $SU(k)$. The $SU(k)$ gauge field is traded for a $SU(k)$ WZW field through a decoupling transformation similar to that of eq.(4).

Using the approach described above we have studied the following cases:

i) Minimal unitary models

These models can be represented as the cosets (Goddard, Kent, Olive (1986))

$$\left(SU(2)_k \times SU(2)_1\right)/SU(2)_{k+1} \ , \tag{8}$$

and the Virasoro central charges lie in the FQS series (Friedan et al. (1984))

$$c = 1 - 6/\left((k+2)(k+3)\right) \ . \tag{9}$$

By making use of eq.(7), one is led to make the identification of the coset (8) with the fermion coset

$$\left[\left(\frac{U(2k)}{SU(k)_2 \times U(1)}\right) \times \left(\frac{U(2)}{U(1)}\right)\right]/SU(2)_{k+1} \ , \tag{10}$$

whose Lagrangian is given by

$$\mathcal{L} = \psi^\dagger \left(\not{\partial} + \not{a} + \not{B} + \not{A} \right) \psi + \bar{\chi}^\dagger \left(\not{\partial} + \not{b} + \not{A} \right) \chi \ , \qquad (11)$$

where all the internal indices are supressed. (The fermions ψ (χ) transform in the fundamental representation of U($2k$) (U(2))). The gauge fields a_μ, b_μ,B_μ,A_μ, in the Lagrangian (11), implement respectively the U(1), U(1), SU$(k)_2$ and SU$(2)_{k+1}$. (The subindices refer to the central charge of the constrained affine currents).

Within this approach one can show that the central charge of these fermionic coset models is given by eq.(9) (Bardakci et al. (1988), Cabra et al. (1990)), all the primaries can be constructed in terms of fermion bilinears (Cabra and Rothe (1995, 1996)), and the correlators can be evaluated in terms of known results on WZW theories (Zamolodchikov, Fateev (1986)).

The point we want to make here is that the fermionic description is more suitable than the bosonic one regarding the construction of primary fields and their operator product algebra. In particular we can identify in a natural way the OD fields as we will show in the examples bellow. There is one interesting example that has been completely described within this scheme, which corresponds to the Ising model ($c = 1/2$, $k = 1$ in eq.(9)). In this case, not only the primary operators, (i.e. the fields with their dimensions in the Kac table), but also the OD algebra has been realized using the gauge invariant fermions (Cabra and Rothe (1995, 1996)).

Gauge invariant fermions

Let us introduce the gauge invariant fermion fields of the theory defined by eq.(11), for $k = 1$,

$$\hat{\psi}^i(x) = \mathrm{e}^{-\mathrm{i}\int_x^\infty \mathrm{d}z_\mu a_\mu} \left(P \mathrm{e}^{-\mathrm{i}\int_x^\infty \mathrm{d}z^\mu A_\mu} \right)_{ij} \psi^j(x) \qquad (12)$$

$$\hat{\chi}^i(x) = \mathrm{e}^{-\mathrm{i}\int_x^\infty \mathrm{d}z_\mu b_\mu} \left(P \mathrm{e}^{-\mathrm{i}\int_x^\infty \mathrm{d}z^\mu A_\mu} \right)_{ij} \chi^j(x) \ . \qquad (13)$$

(Note that for $k = 1$, the B field is not present)

In terms of these fields, the spin (σ) and disorder (μ) fields read

$$\sigma(x) = \hat{\psi}^\dagger \hat{\psi} + \hat{\chi}^\dagger \hat{\chi}, \quad \textit{and} \quad \mu(x) = \hat{\psi}^\dagger \hat{\chi} + \hat{\chi}^\dagger \hat{\psi} \ . \qquad (14)$$

Using the decoupling formulation one can show that they have the correct dimensions (i.e. $h = \bar{h} = 1/16$), and satisfy the OD algebra (Schroer, Truong (1979))

$$\sigma(z_1)\mu(z_2) = \mathrm{e}^{\mathrm{i}\pi\Theta(z_1 - z_2)} \mu(z_2)\sigma(z_1) \ , \qquad (15)$$

where $\Theta(z)$ stands for the Heaviside function.

We will explain how the OD algebra arises in a simpler example later on, but let us point out that while the Schwinger line integrals associated with the non-abelian gauge field A cancel out in the bilinears (14), those associated with the abelian gauge fields a and b will not, and this fact will give rise to the OD algebra.

ii) $\mathbb{Z}_k$ parafermion models

These models are realized as the cosets

$$\mathrm{SU}(2)_k/\mathrm{U}(1) \ , \tag{16}$$

and the Virasoro central charge is given by $c = 2(k-1)/(k+2)$.

In terms of fermionic cosets they correspond to

$$\left(\frac{\mathrm{U}(2k)}{\mathrm{U}(1) \times \mathrm{SU}(k)_2}\right)/\mathrm{U}(1) \ , \tag{17}$$

In order to construct the primary fields we have proceeded in two steps: first we have used the identification of the primary fields of the $\mathrm{SU}(2)_k$ WZW in terms of fermion bilinears made in ref. (Naculich, Schnitzer (1990)), and then, rewritting these bilinears in terms of gif's we have shown that they correspond to OD fields. Other primaries of the $\mathbb{Z}_k$ models, such as parafermion currents, were also constructed as suitable composites of gauge invariant fermions.

$\mathrm{SU}(2)_k$ primaries

The Lagrangian of the fermionic description of the $\mathrm{SU}(2)_k$ WZW model is given by

$$\mathcal{L} = \psi^\dagger \left(\partial\!\!\!/ + a\!\!\!/ + B\!\!\!/\right) \psi \ , \tag{18}$$

where a_μ and B_μ are U(1) and SU(k) gauge fields, implementing the respective constraints .

The fundamental field g and its adjoint $g^\dagger$ of the bosonic $\mathrm{SU}(2)_k$ WZW theory are represented in terms of fermions by the bosonization formulae (Naculich, Schnitzer (1990))

$$g^{ij} = \psi_2^i \psi_2^{j\dagger}, \qquad g^{ij\dagger} = \psi_1^i \psi_1^{j\dagger} \ . \tag{19}$$

All other integrable representations can be constructed as symmetrized normal ordered products of these fundamental fields. Note that these fields are invariant under gauge transformations, both in U(1) and SU(k), as they should be in order to correspond to physical operators.

The theory which we are interested in has one more U(1) constraint (see eq.(17)), which we implement by adding to the Lagrangian (18) the term

$$\Delta L = \psi^\dagger \, b\!\!\!/^3 t^3 \psi \ , \tag{20}$$

where t^3 is an SU(2) generator. This term (after functional integration over b_μ^3) implements the additional U(1) constraint as follows from eqs.(1),(2). It is easy to show, using the approach described above, that the central charge of the resulting model is the correct one.

The fields in (19) vary under gauge transformations associated with the new gauge field, b_μ^3, introduced in eq.(20). In order to ensure invariance also under these transformations, we will define the gauge invariant fermions as

$$\hat{\psi} = \mathrm{e}^{-\mathrm{i}\int_x^\infty \mathrm{d}z^\mu b^3_\mu t^3}\psi \ . \tag{21}$$

In (Cabra and Moreno (1996)) we have shown that all the $\mathbb{Z}_k$ primaries can be built up from these fields. Here we will only discuss an example in some detail, in order to show how OD algebras are realized in terms of gifs.

Using (21) we can construct the gauge invariant version of the g-field and its adjoint in eq.(19) (Cabra and Rothe (1995, 1996))

$$\hat{g}_{ij} = \hat{\psi}_2^i \hat{\psi}_2^{j\dagger}, \qquad \hat{g}^\dagger_{ij} = \hat{\psi}_1^i \hat{\psi}_1^{j\dagger} \ . \tag{22}$$

Let us consider the composites

$$\sigma_1 \equiv \hat{g}_{1,1}, \qquad \mu_1 \equiv \hat{g}^\dagger_{2,1} \ . \tag{23}$$

In the decoupled picture these fields can be rewritten as

$$\begin{aligned} \sigma_1 &\equiv \hat{g}_{1,1} =: \chi_2^{1\alpha} \tilde{U}^{-1\ \alpha\beta} \chi_2^{\dagger 1\beta} :: \mathrm{e}^{2\phi_a} :: \mathrm{e}^{2\phi_b} :, \\ \mu_1 &\equiv \hat{g}^\dagger_{2,1} =: \chi_1^{2\alpha} \tilde{U}^{\alpha\beta} \chi_1^{\dagger 1\beta} :: \mathrm{e}^{-2\phi_a} :: \mathrm{e}^{\varphi_b - \bar{\varphi}_b} : \ . \end{aligned} \tag{24}$$

Here $\tilde{U}$ is the SU(k) WZW field that parametrizes the field B_μ, and ϕ_a, ϕ_b are the U(1) boson fields that parametrize a_μ and b_μ respectively, and φ_b and $\bar{\varphi}_b$ are the chiral (holomorphic and anti-holomorphic) components of the free boson ϕ_b.

The dimensions of these fields are easily evaluated in the decoupled picture and are given by

$$h = \bar{h} = \frac{k-1}{2k(k+2)} \tag{25}$$

and it can be shown that they satisfy the OD algebra (Cabra and Rothe (1995, 1996))

$$\sigma_1(x_1)\mu_1(x_2) = \mathrm{e}^{\frac{\mathrm{i}2\pi}{k}\Theta(x_1 - x_2)} \mu_1(x_2)\sigma_1(x_1) \ . \tag{26}$$

Eqs. (25) and (26) lead one to identify the fields σ and μ defined in (23) with the order and disorder operators in the $\mathbb{Z}_k$ parafermion theory.

Let us stress that the OD algebra has its origin in the particular way in which the holomorphic components of the free boson ϕ_b are combined in eq.(24). This, in turn, is a consequence of the use of the gauge invariant fermions (21). Indeed, one can easily check, using the cannonical commutation rules for ϕ_b that

$$: \mathrm{e}^{2\phi_b(x_1)} :: \mathrm{e}^{(-\varphi_b + \bar{\varphi}_b)(x_2)} := \mathrm{e}^{\frac{\mathrm{i}2\pi}{k}\Theta(x_1 - x_2)} : \mathrm{e}^{(-\varphi_b + \bar{\varphi}_b)(x_2)} :: \mathrm{e}^{2\phi_b(x_1)} : , \tag{27}$$

being the other factors commuting.

This construction can be generalized to all other $\mathbb{Z}_k$ primaries and hence, having identified all the primaries one can pursue the study of perturbations in the Lagrangian approach. We have studied the "thermal" perturbation of $\mathbb{Z}_k$ models within the present approach and reduced the problem to the study of interacting WZW fields. In this scheme, two fixed points of the perturbed system were identified (Cabra and Moreno (1996)).

iii) $SU(N)_k$ WZW models: gif's, spinons and holomorphic factorization

All we have done so far is to study bilinears of the gauge invariant fermions. In this last section we will study the properties of the gauge invariant fermions themselves and show their relevance in the so called "spinon" description of WZW models (Bouwknegt et al. (1995, 1996), Bernard et al. (1994)), as described in ref. (Cabra and Rossini (to app.), Cabra, Moreno, Rossini (in prep.)).

Let us then define the gauge invariant fermion fields (Cabra and Rossini (to app.)) corresponding to the $SU(N)_k$ theory similarly as in (12), (with A_μ replaced by B_μ), that by construction will create the physical excitations, and let us study their properties. For simplicity we will discuss the level one case, although similar considerations apply to the higher level case (Cabra, Moreno, Rossini (in prep.)).

In the decoupled picture, the gifs are given by

$$\begin{array}{ll} \hat{\psi}_1{}^i(z) = \mathrm{e}^{-\varphi(z)}\chi_1^i(z) & \hat{\psi}_1{}^{i\dagger}(\bar{z}) = \chi_1^{i\dagger}(\bar{z})\mathrm{e}^{-\bar{\varphi}(\bar{z})} \\ \hat{\psi}_2^i(\bar{z}) = \mathrm{e}^{\bar{\varphi}(\bar{z})}\chi_2^i(\bar{z}) & \hat{\psi}_2{}^{i\dagger}(z) = \chi_2^{i\dagger}(z)\mathrm{e}^{\varphi(z)} \end{array} \tag{28}$$

where $\varphi(z)$ and $\bar{\varphi}$ are the chiral (holomorphic and anti-holomorphic) components of the free boson ϕ. This fact together with the equation of motion of the free fermions χ ensures that $\hat{\psi}_1^i$ and $\hat{\psi}_2^{i\dagger}$ ($\hat{\psi}_2^i$ and $\hat{\psi}_1^{i\dagger}$) are holomorphic (anti-holomorphic). The conformal dimensions are given by $((N-1)/2N, 0)$ and $(0, (N-1)/2N)$ respectively, thus suggesting that they correspond to the "halves" of the WZW primary $g(z, \bar{z})$.

In fact, it can be shown by the following sequence of identities

$$g^{ij}(z, \bar{z}) = \psi_2^i \psi_2^{j\dagger} = \hat{\psi}_2^i(\bar{z})\hat{\psi}_2^{j\dagger}(z) \ , \tag{29}$$

that the gifs are the holomorphic factors of the WZW primary.

The second important property satisfied by these operators is their OPE which allows us to identify them as the "spinon" operators of ref. (Bouwknegt et al. (1995, 1996), Bernard et al. (1994)):

$$\begin{aligned} \hat{\psi}_1{}^i(z)\hat{\psi}_1{}^j(w) &= \frac{1}{(z-w)^{1/N}}\mathcal{A}(:\chi_1^i(w)\chi_1^j(w):) : \exp 2\varphi(w) : + \ldots \\ &= \frac{1}{(z-w)^{1/N}}\Phi_2^{ij}(w) + \ldots \ , \end{aligned} \tag{30}$$

where $\mathcal{A}$ stands for antisymmetrization and Φ_2^{ij} is the WZW primary with dimension $(N-2)/N$.

As explained in the third reference of (Bouwknegt et al. (1995, 1996), Bernard et al. (1994)) the branch cut in the OPE singularity implies that the excitations created by the spinon fields satisfy generalized statistics with "statistical angle" $\theta = \pi/N$. The chiral Fock space of the $SU(N)_1$ CFT can be constructed in terms of the modes of these fields and this space can be classified into multiplets corresponding to the irreducible representations of the Yangian algebra $Y(sl_N)$ (Bouwknegt et al. (1995, 1996), Bernard et al. (1994)).

In the case of higher levels, the same construction can be done, and in this case, the excitations created by the gauge invariant fermions satisfy non-abelian braiding statistics. The holomorphic factorization of the WZW fields in terms of gifs can also be proved. The results of this investigation will appear elsewhere (Cabra, Moreno, Rossini (in prep.)).

Acknowledgements: I would like to thank to my collaborators E. Moreno, G. Rossini and K.D.Rothe with whom the work presented in this talk has been developed. This work was supported in part by CONICET and Fundación Antorchas, Argentina.

References

Bardakci et al. (1988), Cabra et al. (1990):
Bardakci, K., Rabinovici, E. and Säring, B.(1988): Nucl.Phys. **B299**, 151;
Cabra, D.C., Moreno, E.F. and von Reichenbach, C. (1990): Int.J.Mod.Phys. **A5**, 2313.
Bardakci, K., Crescimanno, M. and Rabinovici, E. (1990, 1991): Nucl.Phys. **B344**, 344; Nucl.Phys. **B349**, 439;
Halpern, M.B. and Obers, N.A. (1996): hep-th/9610081.
Bouwknegt et al. (1995, 1996), Bernard et al. (1994): Bouwknegt, P., Ludwig, A.W.W. and Schoutens, K. (1994):Phys.Lett. **338B**, 448; *ibid* **359B**, 304;
Bouwknegt, P. and Schoutens, K. (1996): hep-th/9607064;
Bernard, D., Pasquier, V. and Serban, D. (1994): Nucl.Phys. **B428**, 612.
Cabra, D.C. and Moreno, E.F. (1996): Nucl.Phys. **B475**, 522.
Cabra, D.C., Moreno, E.F. and Rossini, G.L. (in prep.): "Quasiparticles in $SU(N)_k$ fermion cosets", in preparation.
Cabra, D.C. and Rossini, G.L. (to app.): Mod.Phys.Lett. to appear.
Cabra, D.C. and Rothe, K.D. (1995, 1996): Phys.Rev. D**51**, R2509; Ann.Phys. **251**, 337.
Dagotto, E. and Rice, T.M. (1996): Science **271**, 618;
Sierra, G. (1996): cond-mat/9610057.
Fradkin, E. and Kadanoff, L.P. (1980): Nucl.Phys. **B170**, 1;
Zamolodchikov, A.B. and Fateev, V.A. (1985): Sov.Phys.JETP **62**, 215.
Friedan, D., Qiu, Z. and Shenker, S. (1984): Phys.Rev.Lett. **52**, 1575.
Gawedzki et al. (1989, 1990):
Gawedzki, K. and Kupiainen, A. (1989): Nucl.Phys. **B320**, 624;
Karabali, D., Han Park, Q., Schnitzer, H. and Yang, Z. (1989): Phys.Lett.**216B**, 307;
Karabali, D. and Schnitzer, H. (1990): Nucl.Phys. **B329**, 649.

Ginsparg, P. (1990): in "Fields, Strings and Critical Phenomena", Les Houches School 1988, E.Brezin and J.Zinn-Justin eds. (North-Holland, Amsterdam, 1990).
Goddard, P., Kent, A. and Olive, D. (1986): Int.J.Mod.Phys. **A1**, 303.
Kedem, R., Klassen, T.R., McCoy, B.M. and Melzer, E. (1993): Phys.Lett. **B304**, 263; **B307**, 68.
Knizhnik, V.G. and Zamolodchikov, A.B. (1984): Nucl.Phys. **B247**, 83.
Moore, G. and Read, N. (1991): Nucl.Phys. **B360**, 362;
Block, B. and Wen, X.-G. (1992): Nucl.Phys. **B374**, 615.
Naculich, S.G. and Schnitzer, H.J. (1990): Nucl.Phys. **B347**, 687.
Roskies, R. and Schaposnik, F.A. (1981): Phys.Rev. D**23**, 558;
Polyakov, A. and Wiegmann, P. (1983): Phys.Lett. **131B**, 121;
Gamboa Saraví, R.E., Schaposnik, F.A. and Solomin, J.E. (1981): Nucl. Phys. **B185**, 238.
Schroer, B. and Truong, T.T. (1979): Nucl. Phys. **B154**, 125.
Witten, E. (1984): Commun.Math.Phys. **92**, 455.
Zamolodchikov, A.B. and Fateev, V.A. (1986): Sov.J.Nucl.Phys. **43**, 657.

Exact Solutions in Einstein–Yang–Mills Theories

I.P. Volobuev

Institute of Nuclear Physics, Moscow State University
119 899 Moscow, Russia

D.V. Volkov is famous for his pioneering works on supersymmetry. Strange as it may seem, this fame tends to hide his contributions to the solution of other problems of modern high energy physics. One of such problems is the problem of spontaneous compactification in Einstein–Yang–Mills (EYM) theories.

It is a common knowledge that the extended supergravity with $N \geq 4$ was first constructed by dimensional reduction of supergravity in the space-time of dimension greater than 4. This fact led to the revival of the old idea due to Kaluza and Klein that the space-time may have extra dimensions, which should be treated as physical reality. If one made this assumption, one had to find a dynamical explanation of the fact that the extra dimensions form a compact manifold of extremely small size, i.e. to find solutions of the original multidimensional theory, which would correspond to the factorized structure of multidimensional space-time. This procedure was called spontaneous compactification, and the first solutions of this kind were found in EYM-theories.

D.V. Volkov made a very important contribution to the theory of spontaneous compactification in EYM-theories. He developed the first regular method of solving spontaneous compactification equations for the case of symmetric spaces of extra dimension, based on the parallelizability condition for gauge fields (Volkov, Tkach (1980), Volkov, Tkach (1982), Volkov et al. (1984)). This method revealed the importance of the symmetry group of the space of extra dimensions, and it was the theory of symmetric gauge fields, or invariant connections in fibre bundles over reductive homogeneous spaces which later allowed to find a general solution to the problem of spontaneous compactification in EYM-theories. Application of this theory to EYM-systems with different topological structures of space-time also allowed to find other types of symmetric solutions in EYM-theories, such as generalized Reissner-Nordström solutions and the so called wormhole solutions.

Here we are going to present the general method of finding spontaneous compactification solutions in EYM-theories and to explain the meaning of Volkov's parallelizability condition. Then we will make a brief comment on the other types of exact solutions in EYM-systems.

We take the standard action of EYM-system with gauge group G in the D-dimensional space-time E equipped with the metric g, $\text{sign}\, g = (-, + \cdots +)$

to be

$$S = S_g + S_\Lambda + S_{YM} \ , \tag{1}$$

where

$$\begin{aligned} S_{YM} &= \frac{1}{8\hat{g}^2}\int_E \mathrm{tr}(F_{MN}F^{MN})\mathrm{d}v_E \\ S_g &= \frac{1}{16\pi\hat{\kappa}}\int_E \hat{R}\mathrm{d}v_E \\ S_\Lambda &= -\int_E \Lambda \mathrm{d}v_E \ , \end{aligned} \tag{2}$$

Here $\hat{R}$ and $\mathrm{d}v_E$ are the scalar curvature and the volume element of the space E, canonically calculated from the metric g, and $\hat{g}$, $\hat{\kappa}$ and Λ are multidimensional coupling constants and the cosmological constants.

The corresponding EYM-equations are

$$R_{MN} - \frac{1}{2}g_{MN}\hat{R} = 8\pi\hat{\kappa}\,(T_{MN} - \Lambda g_{MN}) \tag{3}$$
$$\nabla_M F^{MN} + \left[A_M, F^{MN}\right] = \partial_M F^{MN} + \Gamma^M_{LM}F^{LN} + \left[A_M, F^{MN}\right] = 0 \ ,$$

where the Christoffel symbols Γ^L_{MN} are expressed in the standard way in terms of the metric tensor g_{MN}, and the energy-momentum tensor of the gauge field is

$$T_{MN} = -\frac{1}{2\hat{g}^2}\left(\mathrm{tr}(F_{MK}F_N^{\ K}) - \frac{g_{MN}}{4}\mathrm{tr}(F_{KL}F^{KL})\right) \ . \tag{4}$$

As it is usual in the spontaneous compactification theory, we look for the solutions to these equations, which correspond to the direct product structure of the space E, i.e.

$$E = M \times I, \quad g = \eta \oplus \gamma \ , \tag{5}$$

where η is the Minkowski metric on the four dimensional space M, and γ is a metric on the compact homogeneous space I, independent of the coordinates in M. We also assume that $A_\mu = 0 \quad (\mu = 0,1,2,3)$, and the components $A_k \quad (k \geq 4)$ depend only on the coordinates in I.

Under these assumptions the EYM-equations reduce to the spontaneous compactification equations

$$R_{ij} = -\frac{4\pi\hat{\kappa}}{\hat{g}^2}\mathrm{tr}(F_{ik}F_j^{\ k})$$
$$\nabla_i F^{ik} + \left[A_i, F^{ik}\right] = \partial_i F^{ik} + \Gamma^i_{li}F^{lk} + \left[A_i, F^{ik}\right] = 0 \ , \tag{6}$$

where the Ricci tensor R_{ij} and the Christoffel symbols Γ^i_{li} are defined only by the metric γ on K/H. We also get the fine tuning equation for the multidimensional cosmological constant

$$\Lambda = -\frac{1}{8\hat{g}^2}\mathrm{tr}(F_{ik}F^{ik}) \ . \tag{7}$$

Usually, it is assumed that the space I is homogeneous, i.e. $I = K/H$. In this case one can look for K-symmetric solutions. In particular, the metric γ is then K-invariant.

The first regular method to solve the spontaneous compactification equations for the case of symmetric spaces K/H was put forward by D.V. Volkov (Volkov, Tkach (1980), Volkov, Tkach (1982), Volkov et al. (1984)). He proposed that the Yang-Mills equation (6) be replaced by a stronger condition

$$\nabla_i F^{kl} + [A_i, F^{kl}] = \partial_i F^{kl} + \Gamma^k_{mi} F^{ml} + \Gamma^l_{mi} F^{km} + [A_i, F^{kl}] = 0 \ , \tag{8}$$

which he called the parallelizability condition for the gauge field. There is a good reason to believe that this condition was inspired by a similar condition for the curvature tensor

$$\nabla_i R_{klmn} = 0 \ , \tag{9}$$

which is valid for spaces of constant curvature. D.V. Volkov showed that equations (8) and (9) could be solved for arbitrary symmetric spaces K/H. The general solution to the problem of spontaneous compactification was found seven years later and was based on the theory of symmetric gauge fields, or invariant connections (Volobuev, Kubyshin (1987), Volobuev, Kubyshin (1988)). Here we will present the general solution and then discuss, how good was the Volkov's parallelizability condition.

A symmetric gauge field on $E = M \times K/H$ is a field, which is invariant, up to a gauge transformation, under the action of the symmetry group K. In mathematics such a field is described as a K-invariant connection in a principal fibre bundle $P(E, G)$, G being the gauge group. Further on we will assume that the space K/H is reductive, i.e. there exists a decomposition of the Lie algebra $\mathfrak{K} = \mathrm{Lie}(K)$

$$\begin{aligned} &\mathfrak{K} = \mathfrak{H} \oplus \mathfrak{M}, \\ &\mathfrak{H} = \mathrm{Lie}(H), \quad \mathrm{Ad}(H)\mathfrak{M} \subset \mathfrak{M} \ . \end{aligned} \tag{10}$$

We note that the space $\mathfrak{M}$ is naturally identified with the tangent space $T_o K/H$ at $o = [H]$.

Invariant connections in principal fibre bundles over reductive homogeneous spaces are described by the Wang's theorem (Kobayashi, Nomizu (1963)). One can generalize these results to the case of nontransitive group action, which gives that a K-symmetric gauge field A on $E = M \times K/H$ is in a one-to-one correspondence with a triplet $(\tau, \phi, \hat{A})$, where

$$\tau : H \to G \tag{11}$$

is a group homomorphism, ϕ is a function with values in $\mathfrak{M}^* \otimes \mathfrak{G}$, which can also be treated as a linear mapping

$$\phi : \mathfrak{M} \to \mathfrak{G}, \qquad \qquad (12)$$
$$\phi \circ \mathrm{Ad}h = \mathrm{Ad}\tau(h) \circ \phi, \quad h \in H \ ,$$

$\hat{A}$ is a gauge field on M with the gauge group $C = C_G(\tau(H))$, the centralizer of $\tau(H)$ in G.

Let θ be the canonical left-invariant 1-form on K, $s : K/H \to K$ be a section in the H-bundle $K(K/H, H)$ and $\bar{\theta} = s^*\theta$. Then the symmetric field A on $M \times K/H$ can be reconstructed from $(\tau, \phi, \hat{A})$ as follows:

$$A = \hat{A} + \tau'(\bar{\theta}_{\mathfrak{H}}) + \phi(\bar{\theta}_{\mathfrak{M}}) \ . \qquad (13)$$

The corresponding field strength 2-form F is given by

$$F = \hat{F} + (D\phi)(\bar{\theta}_{\mathfrak{M}}) + \frac{1}{2}[\phi(\bar{\theta}_{\mathfrak{M}}), \phi(\bar{\theta}_{\mathfrak{M}})] - $$
$$- \frac{1}{2}\phi([\bar{\theta}_{\mathfrak{M}}, \bar{\theta}_{\mathfrak{M}}]_{\mathfrak{M}}) - \frac{1}{2}\tau'([\bar{\theta}_{\mathfrak{M}}, \bar{\theta}_{\mathfrak{M}}]_{\mathfrak{H}}) \ , \qquad (14)$$

where

$$(D\phi)(\bar{\theta}_{\mathfrak{M}}) = d\phi \wedge \bar{\theta}_{\mathfrak{M}} + [\hat{A}, \phi(\bar{\theta}_{\mathfrak{M}})].$$

The identification of $\mathfrak{M}$ and T_oK/H allows us to introduce a basis $\{u_a\}$ in $\mathfrak{M}$, $(a = 1, 2 \cdots \dim K/H)$, orthonormal with respect to the metric γ, and to define component scalar fields

$$\phi_a(x) \equiv \phi(u_a).$$

Then the action of YM field in E can be reduced to an action in M:

$$S_{YM} = \frac{v(K/H)}{8\hat{g}^2} \int_M \{\mathrm{tr}(F_{\mu\nu}F^{\mu\nu}) + 2\sum_a \mathrm{tr}(D_\mu\phi_a D^\mu\phi_a) - V(\phi)\} dv_M \ , \qquad (15)$$

where $v(K/H)$ is the volume of K/H calculated with the metric γ, and the covariant derivative and the scalar field potential are given by

$$D_\mu\phi_a = \partial_\mu\phi_a(x) + [A_\mu, \phi_a(x)]$$
$$V(\phi) = -\sum_{ab} \mathrm{tr}(F_{ab}F_{ab}) \geq 0 \qquad (16)$$
$$F_{ab} = 2F(u_a, u_b) = [\phi_a(x), \phi_b(x)] - \phi([u_a, u_b]_{\mathfrak{M}}) - \tau'([u_a, u_b]_{\mathfrak{H}}) \ .$$

Thus we see that the reduced theory is a gauge theory, which includes scalar fields minimally coupled to the reduced gauge field. This fact was used in many papers for constructing four-dimensional models by dimensional reduction method (Forgacs, Manton (1980), Kubyshin et al. (1989)).

It is easy to see that the YM-equations (6) on K/H can be obtained from the action

$$S_{\mathrm{eff}} = \frac{1}{8\hat{g}^2} \int_{K/H} \mathrm{tr}(F_{ik}F^{ik}) dv_{K/H} \ . \qquad (17)$$

When taken on symmetric gauge fields, this action is

$$S_{\text{eff}} = \frac{v(K/H)}{8\hat{g}^2} V(\phi) \quad (!!!) \tag{18}$$

As it was shown in (Palais (1979)), the extrema in the class of symmetric fields are extrema in the class of all fields, and therefore the symmetric solutions of YM-equations are the extrema of the potential of scalar fields of the reduced theory.

It is also known (Kobayashi, Nomizu (1963)) that K-invariant metrics γ on K/H are in one-to-one correspondence with H-invariant scalar products in $\mathfrak{M}$ and depend on a finite number of parameters $\{M_K\}$. Therefore, we have

$$V(\phi) = V(\phi, M_k), \quad R_{K/H} = R(M_k) \ . \tag{19}$$

Calculating the scalar curvature of $E = M \times K/H$ with the metric $g = \eta \oplus \gamma$ and the gauge field Lagrangian with $\hat{A} = 0$, $\phi = const$, substituting them into the action (1) and integrating over K/H, we get the following expression

$$v(K/H) \left(\frac{1}{16\pi\hat{\kappa}} R(M_k) - \frac{1}{8\hat{g}^2} V(\phi, M_k) - \Lambda \right) \ , \tag{20}$$

which has the meaning of the cosmological constant in four dimensional theory and therefore should be put equal to zero. Then one can easily show that the spontaneous compactification equations coincide with the equations for the extrema of an effective potential

$$U(\phi, M_k) = \frac{1}{16\pi\hat{\kappa}} R(M_k) - \frac{1}{8\hat{g}^2} V(\phi, M_k) \ , \tag{21}$$

the Einstein equations being

$$R_k = \frac{4\pi\hat{\kappa}}{\hat{g}^2} Q_K \qquad \text{with} \tag{22}$$

$$R_k = \frac{\partial R_{K/H}}{\partial M_k}, \quad Q_k = \frac{1}{2}\frac{\partial V}{\partial M_k} \ .$$

Thus the recipe for solving the spontaneous compactification equations can be formulated as follows:

1. Solve the constraint equation for ϕ and find the basic intertwining operators ϕ_A and the unconstrained scalar fields f^A, so that

$$\phi = \sum_A f^A \phi_A .$$

2. Calculate the scalar field potential $V(\phi, M_k) = V(f^A, M_k)$ and find its extrema
$\tilde{f}^A = f^A(M_k)$ for all possible sets $\{M_k\}$.

3. Calculate $R(M_k)$, derive the Einstein equations and insert $f^A(M_k)$ into them. Then one gets a system of algebraic equations for the parameters $\{M_k\}$.

In the case of symmetric spaces K/H γ is one-parametric and the potential is always of the Higgs type (Kubyshin et al. (1989))

$$V(\phi, M) = M^4(\lambda(|\phi|^2 - \mu^2)^2 + v_0) \ , \tag{23}$$

its extrema being $\tilde{\phi} = 0, \quad |\tilde{\phi}| = \mu$. The scalar curvature for symmetric spaces is given by $R_{K/H} = M^2 r$, where r is a function of their dimension, and the Einstein equation reads

$$\begin{aligned} R(M) &= \frac{4\pi\hat{\kappa}}{\hat{g}^2} V(\phi, M) \\ M^2 r &= \frac{4\pi\hat{\kappa}}{\hat{g}^2} M^4(\lambda(|\phi|^2 - \mu^2)^2 + v_0) \ . \end{aligned} \tag{24}$$

These equations have two solutions, corresponding to two extrema of the Higgs potential. They are

$$M_1^2 = \frac{\hat{g}^2}{4\pi\hat{\kappa}} \frac{r}{\lambda\mu^4 + v_o}, \qquad A = \tau'(\bar{\theta}_{\mathfrak{H}}) \tag{25}$$

$$M_2^2 = \frac{\hat{g}^2}{4\pi\hat{\kappa}} \frac{r}{v_o}, \qquad A = \tau'(\bar{\theta}_{\mathfrak{H}}) + \tilde{\phi}(\bar{\theta}_{\mathfrak{M}}) \ . \tag{26}$$

The parallelizability condition written in terms of ϕ looks like

$$[\phi(u_a), [\phi(u_b), \phi(u_c)] - \tau'([u_b, u_c]] = 0.$$

Now that we have found the general solution, it is easy to check that the parallelizability condition is equivalent to the YM-part of the spontaneous compactification equations in the following cases:

1. $\mathfrak{H}$ is simple.
2. $\mathfrak{H} = \mathfrak{H}_1 \oplus \mathfrak{H}_2$ is semisimple, and the indices of $\tau'(\mathfrak{H}_1)$ and $\tau'(\mathfrak{H}_2)$ in $\mathfrak{G}$ are equal.
3. $\mathfrak{H}$ contains a one-dimensional abelean ideal, $\mathfrak{M}^c = T \oplus \bar{T}$, and $[\phi(T), \phi(T)] = 0$.

Otherwise the parallelizability condition has only the trivial solution $\phi = 0$, which corresponds to the canonical invariant connection in $K(K/H, H)$.

Thus we see that the Volkov's parallelizability condition is a good approximation in the case of symmetric space of extra dimensions, which is almost always equivalent to the exact YM-equations.

The theory of symmetric gauge fields proved to be a very useful tool for finding other types of exact solutions in Einstein–Yang–Mills systems. For example, if the space-time E is Euclidean and of the form $E = R \times S^3$, the method of dimensional reduction of symmetric gauge fields allows to obtain the so called wormhole solutions for arbitrary gauge group G (Bertolami et al. (1991)). If one assumes the space-time E to be four-dimensional and topologically of the form $E = M^2 \times S^2$, one can get the generalized Reissner-Nordstöm solutions (Rudolph (1996)).

The author acknowledges partial support under the INTAS-93-1630-EXT project.

References

Bertolami, O., Mourão, J.M., Picken, R.F., Volobuev, I.P. (1991): Int. Jour. Mod. Phys. **A6**, 4149.
Forgacs, P., Manton, N.S. (1980): Comm. Math. Phys. 72 15
Kobayashi, S., Nomizu, K. (1963): Foundations of Differential Geometry I / II, John Willey & Sons New York London.
Kubyshin, Yu.A., Mourao, J.M., Rudolph, G., Volobuev, I.P. (1989): Lecture Notes in Physics, 349.
Palais, R.S. (1979): Commun.Math.Phys. **69**, 19.
Rudolph, G., Tok, T., Volobuev, I.P. (1996): Int. J. Mod. Phys. **A11**, 4999.
Volkov, D.V., Tkach, V.I. (1980): JETP Lett. **32**, 681.
Volkov, D.V., Tkach, V.I. (1982): Teor. Mat. Fiz. **51**, 171.
Volkov, D.V., Sorokin, D.P., Tkach V.I. (1984): Teor. Mat. Fiz. **61**, 241.
Volobuev, I.P., Kubyshin, Yu.A. (1987): JETP Lett. **45**, P.581.
Volobuev, I.P., Kubyshin, Yu.A. (1988): Teor. Mat. Fiz. **75**, 255.

Higher Massless Irreducible Spins in the BRST Approach

A. Pashnev[1] and M. Tsulaia[2]

[1] JINR–Bogoliubov Theoretical Laboratory, 141980 Dubna, Moscow Region, Russia
E-mail: pashnev@thsun1.jinr.dubna.su
[2] JINR–Bogoliubov Theoretical Laboratory, 141980 Dubna, Moscow Region, Russia
E-mail: tsulaia@thsun1.jinr.dubna.su

Abstract. The BRST -approach is applied to the description of irreducible massless representations of the Poincare group with higher spins.

1 Introduction

The lagrangians, describing irreducible higher spins and guaranteeing the absence of unphysical degrees of freedom must have very special structure and admit some gauge invariance both in the massive and massless cases (Fradkin (1950), Singh, Hagen (1974) ,Chang (1967),Fronsdal (1978),Fronsdal, Fang (1978),de Witt, Freedman (1980), Curtright (1979), Vasiliev (1980)Aragone, Deser (1980)). In some sense the massless case is simpler and, hence, more investigated than the massive one. The hope is that with the help of some Higgs-type effect some of the interacting particles acquire nonzero values of the mass. The alternative way is the dimensional reduction of massless theory leading in general to the arbitrary dependence of mass on spin, at least in the free field theory (Pashnev (1989), Pashnev, Tsulaia (1996),Pashnev, Tsulaia (1997)).

One of the most economic and straightforward method of construction of lagrangians in gauge theories is the method using BRST-charge of the corresponding firstly quantized theory. With the help of BRST-charge the lagrangians for the free field theory for infinite tower of massless higher spin particles (Ouvry, Stern (1986)) and for particular values of spins (Meurice (1987)) were constructed. The lagrangian of (Ouvry, Stern (1986)) describes the system which is infinitely degenerated on each spin level and only with the help of additional constraints we can delete all the extra states having precisely one particle of each spin . These additional constraints are of the second class and their inclusion in the BRST- construction is nontrivial.

The methods of such construction were discussed in (Faddeev, Shatashvili (1986),Egoryan, Manvelyan (1993)). With the help of additional variables one can modify the second class constraints in such a way that they become commuting, i.e the first class . At the same time the number of physical degrees of freedom for both systems of constraints is the same if the number of additional variables coincides with the number of second class constraints.

In the second part of the paper we describe the simple modification of the method of (Faddeev, Shatashvili (1986), Egoryan, Manvelyan (1993)). We show that with the help of additional variables two second class constraints can be transform into three first class constraints with the same number of physical degrees of freedom. The same procedure is used in the third part of the paper for the BRST - description of massless irreducible representations of Poincare group.

2 The Simple Example

To describe all higher spins simultaneously it is convenient to introduce auxiliary Fock space generated by creation and annihilation operators a_μ^+, a_μ with vector Lor
entz index $\mu = 0, 1, 2, ...D-1$, satisfying the following commutation relations

$$[a_\mu, a_\nu^+] = -g_{\mu\nu}, \; g_{\mu\nu} = \mathrm{diag}(1, -1, -1, ..., -1) \; . \tag{1}$$

The general state of the Fock space

$$|\Phi\rangle = \sum \Phi^{(n)}_{\mu_1\mu_2\cdots\mu_n}(x) a^+_{\mu_1} a^+_{\mu_2} \cdots a^+_{\mu_n} |0\rangle \tag{2}$$

depends on space-time coordinates x_μ and its components $\Phi^{(n)}_{\mu_1\mu_2\cdots\mu_n}(x)$ are tensor fields of rank n in the space-time of arbitrary dimension D. The norm of states in this Fock space is not positively definite due to the minus sign in the commutation relation (2.1) for time components of creation and annihilation operators. It means that physical states must satisfy some constraints to have positive norm.

To describe the irreducible massless higher spins we must take into account the following constraints (Pashnev, Tsulaia (1997)):

$$L_0 = -p_\mu^2 \tag{3}$$

$$L_1 = p_\mu a_\mu \;\; L_1^+ = p_\mu a_\mu^+ \tag{4}$$

$$L_2 = \frac{1}{2} a_\mu a_\mu \;\; L_2^+ = \tfrac{1}{2} a_\mu^+ a_\mu^+ \tag{5}$$

$$\tag{6}$$

They form the algebra:

$$[L_1 \, , \, L_{-2}] = -L_{-1} \; [L_{-1} \, , \, L_2] = L_1 \tag{7}$$

$$\tag{8}$$

$$[L_1 \, , \, L_{-1}] = L_0 \; [L_2 \, . \, L_{-2}] = -a_\mu^+ a_\mu + \frac{D}{2} \equiv G_0 \tag{9}$$

It means that the second class constraints $L_{\pm 2}$ must be included in the BRST charge. Following the line of (Faddeev, Shatashvili (1986) , Egoryan, Manvelyan (1993)) we can transform these constraints into commuting ones by introducing additional degrees of freedom like b, b^+. This procedure is rather simple for the classical case when Poisson brackets are used instead of commutators. It leads to the finite system of differentional equations which can be solved without troubles. In the quantum case the corresponding system of equations is infinite due to accounting of repeated commutators.

In this section we describe the modification of the procedure of (Faddeev, Shatashvili (1986) , Egoryan, Manvelyan (1993)) which works well in the case of two second class constraints L_2 and L_{-2} (5) (Pashnev, Tsulaia (1996)). We modify this system of constraints by introduction of *two* additional operators b_1, b_2 together with their conjugates b_1^+, b_2^+: $[b_1, b_1^+] = -1$, $[b_2, b_2^+] = 1$. New constraints are

$$\tilde{L}_2 = L_2 + b_1^+ b_2, \tag{10}$$
$$\tilde{L}_{-2} = L_{-2} + b_2^+ b_1 \ . \tag{11}$$

Together with new operator

$$\tilde{G}_0 = G_0 + b_2^+ b_2 + b_1^+ b_1 \tag{12}$$

they form an $SU(2)$ algebra

$$[\tilde{L}_2 \ , \ \tilde{L}_{-2}] = \tilde{G}_0, \tag{13}$$
$$[\tilde{G}_0 \ , \ \tilde{L}_{\pm 2}] = \mp 2\tilde{L}_{\pm 2} \ . \tag{14}$$

and can be considered as first class constraints. The counting of physical degrees of freedom shows equal number for both systems of constraints. Indeed, we have introduced four additional variables b_i, b_k^+, but instead of two second class constraints each killing one degree of freedom we get three first class constraints which kill together six degrees of freedom.

To illustrate the BRST - approach to this simple system we introduce additional set of anticommuting variables η_0, η_2, η_2^+ having ghost number one and corresponding momenta $\mathcal{P}_0, \mathcal{P}_2^+, \mathcal{P}_2$ with commutation relations:

$$\{\eta_0, \mathcal{P}_0\} = \{\eta_2, \mathcal{P}_2^+\} = \{\eta_2^+, \mathcal{P}_2\} = 1 \ . \tag{15}$$

The standard prescription gives the following nilpotent BRST - charge

$$Q = \eta_2^+ \tilde{L}_2 + \eta_2 \tilde{L}_2^+ + \eta_0 \tilde{G}_0 + \eta_2 \eta_2^+ \mathcal{P}_0 + 2\eta_0 \eta_2^+ \mathcal{P}_2 - 2\eta_0 \eta_2 \mathcal{P}_2^+ \ . \tag{16}$$

Consider the total Fock space generated by creation operators $a_\mu^+, b_i^+, \eta_2^+, \mathcal{P}_2^+$.

The BRST - invariant lagrangian in such Fock space can be written as

$$L = \int d\eta_0 \langle \chi | Q | \chi \rangle \ , \tag{17}$$

$$|\chi\rangle = |S_1\rangle + \eta_2^+ \mathcal{P}_2^+ |S_2\rangle + \eta_0 \mathcal{P}_2^+ |S_3\rangle \ , \tag{18}$$

with vectors $|S_i\rangle$ having ghost number zero and depending only on bosonic creation operators a_μ^+, b_i^+

$$|S_i\rangle = \sum \phi^{n_1,n_2}_{i;\mu_1,\mu_2,...\mu_n} a^+_{\mu_1} a^+_{\mu_2}...a^+_{\mu_n} (b_1^+)^{n_1} (b_2^+)^{n_2} |0\rangle \ . \tag{19}$$

In general the vawefunction $\phi^{n_1,n_2}_{i;\mu_1,\mu_2,...\mu_n}$ can depend on other physical variables of the theory such as space - time coordinates etc.

After the η_0 integration we get the following lagrangian in terms of $|S_i\rangle$

$$\begin{aligned} L = \langle S_1|\tilde{G}_0|S_1\rangle - 2\langle S_1||S_1\rangle - \langle S_2|\tilde{G}_0|S_2\rangle - 2\langle S_2||S_2\rangle + \langle S_3||S_3\rangle - \\ -\langle S_1|\tilde{L}_2^+|S_3\rangle - \langle S_3|\tilde{L}_2|S_1\rangle + \langle S_2|\tilde{L}_2|S_3\rangle + \langle S_3|\tilde{L}_2^+|S_2\rangle \ . \end{aligned} \tag{20}$$

Owing to the nilpotency of the BRST – charge - $Q^2 = 0$, the lagrangian (17) is invariant under the transformation

$$\delta|\chi\rangle = Q|\Lambda\rangle \tag{21}$$

with $|\Lambda\rangle = \mathcal{P}_2^+|\lambda\rangle$. In turn it means the invariance of the lagrangian (20) under the following transformation

$$\delta|S_1\rangle = \tilde{L}_2^+|\lambda\rangle, \tag{22}$$
$$\delta|S_2\rangle = \tilde{L}_2|\lambda\rangle, \tag{23}$$
$$\delta|S_3\rangle = \tilde{G}_0|\lambda\rangle \ . \tag{24}$$

The field $|S_3\rangle$ is auxiliary. Using its equation of motion

$$|S_3\rangle = \tilde{L}_2|S_1\rangle - \tilde{L}_2^+|S_2\rangle \ , \tag{25}$$

the lagrangian (20) can be rewritten in terms of two fields $|S_1\rangle$ and $|S_2\rangle$

$$\begin{aligned} L = \langle S_1|(\tilde{G}_0 - 2 - \tilde{L}_2^+\tilde{L}_2)|S_1\rangle - \langle S_2|(\tilde{G}_0 + 2 + \tilde{L}_2\tilde{L}_2^+)|S_2\rangle + \\ +\langle S_1|\tilde{L}_2^+\tilde{L}_2^+|S_2\rangle + \langle S_2|\tilde{L}_2\tilde{L}_2|S_1\rangle \end{aligned} \tag{26}$$

with the following equations of motion:

$$(\tilde{G}_0 - 2 - \tilde{L}_2^+\tilde{L}_2)|S_1\rangle + \tilde{L}_2^+\tilde{L}_2^+|S_2\rangle = 0, \tag{27}$$
$$(\tilde{G}_0 + 2 + \tilde{L}_2\tilde{L}_2^+)|S_2\rangle - \tilde{L}_2\tilde{L}_2|S_1\rangle = 0 \ . \tag{28}$$

One can show that the gauge freedom (22)-(24) is sufficient to eliminate the fields $|S_2\rangle$ and $|S_3\rangle$ and to kill the b_2^+ dependence in $|S_1\rangle$. After elimination of the field $|S_3\rangle$ there will be residual gauge invariance with under the condition

$$\tilde{G}_0|\lambda\rangle' = 0 \tag{29}$$

Eliminating of field $|S_2\rangle$ with the help of the help of the equation $|S_2\rangle + \tilde{L}_2|\lambda\rangle' = 0$, which is consistent with (29) and equations of motion, the new parameter $|\lambda\rangle''$ will satisfy two conditions $\tilde{G}_0|\lambda\rangle'' = \tilde{L}_1|\lambda\rangle'' = 0$. With the help of this parameter all the fields $|S_{1nm}\rangle$, exept $|S_{1n0}\rangle$ can be eliminated as well.It means that one obtains the following conditions on the reduced field $|\tilde{S}_1\rangle$:

$$\frac{1}{2}a_\mu a_\mu|\tilde{S}_1\rangle = 0,\ (-a_\mu^+ a_\mu + \frac{D}{2} + b_1^+ b_1 - 2)|\tilde{S}_1\rangle = 0\ . \tag{30}$$

The first of the conditions (30) means tracelessness of the wavefunctions $\phi^{n_1,0}_{\mu_1,\mu_2,\dots\mu_n}$. The second one connects n and n_1: $n_1 = n + \frac{D}{2} - 2$ killing effectively the b_1^+ dependence in $|S_1\rangle$. So, the system contains no additional degrees of freedom and describes traceless wavefunctions.

3 Irreducible Massless Higher Spins

Following the line of the previous section, we consider the system of constraints (3) with the additional operator G_0. This operator satisfies the following commutation relations:

$$[L_1\ ,\ G_0] = L_1\ [L_{-1}\ ,\ G_0] = -L_{-1} \tag{1}$$
$$\tag{2}$$

$$[L_2\ ,\ G_0] = 2L_2\ [L_{-2}\ ,\ G_0] = -2L_{-2} \tag{3}$$
$$\tag{4}$$

All other commutaion relations are zero. Again, we modify the system by introduction of *two* additional operators b_1, b_2 together with their conjugates b_1^+, b_2^+: $[b_1, b_1^+] = -1$,$[b_2, b_2^+] = 1$. Because of "unnusial" commutation relation between b_1 and b_1^+ operators, L_0 constraint has to be modified too, in order to get the correct sign for the cinematic term in lagrangian, constructed bellow. We consider the following system: $\tilde{L}_0 = L_0(-1)^{b_2^+b_2 - b_1^+b_1}$, $\tilde{L}_{\pm 1} = \tilde{L}_{\pm 1}$, $\tilde{L}_2 = L_2 + b_1^+ b_2$, $\tilde{L}_{-2} = L_{-2} + b_2^+ b_1$ and $\tilde{G}_0 = G_0 + b_2^+ b_2 + b_1^+ b_1$ The modified operators obey the same algebra as the previous ones. Since we have the system of the first class constaints, the construction of the corresponding BRST charge is straightforward. Let us introduce additional set of anticommuting variables $\eta_0, \eta, \eta_1, \eta_1^+ \eta_2, \eta_2^+$ having ghost number one and corresponding momenta $\mathcal{P}_0, \mathcal{P}, \mathcal{P}_1^+, \mathcal{P}_1, \mathcal{P}_2^+, \mathcal{P}_2$ with commutation relations:

$$\{\eta_0, \mathcal{P}_0\} = \{\eta_1, \mathcal{P}_1^+\} = \{\eta_1^+, \mathcal{P}_1\} = \{\eta, \mathcal{P}\} = \{\eta_2, \mathcal{P}_2^+\} = \{\eta_2^+, \mathcal{P}_2\} = 1\ . \tag{5}$$

The corresponding nilpotent BRST charge has the following form:

$$Q = \eta_0 \tilde{L}_0 + \Omega + \eta N + \mathcal{P}_0 T_1 + \mathcal{P} T_2 \tag{6}$$

where

$$\Omega = \eta_1^+ \tilde{L}_1 + \eta_1 \tilde{L}_1^+ + \eta_2^+ \tilde{L}_2 + \eta_2 \tilde{L}_2^+ - \eta_1 \eta_2^+ \mathcal{P}_1 - \eta_2 \eta_1^+ \mathcal{P}_1^+ \tag{7}$$

$$N = \tilde{G}_0 + \eta_1^+ \mathcal{P}_1 - \eta_1 \mathcal{P}_1^+ + 2\eta_2^+ \mathcal{P}_2 - 2\eta_2 \mathcal{P}_2^+ \quad (8)$$

$$T_1 = \eta_1 \eta_1^+ \quad (9)$$

$$T_2 = \eta_2 \eta_2^+ \quad (10)$$

Since we have to integrate over two ghost zero-modes η_0 and η, we modify the lagrangian to the form (Meurice (1987))

$$L = \frac{1}{2} \int d\eta d\eta_0 (\langle \chi | Q \mathcal{P} | \chi \rangle + h.c.) \quad (11)$$

were the vector $|\chi\rangle$ is expanded in terms of η_0 and η as

$$|\chi\rangle = |\chi_1\rangle + \eta_0 |\chi_2\rangle + \eta |\chi_3\rangle + \eta_0 \eta |\chi_4\rangle \quad (12)$$

The lagrangian is invariant under the transformation with a parameter $|\Lambda\rangle$

$$\delta|\chi\rangle = Q|\Lambda\rangle, \quad (13)$$

provided

$$N|\Lambda\rangle = 0 \quad (14)$$

In this case the vector $|\chi\rangle$ has ghost number 1, while the gauge parameter $|\Lambda\rangle$ has ghost number zero and also can be expressed as a function of two real grassmann variables

$$|\Lambda\rangle = |\Lambda_1\rangle + \eta_0 |\Lambda_2\rangle + \eta |\Lambda_3\rangle + \eta_0 \eta |\Lambda_4\rangle \quad (15)$$

Therefore, we can write the BRST gauge condition for the component fields $N|\Lambda_i\rangle = 0$. After integration the lagrangian takes the form:

$$L = \frac{1}{2}((\langle \chi_1 | N | \chi_4 \rangle + \langle \chi_2 | N | \chi_3 \rangle + \langle \chi_4 | N | \chi_1 \rangle + \langle \chi_3 | N | \chi_2 \rangle) - \\ - \langle \chi_3 | \tilde{L}_0 | \chi_3 \rangle + \langle \chi_4 | T_1 | \chi_4 \rangle - \langle \chi_3 | \Omega | \chi_4 \rangle - \langle \chi_4 | \Omega | \chi_3 \rangle \quad (16)$$

The corresponding equations of motion are:

$$N|\chi_4\rangle = 0 \quad (17)$$

$$N|\chi_3\rangle = 0 \quad (18)$$

$$2\tilde{L}_0 |\chi_3\rangle + 2\Omega |\chi_4\rangle - N|\chi_2\rangle = 0 \quad (19)$$

$$-2T_1 |\chi_4\rangle + 2\Omega |\chi_3\rangle - N|\chi_1\rangle = 0 \ , \quad (20)$$

while the gauge transformation $\delta|\chi\rangle = Q|\Lambda\rangle$ has the following component form:

$$\delta|\chi_1\rangle = \Omega|\Lambda_1\rangle + T_1|\Lambda_2\rangle + T_2|\Lambda_3\rangle \quad (21)$$

$$\delta|\chi_2\rangle = -\Omega|\Lambda_2\rangle + \tilde{L}_0|\Lambda_1\rangle - T_2|\Lambda_4\rangle \quad (22)$$

$$\delta|\chi_3\rangle = -\Omega|\Lambda_3\rangle + T_1|\Lambda_4\rangle \quad (23)$$

$$\delta|\chi_4\rangle = \Omega|\Lambda_4\rangle + \tilde{L}_0|\Lambda_3\rangle \tag{24}$$

The most general expressions for the fields $|\chi_i\rangle$ and $|\Lambda_i\rangle$ in terms of ghosts η_1^+, η_2^+ and antighosts $\mathcal{P}_1^+, \mathcal{P}_2^+$ are rather complicated

$$|\chi_1\rangle = \eta_1^+|A_1\rangle + \eta_2^+|A_2\rangle + \eta_1^+\eta_1^+\mathcal{P}_1^+|A_3\rangle + \eta_1^+\eta_1^+\mathcal{P}_2^+|A_4\rangle \tag{25}$$

$$|\chi_2\rangle = |A_5\rangle + \eta_1^+\mathcal{P}_1^+|A_6\rangle + \eta_2^+\mathcal{P}_2^+|A_7\rangle + \eta_1^+\mathcal{P}_2^+|A_8\rangle + +\eta_2^+\mathcal{P}_1^+|A_9\rangle + \eta_1^+\eta_2^+\mathcal{P}_1^+\mathcal{P}_2^+|A_{10}\rangle \tag{26}$$

$$|\chi_3\rangle = |S_1\rangle + \eta_1^+\mathcal{P}_1^+|S_2\rangle + \eta_2^+\mathcal{P}_2^+|S_3\rangle + \eta_1^+\mathcal{P}_2^+|S_4\rangle + +\eta_2^+\mathcal{P}_1^+|S_5\rangle + \eta_1^+\eta_2^+\mathcal{P}_1^+\mathcal{P}_2^+|S_6\rangle \tag{27}$$

$$|\chi_4\rangle = \mathcal{P}_1^+|S_7\rangle + \mathcal{P}_2^+|S_8\rangle + \eta_1^+\mathcal{P}_1^+\mathcal{P}_2^+|S_9\rangle + \eta_2^+\mathcal{P}_1^+\mathcal{P}_2^+|S_{10}\rangle \tag{28}$$

$$|\Lambda_1\rangle = |\rho_1\rangle + \eta_1^+\mathcal{P}_1^+|\rho_2\rangle + \eta_2^+\mathcal{P}_2^+|\rho_3\rangle + \eta_1^+\mathcal{P}_2^+|\rho_4\rangle + +\eta_2^+\mathcal{P}_1^+|\rho_5\rangle + \eta_1^+\eta_2^+\mathcal{P}_1^+\mathcal{P}_2^+|\rho_6\rangle \tag{29}$$

$$|\Lambda_2\rangle = \mathcal{P}_1^+|\rho_7\rangle + \mathcal{P}_2^+|\rho_8\rangle + \eta_1^+\mathcal{P}_1^+\mathcal{P}_2^+|\rho_9\rangle + \eta_2^+\mathcal{P}_1^+\mathcal{P}_2^+|\rho_{10}\rangle \tag{30}$$

$$|\Lambda_3\rangle = \mathcal{P}_1^+|\sigma_1\rangle + \mathcal{P}_2^+|\sigma_2\rangle + \eta_1^+\mathcal{P}_1^+\mathcal{P}_2^+|\sigma_3\rangle + \eta_2^+\mathcal{P}_1^+\mathcal{P}_2^+|\sigma_4\rangle \tag{31}$$

$$|\Lambda_4\rangle = \mathcal{P}_1^+\mathcal{P}_2^+|\sigma_5\rangle \tag{32}$$

where the vectors $|A_i\rangle, |S_i\rangle, |\rho_i\rangle$ and $|\sigma_i\rangle$ depend only on bosonic creation operators a_μ, b_1^+, b_2^+ and have ghost number zero. Now we are going to prove, that there is a special gauge,in which mass-shell, transvercality and tracelessness conditions are imposed on the physical field $|S_1\rangle$.

Firstly we solve the equation

$$|\chi_4\rangle + \Omega|\Lambda_4\rangle + \tilde{L}_0|\Lambda_3\rangle = 0 \ , \tag{33}$$

which is consistent with the conditions $N|\lambda_i\rangle = 0$ by vertue of the equation of motion $N|\chi_4\rangle = 0$ and the identities

$$[N, \Omega] = [N, T_1] = [N, T_2] = [N, \tilde{L}_0] = 0 \tag{34}$$

We can choose $|\Lambda_3\rangle$ to gauge away $|\chi_4\rangle$ field. For compliteness we present (33) in terms of the component fields

$$|S_7\rangle - \tilde{L}_2^+|\sigma_5\rangle + \tilde{L}_0|\sigma_1\rangle = 0 \tag{35}$$

$$|S_8\rangle + \tilde{L}_1^+|\sigma_5\rangle + \tilde{L}_0|\sigma_2\rangle = 0 \tag{36}$$

$$|S_9\rangle + \tilde{L}_1|\sigma_5\rangle + \tilde{L}_0|\sigma_3\rangle = 0 \tag{37}$$

$$|S_{10}\rangle + \tilde{L}_2|\sigma_5\rangle + \tilde{L}_0|\sigma_4\rangle = 0 \tag{38}$$

The residual gauge parameter $|\Lambda_3\rangle'$, satisfies two conditions

$$N|\lambda_3\rangle' = \tilde{L}_0|\lambda_3\rangle' = 0 \tag{39}$$

Now consider the equations

$$|\chi_1\rangle + \Omega|\Lambda_1\rangle + T_1|\Lambda_2\rangle + T_2|\Lambda_3\rangle' = 0 \tag{40}$$

$$|\chi_2\rangle - \Omega|\Lambda_2\rangle + \tilde{L}_0|\Lambda_1\rangle - T_2|\Lambda_4\rangle' = 0 \tag{41}$$

which are also consistent with the equations of motion and constraint condition on gauge parameters. Using $|\Lambda_1\rangle$ and $|\Lambda_2\rangle$ we gauge away both of these fields.Indeed after writing (40)-(41) in the component form

$$|A_1\rangle + \tilde{L}_1|\rho_1\rangle - \tilde{L}_1^+|\rho_2\rangle - \tilde{L}_2^+|\rho_4\rangle - |R_7\rangle = 0 \tag{42}$$
$$|A_2\rangle - \tilde{L}_1^+|\rho_5\rangle + \tilde{L}_2|\rho_1\rangle - \tilde{L}_2^+|\rho_3\rangle + |R_2\rangle - |\sigma_2\rangle' = 0 \tag{43}$$
$$|A_3\rangle + \tilde{L}_1|\rho_5\rangle - \tilde{L}_2|\rho_2\rangle - \tilde{L}_2^+|\rho_6\rangle - |R_3\rangle + |\sigma_3\rangle' = 0 \tag{44}$$
$$|A_4\rangle + \tilde{L}_1|\rho_3\rangle + \tilde{L}_1^+|\rho_6\rangle - \tilde{L}_2|\rho_4\rangle + |R_{10}\rangle = 0 \tag{45}$$

$$|A_5\rangle - \tilde{L}_1^+|\rho_7\rangle - \tilde{L}_2^+|\rho_8\rangle + \tilde{L}_0|R_1\rangle = 0 \tag{46}$$
$$|A_6\rangle - \tilde{L}_1|\rho_7\rangle - \tilde{L}_2^+|\rho_9\rangle + |R_8\rangle + \tilde{L}_0|\rho_2\rangle = 0 \tag{47}$$
$$|A_7\rangle + \tilde{L}_1^+|\rho_{10}\rangle - \tilde{L}_2|\rho_8\rangle - |R_9\rangle + \tilde{L}_0|\rho_3\rangle = 0 \tag{48}$$
$$|A_8\rangle - \tilde{L}_1|\rho_8\rangle + \tilde{L}_1^+|\rho_9\rangle + \tilde{L}_0|R_4\rangle = 0 \tag{49}$$
$$|A_9\rangle - \tilde{L}_2|\rho_7\rangle - \tilde{L}_2^+|\rho_{10}\rangle + \tilde{L}_0|R_5\rangle - |\sigma_5\rangle' = 0 \tag{50}$$
$$|A_{10}\rangle - \tilde{L}_1|\rho_{10}\rangle + \tilde{L}_2^|\rho_9\rangle + \tilde{L}_0|R_6\rangle = 0 \tag{51}$$

one can choose the gauge, which eliminates all $|A_i\rangle$ vectors (we have denoted corresponding parameters with the capital letters). Finally consider the equation

$$|\chi_3\rangle - \Omega|\Lambda_3\rangle' + T_1|\Lambda_4{}'\rangle = 0 \ , \tag{52}$$

or in component form

$$|S_1\rangle - \tilde{L}_1^+|\sigma_1\rangle' - \tilde{L}_2^+|\sigma_2\rangle' = 0 \tag{53}$$
$$|S_2\rangle - \tilde{L}_1|\sigma_1\rangle' - \tilde{L}_2^+|\sigma_3\rangle' - |\sigma_2\rangle' = 0 \tag{54}$$
$$|S_3\rangle - \tilde{L}_2|\sigma_2\rangle' - \tilde{L}_1^+|\sigma_4\rangle' - |\sigma_3\rangle' = 0 \tag{55}$$
$$|S_4\rangle - \tilde{L}_1|\sigma_2\rangle' + \tilde{L}_1^+|\sigma_3\rangle' - |\sigma_5\rangle' = 0 \tag{56}$$
$$|S_5\rangle - \tilde{L}_2|\sigma_1\rangle' - \tilde{L}_2^+|\sigma_4\rangle' = 0 \tag{57}$$
$$|S_6\rangle + \tilde{L}_2|\sigma_3\rangle' - \tilde{L}_1|\sigma_4\rangle' = 0 \tag{58}$$

Using $|\sigma_3\rangle', |\sigma_5\rangle', |\sigma_4\rangle'$ we can gauge away $|S_3\rangle, |S_4\rangle, |S_6\rangle$. Using $|\sigma_1\rangle'$ we can gauge away $|S_2\rangle$ and impose an additional condition

$$\tilde{L}_1|\sigma_1\rangle'' = 0 \tag{59}$$

Then using this parameter we can gauge away $|S_5\rangle$. Indeed the additional constraint is consistent with the equations of motion. As it is shown in the previous section, we can use $|\sigma_2\rangle'$ to gauge away the b_2^+ dependence in $|S_1\rangle$ And finally we are left with the conditions

$$L_0|S_1\rangle = L_1|S_1\rangle = L_2|S_1\rangle = 0 \tag{60}$$

which mean irreducibility of massless representation of the Poincare group for any value of higher spin, and

$$(\tilde{G}_0 - 3)|S_1\rangle = 0 \tag{61}$$

As it can be seen from the last formula, in a view of the expression (12) our approach is valid only for even dimmensions, with $D \geq 6$.For $D = 4$ condition (61) excludes scalar, and for $D = 2$ - scalar and vector fields.

4 Conclusions

In this paper we have applied the BRST approach to the description of irreducible massless higher spins. We have described the simple modification of the conversion procedure for the second class constraints which can be used for elimination of lower spins, connected with trace parts of the basic field. We conjecture, that the dimensional reduction in the framework of the BRST approach (Pashnev, Tsulaia (1997)) gives the possibility to apply the same procedure to the description of single Regge trajectory

Acknowledgments. This investigation has been supported in part by the Russian Foundation of Fundamental Research, grants 96-02-17634 and 96-02-18126, joint grant RFFR-DFG 96-02-00186G, and INTAS, grant 94-2317 and grant of the Dutch NWO organization.

References

Aragone, C., Deser, S. (1980): Nucl.Phys.,B170, 329
Chang, S.J. (1967): Phys.Rev., 161, 1308
Curtright, T. (1979): Phys.Lett., B85, 219
Egoryan, E.T., Manvelyan, R.P. (1993): Theor. Math.Phys., 94, 241
Faddeev, L.D., Shatashvili, S.L. (1986): Phys.Lett., B167, 225
Fradkin, E.S. (1950): JETP, 20, 27
Fronsdal, C. (1978): Phys.Rev., D18, 3624
Fang, J., Fronsdal, C. (1978): Phys.Rev., D18, 3630
Meurice, Y. (1987): Phys.Lett., B186, 189
Ouvry, S., Stern, J. (1986): Phys.Lett., B177, 335
Pashnev, A.I. (1989): Theor.Math.Phys., 78, 424
Pashnev, A., Tsulaia, M. (1996): Preprint JINR-E2-96-408, Dubna,. hep-th/9611022
Pashnev, A., Tsulaia, M. (1997): Preprint JINR-E2-97-63, hep-th/9703010
Singh, L.P.H., Hagen, C.R. (1974): Phys.Rev., D9, 898; Ibid D9, 910
Vasiliev, M.A. (1980): Sov.J.Nucl.Phys., 32, 855 (439 in english translation)
de Wit, B., Freedman, D.Z. (1980): Phys.Rev., D21, 358

Sonoluminescence and Black Holes as Sources of Squeezed Light

Yu.P. Stepanovsky

National Science Center KPTI,
310108 Kharkov, Ukraine

Abstract. The phenomenon of coherent sonoluminescence is considered as a physical vacuum excitation. The state of the light emitted by sonoluminescence and by black hole evaporation is investigated. It is shown that the light state is a squeezed vacuum state. The question is discussed: is the black hole radiation a thermal radiation?

> My work always tried to unite the true with the beautiful, but when I had to choose one over the other I usually chose the beautiful.
> H.Weyl, (Gross (1988)).

We don't know, does God exist, but nevertheless we know that if He exists, He created the Universe as a pure mathematician. So we believe that without mathematics we can not to understand the Divine Mind in creating the Universe. We believe that abstract patterns, born in the minds of mathematicians, beautifully mesh with the physical structure of the Universe. D.V.Volkov, as you and I do, liked mathematics, but he was very distressed when theoretical physics was replaced by mathematics and pure mathematics in the sense of B.Russel (Bell (1989)): "Mathematics may be defined as a subject in which we newer know what we are talking about, nor whether what we are saying is true." And: "Pure mathematics consists entirely of such asseverations as that, if such proposition is true of *anything*, then such and such another proposition is true. It is essential not to discuss whether the first proposition is really true, and not to mention what the anything is of which it supposed to be true." D.V.Volkov liked another mathematics, mathematics of C.F.Gauss and G.F.B.Riemann, E.Cartan and E.A.Noeter, that possesses great truth and supreme beauty, about which H.Hertz once remarked: "One cannot escape the feeling that these mathematical formulas have an independent existence and intelligence of their own, that they are wiser than we are, wiser even than their discoverers, that we get more out of them than was originally put into them" (Bell (1937)).

Unfortunately my report will not contain much mathematics but I hope that according to uncertainties relation (Gross (1988))

$$\Delta\text{Mathematics} \times \Delta\text{Physics} \geq \text{Constant} \tag{1}$$

it will contain some physics of any interest to you.

Sonoluminescence as a physical vacuum excitation. Sonoluminescence is a convertation of sound energy, in the form of a beam of ultrasonic waves, into light energy. Light is emitted when acoustical energy is focused on a bubble of air trapped in water. Sound having a Mach number of order 10^{-5} is able to create photons with energies of several eV. Radiation occurs only during an interval of less than 50 ps within each cycle of the sound field, which has a period of a few tens of microseconds. The light's spectrum implies that the source of the radiation is similar to a black body at a temperature of tens of thousands of Kelvin (Hiller et al. (1992), Crum (1994), Putterman (1995)).

A beautiful explanation of remarkable phenomenon of sonoluminescence was proposed by J.Schwinger (Schwinger (1992), (1993), (1994)).On putting the question: "How does a macroscopic, classical, hydromechanical system, driven by a macroscopic acoustical force, generate an astonishingly short time scale and an accompanying high electromagnetic frequency, one that is at atomic level?" Schwinger suggested entirely unexpected solution of the problem: sonoluminescence is a result of quantum mechanical excitation of physical vacuum due to the abrupt slowing of the collapse of the bubble of air on which acoustical energy is focused. But why the collapse of the bubble is abruptly slowed? Schwinger suggested: "*The collapse of the cavity is slowed abruptly by the pressure of the light that is created by the abrupt slowing of the collapse.*" Schwinger's theory of sonoluminescence is based on the following phenomenon: the variation in the time of dielectric constant ε of medium is accompanied by emission of photons. Schwinger obtained simple formulas qualitatively describing experimental data. Schwinger's explanation of sonoluminescence was recently discussed by C.Eberlein (Eberlein (1996)), who pointed out on similarity of sonoluminescence and black hole evaporation: both are due to excitation of physical vacuum.

Sonoluminescence as a source of squeezed light. Let scalar field $\Phi(t)$ be the solution of the following equation

$$\left(\frac{1}{c^2}\frac{\partial^2}{\partial t^2} - \frac{1}{\varepsilon}\Delta\right)\Phi(t) = 0 \tag{2}$$

where ε is dielectric constant of medium. Disregarding the spin of photons we suppose that the field $\Phi(t)$ describes the light quanta. Let the solution of the equation (2) be the plane wave

$$\Phi(t) = \mathrm{e}^{\mathrm{i}\mathbf{kx}}\psi(t) \tag{3}$$

The function $\psi(t)$ satisfy the equation

$$\ddot{\psi} + \omega^2(t)\psi = 0 \tag{4}$$

where

$$\omega^2(t) = \frac{c^2\mathbf{k}^2}{\varepsilon(t)} \tag{5}$$

According to Schwinger let us assume that the $\varepsilon(t)$ is jump function of time with the jump from $\varepsilon \neq 1$ to $\varepsilon = 1$. The jump of dielectric constant leads to the transition

$$e^{-i\omega_1 t} \to \frac{\sqrt{\varepsilon}+1}{2} e^{-i\omega_2 t} + \frac{\sqrt{\varepsilon}-1}{2} e^{i\omega_2 t} . \tag{6}$$

Any transition of positive frequency solution to negative frequency solution

$$e^{-i\omega_1 t} \to \alpha e^{-i\omega_2 t} + \beta e^{i\omega_2 t} \tag{7}$$

means the creation of particles, because the quantum field

$$\psi = \frac{1}{\sqrt{\omega_1}} (a_1 e^{-i\omega_1 t} + a_1^+ e^{i\omega_1 t}) \tag{8}$$

will get over in new quantum field

$$\begin{aligned} &\tfrac{1}{\sqrt{\omega_1}} (a_1 e^{-i\omega_1 t} + a_1^+ e^{i\omega_1 t}) \to \\ &\to \tfrac{1}{\sqrt{\omega_2}} \left(\sqrt{\tfrac{\omega_2}{\omega_1}} (\alpha a_1 + \beta^* a_1^+) e^{-i\omega_2 t} + \sqrt{\tfrac{\omega_2}{\omega_1}} (\alpha^* a_1^+ + \beta a_1) e^{i\omega_2 t} \right) = \\ &= \tfrac{1}{\sqrt{\omega_2}} (a_2 e^{-i\omega_2 t} + a_2^+ e^{i\omega_2 t}) \end{aligned} \tag{9}$$

where a_2^+ and a_2 are new creation and annihilation operators.

An overdetermination of creation and annihilation operators means that vacuum with zero number of photons n_1 creates the photons in amounts n_2,

$$n_1 = \langle 0 | a_1^+ a_1 | 0 \rangle = 0 , \tag{10}$$

$$n_2 = \langle 0 | a_2^+ a_2 | 0 \rangle = \frac{\omega_2}{\omega_1} |\beta|^2 = \frac{(\sqrt{\varepsilon}-1)^2}{4\sqrt{\varepsilon}} . \tag{11}$$

Formula (11) gives the number of photons with definite wave vector $\mathbf{k}$. The total number of photons created in the volume V according to Schwinger (Schwinger (1992), (1993), (1994)) is presented as

$$N = \int \frac{(\sqrt{\varepsilon}-1)^2}{4\sqrt{\varepsilon}} \frac{d^3 k}{(2\pi)^3} V . \tag{12}$$

Disregarding the dispersion Schwinger obtained the following expression for the energy emitted in unite volume,

$$\frac{E}{V} = \frac{(\sqrt{\varepsilon}-1)^2}{4\sqrt{\varepsilon}} \int^{k_{\max}} \hbar\omega \frac{d^3 k}{(2\pi)^3} = 2\pi^2 \frac{(\sqrt{\varepsilon}-1)^2}{4\sqrt{\varepsilon}} \frac{\hbar c}{\lambda_{\min}^4} \tag{13}$$

where k_{max} is cut-off wave-number and $\lambda_{min} = 2 \times 10^{-5}$ cm for temperature 3^0C. As remarked Schwinger, this value is strikingly close to the edge of transparency of the water.

Let us assume now that $\varepsilon(t)$ varies in time as

$$\varepsilon(t) = \frac{2\varepsilon}{\varepsilon + 1 + (\varepsilon - 1)\mathrm{th}(t/\tau)} \ . \tag{14}$$

In this case the transition from $\varepsilon \neq 1$ to $\varepsilon = 1$ takes place in the course of the characteristic time τ. For the number n_2 of created photons using (Eckart (1930), Flügge (1971)) we obtain the exact result,

$$n_2 = \frac{1}{\rho - 1} \tag{15}$$

where

$$\rho = \frac{\mathrm{sh}^2(\pi\omega\tau(1 + \frac{1}{\sqrt{\varepsilon}}))}{\mathrm{sh}^2(\pi\omega\tau(1 - \frac{1}{\sqrt{\varepsilon}}))}, \qquad \omega = ck \ . \tag{16}$$

If $\omega\tau \to 0$ from (15) and (16) we obtain (11). If $\omega\tau \gg 1$,

$$n_2 = \frac{1}{\mathrm{e}^{\hbar\omega/kT} - 1} \tag{17}$$

where

$$kT = \frac{\hbar\sqrt{\varepsilon}}{4\pi\tau} \ . \tag{18}$$

Thus we see that our radiation is characterized by Plank's spectrum (17) with a temperature defined by (18). But this radiation is not black radiation. It is well known in quantum optics as a squeezed vacuum (Fabre(1992)). The state of squeezed vacuum is defined by following relation,

$$(\alpha^* a_2 - \beta^* a_2^+)|\psi\rangle = 0 \ , \tag{19}$$

and is according to (9) ordinary vacuum state if we use creation and annihilation operators a_1^+ and a_1,

$$a_1|\psi\rangle = 0 \ . \tag{20}$$

The bilinear combinations of creation and annihilation operators are infinitesimal operators of O(2,1)-Lorentz group (Schwinger (1960)). This property of operators a_1^+ and a_1 helps to find easily the solution of the equation (19) in the form

$$|\psi\rangle = \mathrm{e}^{\gamma(a_2^+)^2}|0\rangle \tag{21}$$

where γ is some constant (Fabre(1992)). Thus we see that possible photon numbers n for the state of squeezed vacuum are 0, 2, 4, 6, ... and drastically differ from photon numbers for the black radiation state 0, 1, 2, 3, ...

Black hole radiation, is it black? Similar problems arise in the case of black hole radiation. As was shown by S.Hawking in his classical paper (Hawking (1975)) the black hole radiation is characterized by Plank's spectrum with some temperature. But Hawking's calculation, been technically more complicated, fundamentally is the same as our calculation. Existence of horizon, although very important for understanding the quantum theory of the black hole radiation, doesn't alter neither relations between creation and annihilation operators a_1^+, a_1 and a_2^+, a_2 nor formula for number of emitted photons n_2. So the photon state in the theory of black hole radiation is the squeezed vacuum as in the case of sonoluminescence.

We know that, when S.Hawking found that nonrotating black holes should create and emit a radiation, at first he thought that his calculation was not valid. But what finally convinced him that the radiation was real was that the spectrum of radiation was exactly that which would be emitted by a black body (Hawking (1988)). It was so beautiful! And just it disturbs the judgement! In spite of black holes emit the radiation with Plank spectrum, this radiation is squeezed radiation to be distinguished in quantum optics from thermal radiation. But maybe we need to overdeterminate what black radiation is, who knows.

Alas!

Vita brevis, ars longa, tempus praeceps, experimentum periculosum, iudicium difficile! (Senecae (1981))

Life is brief, art is long, time is precipitous, experiment is perilous, judgment is difficult!

Last year D.V.Volkov passed away, a great scientist and a great man, and we all are the poorer for this loss. Dmitrij Vasiljevich was a hero to us and we are grateful to our good fortune to associate with him. The memory of D.V.Volkov makes us to try to be better, than we are, and helps us to resist when we are forced now to drag out a life, worthless and disgraceful,

Lives of great men all remind us
We can make our lives sublime,
And, departing, leave behind us
Footprints on the sands of time...
H.W.Longfellow, Psalm of Life (Longfellow (1910))

References

Bell, E.T. (1937): Men of Mathematics.— New York, Simon and Schuster.
Bell, E.T. (1989): Mathematics: Queen and Servant of Science. — Washington, Mathematical Association of America.
Crum, L.A. (1994): Phys. Today. **47** 22.
Eberlein, C. (1996): Phys. Rev. Lett. **76** 3842.
Eckart, C. (1930): Phys. Rev. **35** 1303.
Fabre, C.(1992): Physics Reports **219** 215.
Flügge, S. (1971): Practical Quantum Mechanics I. — Berlin Heidelberg New York, Springer Verlag.
Gross, D.J. (1988): Proc. Nat. Acad. Sci. USA. **85** 8371.
Hiller, R., Putterman, S., Barber, B. (1992): Phys. Rev. Lett. **69** 1182.
Hawking, S.W. (1975): Commun. Math. Phys. **43** 199.

Hawking, S.W. (1988): A Brief History of Time: from the Big Bang to Black Holes. — London New York Toronto Sydney Auckland: Bantam Press.

Longfellow, H.W. (1910): The Poetical Works. — London New York Toronto Melbourne: Oxford University Press.

Putterman, S. (1995): Scientific American, February, 47.

Schwinger, J. (1960): On Angular Momentum, 1952, **in** "Quantum Theory of Angular Momentum", New York: Academic Press.

Schwinger, J. (1970): Quantum Kinematics and Dynamics. — New York: W.A.Benjamin, Inc.

Schwinger, J. (1979): Annals of Physics **119** 192.

Schwinger, J. (1992): Proc. Nat. Acad. Sci. USA. **89** PP. 4091, 1118;
Schwinger, J. (1993): Proc. Nat. Acad. Sci. USA. **90** PP. 958, 2105, 4505, 7285;
Schwinger, J. (1994): Proc. Nat. Acad. Sci. USA. **91** 6473.

Senecae, L.A. (1981): De brevitate vita. — München, F.P. Weiblinger.

Volkov, D.V. (1996): Supergravity before 1976. Proceedings of International Conference "History of Original Ideas and Basic Discoveries in Particle Physics", Plenum Publishing Corporation. P. 663.

Remark Concerning Integrable Hamilton Systems

V.A. Soroka

Kharkov Institute of Physics and Technology
310108, Kharkov, Ukraine

Abstract. Applications of the odd Poisson bracket to the problem of classification of the integrable Hamilton system is discussed.

1. In this report we discuss a possibility of the receipt and classification of integrable Hamilton systems on the basis of the result obtained by author (Volkov, Pashnev, Soroka, Tkach (1986, 1989), Soroka (1989)) concerning the reformulation of the same Hamilton equation of motion with the help of the Hamiltonian-bracket pairs having different Grassmann grading.

The report is organized as follows. The main properties of the odd bracket are presented in Sect. 2. In Sect. 3 it is shown that Hamilton's equations of motion obtained by means of the even Poisson bracket with the help of the even Hamiltonian can be reproduced by the odd bracket using the equivalent odd Hamiltonian. The problem of classification integrable Hamilton systems is discussed in Sect. 4.

2. First, we recall the necessary properties of various graded Poisson brackets. The even and odd brackets in terms of the real even $y_i = (q^a, p_a)$ and odd $\eta^i = \theta^\alpha$ canonical variables have, respectively, the form

$$\{A,B\}_o = A\left[\sum_{a=1}^{n}\left(\overleftarrow{\partial}_{q^a}\overrightarrow{\partial}_{p_a} - \overleftarrow{\partial}_{p_a}\overrightarrow{\partial}_{q^a}\right) - i\sum_{\alpha=1}^{2n}\overleftarrow{\partial}_{\theta^\alpha}\overrightarrow{\partial}_{\theta^\alpha}\right]B\ ; \tag{1}$$

$$\{A,B\}_1 = A\sum_{i=1}^{N}\left(\overleftarrow{\partial}_{y_i}\overrightarrow{\partial}_{\eta^i} - \overleftarrow{\partial}_{\eta^i}\overrightarrow{\partial}_{y_i}\right)B\ , \tag{2}$$

where $\overleftarrow{\partial}$ and $\overrightarrow{\partial}$ are the right and left derivatives, and the notation $\partial_x = \frac{\partial}{\partial x}$ is introduced. By introducing apart from the Grassmann grading $g(A)$ of any quantity A its corresponding bracket grading $g_\epsilon(A) = g(A) + \epsilon \pmod 2$ $(\epsilon = 0, 1)$, the grading and symmetry properties, the Jacobi identities and the Leibnitz rule are uniformly expressed for the both brackets (1,2) as

$$g_\epsilon(\{A,B\}_\epsilon) = g_\epsilon(A) + g_\epsilon(B) \pmod 2\ , \tag{3a}$$

$$\{A,B\}_\epsilon = -(-1)^{g_\epsilon(A)g_\epsilon(B)}\{B,A\}_\epsilon\ , \tag{3b}$$

$$\sum_{(ABC)}(-1)^{g_\epsilon(A)g_\epsilon(C)}\{A,\{B,C\}_\epsilon\}_\epsilon = 0\ , \tag{3c}$$

$$\{A, BC\}_\epsilon = \{A, B\}_\epsilon \, C + (-1)^{g_\epsilon(A) g(B)} \, B\{A, C\}_\epsilon \ , \tag{3d}$$

where (3a)–(3c) have the shape of the Lie superalgebra relations in their canonical form (Berezin (1983)) with $g_\epsilon(A)$ being the canonical grading for the corresponding bracket.

In terms of arbitrary real dynamical variables $x^M = (x^m, x^\alpha) = x^M(y, \eta)$ with the same number of Grassmann even x^m and odd x^α coordinates the odd bracket (2) takes the form

$$\{A, B\}_1 = A \overleftarrow{\partial}_M \, \bar{\omega}^{MN}(x) \, \overrightarrow{\partial}_N \, B \ . \tag{4}$$

The matrix $\bar{\omega}_{MN}$, inverse to $\bar{\omega}^{MN}$

$$\bar{\omega}_{MN} \bar{\omega}^{NL} = \delta^L_M \ , \tag{5}$$

and consisting of the coefficients of the odd closed 2-form, in view of the odd bracket properties (3a)-(3c) can be represented in the form of the grading strength

$$\bar{\omega}_{MN} = \partial_M \bar{A}_N - (-1)^{g(M)g(N)} \partial_N \bar{A}_M \ , \tag{6}$$

where $g(M) = g(x^M)$ and $\partial_M = \partial/\partial x^M$. The coefficients of the 1-form $\bar{A}(d) = \mathrm{d}x^M \bar{A}_M$ satisfy the conditions

$$g(\bar{A}_M) = g(M) + 1 \ , \qquad (\bar{A}_M)^+ = \bar{A}_M \ . \tag{7}$$

As can be seen from (6) $\bar{\omega}_{MN}$ is invariant under gauge transformations

$$\bar{A}'_M = \bar{A}_M + \partial_M \bar{\chi} \tag{8}$$

with functions $\bar{\chi}$ as parameters.

3. Let us consider the Hamilton system containing an equal number n of pairs of even and odd with respect to the Grassmann grading real canonical variables. We require that the equations of motion of the system be reproduced both in the even Poisson-Martin bracket (1) with the help of the even Hamiltonian H and in the odd bracket (2) with the Grassmann-odd Hamiltonian $\bar{H}$, that is (Volkov, Pashnev, Soroka, Tkach (1986, 1989), Soroka (1989)),

$$\frac{\mathrm{d}x^M}{\mathrm{d}t} = \{x^M, H\}_0 = \{x^M, \bar{H}\}_1 \ , \tag{9}$$

where t is the proper time. Using definitions (4) and (1) together with (5),(6) the relations (9) can be represented as the equations

$$(\partial_M \bar{A}_N - (-1)^{g(M)g(N)} \partial_N \bar{A}_M) \omega^{NL} \partial_L H = \partial_M \bar{H} \tag{10}$$

to derive the unknown $\bar{H}$ and $\bar{A}_M$ under the given H and the even matrix ω^{MN} corresponding to the even bracket (1).

In order to solve Eqs. (10) it is convenient to use such real canonical in the even bracket (1) coordinates x^M which contain among canonically conjugate pairs the pair consisting of the proper time t and the Hamiltonian H. It follows from Eqs.(9) that the rest of the canonical quantities z^M would be the integrals of motion for the system considered: even $I_1, ..., I_{2(n-1)}$ and odd $\Theta^1, ..., \Theta^{2n}$. In terms of these coordinates x^M Eqs. (10) take the form

$$(\partial_M \bar{A}_t - \partial_t \bar{A}_M) = \partial_M \bar{H} \ . \tag{11}$$

The quantities $\bar{A}_M, \bar{\chi}$ and $\bar{H}$ can be expanded in powers of the Grassmann variables Θ^α as

$$\bar{A}_m = \sum_{k=1}^{n} \frac{\mathrm{i}^{(k-1)(2k-1)}}{(2k-1)!} A_{m\alpha_1 \ldots \alpha_{2k-1}} \Theta^{\alpha_1} \ldots \Theta^{\alpha_{2k-1}} \ , \tag{12a}$$

$$\bar{A}_\alpha = \sum_{k=0}^{n} \frac{\mathrm{i}^{k(2k+1)}}{(2k)!} B_{\alpha\alpha_1 \ldots \alpha_{2k}} \Theta^{\alpha_1} \ldots \Theta^{\alpha_{2k}} \ , \tag{12b}$$

$$\bar{\chi} = \sum_{k=1}^{n} \frac{\mathrm{i}^{(k-1)(2k-1)}}{(2k-1)!} \chi_{\alpha_1 \ldots \alpha_{2k-1}} \Theta^{\alpha_1} \ldots \Theta^{\alpha_{2k-1}} \ , \tag{12c}$$

$$\bar{H} = \sum_{k=1}^{n} \frac{\mathrm{i}^{(k-1)(2k-1)}}{(2k-1)!} h_{\alpha_1 \ldots \alpha_{2k-1}} \Theta^{\alpha_1} \ldots \Theta^{\alpha_{2k-1}} \ . \tag{12d}$$

The Θ^α coefficients are the real Grassmann-even functions of the even variables $x^m = (t, H, I_1, \ldots, I_{2(n-1)})$ and are chosen to be antisymmetric in the indices contracted with Θ^α. In terms of these functions the gauge transformations (8) have the form

$$\begin{aligned} A'_{m\alpha_1 \ldots \alpha_{2k-1}} &= A_{m\alpha_1 \ldots \alpha_{2k-1}} + \partial_m \chi_{\alpha_1 \ldots \alpha_{2k-1}} \ , (k = 1, \ldots, n) \ ; \\ B'_{[\alpha\alpha_1 \ldots \alpha_{2k}]} &= B_{[\alpha\alpha_1 \ldots \alpha_{2k}]} + \chi_{\alpha\alpha_1 \ldots \alpha_{2k}} \ , (k = 0, 1, \ldots, n-1) \ ; \\ B'_{(\alpha\alpha_j)\alpha_1 \ldots \alpha_{j-1}\alpha_{j+1} \ldots \alpha_{2k}} &= B_{(\alpha\alpha_j)\alpha_1 \ldots \alpha_{j-1}\alpha_{j+1} \ldots \alpha_{2k}} \ , \begin{pmatrix} k = 0, 1, \ldots, n \\ j = 1, \ldots, 2k \end{pmatrix} \ ; \end{aligned} \tag{13}$$

where the expansion in the components with different symmetries of the indices has been used for the tensor antisymmetric in all indices but the first

$$B_{\alpha\alpha_1 \ldots \alpha_N} = B_{[\alpha\alpha_1 \ldots \alpha_N]} + \frac{2}{N+1} \sum_{j=1}^{N} (-1)^{j-1} B_{(\alpha\alpha_j)\alpha_1 \ldots \alpha_{j-1}\alpha_{j+1} \ldots \alpha_N} \ .$$

The additive character of the transformations for the functions $B_{[\alpha\alpha_1 \ldots \alpha_{2k}]}$ $(k = 0, 1, \ldots, n-1)$ allows us to put them equal to zero in the expression (12b) for $\bar{A}_\alpha$ by choosing $\chi_{\alpha\alpha_1 \ldots \alpha_{2k}} = -B_{[\alpha\alpha_1 \ldots \alpha_{2k}]}$. This gauge choice amounts to the following gauge condition

$$\Theta^\alpha \bar{A}_\alpha = 0 \ .$$

Using this condition and Eqs.(11), we obtain the equality

$$\bar{H} = \bar{A}_t \ . \tag{14}$$

which, being substituted again into Eqs.(11), leads to the simple equations

$$\partial_t \bar{A}_M = 0 \ . \tag{15}$$

Thus, in consequence of (15), the solution of Eqs.(11) for $\bar{A}_M$ and $\bar{H}$ in the chosen gauge resides in that the nonzero coefficients $A_{m\alpha_1\ldots\alpha_{2k-1}}$ and $B_{(\alpha\alpha_j)\alpha_1\ldots\alpha_{j-1}\alpha_{j+1}\ldots\alpha_{2k}}$ in expansions (12a,b) for $\bar{A}_M$ are the arbitrary functions (denoted as $a_{m\alpha_1\ldots\alpha_{2k-1}a_{m\alpha_1\ldots\alpha_{2k-1}}}$ and $b_{(\alpha\alpha_j)\alpha_1\ldots\alpha_{j-1}\alpha_{j+1}\ldots\alpha_{2k}}$, respectively) of all, except the proper time t, even variables H and $I_1, \ldots, I_{2(n-1)}$, and the odd Hamiltonian is expressed in terms of these functions with the help of Eq.(14).

Using the gauge transformations (13) with the arbitrary functions $\chi_{\alpha_1\ldots\alpha_{2k-1}}(t, H, I)$, we obtain the general solution of Eqs.(11) in the arbitrary gauge:

$$\begin{aligned}
A_{m\alpha_1\ldots\alpha_{2k-1}} &= a_{m\alpha_1\ldots\alpha_{2k-1}}(H, I) + \partial_m \chi_{\alpha_1\ldots\alpha_{2k-1}}(t, H, I) \ ; \\
B_{\alpha\alpha_1\ldots\alpha_{2k}} &= \frac{2}{2k+1}\sum_{j=1}^{2k}(-1)^{j-1} b_{(\alpha\alpha_j)\alpha_1\ldots\alpha_{j-1}\alpha_{j+1}\ldots\alpha_{2k}}(H, I) + \\
&\quad + \chi_{\alpha\alpha_1\ldots\alpha_{2k}}(t, H, I) \ ; \\
h_{\alpha_1\ldots\alpha_{2k-1}} &= a_{t\alpha_1\ldots\alpha_{2k-1}}(H, I) \ .
\end{aligned}$$

Note that the solution of the analogous problem of finding the even brackets and the corresponding even Hamiltonians, which lead to the same equations of motion

$$\frac{\mathrm{d}x^M}{\mathrm{d}t} = \{x^M, H\}_0 = \{x^M, \tilde{H}\}_{\tilde{0}} \ , \tag{16}$$

has a similar structure but with the difference that the odd quantities $\bar{A}_M, \bar{\chi}$ and $\bar{H}$ has to be replaced by the even ones.

4. Thus, we extended the notion of the bi-Hamiltonian systems (see, for example, (Daf, Okubo (1988))) onto the case when the pairs of the Hamiltonian-bracket, giving the same equations of motion, have an opposite Grassmann grading.

The results obtained can be used for the classification of integrable Hamilton systems. Indeed, if in canonical coordinates for the bracket $\{..., ...\}_0$ Hamiltonian H corresponds to the integrable system, then its equation of motion can be reproduced in accordance with the relations (9) and (16) with the help of Hamiltonians $\bar{H}$ and $\tilde{H}$ by means of the corresponding brackets $\{\ldots, \ldots\}_1$ and $\{\ldots, \ldots\}_{\tilde{0}}$, which, however, written in noncanonical coordinates. Under transition to the canonical coordinates in every bracket $\{\ldots, \ldots\}_1$ and $\{\ldots, \ldots\}_{\tilde{0}}$ the property of integrability apparently does not vanish, but the equations of motion become different from initial ones.

This work was supported in part by the Ukrainian Ministry of Science and Technology, Grant N 2.5.1/54, by Grants INTAS N 93-127 (Extension) and INTAS N 93-633 (Extension).

References

Berezin F.A. (1983): ***Introduction in algebra and analysis with anticommuting variables*** (Moscow State University, 1983) (in Russian).
Daf A., Okubo S. (1988): *Phys. Lett.* **209B**, 311.
Soroka V.A. (1989): *Lett. Math. Phys.* **17**, 201.
Volkov D.V., Pashnev A.I., Soroka V.A. and Tkach V.I. (1986, 1989):
JETP Lett. **44**, 70;
Teor. Mat. Fiz. **79**, 117 (in Russian).

Part II

Quantum Symmetries and q-Deformed Groups

Part II

Quantum Symmetries and q-Deformed Groups

q-Deformed Heisenberg Algebra

J. Wess

Sektion Physik der Ludwig-Maximilians-Universität
Theresienstr. 37, D-80333 München
and
Max-Planck-Institut für Physik
(Werner-Heisenberg-Institut)
Föhringer Ring 6, D-80805 München

In general we describe a physical system by position variables on some manifold and their conjugate momenta. The system is then quantized by imposing canonical commutation relations. We try to generalize this procedure by starting from a noncommutative space. To give some structure to this space we demand it to be a co-module of a quantum group. The conjugate variables are then related to derivatives on the noncommutative space. The "canonical" commutation relations are defined such that they are consistent with selfadjointness of the variables associated with observables. This leads to well defined algebraic structures, depending on the quantum group chosen. For a physical interpretation we study Hilbertspace representations of the algebra and then follow the usual rules of quantum mechanics. The interesting feature of such a system is that the spectrum of the position variables shows a q-lattice structure. The "physical" system lives on a lattice and if the Hamiltonian is expressed in terms of the q-deformed momentum and position variables, then the dynamics of the system will respect the q-lattice structure. Moreover, the q-deformed variables can be expressed in terms of the usual quantum mechanical variables and the usual quantum mechanical Hilbertspace can be decomposed into the irreducible representations of the q-deformed algebra. The Hamiltonian via this procedure is now a Hamiltonian in terms of the usual canonical variables and it can choose a lattice on which all the dynamics takes place.

To illustrate this idea we present the simplest such system - it is one-dimensional. The simplest example is based on the algebra

$$\partial x = 1 + qx\partial \tag{1}$$

where q is a real ($q \in \mathbb{R}$).

More precisely, we study the free algebra generated by the elements x and ∂ and divided by the ideal generated by (1).

If we assume x to be selfadjoint

$$\overline{x} = x \tag{2}$$

we see that this cannot be the case for $i\partial$ because we find from (1) that:

$$\overline{\partial} x = x - \frac{1}{q} + \frac{1}{q} x \overline{\partial} \tag{3}$$

Thus we could study the algebra generated by x, ∂ and $\overline{\partial}$ and divide by (1), (3) and an ideal generated by $\partial\overline{\partial}$ relations. We compute from (1) and (3) $\partial\overline{\partial}x$ and $\overline{\partial}\partial x$ and find that

$$\overline{\partial}\partial = q\partial\overline{\partial} \tag{4}$$

is consistent with these calculations. If we now try to define an operator $\hat{p}$ by $\hat{p} = -\frac{i}{2}(\partial - \overline{\partial})$ we find that the $x, \hat{p}$ relations do not close. The real part of ∂ has to be introduced as well. Thus our Heisenberg algebra would have one space and two momentum operators - a system that will hardly find a physical interpretation.

It turns out that $\overline{\partial}$ can be related to ∂ and x in a nonlinear way. This relation involves the scaling operator Λ:

$$\begin{aligned} &\Lambda \equiv q^{\frac{1}{2}}\,(1 + (q-1)x\partial) \\ &\Lambda x = qx\Lambda \\ &\Lambda\partial = q^{-1}\partial\Lambda \end{aligned} \tag{5}$$

The scaling property follows from (1).

We now define

$$\tilde{\partial} = -q^{-\frac{1}{2}}\Lambda^{-1}\partial \tag{6}$$

Λ^{-1} is defined by an expansion in $(q-1)$. We find

$$\begin{aligned} &\tilde{\partial} x = -\tfrac{1}{q} + \tfrac{1}{q} x\tilde{\partial} \\ &\tilde{\partial}\partial = q\partial\tilde{\partial} \end{aligned} \tag{7}$$

Comparing this with (3) and (4) it follows from (6) and (7) that conjugation in the x, ∂ algebra can be defined by

$$\overline{x} = x \quad , \quad \overline{\partial} = -q^{-\frac{1}{2}}\Lambda^{-1}\partial \tag{8}$$

Conjugating Λ and using (8) shows that

$$\overline{\Lambda} = \Lambda^{-1} \tag{9}$$

Λ is a unitary element of the algebra, this justifies the factor $q^{\frac{1}{2}}$ in the definition of Λ.

The existence of a scaling operator Λ and the definition of the conjugation (8) seems to be very specific for the x, ∂ algebra (1). It is however generic in the sense that a scaling operator and a definition of conjugation based on it can be found for all the quantum planes defined by $SO_q(n)$ and $SO_q(1,n)$.

The definition of the q-deformed Heisenberg algebra will now be based on the definition of the momentum:

$$p = -\frac{i}{2}(\partial - \overline{\partial}) \tag{10}$$

It is selfadjoint. From the x, ∂ algebra and the definition of $\overline{\partial}$ follows

$$\begin{aligned} & q^{\frac{1}{2}} xp - q^{-\frac{1}{2}} px = \mathrm{i}\Lambda^{-1} \\ & \Lambda x = qx\Lambda \qquad \Lambda p = q^{-1} p\Lambda \end{aligned} \tag{11}$$

and

$$\overline{p} = p \quad , \quad \overline{x} = x \quad , \quad \overline{\Lambda} = \Lambda^{-1} \tag{12}$$

These algebraic relations can be verified in the x, ∂ representation where the ordered x, ∂ monomials form a basis. We shall take (11) and (12) as the defining relations for the q-deformed Heisenbergalgebra without making further reference to its x, ∂ representation.

Let us discuss the representations of (11). We assume p to be represented by an essentially selfadjoint operator and therefore diagonalyzable. We assume p to be diagonal. Let p_0 be an eigenvalue and $\mid 0\rangle$ its eigenstate.

$$p \mid 0\rangle = p_0 \mid 0\rangle \tag{13}$$

From (11) follows that $\Lambda^n \mid 0\rangle$ will have the eigenvalue $q^n p_0$

$$\begin{aligned} & p \mid n\rangle = q^n p_0 \mid n\rangle \\ & \Lambda \mid n\rangle = \mid n+1\rangle \end{aligned} \tag{14}$$

As Λ is unitary and p selfadjoint we find:

$$\langle m \mid n\rangle = \delta_{mn} \tag{15}$$

As Hilbertspace H_{p_0} we take the completion of the linear span D_{p_0}

$$D_{p_0} = \left\{ \sum_{\text{finite}} c_n \mid n\rangle \right\} \tag{16}$$

$$H_{p_0} = \left\{ \sum_{n=-\infty}^{\infty} c_n \mid n\rangle \quad , \quad \sum_{n=-\infty}^{\infty} \mid c_n \mid^2 < \infty \right\}$$

On this Hilbertspace x can be represented as follows:

$$x \mid n\rangle = \frac{\mathrm{i}}{p_0} \frac{q^{-n}}{q - \frac{1}{q}} \left(q^{\frac{1}{2}} \mid n-1\rangle - q^{-\frac{1}{2}} \mid n+1\rangle \right) \tag{17}$$

The eqns (14) and (17) define a representation of the algebra (11). In this representation p is essentially selfadjoint and Λ is unitary. The operator x is symmetric:

$$\overline{\langle m \mid x \mid n\rangle} = \langle n \mid x \mid m\rangle \tag{18}$$

It follows from (11) that another representation of the algebra can be otained by defining

$$p^s = \pm sp \quad , \quad x^s = \pm s^{-1} x \quad , \quad \Lambda^s = \Lambda \tag{19}$$

where s is a real number. For $1 \leq s < q$ (we now consider the case $q \geq 1$) the representations obtained that way will be inequivalent.

We now show that x cannot be essentially selfadjoint in these representations. We assume that it is and thus can be diagonalized. Let $| x_0\rangle$ be an eigenvector with eigenvalue x_0. The operator Λ lowers and Λ^{-1} raises this eigenvalue. The states obtained that way are orthogonal (x essentially selfadjoint).

$$x \mid x_0\rangle = x_0 \mid x_0\rangle \tag{20}$$

$$| x_0\rangle = \sum_{n=-\infty}^{\infty} c_n \mid n\rangle \quad , \quad \sum_{n=-\infty}^{\infty} | c_n |^2 = 1$$

From the algebra follows

$$\langle x_0 \mid p \mid x_0\rangle x_0 (q^{\frac{1}{2}} - q^{-\frac{1}{2}}) = 0 \tag{21}$$

$$s p_0 \sum_{n=-\infty}^{\infty} q^n \mid c_n |^2 = 0$$

This contradicts (20).

The story is that x is symmetric but not essentially selfadjoint. It can be shown that x has exactly one eigenvector for each of the eigenvalues $\pm$i. Thus it has a one-parameter family of selfadjoint extensions. But none of these extensions will represent the algebra. This also follows from the argument given above. It is however possible to find reducible representations of the algebra (11) where the selfadjoint extension of x will satisfy the algebra (11). It is obvious from (21)that this is only possible if p_0 is allowed to have both signs. This will be discussed in the next section.

We now show that the algebra (11) can also be represented in the usual "quantum mechanical" Hilbertspace. With the notation

$$\hat{z} = -\frac{\mathrm{i}}{2}(\hat{p}\hat{x} + \hat{x}\hat{p}) = -\mathrm{i}\hat{p}\hat{x} + \frac{1}{2} = -\mathrm{i}\hat{x}\hat{p} - \frac{1}{2} \tag{22}$$

and, for any quantity A:

$$[A] = \frac{q^A - q^{-A}}{q - q^{-1}} \tag{23}$$

we define

$$p = \hat{p} \tag{24}$$

$$x = \frac{[\hat{z} + \frac{1}{2}]}{(\hat{z} + \frac{1}{2})}\hat{x}$$

$$\Lambda = q^{-\hat{z}}$$

and show that p, x satisfy the algebraic relations (11) if $\hat{p}$ and $\hat{x}$ satisfy the canonical commutation relations. We start from:

$$\hat{x}\hat{z} = (\hat{z} + 1)\hat{x} \quad , \quad \hat{p}\hat{z} = (\hat{z} - 1)\hat{p} \tag{25}$$

and conclude:

$$\hat{x}f(\hat{z}) = f(\hat{z}+1)\hat{x} \quad , \quad \hat{p}f(\hat{z}) = f(\hat{z}-1)\hat{p} \tag{26}$$

Let us now verify the first relation of (11).

$$q^{\frac{1}{2}}\frac{[\hat{z}+\frac{1}{2}]}{(\hat{z}+\frac{1}{2})}\hat{x}\hat{p} - q^{-\frac{1}{2}}\hat{p}\frac{[\hat{z}+\frac{1}{2}]}{(\hat{z}+\frac{1}{2})}\hat{x} = \tag{27}$$

$$q^{\frac{1}{2}}\frac{[\hat{z}+\frac{1}{2}]}{(\hat{z}+\frac{1}{2})}\hat{x}\hat{p} - q^{-\frac{1}{2}}\frac{[\hat{z}-\frac{1}{2}]}{(\hat{z}-\frac{1}{2})}\hat{p}\hat{x}$$

$$= \mathrm{i}q^{\frac{1}{2}}[\hat{z}+\frac{1}{2}] - q^{-\frac{1}{2}}[\hat{z}-\frac{1}{2}] = \mathrm{i}q^{\hat{z}} = \mathrm{i}\Lambda^{-1}$$

The other relations are verified in the same way. That p is selfadjoint and Λ unitary if $\hat{p}$ and $\hat{x}$ are selfadjoint is obvios. Let us show that x is selfadjoint as well:

$$\bar{\hat{x}} = \hat{x}\frac{[-\hat{z}+\frac{1}{2}]}{(-\hat{z}+\frac{1}{2})} = \hat{x}\frac{[\hat{z}-\frac{1}{2}]}{(\hat{z}-\frac{1}{2})} = \frac{[\hat{z}+\frac{1}{2}]}{(\hat{z}+\frac{1}{2})}\hat{x} = x \tag{28}$$

The x, p, Λ algebra does not change if $\hat{x}\,\hat{p}$ are changed by a canonical transformation. A class of such canonical transformations is:

$$\tilde{p} = f(\hat{z})\hat{p} \quad , \quad \tilde{x} = \hat{x}f^{-1}(\hat{x}) \tag{29}$$

We immediately see that

$$\tilde{x}\tilde{p} - \tilde{p}\tilde{x} = \mathrm{i}$$

The elements $\tilde{p}, \tilde{x}$ will be selfadjoint if

$$\overline{f}(\bar{\hat{z}}) = f(\hat{z}+1) \tag{30}$$

An example of such an f is:

$$f^{-1}(\hat{z}) = \frac{[\hat{z}-\frac{1}{2}]}{\hat{z}-\frac{1}{2}} \tag{31}$$

With such a canonical transformation we find:

$$p = \frac{[\tilde{z}-\frac{1}{2}]}{(\tilde{z}-\frac{1}{2})}\tilde{p} \tag{32}$$

$$x = \tilde{x}$$

$$\Lambda = q^{-\tilde{z}}$$

This is again a representation of the $x\,p\,\Lambda$ algebra.

Let us see how the representation (24) is related to the representations (14), (17), (19). We start from a representation where $\hat{p}$ is diagonal:

$$\hat{p} \mid k_0\rangle = k_0 \mid k_0\rangle \quad \langle k_0' \mid k_0\rangle = \delta(k' - k_0) \tag{33}$$

Eqn.(19) suggests the notation:

$$p^s \mid n,\sigma\rangle \mid s\rangle = \sigma s q^n p_0 \mid n,\sigma\rangle \mid s\rangle \tag{34}$$

$$x^s \mid n,\sigma\rangle \mid s\rangle = \sigma s^{-1} p_0^{-1} q^{-n} \frac{\mathrm{i}}{q-\frac{1}{q}} (q^{\frac{1}{2}} \mid n-1,\sigma\rangle \mid s\rangle - q^{-\frac{1}{2}} \mid n+1,\sigma\rangle \mid s\rangle)$$

$$\langle s \mid s'\rangle = \delta(s-s')$$

$$\sigma = \pm 1 \quad , \quad 1 \leq s < q \ .$$

The representation (24) can be reduced:

$$\mid n,\sigma\rangle \mid s\rangle = \int \mathrm{d}k_0 q^{\frac{n}{2}} \delta(k_0 - \sigma s q^n) \mid k_0\rangle \tag{35}$$

Normalization:

$$\begin{aligned} &\langle n',\sigma' \mid \langle s' \mid\mid n,\sigma\rangle \mid s\rangle = \qquad (36)\\ &= \int \mathrm{d}k_0 \mathrm{d}k_0' q^{\frac{1}{2}(n+n')} \delta(k_0 - k_0')\delta(k_0 - \sigma s q^n)\delta(k_0' - \sigma' s' q^{n'})\\ &= \int \mathrm{d}k_0 q^{\frac{1}{2}(n+n')} \delta(k_0 - \sigma s q^n)\delta(k_0 - \sigma' s' q^{n'}) =\\ &= q^{\frac{1}{2}(n+n')} \delta(\sigma' s' q^{n'} - \sigma s q^n) = q^{\frac{1}{2}(n+n')} \frac{1}{q^n} \delta_{n,n'}\delta_{\sigma,\sigma'}\delta(s-s')\\ &= \delta_{n,n'}\delta_{\sigma,\sigma'}\delta(s-s') \quad . \end{aligned}$$

The transformation (35) can be inverted:

$$\mid k_0\rangle = \int_1^s \mathrm{d}s \sum_{n=-\infty}^{\infty} \sum_{\sigma=+,-} q^{\frac{n}{2}} \delta(k_0 - \sigma s q^n) \mid n,\sigma\rangle \mid s\rangle \tag{37}$$

Let us now act with Λ as defined by (24) on (35). On a wavefunction in the k_0 representation (33) $\hat{x}$ acts as $\mathrm{i}\frac{\partial}{\partial k_0}$. The action of Λ then is

$$\Lambda f(k_0) = q^{-\frac{1}{2}} q^{-k_0 \frac{\partial}{\partial k_0}} f(k_0) = q^{-\frac{1}{2}} f(q^{-1} k_0) \tag{38}$$

$$\Lambda q^{\frac{n}{2}} \delta(k_0 - \sigma s q^n) = q^{-\frac{1}{2}} q^{\frac{n}{2}} \delta(q^{-1} k_0 - \sigma s q^n) = q^{\frac{1}{2}(n+1)} \delta(k_0 - \sigma s q^{(n+1)}) \quad ,$$

$$\Lambda \mid n\sigma\rangle \mid s\rangle = \mid n+1,\sigma\rangle \mid s\rangle$$

The action of x can be otained from (38) by

$$x = \frac{\mathrm{i}}{q-\frac{1}{q}} (q^{\frac{1}{2}} \Lambda^{-1} - q^{-\frac{1}{2}} \Lambda) p^{-1} \tag{39}$$

Its action on $\mid n,\sigma\rangle \mid s\rangle$ is identical with (34). This shows that the representation (24) decomposes into a direct integral of representations of the type (17) and (19).

Supersymmetric Reflection Matrices

M. Moriconi[1] and K. Schoutens[2]

[1] High Energy Section, ICTP
Strada Costiera 11, Trieste, 34100, Italy
e.mail: moriconi@ictp.trieste.it
[2] Institute for Theoretical Physics, University of Amsterdam
Valckenierstraat 65, 1018 XE Amsterdam, The Netherlands
e.mail: kjs@phys.uva.nl

Abstract. We briefly review the general structure of integrable particle theories in $1+1$ dimensions having $N=1$ supersymmetry. Examples are specific perturbed superconformal field theories (of Yang-Lee type) and the $N=1$ supersymmetric sine-Gordon theory. We comment on the modifications that are required when the $N=1$ supersymmetry algebra contains non-trivial topological charges.

1 Introduction

Quantum field theory (QFT) provides a powerful and unifying language to understand a variety of physical phenomena. In general we may define a QFT by choosing a set of fields that transform according to some irreducible representation of the Poincaré group, together with a prescription (for example, coming from a lagrangian) that gives us the dynamics of these fields. In the context of Particle Physics, it is strongly believed that the Poincaré group is a true symmetry of the world and therefore we always take it for granted.

In general, a QFT becomes more tractable if it possesses additional symmetries beyond Poincaré invariance. Via Ward identities, extra symmetries provide strong constraints on correlation functions and so make possible a more thorough analytical treatment. To introduce new symmetries in a QFT is rather delicate: we do not want to oversimplify the specific models we are looking at but we would like to have enough symmetry to improve the physical properties and to gain the upper hand in controlling the theory. Among the possible symmetries we may consider, *supersymmetry* stands out as a very special one. Supersymmetry unifies the apparently incompatible concepts of bosons and fermions, and often improves the physical properties of specific field theories. At the same time, supersymmetric theories are usually easier to analyze. This last remark applies to supersymmetric Yang-Mills theory (in 4 dimensions), to string models (in 10 dimensions) and to models of QFT in 2 dimensions.

Focussing on QFT's in $1+1$ dimensions, we may further specialize to models that are *integrable*. By this we mean that we assume the existence of 'enough' (meaning an infinite number of) charges in involution.[1] For a QFT describing (massive) particles, integrability implies the factorizability of the scattering matrix: the S-matrix for n-particle scattering factorizes into two-body S-matrices, there is no particle production and the individual momenta of the particles are conserved. By assuming both supersymmetry and integrability, we thus arrive at very simple supersymmetric particle theories which can be analyzed in closed form, and which may serve as prototypes for supersymmetric particle theories in higher dimensions.

In addition to the motivation we have given so far, there are more direct reasons for considering QFT's in two dimensions. For one thing, such theories directly apply to the analysis of either classical systems of statistical mechanics in 2 dimensions, or quantum mechanical systems in 1+1 dimensions (that is, on a line). In addition, there are applications to problems in $3+1$ dimensions, where the essential physics takes place in the radial direction. In the latter type of applications, the models live on a half-line and the behavior at the boundary[2] is important. Examples are the Callan-Rubakov effect (the catalysis of baryon decay in the field of a magnetic monopole), the Hawking effect (quantum black-hole evaporation), the Kondo problem (magnetic impurities coupling to conduction electrons) or edge current tunneling in the quantum Hall effect.

In view of these applications, it is an interesting problem to consider integrable supersymmetric particle theories in 1+1 dimensions in the presence of a boundary. In a recent paper Moriconi and Schoutens (1997) we obtained a general form for boundary reflection matrices in $N = 1$ supersymmetric theories, and we worked out a number of examples.

In this note we outline the general structure of integrable supersymmetric QFT's in 1+1 dimensions, paying particular attention to their boundary scattering. We shall briefly introduce specific examples (which are perturbations of supersymmetric Yang-Lee-type conformal field theories and the breathers in the supersymmetric sine-Gordon theory). We shall also comment on the extension of these results to the case of $N = 1$ supersymmetry with non-zero topological charge.

2 S-Matrices and Reflection Matrices: General

Given a bulk integrable field theory, one may start the analysis by determining the two-body scattering matrix. This is a key ingredient for the understanding of the physics of the model and the first step towards computing correlation functions using the form-factor approach. From now on the term "S-matrix" will be used for the two-body S-matrix unless stated otherwise explicitly. We

[1] To say that charges are in involution means (in classical mechanics) that the Poisson brackets of any two of them vanish or (in QFT) that they all mutually commute.

[2] the origin of space in the original three-dimensional formulation

will use the rapidity variable θ, which parametrizes the on-shell momenta of the particles by $p_0 = m\cosh(\theta)$ and $p_1 = m\sinh(\theta)$. The S-matrix between particles 1 and 2 can be written as $S_{a_1a_2}^{b_1b_2}(\theta_{12}))$, where $\theta_{12} = \theta_1 - \theta_2$ is the difference of the rapidities of the incoming particles.

Let us now briefly outline the general strategy for obtaining the S-matrix for an integrable model with some non-trivial symmetries. One starts by writing down the most general S-matrix compatible with the unbroken symmetries and then requires integrability. This is done by imposing the famous Yang-Baxter equation (YBE). The YBE is shown in figure 1.

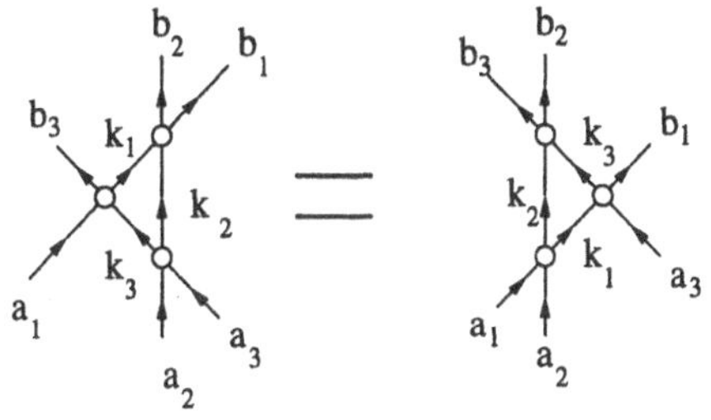

Fig. 1 Pictorial representation of the YBE

In a formula the YBE reads

$$\sum_k S_{a_1a_2}^{k_1k_2}(\theta_{12})S_{k_1a_3}^{b_1k_3}(\theta_{13})S_{k_2k_3}^{b_2b_3}(\theta_{23}) =$$

$$= \sum_k S_{a_2a_3}^{k_2k_3}(\theta_{23})S_{a_1k_3}^{k_1b_3}(\theta_{13})S_{k_1k_2}^{b_1b_2}(\theta_{12}) \ . \tag{1}$$

Once the YBE is solved we impose the usual constraints from general S-matrix theory, that is, analyticity, crossing-symmetry and unitarity. After we managed to do all that (it can be done in many cases!) we have to impose the so-called *bootstrap principle*: bound states are to be treated on the same footing as asymptotic states. This, together with integrability as encoded in the YBE, provides a very restrictive set of equations that greatly constrain the initial S-matrix. Once we reach a self-consistent spectrum we will have found the minimal S-matrix for our model. Of course this can not be the whole story, since different models with the same symmetry and same spectrum may correspond to quite different lagrangians, say. This ambiguity is indeed present and it is called CDD ambiguity, after the work of Castillejo, Dalitz and Dyson (Castillejo et al. (1956)).

One of the ways to test a conjectured S-matrix is through the thermodynamic Bethe Ansatz (TBA). This is a general procedure where we start with the S-matrix as input and compute some ultraviolet physical properties such as the ground state energy (central charge) and scaling dimensions of the underlying QFT. Comparing these with ultraviolet data obtained from a lagrangian or from conformal field theory, we have a non-trivial check on the conjectured S-matrix.

Next we go from the bulk theory to a theory defined on half-line. We will have to specify the boundary action or simply assume that the boundary action is such as to preserve integrability and the extra symmetries of the theory. The theory is then described in the bulk by the same S-matrix as before but now we have to find reflection matrices $R_a^b(\theta)$, which tell us how particles scatter off the boundary. We will assume that the boundary does not change the particle species, so that $R_a^b(\theta) = \delta_a^b R_a(\theta)$. Integrability is imposed now via the boundary Yang-Baxter equation (BYBE), which can be represented in this case as

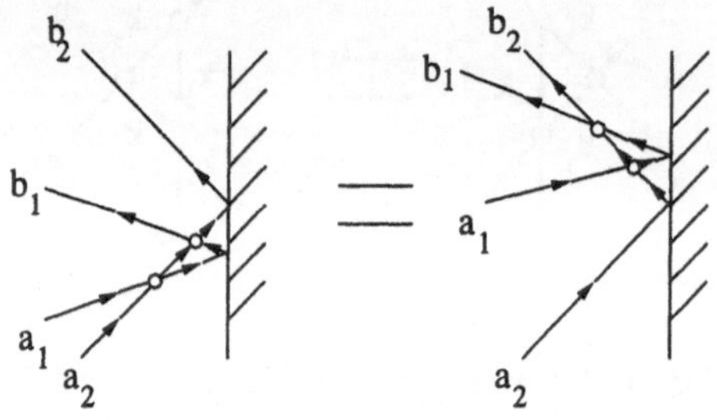

Fig. 2 Pictorial representation of the BYBE

In a formula[3],

$$R_{a_2}(\theta_2) S_{a_1 a_2}^{c_1 d_2}(\theta_1 + \theta_2) R_{c_1}(\theta_1) S_{d_2 c_1}^{b_2 b_1}(\theta_1 - \theta_2) =$$

$$S_{a_1 a_2}^{c_1 c_2}(\theta_1 - \theta_2) R_{c_1}(\theta_1) S_{c_2 c_1}^{b_2 b_1}(\theta_1 + \theta_2) R_{b_2}(\theta_2) \ . \qquad (2)$$

The general idea is clear now. Once we found an exact (bulk) S-matrix we can study boundary versions of the same theory by solving the BYBE and the analogous requirements on the reflection matrix (analyticity, unitarity, boundary crossing-unitarity). Note that the introduction of a boundary necessarily changes the structure of conserved charges. Some of the charges that were conserved in the bulk are not conserved anymore, e.g., linear momentum.

3 $N = 1$ Supersymmetry Without Topological Charges

Supersymmetry may sound as an odd concept in $1+1$ dimensions, since the very definition of bosons and fermions is based on the behavior of the matter field under the rotation group. Since we do not have a rotation group in $1+1$ dimensions one may feel uneasy with such concepts. One may realize,

[3] no sum over a_1, a_2, b_1, b_2

however, that the Lorentz group (which is non-trivial in 1+1 dimension) suffices to define the notions of bosonic fields (of integer Lorentz spin) and fermionic fields (of half-integer Lorentz spin). Supersymmetry comes with a parity operator Q_L, which has eigenvalue $+1$ on bosonic states and -1 on fermionic states. The $N = 1$ supersymmetry algebra takes the form

$$\begin{gathered} \{Q_L, Q_\pm\} = 0 \\ {Q_+}^2 = p_0 + p_1, \qquad {Q_-}^2 = p_0 - p_1 \\ \{Q_+, Q_-\} = 0 \,. \end{gathered} \tag{3}$$

The anticommutator $\{Q_+, Q_-\}$ could have been non-zero, and equal to a real number $Z \leq 2$, which would correspond to a topological charge Z. We will initially consider the situation $Z = 0$ and later show how it can be generalized to $Z \neq 0$. We will use the following realization

$$\begin{aligned} &Q_+(\theta) = \sqrt{m}\, e^{\frac{\theta}{2}} \begin{pmatrix} 0 & 1 \\ 1 & 0 \end{pmatrix}, \quad Q_-(\theta) = \sqrt{m}\, e^{-\frac{\theta}{2}} \begin{pmatrix} 0 & -i \\ i & 0 \end{pmatrix}, \\ &Q_L(\theta) = \begin{pmatrix} 1 & 0 \\ 0 & -1 \end{pmatrix} \,. \end{aligned} \tag{4}$$

We can then define how these operators act on multi-particle states (all we need are two-particle states) and impose that the S-matrix commutes with the supersymmetry generators (Schoutens (1990)). There is one point here that we should stress: to date all exact $N = 1$ S-matrices are of a special form, given by

$$S^{[ij]}(\theta) = S_B^{[ij]}(\theta) S_{BF}^{[ij]}(\theta) \,, \tag{5}$$

where $S_B^{[ij]}(\theta)$ is the S-matrix of the bosonic projection of the theory and $S_{BF}^{[ij]}(\theta)$ is a universal S-matrix that mixes bosons and fermions in such a way that the final non-diagonal S-matrix commutes with the supersymmetry charges. The bosonic factor $S_B^{[ij]}(\theta)$, describing the scattering of bosons b_i and b_j, is assumed to be diagonal. All the physical (bound-state) poles of the total S-matrix are contained in this factor. The fermions are labeled by f_i with $i = 1, 2, \ldots, n$. The particles b_i and f_i have the same mass and form a supermultiplet under the $N = 1$ supersymmetry. The general $S_{BF}^{[ij]}$ matrix was proposed by one of us in Schoutens (1990) and we refer to that paper for a more complete discussion.

In Schoutens (1990) it was found that integrability and supersymmetry alone fix the form of $S_{BF}^{[ij]}(\theta)$ up to one constant α. To fix this constant we have to look at the bootstrap relations. It can be shown that if the particles b_i and b_j for a bound state b_k then we have the following relation

$$\alpha = -\frac{(2m_i^2 m_j^2 + 2m_i^2 m_k^2 + 2m_j^2 m_k^2 - m_i^4 - m_j^4 - m_k^4)^{\frac{1}{2}}}{2m_i m_j m_k} \,. \tag{6}$$

Another consequence of supersymmetry is that once we have a three-point coupling between particles b_i, b_j and b_k we will also have three-point couplings for (f_i, f_j, b_k), (f_i, b_j, f_k) and (b_i, f_j, f_k). The following ratio is then obtained

$$\frac{f_{f_i f_j b_k}}{f_{b_i b_j b_k}} = \left(\frac{m_i + m_j - m_k}{m_i + m_j + m_k} \right)^{\frac{1}{2}} . \tag{7}$$

From conditions such as (6) it is clear that bosonic theories that can be supersymmetrized in this simple manner have to be rather special. In all known examples, the masses m_i of the particles b_i come out as

$$m_i = \frac{\sin(i\beta\pi)}{\sin(\beta\pi)}, \qquad \beta = \frac{1}{2n+1} , \tag{8}$$

for $i = 1, 2, \ldots, n$, $\alpha = -\sin(\beta\pi)$, and there is a specific bound state structure, related to $A_{2n}^{(2)}$ group theory. In section V we shall present some explicit examples.

4 Supersymmetric Reflection Matrices

In this section we will explain how to obtain boundary reflection matrices for $N = 1$ supersymmetric theories by circumventing some of the typical difficulties of boundary integrable models. The basic assumption is that the boundary action is such that *integrability and supersymmetry are both preserved.* On top of that we assume that the reflection matrix can be factorized in a similar fashion to the bulk S-matrix,

$$R(\theta) = R_B(\theta)\, R_{BF}(\theta) . \tag{9}$$

The $R_B(\theta)$ factor is the reflection matrix for the bosonic projection of the theory, and the $R_{BF}(\theta)$ is the "supersymmetric" piece. This factor has the following representation in a $|b\rangle$, $|f\rangle$ basis

$$R_{BF}(\theta) = \begin{pmatrix} R_{bb}(\theta) & R_{bf}(\theta) \\ R_{fb}(\theta) & R_{ff}(\theta) \end{pmatrix} . \tag{10}$$

As we will see now, this will be enough to fix almost completely the reflection matrices, in a similar way to what happens in the bulk case.

If we assume that supersymmetry is preserved by the boundary action we have to impose the "commutation" relation between the reflection matrix and some linear combination of the two bulk supercharges

$$\mathcal{Q}(\theta)R(\theta) = R(\theta)\mathcal{Q}(-\theta) , \tag{11}$$

where $\mathcal{Q}(\theta) = aQ_+(\theta) + bQ_-(\theta)$, a and b some arbitrary real numbers. It is easy to see that the only solutions for (11) are $a = \pm b$ and

$$R^{\pm}_{BF}(\theta) = Z^{\pm}(\theta) \begin{pmatrix} \cosh(\frac{\theta}{2} \pm i\frac{\pi}{4}) & e^{i\frac{\pi}{4}} Y(\theta) \\ e^{-i\frac{\pi}{4}} Y(\theta) & \cosh(\frac{\theta}{2} \mp i\frac{\pi}{4}) \end{pmatrix} . \tag{12}$$

By imposing BYBE we find that $Y(\theta) = 0$. This means that the boundary can not change the fermion number of the incoming particle.

The mass (supermultiplet) dependence of $R^{(\pm)}_{BF}$ is encoded in the prefactor $Z^{(\pm)}$. So in order to have a complete description of boundary scattering we have to fix these functions, by imposing unitarity and boundary crossing-symmetry. This was done in Moriconi and Schoutens (1997) and we refer to that paper for details. On the other hand, without any further work we can see immediately that the ratio R_b/R_f is universal

$$\frac{R^{\pm}_b(\theta)}{R^{\pm}_f(\theta)} = \frac{\cosh(\frac{\theta}{2} \pm i\frac{\pi}{4})}{\cosh(\frac{\theta}{2} \mp i\frac{\pi}{4})} . \tag{13}$$

These results follows directly from supersymmetry, the factorization Ansatz (9) and the specific realization of the superalgebra that we are using.

5 Examples: Supersymmetric Yang-Lee Models and Breathers in the Supersymmetric sine-Gordon Model

In this section we give some examples of theories that realize the general structure presented in sections III and IV.

The first series of examples are the so-called supersymmetric generalized Yang-Lee models. They are obtained as integrable deformations of specific $N = 1$ superconformal field theories of central charges $c = -3n(4n+3)/(2n+2)$, $n = 1, 2, \ldots$, where the perturbing field is the bottom component of the Neveu-Schwarz field labeled as $\phi_{(1,3)}$. The spectrum of the massive deformation is as in (8). The first model of this series corresponds to the supersymmetrization of the Yang-Lee model.

The supersymmetric sine-Gordon theory is defined by the following action in euclidean space-time

$$\begin{aligned} S_{ssG} &= \int_{-\infty}^{\infty} \mathrm{d}y \int_{-\infty}^{\infty} \mathrm{d}x \left\{ \frac{1}{2}(\partial_x \phi)^2 + \frac{1}{2}(\partial_y \phi)^2 - \bar{\psi}(\partial_x - i\partial_y)\bar{\psi} + \right. \\ &\left. + \psi(\partial_x + i\partial_y)\psi - \frac{m^2}{\beta^2_{ssG}} \cos(\beta_{ssG}\phi) - 2m\bar{\psi}\psi \cos(\frac{\beta_{ssG}\phi}{2}) \right\} , \end{aligned} \tag{14}$$

where ϕ is the bosonic field and ψ and $\bar{\psi}$ are the components of a Majorana fermion. The spectrum of the full quantum theory contains (anti-)soliton multiplets and bound state multiplets (b_j, f_j), $j = 1, 2, \ldots < \lambda$, $\lambda = 2\pi \left(1 - (\beta^2_{ssG}/4\pi)\right) / \beta^2_{ssG}$, of masses (8) with $\beta = 1/(2\lambda)$.

In all these examples, the bulk S-matrices and boundary R-matrices are of the general form discussed in sections III and IV (Schoutens (1990), Ahn (1991), Moriconi and Schoutens (1997)). The detailed form of the reflection matrices was worked out in our recent paper (Moriconi and Schoutens (1997)).

We already mentioned that the thermodynamic Bethe Ansatz (TBA) is a very effective way to test the validity of conjectured S-matrices. While this analysis is more or less routine for diagonal S-matrices, the analysis for non-diagonal S-matrices is non-trivial and has to be studied on a case by case basis. Fortunately, $N = 1$ supersymmetric integrable models can be mapped into the eight-vertex model at a special point, where they satisfy the so-called "free-fermion" condition, which allows to complete the TBA program. This was done in Ahn (1994) for the super Yang-Lee case and in Moriconi and Schoutens (1996) for the more general perturbed superconformal field theories discussed in this section.

In the case of the supersymmetric sine-Gordon theory, we have been able to propose exact reflection matrices without knowing the boundary action. An interesting problem is then to find the boundary actions that correspond to these matrices. Inami, Odake and Zhang (1995) have proposed two possible boundary actions that preserve integrability and supersymmetry. Their proposal is based on the study of the conserved charges, at the classical level, in the presence of a boundary. In Moriconi and Schoutens (1997) we established a connection between this proposal and our reflection matrices by looking at the weak coupling limit of the supersymmetric sine-Gordon model (Moriconi and Schoutens (1997)).

6 Topological Charges

In the presence of topological charges, the anticommutator $\{Q_+, Q_-\}$ changes to

$$\{Q_+, Q_-\} = Z \ , \tag{15}$$

with $Z \leq 2$.[4] We will adopt the following realization

$$Q_+(\theta) = \sqrt{m}\, e^{\frac{\theta}{2}} \begin{pmatrix} 0 & 1 \\ 1 & 0 \end{pmatrix} , \quad Q_-(\theta) = \sqrt{m}\, e^{-\frac{\theta}{2}} \begin{pmatrix} 0 & e^{i\alpha} \\ e^{-i\alpha} & 0 \end{pmatrix} ,$$

$$Q_L(\theta) = \begin{pmatrix} 1 & 0 \\ 0 & -1 \end{pmatrix} , \tag{16}$$

where $\cos(\alpha) = Z/2$. Note that the case $Z = 0$ is obtained when $\alpha = -\pi/2$. Similarly to the case without topological charges we assume that the reflection matrices will be of the same factorized form as in (9).

[4] This Z should not be confused with the prefactors $Z^{(\pm)}$.

Following the same approach as in the case without topological charges it is easy to see that the "supersymmetric" part of the reflection matrix will have the following form

$$R_{BF}^{(\pm)}(\theta) = Z^{(\pm)}(\theta) \begin{pmatrix} \cosh(\frac{\theta}{2} + i(\frac{\pi}{4} + \frac{\alpha}{2}) \pm i\frac{\pi}{4}) & 0 \\ 0 & \cosh(\frac{\theta}{2} + i(\frac{\pi}{4} + \frac{\alpha}{2}) \mp i\frac{\pi}{4}) \end{pmatrix} . \tag{17}$$

At $Z = 0$ this reduces to the reflection matrix in (13). Again we recall that this is the reflection matrix obtained by imposing that the boundary action preserves supersymmetry and integrability. Notice that we again have a universal ratio

$$\frac{R_b^{\pm}}{R_f^{\pm}} = \frac{\cosh(\frac{\theta}{2} + i(\frac{\pi}{4} + \frac{\alpha}{2}) \pm i\frac{\pi}{4})}{\cosh(\frac{\theta}{2} + i(\frac{\pi}{4} + \frac{\alpha}{2}) \mp i\frac{\pi}{4})} . \tag{18}$$

Recently, Hollowood and Mavrikis(1997) have proposed exact $N = 1$ supersymmetric S-matrices for theories with non-zero topological charges. We expect that the reflection matrices (17) can consistently be combined with these new S-matrices, in the sense that together they form a solution of the BYBE.

Acknowledgements

We would like to thank Roland Köberle and Andreas Fring for useful discussions. One of us (MM) would like to thank the University of Amsterdam, where part of this work was done, for the warm hospitality. The research of KS was supported in part by the foundation FOM of the Netherlands.

References

Ahn, C. (1991): Nucl. Phys. **B354**, 57-84.
Ahn, C. (1994): Nucl. Phys. **B422**, 449-475, hep-th/9306146.
Castillejo, L., Dalitz, R.H. and Dyson, F.J. (1956): Phys. Rev. **101**, 453-458.
Hollowood, T. and Mavrikis, E. (1997): Nucl. Phys. **B484**, 631-652, hep-th/9606116.
Inami, T., Odake, S. and Zhang, Y.-Z. (1995): Phys. Lett. **B359**, 118-124, hep-th/9506157.
Moriconi, M. and Schoutens, K. (1996): Nucl. Phys. **B464**, 472-491, hep-th/9511008.
Moriconi, M. and Schoutens, K. (1997): Nucl. Phys. **B487**, 756-778, hep-th/9605219.
Schoutens, K. (1990): Nucl. Phys. **B344**, 665-695.

Deformed Oscillator Algebras and Higher-Spin Gauge Interactions of Matter Fields in 2+1 Dimensions

M.A. Vasiliev

I.E.Tamm Department of Theoretical Physics,
Lebedev Physical Institute
Leninsky prospect 53, 117924, Moscow, Russia

Abstract. We formulate a non-linear system of equations which describe higher-spin gauge interactions of massive matter fields in 2+1 dimensional space-time and explain some properties of the deformed oscillator algebra which underlies this formulation. In particular we show that the parameter of mass M of matter fields is related to the deformation parameter in this algebra.

1 Introduction

Dmitrij Vassilievich Volkov was a brilliant scientist who had made a great contribution to theoretical physics and created a remarkable scientific school in Kharkov. His most famous results are related to the creation of supersymmetric theories. The original approach invented by Volkov and collaborators was based on the application of invariant connection forms (Volkov et al. (1974a), (1974b), (1975)). A further development of these geometric ideas in the modern field theory was extremely fruitful. In this talk we argue that a proper generalization of the Volkov's ideas leads to a universal method of description of relativistic dynamics in terms of certain zero-curvature conditions supplemented with appropriate constraints. We will illustrate this by considering an example of matter fields interacting through higher-spin (HS) gauge fields in 2+1 dimensions.

2 Higher-Spin Symmetries in 2+1 Dimensions and Deformed Oscillator Algebras

HS algebras in d space-time dimensions are certain infinite-dimensional extensions of space-time symmetry algebras s_d (Fradkin and Vasiliev (1987), Vasiliev (1988)), which act on appropriate physical fields. HS symmetries can be gauged by virtue of introducing appropriate HS gauge fields. In 2+1 dimensions, HS gauge fields do not propagate rather mediating interactions of matter sources analogously to the case of the gravitational field in 2+1 dimensions (Achucarro and Townsend (1986), Witten (1989)). This is a greatly simplifying property compared to the HS dynamics in four and higher dimensions. HS symmetries in 2+1 dimensions are still non-trivial as well as HS matter multiplets, i.e. the multiplets of fields on which the HS symmetries are realized. They are however very simple: ordinary scalar and spinor fields of an arbitrary mass. The analysis of HS interactions of relatively simple lower dimensional models sheds some light on general properties of HS models.

It is most useful to start with the space-time symmetries of (anti) - de Sitter type $s_d = o(d-1,2)$ analyzing a possibility of taking a flat limit afterwards. The case of $d=3$ is special because $s_3 = o(1,2) \oplus o(1,2) = sp(2) \oplus sp(2)$ is not simple. Originally it was conjectured (Blencowe (1989)) that a 3d HS algebra is the direct sum of two Heisenberg-Weyl algebras (more precisely, of their Lie supercommutator superalgebras), each constructed from the ordinary oscillators [1] $[y_\alpha, y_\beta] = 2i\epsilon_{\alpha\beta}$. Because 3d HS gauge fields are not propagating one can write the Chern-Simons action for the pure gauge HS system, $S = \int_{M_3} \mathrm{str}(A\wedge dA + \frac{2}{3}A\wedge A\wedge A)$ with the gauge fields A taking values in the HS algebra. In (Bergshoeff et al. (1990), Bordemann et al. (1989), Vasiliev (1989), (1991)) it was shown that there exists a one-parametric class of infinite-dimensional algebras which we denote $hs(2;\nu)$ (ν is an arbitrary real parameter), all containing sp(2) as a subalgebra. This allows one to define a class of HS algebras $g = hs(2;\nu) \oplus hs(2;\nu)$. The supertrace operation was defined in (Vasiliev (1989), (1991)) where also a useful realization of the supersymmetric extension of $hs(2;\nu)$ was given, based on a certain deformed oscillator algebra. Since this construction will be used below and also gets interesting applications in a number of different physical problems let us explain its properties in somewhat more details.

Consider an associative algebra $Aq(2;\nu)$ with a general element of the form

$$f(q,K) = \sum_{n=0}^{\infty} \sum_{A=0,1} \frac{1}{n!} f^{A\,\alpha_1\ldots\alpha_n}(K)^A q_{\alpha_1}\ldots q_{\alpha_n} \ , \tag{1}$$

under condition that the coefficients $f^{A\,\alpha_1\ldots\alpha_n}$ are symmetric with respect to the indices $\alpha_j = 1,2$ and that the generating elements q_α satisfy the relations

$$[q_\alpha, q_\beta] = 2i\epsilon_{\alpha\beta}(1+\nu K)\,, \quad Kq_\alpha = -q_\alpha K\,, \quad K^2 = 1\ , \tag{2}$$

where ν is an arbitrary parameter. In other words, $Aq(2;\nu)$ is the enveloping algebra for the relations (2), the deformed oscillator algebra.

An important property of this algebra is that for all ν the bilinears

$$T_{\alpha\beta} = \frac{1}{4i}\{q_\alpha, q_\beta\} \tag{3}$$

[1] The indices α, β, $\gamma = 1,2$ are treated as spinor indices in 2+1 dimensions. These are lowered and raised with the aid of the symplectic form $\epsilon_{\alpha\beta} = -\epsilon_{\beta\alpha}$, $\epsilon_{12} = \epsilon^{12} = 1$, $A^\alpha = \epsilon^{\alpha\beta}A_\beta$, $A_\alpha = A^\beta\epsilon_{\beta\alpha}$.

have sp(2) commutation relations and rotate q_α as a sp(2) vector

$$[T_{\alpha\beta}, T_{\gamma\eta}] = (\epsilon_{\alpha\gamma}T_{\beta\eta} + \epsilon_{\beta\gamma}T_{\alpha\eta} + \epsilon_{\alpha\eta}T_{\beta\gamma} + \epsilon_{\beta\eta}T_{\alpha\gamma}), \tag{4}$$

$$[T_{\alpha\beta}, q_\gamma] = \epsilon_{\alpha\gamma}q_\beta + \epsilon_{\beta\gamma}q_\alpha. \tag{5}$$

The deformed oscillators described above have a long history and were originally discovered by Wigner (Wigner (1950)) who addressed a question whether it is possible to modify the commutation relations for the normal oscillators $a^\pm$ in such a way that the basic commutation relations $[H, a^\pm] = \pm a^\pm$, $H = \frac{1}{2}\{a^+, a^-\}$ remain valid. By analyzing this problem in the Fock-type space Wigner found a one-parametric deformation of the standard commutation relations which corresponds to a particular realization of the commutation relations (2) with the identification $a^+ = q_1$, $a^- = \frac{1}{2i}q_2$, $H = T_{01}$ and $K = (-1)^N$ where N is the particle number operator. These commutation relations were discussed later by various authors in particular in the context of parastatistics (see e.g. (Yang (1951), Boulware and Deser (1963), Mukunda et al. (1980)).

According to (3) and (5) the sp(2) symmetry generated by $T_{\alpha\beta}$ extends to osp(1,2) supersymmetry by identifying the supergenerators with q_α. In fact, as shown in (Bergshoeff et al. (1991)), one can start from the osp(1,2) algebra to derive the deformed oscillator commutation relations. Since this construction is instructive in many respects we reproduce it here.

One starts with the (super)generators $T_{\alpha\beta}$ and q_α which by definition of osp(1,2) satisfy the commutation relations (3)-(5). Since α and β take only two values one can write

$$[q_\alpha, q_\beta] = 2i\epsilon_{\alpha\beta}(1+Q) \ , \tag{6}$$

where Q is some new "operator" while the unit term is singled out for convenience. Inserting this back into (5) with the substitution of (3) and completing the commutations one observes that (5) is true if and only if Q anticommutes with q_α,

$$Qq_\alpha = -q_\alpha Q. \tag{7}$$

The relation (4) does not add anything new since it is a consequence of (3) and (5). As a result we arrive (Bergshoeff et al. (1991)) at the following important

Corollary: *The enveloping algebra of* osp(1,2), $U(\mathrm{osp}(1,2))$, *is isomorphic to the enveloping algebra of the deformed oscillator relations (6) and (7).*

In other words, the associative algebra with the generating elements q_α and Q subject to the relations (6) and (7) is the same as the associative algebra with the generating elements q_α and $T_{\alpha\beta}$ subject to the osp(1,2) commutation relations (3)-(5).

Computing the quadratic Casimir operator of osp(1,2)

$$C_2 = -\frac{1}{2}T_{\alpha\beta}T^{\alpha\beta} - \frac{i}{4}q_\alpha q^\alpha . \tag{8}$$

one easily derives using (6) that

$$C_2 = -\frac{1}{4}(1 - Q^2) . \tag{9}$$

Let us now consider the factor algebra of $U(\mathrm{osp}(1,2))$ over its ideal $I_{(C_2+\frac{1}{4}(1-\nu^2))}$ generated by the element $(C_2+\frac{1}{4}(1-\nu^2))$ where ν is an arbitrary number. In other words we assume that every element of $U(\mathrm{osp}(1,2))$ which is of the form $\left(C_2 + \frac{1}{4}(1-\nu^2)\right) a$, $\forall a \in U(\mathrm{osp}(1,2))$ is equivalent to zero. This factorization can be achieved in terms of the deformed oscillators (6), (7) by setting [2]

$$Q = \nu K , \qquad K^2 = 1 \qquad K q_\alpha = -q_\alpha K . \tag{10}$$

Thus, it is shown (Bergshoeff et al. (1991)) that the algebra $Aq(2,\nu)$ introduced in (Vasiliev (1989), (1991)) is isomorphic to $U(\mathrm{osp}(1,2))/I_{(C_2+\frac{1}{4}(1-\nu^2))}$. This fact has a number of simple but important consequences. For example, any representation of the superalgebra osp(1,2) with $C_2 = -\frac{1}{4}(1-\nu^2)$ forms a representation of $Aq(2,\nu)$ ($\nu \neq 0$) and vice versa (for all ν including $\nu = 0$). In particular this is the case for finite-dimensional representations corresponding to the values $\nu = 2l+1$, $l \in \mathbf{Z}$ with $C_2 = l(l+1)$. This fact has been used in (Brink and Vasiliev (1995)) for the construction of the generalized Toda field theories interpolating between ordinary finite-component Toda field theories. Let us note that the even subalgebra of $Aq(2;\nu)$ spanned by the elements of the form (1) with $f(q,K) = f(-q,K)$ decomposes into a direct sum of two subalgebras $Aq^E_\pm(2;\nu)$ spanned by the elements $P_\pm f(q,K)$ with $f(-q,K) = f(q,K)$, $P_\pm = \frac{1}{2}(1 \pm K)$. These algebras can be shown to be isomorphic to the factor algebras $U(\mathrm{sp}(2))/I_{(C_2+\frac{3\pm 2\nu-\nu^2}{4})}$ where $C_2 = -\frac{1}{2}T_{\alpha\beta}T^{\alpha\beta}$ is the quadratic Casimir operator of sp(2) and can be interpreted as (infinite-dimensional) algebras interpolating between the ordinary finite-dimensional matrix algebras. Such interpretation of $U(\mathrm{sp}(2))/I_{(C_2-c)}$ was given by Feigin in (Feigin (1988)).

A very important property of $Aq(2;\nu)$ is that it admits (Vasiliev (1989), (1991)) a uniquely defined supertrace operation

$$\mathrm{str}(f) = f^0 - \nu f^1 , \tag{11}$$

such that $\mathrm{str}(fg) = (-1)^{\pi_f\pi_g}\mathrm{str}(gf)$, $\forall f,g$ having a definite parity, $f(-q,K) = (-1)^{\pi_f} f(q,K)$ (i.e. $\mathrm{str}(1) = 1$, $\mathrm{str}(K) = -\nu$ while all higher monomials of q_α in (1) do not contribute under the supertrace). This supertrace reduces (Brink and Vasiliev (1995)) to the ordinary supertrace of finite-dimensional algebras for the special values of the parameter $\nu = 2l+1$ which correspond to the values of the Casimir operator related to finite-dimensional

[2] The point $\nu = 0$ is special since one can consider a case(s) with $Q^2 = 0$, $Q \neq 0$.

representations of osp(1,2) (sp(2) in the bosonic case). This property allows one to handle the algebras $Aq(2;\nu)$ very much the same way as ordinary finite-dimensional (super)matrix algebras. What happens for special values of $\nu = 2l+1$ is that $Aq(2;\nu)$ acquires ideals I_l such that $Aq(2;\nu)/I_l$ amounts to appropriate (super)matrix algebras. These ideals were described in (Vasiliev (1989), (1991)) as null vectors of the invariant bilinear form $\mathrm{str}(ab)$, $a, b \in Aq(2;\nu)$.

The identification of $Aq(2;\nu)$ with $U(\mathrm{osp}(1,2))/I_{(C_2+\frac{1}{4}(1-\nu^2))}$ makes transparent such properties of the deformed oscillator algebra as relationship of the representations of $Aq(2;\nu)$ with those of osp(1,2) (including its finite-dimensional representations for special values of $\nu = 2l+1, \forall l \in \mathbf{Z}$) and $N=1$ supersymmetry (as inner osp(1,2) automorphisms). A more interesting property (Bergshoeff et al. (1991)) is that $Aq(2;\nu)$ admits $N=2$ supersymmetry osp(2,2) with the generators

$$T_{\alpha\beta} = \frac{1}{4i}\{q_\alpha\,, q_\beta\}\,, \quad Q_\alpha = q_\alpha\,, S_\alpha = q_\alpha K\,, \quad J = K+\nu\,. \tag{12}$$

These properties find interesting applications (see, e.g., Plyushchay (1997) and references therein).

As we demonstrate below the deformed oscillator algebras serve as a main tool for the description of the d=3 HS dynamics. The reason is that they are related to the enveloping algebras of the space-time symmetries and allow us to formulate a non-linear dynamics with explicit local Lorentz symmetry. In its turn, the analysis of the HS dynamics presented below is interesting in the context of the deformed oscillator algebra itself because algebraically it reduces to the construction of its embedding into a direct product of two ordinary Heisenberg-Weyl (i.e. oscillator) algebras equipped with certain twist operators.

Coming back to the HS problem in 2+1 dimensions we note that to describe a doubling of the elementary algebras in $g = hs(2;\nu) \oplus hs(2;\nu)$ it suffices to introduce an additional central involutive generating element ψ: $[\psi, q_\alpha] = 0$, $[\psi, K] = 0$, $\psi^2 = 1$. The two simple subalgebras of g are singled out by the projection operators $\Pi_\pm = \frac{1}{2}(1 \pm \psi)$. The full set of HS gauge fields in 2+1 dimensions, the gauge fields of g, thus is

$$A(q,K,\psi|x){=}dx^\nu \sum_{n=0}^{\infty} \sum_{B=0,1} \frac{1}{2i\,n!}(\omega_\nu^{B\alpha_1\ldots\alpha_n}(x)$$

$$+\psi h_\nu^{B\alpha_1\ldots\alpha_n}(x))\,(K)^B\, q_{\alpha_1}\cdots q_{\alpha_n}\,. \tag{13}$$

The field strengths and gauge transformation laws are defined in the usual way

$$R(q,K,\psi|x) = dA(q,K,\psi|x) + A(q,K,\psi|x)\wedge A(q,K,\psi|x)\,, \tag{14}$$

$$\delta A(q,K,\psi|x) = \mathrm{d}\epsilon(q,K,\psi|x) + [A(q,K,\psi|x)\,, \epsilon(q,K,\psi|x)]\,, \tag{15}$$

where $\mathrm{d} = dx^\nu \frac{\partial}{\partial x^\nu}$. The gravitational fields

$$A^{gr} = \frac{1}{4i}(\omega^{\alpha\beta} + h^{\alpha\beta}\psi)q_\alpha q_\beta \tag{16}$$

take values in the subalgebra sp(2) ⊕ sp(2). The pure gauge Chern-Simons HS action reduces to the Witten gravity action (Witten (1989)) in the spin 2 sector and to the Blencowe's HS action (Blencowe (1989)) in the case of $\nu = 0$.

3 Unfolded Equations

In this report which is based on a recent papers (Barabanschikov et al. (1997), Prokushkin and Vasiliev (in preparation)) we answer to the question how to introduce interactions of HS gauge fields with propagating matter fields at the level of equations of motion using an approach which we call "unfolded formulation" (Vasiliev (1994)). It consists of reformulation of dynamical equations in a form of certain zero-curvature conditions and covariant constancy conditions

$$\mathrm{d}\omega + \omega \wedge \omega \ = 0,, \qquad \mathrm{d}B^A + \omega^i t_i{}^A{}_B B^B = 0 \ , \tag{17}$$

supplemented with some gauge invariant constraints

$$\chi(B) = 0 \tag{18}$$

which do not contain space-time derivatives. Here $\omega(x) = dx^\nu \omega^i_\nu(x) T_i$ is a gauge field taking values in some Lie superalgebra l ($T_i \in l$), and $B^A(x)$ is a set of 0-forms which take values in the representation space of some representation $(t_i)^B{}_A$ of l.

An interesting property of this form of equations is that their dynamical content is hidden in the constraints (18). Indeed, locally one can integrate out (17) explicitly as $\omega = g(x)\mathrm{d}g^{-1}(x)$, $B(x) = t_{g(x)}(B_0)$ where $g(x)$ is an arbitrary invertible element while B_0 is an arbitrary x - independent representation element and $t_{g(x)}$ is the exponential of the representation t of l. Since the constraints $\chi(B)$ are gauge invariant one is left with the only condition $\chi(B_0) = 0$. Let $g(x_0) = I$ for some point of space-time x_0. Then $B_0 = B(x_0)$.

Such a formulation can in principle be applied to an arbitrary dynamical system provided that a representation t is infinite-dimensional. Being based on zero-curvature conditions it has deep similarities with the original approach by Volkov and collaborators (citev-v). A crucial feature of our approach is that the set of 0-forms B has to be reach enough to describe all space-time derivatives of the dynamical fields while the constraints (18) effectively impose all restrictions on the space-time derivatives required by the dynamical equations under consideration. Given solution of (18) one knows all derivatives of the dynamical fields compatible with the field equations and can therefore reconstruct these fields by analyticity in some neighborhood of x_0. The specificity of the HS dynamics which makes such an approach adequate is that HS symmetries mix all orders of derivatives which therefore have to be contained in a representation t of HS symmetries.

Let us illustrate this by the example of a scalar field ϕ obeying the massless Klein-Gordon equation $\Box\phi = 0$ in a flat space-time of an arbitrary dimension d. Here l is identified with the Poincare algebra $iso(d-1,1)$ which gives rise to the gauge fields $\omega_\nu = (h_\nu{}^a, \omega_\nu{}^{ab})$ $(a, b = 0 - (d-1))$. The zero curvature conditions of $iso(d-1,1)$, $R_{\nu\mu}{}^a = 0$ and $R_{\nu\mu}{}^{ab} = 0$, imply that the vierbein $h_\nu{}^a$ and Lorentz connection $\omega_\nu{}^{ab}$ describe the flat geometry. Fixing the local Poincare gauge transformations one can set

$$h_\nu{}^a = \delta_\nu^a\,, \quad \omega_\nu{}^{ab} = 0\,. \tag{19}$$

To describe dynamics of a spin zero massless field $\phi(x)$ let us introduce an infinite collection of 0-forms $\phi_{a_1\ldots a_n}(x)$ which are totally symmetric traceless tensors

$$\eta^{bc}\phi_{bca_3\ldots a_n} = 0\,, \tag{20}$$

where η^{bc} is the flat Minkowski metrics. The "unfolded" version of the Klein-Gordon equation has a form of the following infinite chain of equations

$$\partial_\nu\phi_{a_1\ldots a_n}(x) = h_\nu{}^b\phi_{a_1\ldots a_n b}(x)\,, \tag{21}$$

where we have replaced the Lorentz covariant derivative by the ordinary flat derivative ∂_ν using the gauge condition (19). The tracelessness condition (20) is a specific realization of the constraints (18) while the system of equations (21) is a particular example of the equations (17). It is easy to see that this system is formally consistent.

To show that the system (21) is equivalent to the free massless field equation $\Box\phi(x) = 0$ let us identify the scalar field $\phi(x)$ with the $n = 0$ member of the tower of 0-forms $\phi_{a_1\ldots a_n}$. Then the first two equations (21) read $\partial_\nu\phi = \phi_\nu$ and $\partial_\nu\phi_\mu = \phi_{\mu\nu}$, respectively. The former tells us that ϕ_ν is a first derivative of ϕ. The latter implies that $\phi_{\nu\mu}$ is a second derivative of ϕ. However, because of the tracelessness condition (20) it imposes the Klein-Gordon equation $\Box\phi = 0$. It is easy to see that all other equations in (21) express highest tensors in terms of the higher-order derivatives $\phi_{\nu_1\ldots\nu_n} = \partial_{\nu_1}\ldots\partial_{\nu_n}\phi$ and impose no additional conditions on ϕ. The tracelessness conditions are all satisfied once the Klein-Gordon equation is true.

4 Free Fields in 2+1 AdS Space

Let us now confine ourselves to the 2+1 dimensional case and generalize the above analysis of the scalar field dynamics to the AdS geometry. The gauge fields of the AdS algebra $o(2,2) \sim \mathrm{sp}(2) \oplus \mathrm{sp}(2)$ are identified with the gravitational fields, $A_\nu = (\lambda h_{\nu\,,\alpha\beta}\,;\omega_{\nu\,,\alpha\beta})$. The zero-curvature conditions $R_{\nu\mu} = 0$ for the AdS algebra in its orthogonal realization take a form

$$R_{\nu\mu\,,ab} = \lambda^2(h_{\nu a}h_{\mu b} - h_{\nu b}h_{\mu a})\,, \quad R_{\nu\mu\,,a} = 0 \tag{22}$$

$(\nu\,,\mu\ldots;a\,,b\ldots = 0-2)$, where $R_{\nu\mu\,,ab}$ and $R_{\nu\mu\,,a}$ are the Riemann and torsion tensors, respectively. From (22) one concludes that the zero curvature equations for the algebra $o(2,2)$ on a 3d manifold do indeed describe the AdS space provided that $h_\nu{}^a$ is identified with a dreibein and is invertible.

It is an important property of the 3d geometry that one can resolve the tracelessness conditions (20) by using the formalism of two-component spinors: a totally symmetric traceless tensor $\phi_{a_1\ldots a_n}$ is equivalent to a totally symmetric multispinor $C_{\alpha_1\ldots\alpha_{2n}}$. Let us now address the question what is a general form of the equations analogous to (21) such that their integrability conditions reduce to (22). The result is that, up to a freedom in field redefinitions, these are equations of the form (Vasiliev (1994))

$$DC_{\alpha_1\ldots\alpha_{2n}} = h^{\beta\gamma}C_{\alpha_1\ldots\alpha_{2n}\beta\gamma} + 2n(2n-1)e(2n,\lambda,M)h_{\{\alpha_1\alpha_2}C_{\alpha_3\ldots\alpha_{2n}\}_\alpha} \tag{23}$$

where D is the Lorentz covariant derivative, $DB_\alpha \equiv \mathrm{d}B_\alpha + \omega_\alpha{}^\beta B_\beta$, and

$$e(l,\lambda,M) = \frac{1}{4}\lambda^2 - \frac{1}{2}\frac{M^2}{l^2-1} \qquad (l \geq 2)\ . \tag{24}$$

One can see that the freedom in an arbitrary parameter M is just the freedom of the relativistic field equations in the parameter of mass.

Thus the equations (23) describe a scalar field of an arbitrary mass in 2+1 dimensions. Now let us show how these equations can be generated with the aid of the generalized oscillators (2). To this end we introduce the generating function

$$C(q_\alpha,K|x) = \sum_{n=0}^{\infty}\sum_{A=0,1}\frac{1}{n!}C_{\alpha_1\ldots\alpha_n}(x)(K)^A q^{\alpha_1}\ldots q^{\alpha_n}\ . \tag{25}$$

The relevant equations acquire then the following simple form

$$DC(q_\alpha,K|x) = \frac{1}{4i}\{h^{\alpha\beta}q_\alpha q_\beta, C(q,K|x)\} \tag{26}$$

(from now on we use the dimensionless units with a unit AdS radius, $\lambda = 1$).

To see that the integrability conditions for (26) reduce to the zero-curvature conditions for $\mathrm{sp}(2)\oplus\mathrm{sp}(2)$ one observes that there is an automorphism of the AdS algebra which changes a sign of the AdS translations. This automorphism allows one to introduce a "twisted representation" of the AdS algebra with the anticommutator instead of commutator in the translational part of the AdS algebra. This twisted representation just leads to the covariant constancy equations (26).

Since the terms in (26) which depend on the background gravitational fields only contain even combinations of the oscillators q_α the full system of equations decomposes into four independent subsystems which can be singled out by virtue of the projection operators $P_\pm = \frac{1}{2}(1 \pm K)$ either in the boson or in the fermion sectors (even (odd) functions $C(q_\alpha, K|x)$ of q_α describe bosons (fermions)). The explicit calculation which involves some reorderings of q_α and rescalings of fields then shows that the irreducible boson subsystems projected out by $P_\pm$ indeed reduce to the equations of motion of the form (23) for a massive scalar field of mass $M^2 = \frac{1}{2}\nu(\nu \mp 2)$. Remarkably, the same equations in the fermion sector describe spin $\frac{1}{2}$ fermion fields of the mass $M^2 = \frac{1}{2}\nu^2$.

An important achievement of the reformulation of the free field equations in the form (26) is that this form suggests that the global HS symmetry algebra realized on the matter fields of mass $M(\nu)$ is $g = hs(2;\nu) \oplus hs(2;\nu)$ with the gauge fields (2). To simplify the formulation it is convenient to introduce two Clifford variables $\{\psi_i, \psi_j\} = 2\delta_{ij}$ $(i,j = 1,2)$ instead of ψ. One then introduces the full set of HS gauge fields as $W_\nu(q_\alpha, K, \psi_{1,2}|x)$ and realizes the gravitational fields as

$$W_\nu^{gr} = \frac{1}{4i}(\omega_\nu{}^{\alpha\beta} + h_\nu{}^{\alpha\beta}\psi_1)q_\alpha q_\beta \ . \tag{27}$$

The generating function for 0-forms is

$$C(q, K, \psi_{1,2}|x) = C^{mat}(q, K, \psi_1|x)\psi_2 + C^{aux}(q, K, \psi_1|x) \ . \tag{28}$$

Now let us consider the zero curvature equations

$$0 = R \equiv dW(q, K, \psi|x) + W(q, K, \psi|x) \wedge W(q, K, \psi|x) \tag{29}$$

along with the covariant constancy conditions in the adjoint representation of the HS algebra

$$\begin{aligned} 0 = dC(q, K, \psi|x) &+ W(q, K, \psi|x)C(q, K, \psi|x) \\ &- C(q, K, \psi|x)W(q, K, \psi|x) \ . \end{aligned} \tag{30}$$

Due to the factor of ψ_2 in front of C^{mat} the equations for C^{mat} turn out to be equivalent to the equations (26) in the gauge in which only the gravitational part (27) of the vacuum HS gauge fields is non-vanishing. The fields C^{aux} can be shown (Vasiliev (1994)) to be of a topological type so that each irreducible subsystem in this sector can describe at most a finite number of degrees of freedom and trivializes in a topologically trivial situation. Thus the effect of introducing a second Clifford element consists of addition of some topological fields.

5 Non-linear Dynamics

To describe non-linear HS dynamics of matter fields in 2+1 dimensions we start with a system of equations which is very close to that introduced in (Vasiliev (1992)) for a particular case of massless matter fields. We introduce three types of the generating functions $dx^\nu W_\nu(z_\alpha, y_\beta, K, \psi_i|x)$, $s_\gamma(z_\alpha, y_\beta, K, \psi_i|x)$ and $B(z_\alpha, y_\beta, K, \psi_i|x)$ which depend on the space-time variables x^μ and auxiliary variables (z_α,y_β,K,ψ_i) such that the two Clifford elements ψ_i commute to all other variables, while the bosonic spinor variables z_α and y_β commute to each other but anticommute with K , *i.e.* $\{K, z_\alpha\} = \{K, y_\alpha\} = 0$, $K^2 = 1$. Their physical content is as follows: $dx^\nu W_\nu$ is the generating function for HS gauge fields, B contains physical matter degrees of freedom along with some auxiliary variables, and s_γ is entirely auxiliary variable which allows one to formulate the full system of equations in a compact form.

This formulation is based on the following star-product law which endows the space of functions $f(z,y)$ with a structure of associative algebra

$$(f * g)(z,y) = (2\pi)^{-2} \int d^2u\, d^2v\, f(z+u, y+u) \times g(z-v, y+v) \exp i(u_\alpha v^\alpha) \ . \tag{31}$$

This product law provides a particular symbol realization of the Heisenberg–Weyl algebra. In particular one finds that $[y_\alpha, y_\beta]_* = -[z_\alpha, z_\beta]_* = 2i\epsilon_{\alpha\beta}$. The full system of equations has the form:

$$dW + W * \wedge W = 0\,, \quad ds_\alpha + \tilde{W} * s_\alpha - s_\alpha * W = 0\,,$$

$$dB + W * B - B * W = 0 \ , \tag{32}$$

and

$$\tilde{s}_\alpha * s_\beta - \tilde{s}_\beta * s_\alpha = -2i\epsilon_{\alpha\beta}(1 + \kappa * B)\,, \qquad \tilde{B} * s_\alpha - s_\alpha * B = 0 \ , \tag{33}$$

where

$$\tilde{a}(z,y,K,\psi_i|x) = a(z,y,-K,\psi_i|x) \quad \forall a \tag{34}$$

and $\kappa = K \exp i(z_\alpha y^\alpha)$ is a central element of the algebra which has vanishing star commutators with y_α, z_α, K and ψ_i.

The system of equations (32),(33) is explicitly invariant under the general coordinate transformations and the HS gauge transformations of the form

$$\delta W = d\epsilon + W * \epsilon - \epsilon * W, \quad \delta B = B * \epsilon - \epsilon * B, \quad \delta s_\alpha = s_\alpha * \epsilon - \tilde{\epsilon} * s_\alpha \tag{35}$$

To elucidate its physical content one has to analyze this system perturbatively near some vacuum solution. In the massless case the appropriate vacuum solution (Vasiliev (1992)) is

$$B_0 = 0\,, \quad s_{0\alpha} = z_\alpha\,, \quad W_0 = \omega(y, K, \psi_{1,2}) \tag{36}$$

with the vacuum gauge field ω satisfying the zero curvature condition $d\omega + \omega * \wedge\omega = 0$. It can be shown along the lines of (Vasiliev (1992)) that the system of equations (32),(33) expanded near this vacuum solution properly describes dynamics of massless matter fields on the free field level and beyond.

The main result of this report consists (Prokushkin and Vasiliev (in preparation)) of the observation that the same system (32), (33) expanded near another vacuum solution describes dynamics of matter fields with an arbitrary mass. This is a solution with

$$B_0 = \nu \ , \tag{37}$$

where ν is an arbitrary constant. For a constant field B_0 only the first of the equations (33) remains non-trivial. Remarkably it turns out to be possible to find its explicit solution

$$s_{0\alpha} = z_\alpha + \nu(z_\alpha - y_\alpha) \int_0^1 dt\ te^{it(z_\beta y^\beta)} K \tag{38}$$

(it is not too difficult to check that (38) satisfies (33) by a direct substitution). Now let us turn to the equations (32). The third of these equations is trivially satisfied. The second one reads

$$\tilde{W} * s_{0\alpha} - s_{0\alpha} * W = 0 \ , \tag{39}$$

where we have taken into account that $ds_{0\alpha} = 0$. Eq.(39) is a complicated integral equation. The key observation however is that it admits the following two particular solutions: $W_0 = q_\alpha$ $(\alpha = 1, 2)$,

$$q_\alpha = y_\alpha + \nu K(z_\alpha - y_\alpha) \int_0^1 dt\ (1-t) e^{itz_\beta y^\beta} \ . \tag{40}$$

Taking into account that $*$-product is associative it allows us to describe a general solution of (39) as an arbitrary element $W_0 = \omega(q_\alpha, K, \psi_{1,2}|x)$ whose arguments are treated as some non-commutative elements of the star-product algebra.

To make contact with the previous consideration it remains to check by explicit computation that the elements q_α indeed obey (2). Thus, the vacuum solution with a constant field (37) leads automatically to the deformed oscillator algebra with the deformed oscillators realized as some functions of z, y and K, i.e. as elements of the tensor product of two Heisenberg-Weyl algebras (equipped with the operator K). Finally, it remains to observe that the first of the equations (32) reduces to the zero curvature equation which describes the AdS background space. Since, as argued in the section 4, ν governs the parameter of mass of matter fields we arrive at the conclusion that a particular value of the parameter of mass is determined by a vacuum value of the field B.

Next one can analyze the full system of equations perturbatively by inserting the expansions of the form: $W = W_0 + W_1 + \ldots$, $B = B_0 + B_1 + \ldots$, $s_\alpha = s_{0\alpha} + s_{1\alpha} + \ldots$. In particular one can derive in the lowest orders that

$$B_1(z, y, K, \psi|x) = C(q, K, \psi|x) \ , \tag{41}$$

$$W_1(z, y, K, \psi|x) = \omega(q, K, \psi|x) + \Delta W_1(C) , \qquad s_{1\alpha} = s_{1\alpha}(C) \ , \tag{42}$$

where $s_{1\alpha}(C)$ and $\Delta W_1(C)$ are some functionals of the field C which remains arbitrary and has to be identified with generating function (28). Inserting this back into (32) one obtains the free field equations for C in the linearized approximation from the third equation and the equations of the form $d\omega + \omega * \wedge \omega - J(\omega, C^2) = 0$ from the first one where $J(\omega, C^2)$ is expected to describe HS currents (including the gravitational and spin - 1 ones).

6 Lorentz Covariance

After it is argued that the system (32), (33) describes properly HS dynamics in 2+1 dimensions let us explain what a physical principle fixes a particular form of these equations. Remarkably, this is a simple and physically important requirement that local Lorentz symmetry should be a particular symmetry of the equations.

Let us consider the following element of the algebra

$$L^{tot}_{\alpha\beta} = \frac{i}{4} \left(\{z_\alpha, z_\beta\}_* - \{y_\alpha, y_\beta\}_* \right) \ . \tag{43}$$

Infinitesimal local Lorentz transformations with a parameter $\eta^{\alpha\beta}$ are generated as $\delta Q(z, y) = [Q, \eta^{\alpha\beta} L^{tot}_{\alpha\beta}]_*$. Indeed these generators rotate properly the elementary spinor generating elements of the algebra, $\delta z_\alpha = \eta_{\alpha\beta} z^\beta$, $\delta y_\alpha = \eta_{\alpha\beta} y^\beta$, and therefore induce principal sp(2) transformations (automorphisms) of the whole algebra.

Although the argument above proves explicit local Lorentz invariance of the system of equations (32), (33), this symmetry is spontaneously broken due to the first of the constraints (33). Indeed, since the right hand side of this constraint has a non-vanishing vacuum value, s_α itself must have a non-vanishing vacuum value (38). The question therefore is whether there exists another local Lorentz symmetry which rotates properly spinor indices of the dynamical fields leaving invariant a vacuum solution. In fact the existence of such a Lorentz symmetry in all orders in interactions is a highly non-trivial property which fixes the constraints (33).

The point is that according to the analysis of the previous section after the constraints (33) are solved the generating functions for matter fields and gauge fields are described by arbitrary functions of only one spinor variable q_α, $C(q, K, \psi|x)$ and $\omega(q, K, \psi|x)$, respectively. In the linearized approximation the Lorentz generators which rotate properly q_α are $L_{\alpha\beta} = \frac{1}{4i}\{q_\alpha, q_\beta\}_*$.

Thus what we need is a proper generalization of these generators, to all orders in interactions. The constraints (33) indeed guarantee that such Lorentz generators $l_{\alpha\beta}$ can be constructed.

To see this we first observe that the constraints (33) give a particular realization of the deformed oscillator algebra (2). To this end it is convenient to introduce a new auxiliary generating element ρ which has the properties

$$\{\rho, K\} = 0\,, \qquad \rho^2 = 1\,. \tag{44}$$

Let us introduce a new variable $t_\alpha = \rho s_\alpha$. A role of ρ is that it compensates the twiddle operation (34) in (32) and (33) so that the constraints (33) take the form

$$t_\alpha * t_\beta - t_\beta * t_\alpha = -2i\epsilon_{\alpha\beta}(1+\kappa * B),$$

$$B * t_\alpha - t_\alpha * B = 0, \quad \kappa * t_\alpha + t_\alpha * \kappa = 0\,, \tag{45}$$

where $\kappa = K \exp i(z_\alpha y^\alpha)$ is a central element of the original algebra which now anticommutes with t_α due to (44). To make contact with (2) one identifies t_α, B and κ with iq_α, ν, and K, respectively [3]. As a consequence of the general property (5) one concludes that the elements

$$M_{\alpha\beta} = \frac{i}{4}\{t_\alpha, t_\beta\}_* = \frac{i}{4}(\tilde{s}_\alpha * s_\beta + \tilde{s}_\beta * s_\alpha) \tag{46}$$

obey the Lorentz commutation relations and rotate properly t_α. Now one can come back to the original ρ - independent variables arriving at the relations

$$\tilde{M}_{\alpha\beta} * s_\gamma - s_\gamma * M_{\alpha\beta} = \epsilon_{\alpha\gamma} s_\beta + \epsilon_{\beta\gamma} s_\alpha\,. \tag{47}$$

Let us now introduce the generators

$$l_{\alpha\beta} = L^{tot}_{\alpha\beta} - M_{\alpha\beta}\,. \tag{48}$$

From (33) it follows that

$$\delta B = [B\,, \eta^{\alpha\beta} l_{\alpha\beta}]_* = [B\,, \eta^{\alpha\beta} L^{tot}_{\alpha\beta}]_*\,, \tag{49}$$

i.e. $l_{\alpha\beta}$ rotate properly physical fields like $C(q, K, \psi|x)$ in all orders in interactions. Assuming that L^{tot} rotates properly s_α one concludes [4] that

$$\delta s_\alpha = s_\alpha * \eta^{\gamma\beta} l_{\gamma\beta} - \eta^{\gamma\beta}\tilde{l}_{\gamma\beta} * s_\alpha = \frac{\delta s_\alpha}{\delta B}\delta B \tag{50}$$

[3] Note that the vacuum solution $t_{0\alpha} = \rho s_{0\alpha}$ (38) therefore again describes an embedding of the deformed oscillator algebra into a (equipped) direct product of the two Heisenberg-Weyl algebras

[4] There is some gauge ambiguity in the generic solution of the first of the constraints (33) for s_α in terms of B. The assumption above is true when s_α is reconstructed entirely in terms of B without introducing any external constant spinors beyond those in the vacuum solution $s_{0\alpha}$.

and that the gauge transformations induced by $l_{\alpha\beta}$ satisfy sp(2) commutation relations. Also

$$\delta W = D(\eta^{\alpha\beta} l_{\alpha\beta}) = \mathrm{d}(\eta^{\alpha\beta}) l_{\alpha\beta} + [W, \eta^{\alpha\beta} L^{tot}_{\alpha\beta}] \tag{51}$$

because $\mathrm{d}(L^{tot}) = 0$ while $D(M_{\alpha\beta}) = 0$ as a consequence of the second of the equations (32). From (51) one concludes that the gauge field for a true local Lorentz symmetry is

$$W_L = \omega_L^{\alpha\beta} l_{\alpha\beta} \tag{52}$$

while all other gauge fields are rotated properly under the Lorentz transformations.

Let us emphasize that the above analysis guarantees Lorentz symmetry in all orders in interactions. Thus, it is the Lorentz symmetry principle which fixes a form of the equations and enforces appearance of the deformed oscillator algebra in the HS problem.

7 Concluding remarks

The proposed formulation of HS interactions admits interpretation of the parameter of mass as a module of the space of vacuum solutions, i.e. the same equations describe HS interactions of massive multiplets with different masses depending on a chosen vacuum solution. As a result different global HS symmetries of the linearized matter multiplets are different stability subgroups of the full HS symmetry, which leave invariant vacuum solutions. These global HS symmetries turn out to be pairwise non-isomorphic for different values of the parameter of mass. It is worth mentioning that the model under consideration (eq.(26)) possesses $N = 2$ supersymmetry osp(2; 2) with the generators (12). The constraints have a form of the deformed oscillator algebra as a consequence of the requirement that the equations of motion of matter fields interacting with HS fields must possess local Lorentz symmetry which is guaranteed by the properties (4) and (5).

The research described in this report was supported in part by the Russian Foundation for Basic Research, Grant No.96-01-01144 and by the European Community Commission under the contract INTAS, Grant No.93-633-*ext*.

References

Achucarro, A. and Townsend, P.K. (1986): Phys. Lett. **B180**, 89.
Akulov, V. P. and Volkov, D.V. (1974a): Teor. Mat. Fiz. **18**, 39;
Volkov, D.V. and Soroka, V.A. (1974b): Teor. Mat. Fiz. **20**, 291;
Akulov, V.P., Volkov, D.V. and Soroka, V.A. (1975): JETP Lett. **22**, 396.
Barabanschikov, A.V., Prokushkin, S.F. and Vasiliev, M.A. (1997): Teor. Mat. Fiz. **110**, 372-384; hep-th/9609034
Bergshoeff, E., de Wit, B. and Vasiliev, M.A. (1991): Nucl. Phys. **B366**, 315.

Bergshoeff, E., Blencowe, M. and Stelle, K. (1990): Comm. Math. Phys. **128**, 213.
Blencowe, M.P. (1989): Class. Quantum Grav. **6**, 443.
Bordemann, M., Hoppe, J. and Schaller, P. (1989): Phys. Lett. **232**, 199.
Brink, L. and Vasiliev, M. A. (1995): Nucl. Phys. **B457**, 273.
Feigin, B.L. (1988): Uspehi Mat. Nauk, **43**, 169.
Fradkin, E.S. and Vasiliev, M.A. (1987): *Ann. of Phys.* **177**, 63.
Plyushchay, M. (1997): contribution to this volume.
Prokushkin, S.F. and Vasiliev, M.A. (in preparation).
Vasiliev, M.A. (1988): Fortschr. Phys. **36**, 33;
Nucl. Phys. **301**, (1988) 26.
Vasiliev, M.A. (1989): JETP Lett. **50**, 374;
Vasiliev, M.A. (1991): Int. J. Mod. Phys. **A6**, 1115.
Vasiliev, M.A. (1992): Mod. Phys. Lett. **A7**, 3689.
Vasiliev, M.A. (1994): Class. Quant. Grav. **11**, 649.
Wigner, E. (1950): Phys. Rev. **77**, 711.
Witten, E. (1989): Nucl. Phys. **B311**, 46.
Yang, L.M. (1951): Phys. Rev. **84**, 788;
Boulware, D.G. and Deser, S. (1963): Il Nouvo Cimento, **XXX**, 231;
Mukunda, N., Sudarshan, E.C.G., Sharma, J.K. and Mehta, C.L. (1980): J. Math. Phys. **21**, 2386.

Universality of the R-Deformed Heisenberg Algebra

M.S. Plyushchay

Institute for High Energy Physics, Protvino, Moscow Region, 142284 Russia
Departamento de Fisica – ICE, Universidade Federal de Juiz de Fora
36036-330 Juiz de Fora, MG Brazil
E-mail: plyushchay@mx.ihep.su

Abstract. We show that deformed Heisenberg algebra with reflection emerging in parabosonic constructions is also related to parafermions. This universality is discussed in different algebraic aspects and is employed for the description of spin-j fields, anyons and supersymmetry in 2+1 dimensions.

1 Introduction

The R-deformed Heisenberg algebra (RDHA) is given by the generators a^-, a^+, R, and 1 satisfying the (anti)commutation relations

$$[a^-, a^+] = 1 + \nu R, \quad \{a^\pm, R\} = 0, \quad R^2 = 1, \quad [1, a^\pm] = [1, R] = 0 \ , \tag{1}$$

where $\nu \in \mathbb{R}$ is a deformation parameter and R is a reflection operator. It emerges in the context of quantization schemes generalizing bosonic commutation relations as follows. Let us consider a quantum mechanical system having the bosonic-like Hamiltonian, $H = \frac{1}{2}\{a^+, a^-\}$, and bosonic-like equations of motion,

$$\frac{1}{2}[\{a^-, a^+\}, a^\pm] = \pm a^\pm \ , \tag{2}$$

and put the question: what is the most general form of commutation relations for the operators a^+ and a^- which would lead to eqs. (2)? The answer is given by the R-deformed Heisenberg algebra (1) (Ohnuki, Kamefuchi (1982), Plyushchay (1997)). In this way, in fact, the deformed Heisenberg algebra (1) was introduced by Wigner (Wigner (1950)).

Trilinear commutation relations (2) characterize the *parabosonic* system of order $p = 1, 2, \ldots$ in the case when $\nu = p - 1 = 0, 1, \ldots$ (Ohnuki, Kamefuchi (1982), Macfarlane (1994)). Generalization of these trilinear commutation relations to the systems with many degrees of freedom as well as construction of their fermionic modification lead Green and Volkov to the discovery of parafields and parastatistics (Green (1953), Volkov (1959, 1960), Greenberg (1964), Greenberg, Messiah (1965), Govorkov (1983)). Recently the algebra (1) was rediscovered in the context of integrable systems (Polychronakos (1992)) where it was used for solving quantum mechanical Calogero model

(Polychronakos (1992), Brink, Hansson, Konstein, Vasiliev (1993)) (see also ref. (Yang (1951)). It was also employed for bosonization of supersymmetric quantum mechanics (Brzezinski, Egusquiza, Macfarlane (1993), Plyushchay (1994), Plyushchay (1996), Plyushchay (1996a)) and for describing anyons (Plyushchay (1996a), Plyushchay (1994a) within the framework of the group-theoretical approach (Plyushchay (1990), Jackiw, Nair (1991), Plyushchay (1991), Plyushchay (1991a, 1992), Sorokin, Volkov (1993)).

The existence of infinite-dimensional unitary representations of algebra (1) on the half-line $\nu > -1$ and the relationship between trilinear commutation relations and R-deformed Heisenberg algebra mean that the latter can be considered as the algebra supplying us with some generalization of parabosons for the case of non-integer statistical parameter $p = \nu + 1 > 0$ (Macfarlane (1994)). But it turns out that the algebra (1) obtained originally by generalizing *bosonic* commutation relations, has also finite-dimensional representations of the deformed (para)*fermionic* nature (Plyushchay (1997), Plyushchay (1996b)). Thus, the R-deformed Heisenberg algebra reveals some properties of universality which are the subject of the present talk based on recent papers (Plyushchay (1997), Plyushchay (1996b), Plyushchay (1997a)).

2 Aspects of Universality

Infinite-dimensional unitary representations of the algebra (1) taking place for $\nu > -1$ can be realized on the Fock space with complete orthonormal basis of states $|n\rangle = C_n (a^+)^n|0\rangle$, $n = 0, 1, \ldots$, $a^-|0\rangle = 0$, $\langle 0|0\rangle = 1$, $R|0\rangle = |0\rangle$, where $C_n = ([n]_\nu!)^{-1/2}$, $[n]_\nu! = \prod_{l=1}^n [l]_\nu$, $[l]_\nu = l + \frac{1}{2}(1 - (-1)^l)\nu$. The reflection operator is realized as $R = (-1)^N$, $N = \frac{1}{2}\{a^+, a^-\} - \frac{1}{2}(\nu + 1)$, $N|n\rangle = n|n\rangle$, and introduces Z_2-grading structure in the space of states, $R|k\rangle_\pm = \pm|k\rangle_\pm$, $|k\rangle_+ = |2k\rangle$, $|k\rangle_- = |2k + 1\rangle$, $k = 0, 1, \ldots$. The even, '+', and odd, '−', subspaces are separated by the projectors $\Pi_\pm = \frac{1}{2}(1 \pm R)$.

Due to the commutation relation

$$[a^-, (a^+)^n] = \left(n + \frac{1}{2}(1 - (-1)^n)\nu R\right)(a^+)^{n-1} , \qquad (3)$$

at special values of the deformation parameter, $\nu = -(2p + 1)$, $p = 1, 2, \ldots$, the relation $\langle\langle m|n\rangle\rangle = 0$, $|n\rangle\rangle \equiv (a^+)^n|0\rangle$, holds for $n \geq 2p + 1$ and arbitrary m. Therefore, there are $(2p+1)$-dimensional irreducible representations of the algebra (1) with $\nu = -(2p+1)$, in which the relations $(a^+)^{2p+1} = (a^-)^{2p+1} = 0$ are valid. The latter relations are a characteristic property of parafermions of order $2p$, and we arrive at the nilpotent algebra

$$[a^-, a^+] = 1 - (2p + 1)R, \quad (a^\pm)^{2p+1} = 0, \quad \{a^\pm, R\} = 0,$$
$$R^2 = 1, \quad p = 1, 2, \ldots . \qquad (4)$$

Operator a^+ can be interpreted here as a paragrassmann variable θ, $\theta^{2p+1} = 0$, whereas operator a^- can be considered as corresponding differentiation operator ∂ defined by relation (3). Therefore, the algebra (1) at $\nu = -(2p+1)$ can be considered as a paragrassmann algebra of order $2p$ (Filipov, Isaev, Kurdikov (1992, 1993), Fleury, Rausch de Traubenberg (1992, 1994)) with a special differentiation operator. It was called in ref. (Plyushchay (1996b)) the R-paragrassmann algebra.

The finite-dimensional representations can be realized as matrix representations with diagonal operator $R = \mathrm{diag}(+1, -1, +1, \ldots, -1, +1)$, and with operators $a^{\pm}$ realized as $(a^+)_{ij} = A_j \delta_{i-1,j}$, $(a^-)_{ij} = B_i \delta_{i+1,j}$, where $A_{2k+1} = -B_{2k+1} = \sqrt{2(p-k)}$, $k = 0, 1, \ldots, p-1$, $A_{2k} = B_{2k} = \sqrt{2k}$, $k = 1, \ldots, p$. Operators a^+ and a^- are mutually conjugate, $(\Psi_1, a^- \Psi_2)^* = (\Psi_2, a^+ \Psi_1)$, with respect to the indefinite scalar product

$$(\Psi_1, \Psi_2) = \bar{\Psi}_{1n} \Psi_2^n \ , \quad \bar{\Psi}_n = \Psi^{*k} \hat{\eta}_{kn} \ , \tag{5}$$

where $\Psi^n = \langle n | \Psi \rangle$ and $\hat{\eta} = \mathrm{diag}(1, -1, -1, +1, +1, \ldots, (-1)^p, (-1)^p)$ is indefinite metric operator.

Finite-dimensional representations of RDHA can also be described in terms of hermitian conjugate operators $f^+ = a^+$, $f^- = a^- R$ and of positive definite scalar product, $\langle \Psi_1, \Psi_2 \rangle = \Psi_1^{*n} \Psi_2^n$. In terms of these, nilpotent algebra (4) is

$$\{f^+, f^-\} = (2p+1) - R, \quad (f^{\pm})^{2p+1} = 0, \quad \{R, f^{\pm}\} = 0, \quad R^2 = 1 \ . \tag{6}$$

Operators $I_+ = f^+$, $I_- = f^-$ generate a nonlinear deformation of su(2) algebra of the form

$$[I_+, I_-] = 2I_3(-1)^{I_3+p}, \quad [I_3, I_{\pm}] = \pm I_{\pm} \tag{7}$$

involving the reflection operator $R = (-1)^{I_3+p}$. For $p = 1$ one has the relation $I_3(-1)^{I_3+1} = I_3$, and in this particular case the deformed su(2) algebra turns into the standard su(2). Relations (7) can be presented as (Plyushchay (1997))

$$[[f^+, f^-], f^{\pm}] = 2I_3(2I_3 \mp 1)(-1)^{I_3+p} f^{\pm} \tag{8}$$

with $I_3 = C(-1)^{C+p}$, $C = \frac{1}{2}[f^+, f^-]$, that gives a deformation of parafermionic algebra of order $2p$. At $p = 1$ this algebra turns into the standard parafermionic algebra of order 2.

The (anti)commutation relations (6), (7) can be presented as $\{f^+, f^-\} = F(N+1) + F(N)$, $[f^-, f^+] = F(N+1) - F(N)$, with function $F(N) = N(-1)^N + (p + \frac{1}{2})(1 - (-1)^N)$, where $N = I_3 + p$ is the number operator. This means that deformed parafermionic algebra (8) belongs to the class of generalized deformed parafermionic algebras introduced by Quesne (Quesne (1994)). Note also that the deformed su(2) algebra (7) can be realized via the generators of the standard su(2) algebra, $[J_+, J_-] = 2J_3$, $[J_3, J_{\pm}] = \pm J_{\pm}$, taken in $(2p+1)$-dimensional representation, $J_3^2 + \frac{1}{2}\{J_+, J_-\} = p(p+1)$, according to the prescription (Curtright, Zachos (1990), Polychronakos

(1990), Roček (1991), Plyushchay (1997)): $I_3 = J_3$, $I_- = (I_+)^\dagger = J_-\Phi(J_3)$ with $\Phi(J_3) = [2p+1+(-1)^N(2J_3-1)]\cdot[2N(p-J_3+1)]^{-1}$.

The linear combinations of the initial creation-annihilation operators, $\mathcal{L}_1 = \frac{1}{\sqrt{2}}(a^+ + a^-)$ and $\mathcal{L}_2 = \frac{i}{\sqrt{2}}(a^+ - a^-)$, satisfy the commutation relations $[\mathcal{L}_\alpha, \mathcal{L}_\beta] = i\epsilon_{\alpha\beta}(1+\nu R)$, $\epsilon_{\alpha\beta} = -\epsilon_{\beta\alpha}$, $\epsilon_{12} = 1$, and are hermitian in the case of infinite-dimensional representations and self-conjugate with respect to the scalar product (5) in the case of finite dimensional representations of RDHA. These operators together with quadratic operators J_μ, $\mu = 0,1,2$, $J_0 = \frac{1}{4}\{a^+,a^-\}$, $J_1 \pm iJ_2 = J_\pm = \frac{1}{2}(a^\pm)^2$, form the set of generators of osp(1|2) superalgebra: $\{\mathcal{L}_\alpha, \mathcal{L}_\beta\} = 4i(J\gamma)_{\alpha\beta}$, $[J_\mu, J_\nu] = -i\epsilon_{\mu\nu\lambda}J^\lambda$, $[J_\mu, \mathcal{L}_\alpha] = \frac{1}{2}(\gamma_\mu)_\alpha{}^\beta\mathcal{L}_\beta$. Here γ-matrices appear in the Majorana representation, $(\gamma^0)_\alpha{}^\beta = -(\sigma^2)_\alpha{}^\beta$, $(\gamma^1)_\alpha{}^\beta = i(\sigma^1)_\alpha{}^\beta$, $(\gamma^2)_\alpha{}^\beta = i(\sigma^3)_\alpha{}^\beta$. Hence, J_μ are even and $\mathcal{L}_\alpha$ are odd generators of the superalgebra with J_μ forming $so(2,1) \sim \mathrm{sl}(2,R)$ subalgebra and $\mathcal{L}_\alpha$ being an $so(2,1)$ spinor. The osp(1|2) Casimir operator $\mathcal{C} \equiv J^\mu J_\mu - \frac{i}{8}\mathcal{L}^\alpha\mathcal{L}_\alpha$, $J^\mu = \eta^{\mu\nu}J_\nu$, $\eta^{\mu\nu} = \mathrm{diag}(-,+,+)$, $\mathcal{L}^\alpha = \epsilon^{\alpha\beta}\mathcal{L}_\beta$, $\epsilon_{\alpha\gamma}\epsilon^{\gamma\beta} = -\delta_\alpha^\beta$, takes the fixed value $\mathcal{C} = \frac{1}{16}(1-\nu^2)$. Therefore, every infinite- or finite-dimensional representation of RDHA supplies us with the corresponding irreducible representation of osp(1|2) superalgebra revealing its universality in some another but related aspect (Plyushchay (1997)). Every such representation is reducible with respect to the action of the $so(2,1)$ generators J_μ: $J^2 = J^\mu J_\mu = -\hat{\alpha}(\hat{\alpha}-1)$, where $\hat{\alpha} = \frac{1}{4}(1+\nu R)$. Hence, J_μ act irreducibly on even, '+', and odd, '−', subspaces spanned by the states $|k\rangle_+$ and $|k\rangle_-$, $J^2|k>_\pm = -\alpha_\pm(\alpha_\pm - 1)|k>_\pm$, where $\alpha_+ = \frac{1}{4}(1+\nu)$, $\alpha_- = \alpha_+ + 1/2$ and $J_0|k>_\pm = (\alpha_\pm + k)|k>_\pm$, $k = 0,1,\ldots$. For infinite-dimensional representations of RDHA ($\nu > -1$), this gives the direct sum of infinite-dimensional unitary irreducible representations of $\mathrm{sl}(2,R)$, $\mathcal{D}^+_{\alpha_+} \oplus \mathcal{D}^+_{\alpha_-}$, being half-bounded representations of the discrete series characterized by parameters $\alpha_+ > 0$ and $\alpha_- > 1/2$ (Plyushchay (1993)). In the case of finite-dimensional representations of RDHA one gets $J^2|l>_\pm = -j_\pm(j_\pm+1)|l>_\pm$, $j_+ = p/2$, $j_- = (p-1)/2$, where $l = 0,1,\ldots,p$ for $|l>_+$ and $l = 0,1\ldots,p-1$ for $|l>_-$. Thus, we have the direct sum of spin-j_+ and spin-j_- finite-dimensional representations with $so(2,1)$ spin parameter shifted in 1/2, where the operator J_0 has the spectra $j_0 = (-j_+, -j_+ + 1, \ldots, j_+)$ and $j_0 = (-j_-, -j_- + 1, \ldots, j_-)$ (Plyushchay (1996b)).

The described osp(1|2) superalgebraic construction can be extended to the OSp(2|2) supersymmetry. Indeed, defining the operators $\Delta = -\frac{1}{2}(R+\nu)$ and $Q^+ = a^+\Pi_-$, $Q^- = a^-\Pi_+$, $S^+ = a^+\Pi_+$, $S^- = a^-\Pi_-$, we find that operators J_μ and Δ are even generators of osp(2|2) superalgebra, forming its $\mathrm{sl}(2,R)\times u(1)$ subalgebra, whereas $Q^\pm$ and $S^\pm$ are its odd generators (Plyushchay (1996a), Plyushchay (1996b)). In terms of the latter, osp(1|2) odd generators are presented as $\mathcal{L}_1 = \frac{1}{\sqrt{2}}(Q^+ + Q^- + S^+ + S^-)$, $\mathcal{L}_2 = \frac{i}{\sqrt{2}}(Q^+ - Q^- + S^+ - S^-)$. This more broad OSp(2|2) supersymmetry was revealed in ref. (Plyushchay (1996a)) as a dynamical symmetry of bosonized supersymmetric quantum mechanical 2-body Calogero model. Let us also note that the pair of odd generators Q^+ and Q^- together with even generator $H_+ = 2J_0 + \Delta$ form $s(2)$ superalgebra, $Q^{\pm 2} = 0$, $\{Q^+, Q^-\} = H_+$, $[Q^\pm, H_+] = 0$, whereas S^+ and S^- are odd generators of $s(2)$ superalgebra with even generator $H_- = 2J_0 - \Delta$. This latter observation (Brzezinski, Egusquiza, Macfarlane (1993), Plyushchay (1994)) was a starting point for bosonization of Witten supersymmetric quantum mechanics (Witten (1981, 1982)) realized in ref. (Plyushchay (1996)).

To conclude the discussion of the formal algebraic aspects, let us note that RDHA can be presented in the form related to the generalized statistics (Scipioni (1994), de Falco, Mignani, Scipioni (1995)). To this end, let us define the operators $c^- = a^- G_\nu^{-1/2}(R)$, $c^+ = G_\nu^{-1/2}(R)a^+$, where $G_\nu(R) = |1-\nu R|$, $\nu \neq 1$. They give the normalized form of the RDHA (Plyushchay (1996b)), $c^-c^+ - g_\nu c^+c^- = 1$, with $g_\nu = (1-\nu)^R(1+\nu)^{-R}$ in the case $-1 < \nu < 1$, or $c^-c^+ - g_\nu c^+c^- = R$ with $g_\nu = (\nu-1)^R(1+\nu)^{-R}$ for $\nu > 1$ and $g_\nu = p^R(1+p)^{-R}$ for $\nu = -(2p+1)$ (cp. with guon commutation relations (Scipioni (1994), de Falco, Mignani, Scipioni (1995))). The limit $|\nu| \to \infty$ in both cases $\nu > 1$ and $\nu = -(2p+1)$ leads to the algebra

$$c^-c^+ - c^+c^- = R, \quad \{R, c^\pm\} = 0, \quad R^2 = 1 \ . \tag{9}$$

This algebra has two-dimensional irreducible representation with reflection operator realized as $R = 1 - 2c^+c^-$ that reduces eqs. (9) to the standard fermionic anticommutation relations, $\{c^+, c^-\} = 1$, $c^{\pm 2} = 0$. Therefore, the fermionic algebra can be obtained as a limit case of RDHA. The commutation relations (9) can be generalized to the algebra with phase operator (Plyushchay (1996b)),

$$[a, \bar{a}] = \mathcal{R}, \quad \mathcal{R}^p = 1, \quad \mathcal{R}a = qa\mathcal{R}, \quad \mathcal{R}\bar{a} = q^{-1}\bar{a}\mathcal{R} \ , \tag{10}$$

where $q = e^{-i2\pi/p}$, $p = 2, 3, \ldots$. Via the transformation $c = q^{-1/2}a\mathcal{R}^{-1/2}$, $\bar{c} = q^{-1/2}\mathcal{R}^{-1/2}\bar{a}$, the algebra (10) can be related to the q-deformed Heisenberg algebra $c\bar{c} - q\bar{c}c = 1$ (Arik, Coon (1976), Kuryshkin (1980), Macfarlane (1989), Biedenharn (1989)) with deformation parameter q being the primitive root of unity.

3 Universal Spinor Set of Field Equations and 3D SUSY

Now we turn to some applications of the RDHA algebra (Plyushchay (1996b), Plyushchay (1997a)). Namely, we shall employ the described infinite- and finite-dimensional representations for the construction of the universal minimal set of linear differential equations describing either ordinary integer and half-integer $(2+1)$-dimensional spin-j fields or anyons. We shall also generalize the construction for the case of some 3D SUSY field systems and find the corresponding field actions.

So, let us introduce the field $\Psi = \Psi^n(x)$ carrying some irreducible representation of RDHA. The corresponding total angular momentum operator has the form $M_\mu = -\epsilon_{\mu\nu\lambda}x^\nu P^\lambda + J_\mu$, $P_\mu = -i\partial_\mu$, and generates Lorentz transformations, $\Psi(x) \to \Psi'(x') = \exp(\mathrm{i}M^\mu\omega_\mu)\Psi(x)$, which are specified by parameters ω_μ. In the chosen fixed representation of RDHA we introduce the self-conjugate spinor operator

$$\mathcal{Q}_\alpha = \mathcal{Q}_\alpha^\dagger = R\mathcal{P}_\alpha + \epsilon m\mathcal{L}_\alpha \tag{11}$$

with $\epsilon = \pm$, $\mathcal{P}_\alpha = (P\gamma)_\alpha{}^\beta\mathcal{L}_\beta$ and m a mass parameter. Now we postulate the covariant set of field equations,

$$\mathcal{Q}_\alpha\Psi = 0\ . \tag{12}$$

To clarify the physical content of the field Ψ subject to eq. (12), we decompose it into irreducible sl$(2,R)$ components $\Psi_\pm$, $\Psi = \Psi_+ + \Psi_-$, $R\Psi_\pm = \pm\Psi_\pm$. As a consequence of the equations (12) and the (anti)commutation relations

$$\begin{aligned}
&\{\mathcal{Q}_\alpha, \mathcal{Q}_\beta\} = 4\mathrm{i}(P^2+m^2)(J\gamma)_{\alpha\beta} - 8\mathrm{i}(\Delta_+\Pi_+ + \Delta_-\Pi_-)(P\gamma)_{\alpha\beta},\\
&[\mathcal{Q}_\alpha, \mathcal{Q}_\beta] = -(\mathcal{Q}^\rho\mathcal{Q}_\rho)\epsilon_{\alpha\beta},\\
&\mathcal{Q}^\rho\mathcal{Q}_\rho = \mathrm{i}(P^2+m^2)(1+\nu R) + 8\mathrm{i}\epsilon m(\Delta_+\Pi_+ - \Delta_-\Pi_-)\ ,
\end{aligned}$$

$\Delta_\pm = PJ - \epsilon m\frac{1}{4}(\nu \pm 1)$, we find that $\Psi_- = 0$, whereas 'even' component Ψ_+ satisfies Klein-Gordon, $(P^2+m^2)\Psi_+ = 0$, and spin, $(PJ - s_+m)\Psi = 0$, equations, where the spin value is defined by the deformation parameter corresponding to the chosen irreducible representation of RDHA, $s_+ = \epsilon\frac{1}{4}(1+\nu)$. Hence, the spinor set of equations (12) describes ordinary integer, $s_+ = -\epsilon k$, or half-integer, $s_- = -\epsilon(k+1/2)$, spin fields under the choice of $\nu = -(4k+1)$, $k = 1, 2, \ldots$, and $\nu = -(4k+3)$, $k = 0, 1, \ldots$, respectively. On the other hand, in the case of infinite-dimensional representations ($\nu > -1$), the basic equations describe the fields with arbitrary value of spin $s_+ \in \mathbf{R}$, $s_+ \neq 0$. In this case the spin equation, $(PJ - s_+m)\Psi = 0$, is the (2+1)-dimensional analog of the Majorana equation (Plyushchay (1991), Majorana (1932), Rühl (1967)), whereas the basic spinor set of equations (12) is some analog of the 4D Dirac positive-energy set of linear differential equations (Dirac (1971, 1972)).

Therefore, the R-deformed Heisenberg algebra gives a possibility to construct a universal minimal spinor set of linear differential equations (11), (12) describing spin-j fields and anyons (Plyushchay (1997a)).

As we have seen, any irreducible representation of the osp$(1|2)$ superalgebra is the direct sum of two irreducible representations of $so(2,1)$ subalgebra. The latter are specified by the parameters α_+ and α_-, or j_+ and j_- related as $\alpha_- - \alpha_+ = j_+ - j_- = 1/2$. So, one can try to modify the set of linear differential equations (11), (12) in such a way that it would have nontrivial solutions not only in '+' subspace, but also in '−' subspace. If these states will have equal mass but their spin will be shifted for $\Delta s = 1/2(\mathrm{mod}\, n)$, we shall have the states forming a supermultiplet. Such a modification can be realized, but only for two special cases corresponding to $\nu = -5$ and $\nu = -7$. The reason is that the consistent system of Klein-Gordon and spin equations cannot be obtained independently in the odd subspace. But for those two cases one can arrive at the linear spin equation only which itself will generate the Klein-Gordon equation. Such corresponding linear spin equation is the Dirac equation or the equation for topologically massive vector U(1) gauge field (Jackiw, Templeton (1981), Schonfeld (1981), Deser, Jackiw, Templeton (1982)). We present here the final result whereas the details can be found in ref. (Plyushchay (1997a)).

The self-conjugate spinor linear differential operator $\mathcal{D}_\alpha = \mathcal{D}_\alpha^\dagger$ generating the corresponding set of equations

$$\mathcal{D}_\alpha \Psi = 0 \tag{13}$$

is given by

$$\mathcal{D}_\alpha = (p - \frac{1}{2})\epsilon m \mathcal{L}_\alpha - \mathcal{J}_\alpha + \frac{1}{2}(1 - pR)\mathcal{P}_\alpha \ , \tag{14}$$

where $p = 2, 3$ correspond to $\nu = -5, -7$. In these two cases we have supermultiplets with spin content $(s_+ = \epsilon,\ s_- = \frac{1}{2}\epsilon)$ for $\nu = -5$ and $(s_+ = \frac{3}{2}\epsilon,\ s_- = \epsilon)$ for $\nu = -7$. It is interesting to note that in both cases, $p = 2, 3$, the spinor supercharge operator $\mathcal{Q}_\alpha$, $\{\mathcal{Q}_\alpha, \mathcal{D}_\beta\} \approx 0$, which transforms the corresponding supermultiplet components, coincides with operator (11) taken in 5- and 7-dimensional representations. The weak equality means that the left hand side turns into zero on the physical subspace given by eqs. (13), and we have the following typical superalgebraic relations: $\{\mathcal{Q}_\alpha, \mathcal{Q}_\beta\} \approx -4i\epsilon m(p+2)(P\gamma)_{\alpha\beta}$, $[P_\mu, \mathcal{Q}_\alpha] = 0$. If we not require that $\mathcal{D}_\alpha^\dagger = \mathcal{D}_\alpha$, we shall have a possibility to get supermultiplets with the spin content $(s_+ = \epsilon,\ s_- = -\frac{1}{2}\epsilon)$ and with the spin shift $|\Delta s| = |s_+ - s_-| = 3/2$ for $\nu = -5$, and $(s_+ = \frac{3}{2}\epsilon,\ s_- = -\epsilon)$ with $|\Delta s| = 5/2$ for $\nu = -7$. The corresponding operators, having the property $\mathcal{D}_\alpha^\dagger = \mathcal{D}_\alpha + \mathcal{P}_\alpha$, are given by

$$\begin{aligned} \mathcal{D}_\alpha &= -\frac{1}{2}\epsilon m \mathcal{L}_\alpha - \mathcal{J}_\alpha - \frac{1}{2}R\mathcal{P}_\alpha, \quad \nu = -5 \ ; \\ \mathcal{D}_\alpha &= -\frac{1}{2}\epsilon m \mathcal{L}_\alpha - \frac{1}{2}\mathcal{J}_\alpha - \frac{1}{4}(1 + R)\mathcal{P}_\alpha, \quad \nu = -7 \ . \end{aligned} \tag{15}$$

Since the universal equations (12) as well as equations for 3D SUSY field systems (13) are two equations for one multi- or infinite-component field, the corresponding action $\mathcal{A} = \int L d^3x$ has to contain some auxiliary fields. The Lagrangian leading to the equations (12) for basic field Ψ can be chosen as (Plyushchay (1997a))

$$L = (\bar{\Psi} + \bar{\chi}^\alpha \mathcal{Q}_\alpha)(\Psi + \mathcal{Q}_\beta \chi^\beta) - \bar{\Psi}\Psi \ , \tag{16}$$

where $\bar{\Psi} = \Psi^\dagger$, $\bar{\chi}^\alpha = \chi^{\dagger\alpha}$ for $\nu > -1$, whereas $\bar{\Psi} = \Psi^\dagger \hat{\eta}$, $\bar{\chi}^\alpha = \chi^{\dagger\alpha}\hat{\eta}$ for $\nu = -(2p+1)$, that guarantees the reality of the Lagrangian. Besides the basic equations (12), the action leads to the equations $\mathcal{Q}_\alpha \chi^\alpha = 0$ and is invariant with respect to the transformations $\chi^\alpha \to \chi^\alpha + \Pi^{\alpha\beta}\Lambda_\beta$, where $\Pi_{\alpha\beta} = (\mathcal{Q}^\sigma \mathcal{Q}_\sigma)\epsilon_{\alpha\beta} - \mathcal{Q}_\alpha \mathcal{Q}_\beta$, $\mathcal{Q}^\alpha \Pi_{\alpha\beta} = 0$. This means that fields χ^α play the role of auxiliary fields having no independent dynamics. The same equations can be obtained also from the action with the linear form of Lagrangian,

$$L' = \bar{\chi}^\alpha \mathcal{Q}_\alpha \Psi + \bar{\Psi} \mathcal{Q}_\alpha \chi^\alpha \ . \tag{17}$$

The change of $\mathcal{Q}_\alpha$ for $\mathcal{D}_\alpha$ in Lagrangians (16) and (17) will give the action functionals for the special SUSY cases (14) corresponding to $\nu = -5$ and $\nu = -7$. For SUSY systems given by eq. (15), the corresponding quadratic and linear Lagrangians are $L = (\bar{\Psi} + \bar{\chi}^\alpha \mathcal{D}_\alpha)(\Psi + \mathcal{D}^\dagger_\beta \chi^\beta) - \bar{\Psi}\Psi$ and $L' = \bar{\chi}^\alpha \mathcal{D}_\alpha \Psi + \bar{\Psi} \mathcal{D}^\dagger_\alpha \chi^\alpha$.

The described first and second order field Lagrangians can be used as a departing point for realization of the second quantization of the corresponding (2+1)-dimensional field systems. This, in particular, could clarify the question on the spin-statistics relation for the described fractional spin fields.

References

Arik, M. and Coon, D.D. (1976): *J. Math. Phys.* **17**, 524;
Kuryshkin, V. (1980): *Ann. Fond. L. de Broglie* **5**, 111;
Macfarlane, A.J. (1989): *J. Phys. A* **22**, 4581;
Biedenharn, L.C. (1989): *J. Phys. A* **22**, L873.

Brink, L., Hansson, T.H, Konstein, S. and Vasiliev, M.A. (1993): *Nucl. Phys.* **B401**, 591.

Brzezinski, T. Egusquiza, I.L. and Macfarlane, A.J. (1993): *Phys. Lett.* **B311**, 202.

Curtright, T. and Zachos, C. (1990): *Phys. Lett.* **B243**, 237;
Polychronakos, A.P. (1990): *Mod. Phys. Lett.* **A5**, 2325;
Roček, M. (1991): *Phys. Lett.* **B255**, 554.

Dirac, P.A.M. (1971, 1972): *Proc. Roy. Soc. London Ser. A* **322**, 435, **328**, 1572.

Filippov, A.T., Isaev, A.P. and Kurdikov, A.P. (1992, 1993): *Mod. Phys. Lett.* **A7**, 2129; *Teor. Mat. Fiz.* **94**, 213;
Fleury, N. and Rausch de Traubenberg, M. (1992, 1994): *J. Math. Phys.* **33**, 3356; *Adv. Appl. Clif. Alg.* **4**, 123.

Green, H.S. (1953): *Phys. Rev.* **90**, 270.

Greenberg, O.W. (1964): *Phys. Rev. Lett.* **13**, 598;
Greenberg, O.W. and Messiah, A.M.L. (1965): *Phys. Rev.* **B138**, 1155;
Govorkov, A.B. (1983): *Theor. Math. Phys.* **54**, 234.

Jackiw, R. and Nair, V.P. (1991): *Phys. Rev.* **D43**, 1933.

Jackiw, R. and Templeton, S. (1981): *Phys. Rev.* **D23**, 2291;
Schonfeld, J. (1981) *Nucl. Phys.* **B185**, 157;
Deser, S., Jackiw, R. and Templeton, S. (1982): *Phys. Rev. Lett.* **48**, 975.

Macfarlane, A.J. (1994): *Generalized Oscillator Systems and Their Parabosonic Interpretation*, in: Proc. Inter. Workshop on Symmetry Methods in Physics, eds. A. N. Sissakian, G. S. Pogosyan and S. I. Vinitsky (JINR, Dubna, 1994) p. 319.

Majorana, E. (1932): *Nuovo Cim.* **9**, 335;
Rühl, W. (1967): *Comm. Math. Phys.* **6**, 312.
Ohnuki, Y. and Kamefuchi, S. (1982): *Quantum Field Theory and Parastatistics*, University Press of Tokyo, 1982.
Plyushchay, M.S. (1990): *Phys. Lett.* **B248**, 107.
Plyushchay, M.S. (1991): *Phys. Lett.* **B262**, 71; *Nucl. Phys.* **B362**, 54.
Plyushchay, M.S. (1991a, 1992): *Phys. Lett.* **B273**, 250; *Int. J. Mod. Phys.* **A7**, 7045.
Plyushchay, M.S. (1993): *J. Math. Phys.* **34**, 3954.
Plyushchay, M.S. (1994): *Supersymmetry without Fermions*, **hep-th/9404081**.
Plyushchay, M.S. (1994): *Phys. Lett.* **B320**, 91.
Plyushchay, M.S. (1996): *Mod. Phys. Lett.* **A11**, 397.
Plyushchay, M.S. (1996a): *Ann. Phys.* **245**, 339.
Plyushchay, M.S. (1996b): *Mod. Phys. Lett.* **A11**, 2953.
Plyushchay, M.S. (1997): *Nucl. Phys.* **B491**, 619 [**hep-th/9701091**].
Plyushchay, M.S. (1997a): *R-deformed Heisenberg algebra, anyons and* $d = 2 + 1$ *supersymmetry*, **hep-th/9705034**, to appear in *Mod. Phys. Lett.* **A12**.
Polychronakos, A.P. (1992): *Phys. Rev. Lett.* **69**, 703.
Quesne, C. (1994): *Phys. Lett.* **A193**, 245.
Scipioni, R. (1994): *Phys. Lett.* **B327**, 56; *Nuovo Cim.* **B109**, 479; De Falco, L., Mignani, R. and Scipioni, R. (1995): *Nuovo Cim.* **A108**, 1029.
Sorokin, D.P. and Volkov, D.V. (1993): *Nucl. Phys.* **B409**, 547.
Volkov, D.V. (1959, 1960): *Sov. Phys. JETP* **9**, 1107; **9**, 375.
Wigner, E.P. (1950): *Phys. Rev.* **77**, 711.
Witten, E. (1981, 1982): *Nucl. Phys.* **B188**, 513; **B202**, 253.
Yang, L.M. (1951): *Phys. Rev.* **84**, 788.

The Dual Algebra of the Jordanian $GL_{g,h}(2)$

B.L. Aneva, V.K. Dobrev and S.G. Mihov

Bulgarian Academy of Sciences
Institute of Nuclear Research and Nuclear Energy
72 Tsarigradsko Chaussee
1784 Sofia, Bulgaria

1 Introduction

The group GL(2) admits two distinct quantum group deformations with central quantum determinant: $GL_q(2)$ (Drinfeld (1985)) and $GL_h(2)$ (Demidov et al. (1990)), (Zakrzewski (1991)). These are the only possible such deformations (up to isomorphism) (Kupershmidt (1992)). Both may be viewed as special cases of two parameter deformations: $GL_{p,q}(2)$ (Demidov et al. (1990)) and $GL_{g,h}(2)$ (Aghamohammadi (1993)). In the initial years of studies of quantum groups mostly $GL_q(2)$ and $GL_{p,q}(2)$ were studied. More recently started studies on $SL_h(2)$ and its dual quantum algebra $U_h(sl(2))$ (Ohn (1992)). In particular, aspects of differential calculus (Aghamohammadi (1993)), and differential geometry (Karimipour (1994)) were developed, the universal R-matrix for $U_h(sl(2))$ was given in (Vladimirov (1993)), (Khorrami et al. (1995)), (Ballesteros and Herranz (1996)), the representations of $U_h(sl(2))$ were studied in (Dobrev (1996)), (Abdesselam et al. (1996)), (Aizawa (1997)), (Van der Jeugt (1997)). However, there are no studies until now of the two-parameter Jordanian matrix quantum group $GL_{g,h}(2)$. Even the dual of this algebra is not known.

This is the problem discussed in this contribution, which follows mostly (Aneva et al. (1997)). We give the Hopf algebra $\mathcal{U}_{g,h}$ dual to the Jordanian matrix quantum group $GL_{g,h}(2)$. As an algebra it depends on the sum $\tilde{g} = (g+h)/2$ of the two parameters and is split in two subalgebras: $\mathcal{U}'_{g,h}$ (with three generators) and $U(\mathcal{Z})$ (with one generator). The subalgebra $U(\mathcal{Z})$ is a central Hopf subalgebra of $\mathcal{U}_{g,h}$. The subalgebra $\mathcal{U}'_{g,h}$ is not a Hopf subalgebra and its coalgebra structure depends on both parameters. We discuss also two interesting one-parameter special cases: $g = h$ and $g = -h$. The subalgebra $\mathcal{U}'_{h,h}$ is a Hopf algebra and coincides with the algebra introduced by Ohn as the dual of $SL_h(2)$. The subalgebra $\mathcal{U}'_{-h,h}$ is isomorphic to $U(sl(2))$ as an algebra but has a nontrivial coalgebra structure and again is not a Hopf subalgebra of $\mathcal{U}_{-h,h}$.

2 Jordanian Matrix Quantum Group $\mathrm{GL}_{g,h}(2)$

In this Section we recall the Jordanian two parameter deformation $\mathrm{GL}_{g,h}(2)$ of GL(2) introduced in (Aghamohammadi (1993)). One starts with a unital associative algebra generated by four elements a, b, c, d of a quantum matrix $M = \begin{pmatrix} a & b \\ c & d \end{pmatrix}$ with the following relations $(g, h \in \mathbb{C})$:

$$\begin{gathered} [a,c] = gc^2, \qquad [d,c] = hc^2, \qquad [a,d] = gdc - hac \\ [a,b] = h(\mathcal{D} - a^2), \qquad [d,b] = g(\mathcal{D} - d^2), \qquad [b,c] = gdc + hac - ghc^2 \\ \mathcal{D} = ad - bc + hac = ad - cb - gdc + ghc^2 \ , \end{gathered} \tag{1}$$

where $\mathcal{D}$ is a multiplicative quantum determinant which is not central (unless $g = h$).

The above algebra is turned into a bialgebra $A_{g,h}(2)$ with the standard GL(2) co-product δ and co-unit ε :

$$\delta\left(\begin{pmatrix} a & b \\ c & d \end{pmatrix}\right) = \begin{pmatrix} a \otimes a + b \otimes c & a \otimes b + b \otimes d \\ c \otimes a + d \otimes c & c \otimes b + d \otimes d \end{pmatrix},$$

$$\varepsilon\left(\begin{pmatrix} a & b \\ c & d \end{pmatrix}\right) = \begin{pmatrix} 1 & 0 \\ 0 & 1 \end{pmatrix} . \tag{2}$$

Further, we suppose that $\mathcal{D}$ is invertible, i.e., there is an element $\mathcal{D}^{-1}$ which obeys:

$$\mathcal{D}\mathcal{D}^{-1} = \mathcal{D}^{-1}\mathcal{D} = 1_{\mathcal{A}}, \qquad (\mathcal{D}^{-1}) = \mathcal{D}^{-1} \otimes \mathcal{D}^{-1}, \qquad \varepsilon(\mathcal{D}^{-1}) = 1 \ . \tag{3}$$

In this case one defines the left and right inverse matrix of M Aghamohammadi (1993):

$$M^{-1} = \mathcal{D}^{-1} \begin{pmatrix} d + gc & -b + g(d-a) + g^2 c \\ -c & a - gc \end{pmatrix} =$$

$$= \begin{pmatrix} d + hc & -b + h(d-a) + h^2 c \\ -c & a - hc \end{pmatrix} \mathcal{D}^{-1} \ . \tag{4}$$

The quantum group $\mathrm{GL}_{g,h}(2)$ is *defined* as the Hopf algebra obtained from the bialgebra $A_{g,h}(2)$ when $\mathcal{D}^{-1}$ exists and with antipode given by the formula:

$$\gamma(M) \ = \ M^{-1} \ . \tag{5}$$

For $g = h$ one obtains from $\mathrm{GL}_{g,h}(2)$ the matrix quantum group $\mathrm{GL}_h(2) = \mathrm{GL}_{h,h}(2)$, and, if the condition $\mathcal{D} = 1_{\mathcal{A}}$ holds, the matrix quantum group $\mathrm{SL}_h(2)$.

3 The Dual of $\mathrm{GL}_{g,h}(2)$

Two Hopf algebras $\mathcal{U}, \mathcal{A}$ are said to be *in duality* (Abe (1980)) if there exists a doubly nondegenerate bilinear form

$$\langle , \rangle : \mathcal{U} \times \mathcal{A} \longrightarrow \mathbb{C}, \langle , \rangle : (u,a) \mapsto \langle u,a \rangle, u \in \mathcal{U}, a \in \mathcal{A} , \qquad (6)$$

such that, for $u, v \in \mathcal{U}, a, b \in \mathcal{A}$:

$$\begin{aligned} &\langle u, ab\rangle = \langle \delta_{\mathcal{U}}(u), a \otimes b\rangle, \qquad \langle uv, a\rangle = \langle u \otimes v, \delta_{\mathcal{A}}(a)\rangle \\ &\langle 1_{\mathcal{U}}, a\rangle = \varepsilon_{\mathcal{A}}(a), \qquad \langle u, 1_{\mathcal{A}}\rangle = \varepsilon_{\mathcal{U}}(u) \\ &\langle \gamma_{\mathcal{U}}(u), a\rangle = \langle u, \gamma_{\mathcal{A}}(a)\rangle \ . \end{aligned} \qquad (7)$$

It is enough to define the pairing (6) between the generating elements of the two algebras. The pairing between any other elements of $\mathcal{U}$, $\mathcal{A}$ follows then from relations (7) and the standard bilinear form inherited by the tensor product.

The duality between two bialgebras or Hopf algebras may be used also to obtain the unknown dual of a known algebra. For that it is enough to give the pairing between the generating elements of the unknown algebra with arbitrary elements of the PBW basis of the known algebra. Using these initial pairings and the duality properties one may find the unknown algebra. Such an approach was first given by Sudbery (Sudbery (1990)). He obtained $U_q(sl(2)) \otimes U(u(1))$ as the algebra of tangent vectors at the identity of $\mathrm{GL}_q(2)$. The initial pairings were defined through the tangent vectors at the identity. However, such calculations become very difficult for more complicated algebras. Thus, in (Dobrev (1992)) a generalization was proposed in which the initial pairings are postulated to be equal to the classical undeformed results. This generalized method was applied in (Dobrev (1992)) to the standard two-parameter deformation $\mathrm{GL}_{p,q}(2)$, (where also Sudbery's method was used), then in (Dobrev and Parashar (1993a)) to the multiparameter deformation of $\mathrm{GL}(n)$, and in (Dobrev and Parashar (1993b)) to matrix quantum Lorentz group of (Podles, Woronowicz (1990), Carow-Watamura, Schlieker, Scholl, Watamura (1991)). One should note that the dual of $\mathrm{GL}_{p,q}(2)$ was obtained also in (Schirrmacher et al. (1991)) by methods of q-differential calculus.

Here we apply the method of (Dobrev (1992)) to find the dual of $\mathrm{GL}_{g,h}(2)$. Following (Dobrev (1992)) we first need to fix a PBW basis of $\mathrm{GL}_{g,h}(2)$. At first one may be inclined to use a PBW basis as the one introduced in (Ohn (1992)) for the case $g = h$, namely consisting of all monomials $a^k d^\ell b^m c^n$, where $k, \ell, m, n \in \mathbb{Z}_+$. However, it turns out that it would be simpler to work with the PBW basis: $a^k\ d^\ell\ c^n\ b^m$, $k, \ell, m, n \in \mathbb{Z}_+$. Further simplification results if we make the following change of generating elements and parameters:

$$\tilde{a} = \tfrac{1}{2}(a+d), \quad \tilde{d} = \tfrac{1}{2}(a-d), \qquad \tilde{g} = \tfrac{1}{2}(g+h), \quad \tilde{h} = \tfrac{1}{2}(g-h) \ . \qquad (8)$$

With these generating elements and parameters the algebra relations become:

$$\begin{aligned} c\tilde{a} &= \tilde{a}c - \tilde{g}c^2, \qquad c\tilde{d} = \tilde{d}c - \tilde{h}c^2, \qquad \tilde{d}\tilde{a} = \tilde{a}\tilde{d} - \tilde{g}\tilde{d}c + \tilde{h}\tilde{a}c \\ b\tilde{a} &= \tilde{a}b + \tilde{g}cb - 2\tilde{h}\tilde{a}\tilde{d} + 2\tilde{g}\tilde{d}^2 + (\tilde{g}^2 - \tilde{h}^2)\tilde{a}c + \tilde{g}(\tilde{h}^2 - \tilde{g}^2)c^2 \\ b\tilde{d} &= \tilde{d}b - \tilde{h}cb + 2\tilde{g}\tilde{a}\tilde{d} - 2\tilde{h}\tilde{d}^2 + (\tilde{h}^2 - \tilde{g}^2)\tilde{d}c + \tilde{h}(\tilde{g}^2 - \tilde{h}^2)c^2 \\ bc &= cb + 2\tilde{g}\tilde{a}c - 2\tilde{h}\tilde{d}c + (\tilde{h}^2 - \tilde{g}^2)c^2 \\ \mathcal{D} &= \tilde{a}^2 - \tilde{d}^2 - cb + (\tilde{g}^2 - \tilde{h}^2)c^2 - \tilde{g}\tilde{a}c + \tilde{h}\tilde{d}c \ . \end{aligned} \tag{9}$$

Note that these relations are written in anticipation of the PBW basis:

$$f = f_{k,\ell,m,n} = \tilde{a}^k\ \tilde{d}^\ell\ c^n\ b^m, \qquad k,\ell,m,n \in \mathbb{Z}_+ \ . \tag{10}$$

The coalgebra relations become:

$$\delta\left(\begin{pmatrix} \tilde{a} & b \\ c & \tilde{d} \end{pmatrix}\right) = \tag{11}$$

$$= \begin{pmatrix} \tilde{a}\otimes\tilde{a} + \tilde{d}\otimes\tilde{d} + \frac{1}{2}b\otimes c + \frac{1}{2}c\otimes b & \tilde{a}\otimes b + \tilde{d}\otimes b + b\otimes\tilde{a} - b\otimes\tilde{d} \\ c\otimes\tilde{a} + c\otimes\tilde{d} + \tilde{a}\otimes c - \tilde{d}\otimes c & \tilde{a}\otimes\tilde{d} + \tilde{d}\otimes\tilde{a} + \frac{1}{2}b\otimes c - \frac{1}{2}c\otimes b \end{pmatrix}$$

$$\varepsilon\left(\begin{pmatrix} \tilde{a} & b \\ c & \tilde{d} \end{pmatrix}\right) = \begin{pmatrix} 1 & 0 \\ 0 & 0 \end{pmatrix} , \tag{12}$$

$$\begin{aligned} \gamma\left(\begin{pmatrix} \tilde{a} & b \\ c & \tilde{d} \end{pmatrix}\right) &= \\ &= \mathcal{D}^{-1}\begin{pmatrix} \tilde{a} - \tilde{d} + (\tilde{g} + \tilde{h})c & -b - 2(\tilde{g} + \tilde{h})\tilde{d} + (\tilde{g} + \tilde{h})^2 c \\ -c & \tilde{a} + \tilde{d} - (\tilde{g} + \tilde{h})c \end{pmatrix} = \\ &= \begin{pmatrix} \tilde{a} - \tilde{d} + (\tilde{g} - \tilde{h})c & -b + 2(\tilde{h} - \tilde{g})\tilde{d} + (\tilde{g} - \tilde{h})^2 c \\ -c & \tilde{a} + \tilde{d} + (\tilde{h} - \tilde{g})c \end{pmatrix}\mathcal{D}^{-1} \ . \end{aligned} \tag{13}$$

Let us denote by $\mathcal{U}_{g,h} = U_{g,h}(gl(2))$ the unknown yet dual algebra of $\mathrm{GL}_{g,h}(2)$, and by A, B, C, D the four generators of $\mathcal{U}_{g,h}$. Following (Dobrev (1992)) we shall define the pairing $\langle Z, f\rangle$, $Z = A, B, C, D$, f is from (10), as the classical tangent vector at the identity:

$$\langle Z, f\rangle \equiv \varepsilon\left(\frac{\partial f}{\partial y}\right), (Z,y) = (A,\tilde{a}),\ (B,b),\ (C,c),\ (D,\tilde{d}) \ . \tag{14}$$

From this we get the explicit expressions:

$$\begin{aligned} \langle A, f\rangle &= \varepsilon\left(\tfrac{\partial f}{\partial \tilde{a}}\right) = k\delta_{\ell 0}\delta_{m0}\delta_{n0} \\ \langle B, f\rangle &= \varepsilon\left(\tfrac{\partial f}{\partial b}\right) = \delta_{\ell 0}\delta_{m1}\delta_{n0} \\ \langle C, f\rangle &= \varepsilon\left(\tfrac{\partial f}{\partial c}\right) = \delta_{\ell 0}\delta_{m0}\delta_{n1} \\ \langle D, f\rangle &= \varepsilon\left(\tfrac{\partial f}{\partial \tilde{d}}\right) = \delta_{\ell 1}\delta_{m0}\delta_{n0} \ . \end{aligned} \tag{15}$$

First we find the commutation relations between the generators of $\mathcal{U}_{g,h}$. Below we shall need expressions like $e^{\nu B}$ which we define as formal power series $e^{\nu B} = 1_{\mathcal{U}} + \sum_{p\in\mathbb{N}} \frac{\nu^p}{p!} B^p$. We have (Aneva et al. (1997)):

Proposition 1: The commutation relations of the generators A, B, C, D are:

$$\begin{aligned} &[B,C] = D, \quad [D,B] = \tfrac{1}{\tilde{g}}(e^{2\tilde{g}B} - 1_{\mathcal{U}}), \quad [D,C] = -2C + \tilde{g}D^2 - \tilde{g}A \\ &[A,B] = 0, \qquad [A,C] = 0, \qquad [A,D] = 0 \; . \end{aligned} \tag{16}$$

Note that the commutation relations (16) depend only on the parameter $\tilde{g}$ and that the generator A is central. This is similar to the situation of the dual algebra $\mathcal{U}_{p,q}$ of the standard matrix quantum group $\mathrm{GL}_{p,q}$ the commutation relations of which depend only on the combination $q' = \sqrt{pq}$ and also one generator is central (Schirrmacher et al. (1991)), (Dobrev (1992)). Here the central generator appears as a central extension but this is fictitious since this may be corrected by a change of basis. Besides such a change we shall make a change of generating elements of $\mathcal{U}_{g,h}$ in order to bring the commutation relations to a form closer to the algebra of (Ohn (1992)). Thus, we make the following substitutions:

$$D = e^{\mu B}\, H\, e^{\nu B}, \quad C = e^{\mu B}\, Y\, e^{\nu B} - \tfrac{\tilde{g}}{2}\sinh(\tilde{g}B)e^{\tilde{g}B} - \tfrac{\tilde{g}}{2}A, \quad \mu + \nu = \tilde{g} \; , \tag{17}$$

to obtain the following commutation relations instead of (16) :

$$\begin{aligned} &[B,Y] = H, \quad [H,B] = \tfrac{2}{\tilde{g}}\sinh(\tilde{g}B), \quad [H,Y] = -Y\cosh(\tilde{g}B) - \cosh(\tilde{g}B)Y \\ &\qquad\qquad [A,B] = 0, \quad [A,Y] = 0, \quad [A,H] = 0 \; . \end{aligned} \tag{18}$$

The Casimir operator of $\mathcal{U}_{g,h}$ is (f_1, f_2 being arbitrary polynomial functions) :

$$\begin{aligned} \mathcal{C}_2 = f_1(A)\Big(&\tfrac{1}{2}\,(H^2 + \cosh^2(\tilde{g}B)) + \\ &\tfrac{1}{\tilde{g}}\,(Y\sinh(\tilde{g}B) + \sinh(\tilde{g}B)Y)\Big) + f_2(A) \; . \end{aligned} \tag{19}$$

We turn now to the coalgebra structure of $\mathcal{U}_{g,h}$. We have (Aneva et al. (1997)):

Proposition 2: (i) The comultiplication in the algebra $\mathcal{U}_{g,h}$ is given by:

$$\delta_{\mathcal{U}}(A) = A \otimes 1_{\mathcal{U}} + 1_{\mathcal{U}} \otimes A \tag{20}$$

$$\delta_{\mathcal{U}}(B) = B \otimes 1_{\mathcal{U}} + 1_{\mathcal{U}} \otimes B \tag{21}$$

$$\begin{aligned} \delta_{\mathcal{U}}(Y) = \; &Y \otimes e^{-\tilde{g}B} + e^{\tilde{g}B} \otimes Y - \\ &\tfrac{\tilde{h}^2}{\tilde{g}}\,\sinh(\tilde{g}B) \otimes A^2 e^{-\tilde{g}B} + \tilde{h}H \otimes Ae^{-\tilde{g}B} \end{aligned} \tag{22}$$

$$\delta_{\mathcal{U}}(H) = H \otimes e^{-\tilde{g}B} + e^{\tilde{g}B} \otimes H - \tfrac{2\tilde{h}}{\tilde{g}}\,\sinh(\tilde{g}B) \otimes Ae^{-\tilde{g}B} \; . \tag{23}$$

(ii) The co-unit relations in $\mathcal{U}_{g,h}$ are given by:

$$\varepsilon_{\mathcal{U}}(Z) = 0, \qquad Z = A, B, Y, H \ . \tag{24}$$

(iii) The antipode in the algebra $\mathcal{U}_{g,h}$ is given by:

$$\gamma_{\mathcal{U}}(A) = -A \tag{25}$$

$$\gamma_{\mathcal{U}}(B) = -B \tag{26}$$

$$\gamma_{\mathcal{U}}(Y) = -e^{-\tilde{g}B} Y e^{\tilde{g}B} + \tfrac{\tilde{h}^2}{\tilde{g}} \sinh(\tilde{g}B) A^2 + \tilde{h} e^{-\tilde{g}B} H A e^{\tilde{g}B} \tag{27}$$

$$\gamma_{\mathcal{U}}(H) = -e^{-\tilde{g}B} H e^{\tilde{g}B} - \tfrac{2\tilde{h}}{\tilde{g}} \sinh(\tilde{g}B) A \ . \tag{28}$$

Thus, we can state our final result:

Theorem: The Hopf algebra $\mathcal{U}_{g,h}$ dual to $\mathrm{GL}_{g,h}(2)$ is generated by A, B, Y, H, cf. (15), (17). It is given by relations (18), (20), (24), (25). As an algebra it depends only on one parameter $\tilde{g} = (g+h)/2$ and is split in two subalgebras: $\mathcal{U}'_{g,h}$ generated by B, Y, H and $U(\mathcal{Z})$, where the algebra $\mathcal{Z}$ is spanned by A. The subalgebra $U(\mathcal{Z})$ is central in $\mathcal{U}_{g,h}$ and is also a Hopf subalgebra of $\mathcal{U}_{g,h}$. The subalgebra $\mathcal{U}'_{g,h}$ is not a Hopf subalgebra.

4 One-Parameter Cases

It is interesting to discuss one-parameter special cases of the matrix quantum group $\mathrm{GL}_{g,h}(2)$ and its dual. The one-parameter matrix quantum group $\mathrm{GL}_{\tilde{g}}(2)$ (Demidov et al. (1990)), (Zakrzewski (1991)), is obtained from $\mathrm{GL}_{g,h}(2)$ by setting $g = h = \tilde{g}$. Thus the dual algebra $\mathcal{U}_{\tilde{g}} \equiv \mathcal{U}_{\tilde{g},\tilde{g}}$ of $\mathrm{GL}_{\tilde{g}}(2)$ is obtained by setting $\tilde{h} = \frac{1}{2}(g-h) = 0$ in (18), (20), (24), (25). We see that the one-parameter Hopf algebra $\mathcal{U}_{\tilde{g}}$ is split in two Hopf subalgebras $\mathcal{U}'_{\tilde{g}} \equiv \mathcal{U}'_{\tilde{g},\tilde{g}}$ and $U(\mathcal{Z})$ and we may write: $\mathcal{U}_{\tilde{g}} = \mathcal{U}'_{\tilde{g}} \otimes U(\mathcal{Z})$. Thus, now we compare the algebra $\mathcal{U}'_{\tilde{g}}$ with the algebra of (Ohn (1992)). We see that after the identification $B \mapsto X$, $\tilde{g} \mapsto -h$, the algebra $\mathcal{U}'_{\tilde{g}}$ coincides with the algebra of Ohn. We also note that the algebra $\mathcal{U}'_{\tilde{g}}$ in the basis $B, \tilde{C}, D$ coincides for $\tilde{h} = 0$ with the version given in (Ballesteros and Herranz (1996)) after the identification: $(B, \tilde{C}, D; \tilde{g}) \mapsto (A_+, A_-, A; z)$, and by using the opposite coalgebra structure.

Another one-parameter case is: $g = -h = \tilde{h}$, i.e., $\tilde{g} = 0$. From (18), (20), (25), we obtain that the subalgebra $\mathcal{U}'_{\tilde{h},-\tilde{h}}$ is isomorphic to the undeformed $U(sl(2))$ with $sl(2)$ spanned by B, Y, H. However, as in the general case, the coalgebra sector is not classical, and the generators B, Y, H do not close a co-subalgebra.

Acknowledgments

B.L.A. was supported in part by BNFR under contract Ph-404, V.K.D. was supported in part by BNFR under contract Ph-643.

References

Abdesselam, B., Chakrabarti, A. and Chakrabarti, R. (1996): Mod. Phys. Lett. **A11**, 2883.

Abe, E. (1980): *Hopf Algebras*, (Cambridge Univ. Press.

Aghamohammadi, A. (1993): Mod. Phys. Lett. **A8**, 2607.

Aizawa, N. (1997): preprint OWUAM-020, q-alg/9701022.

Aneva, B.L., Dobrev, V.K. and Mihov, S.G. (1997): preprint INRNE-TH/3/97.

Ballesteros, A. and Herranz, F.J. (1996): J. Phys. **A29** L311.

Demidov, E.E., Manin, Yu.I., Mukhin, E.E. and Zhdanovich, D.V. (1990): Progr. Theor. Phys. Suppl. **102**, 203.

Dobrev, V.K. (1992): J. Math. Phys. **33**, 3419.

Dobrev, V.K. and Parashar, P. (1993): J. Phys. **A26**, 6991.

Dobrev, V.K. and Parashar, P. (1993):, Lett. Math. Phys. **29**, 259.

Dobrev, V.K. (1996): ICTP preprint IC/96/14 (January 1996).

Drinfeld, V.G. (1985): Dokl. Akad. Nauk SSSR **283**, 1060; in: Proceedings ICM, (MSRI, Berkeley, 1986) p.798.

Karimipour, V. (1994): Lett. Math. Phys. **30**, 87.

Khorrami, M., Shariati, A., Abolhassani, M.R. and Aghamohammadi, A. (1995): Mod. Phys. Lett. **A10**, 873.

Kupershmidt, B.A. (1992): J. Phys. **A25**, L1239.

Ohn, C. (1992): Lett. Math. Phys. **25** 85.

Podles, P., Woronowicz, S.L. (1990): Comm. Math. Phys. **130**, 381;
Carow-Watamura, U., Schlieker, M., Scholl, M., Watamura, S. (1991): Zeit. Phys. **C48**, 159.

Schirrmacher, A., Wess, J. and Zumino, B. (1991): Z. Phys. **C49**, 317.

Sudbery, A. (1990): in: Proceedings of the Workshop on Quantum Groups, Argonne National Lab , eds. T. Curtright, D. Fairlie and C. Zachos, (World Sci, 1991) p. 33.

Van der Jeugt, J. (1997): q-alg/9703011.

Vladimirov, A.A. (1993): Mod. Phys. Lett. **A8**, 2573.

Zakrzewski, S. (1991): Lett. Math. Phys. **22**, 287.

Supertraces on Some Deformations of Heisenberg Superalgebra

S.E. Konstein

I.E.Tamm Department of Theoretical Physics, P. N. Lebedev Physical Institute, 117924 Leninsky Prospect 53, Moscow, Russia.

In this report I discuss some algebraic properties of the algebra $H_{W(\mathbf{R})}(\nu)$, the algebra of observables of the rational Calogero model (Calogero (1969, 1971)) with harmonic interaction based on the classical root system $\mathbf{R}$ of A_{N-1}, B_N, C_N or D_N types (Olshanetsky, Perelomov (1983)). Namely I show how one can prove the existence of the supertraces on this algebra and count up the number of these supertraces. The algorithm of finding of these supertraces for arbitrary element of the algebra is described briefly. Two- and three-particle cases are considered in more details. Detailed proofs of followed statements are given in (Konstein, Vasiliev (1995, 1996), Konstein (1996)).

1 Definitions and Results

1. The algebra $H_{W(\mathbf{R})}(\nu)$ is the deformations of the associative Heisenberg algebra, which describes naturally the solutions of N-particle rational Calogero model with harmonic part in potential based on the classical root systems. Its elements are the polynomials of N pairs of (deformed) oscillators a_i^α, $\alpha = 0, 1$, $i = 1, ..., N$ and of elements of Weyl group $G = W(\mathbf{R})$ of the root system $\mathbf{R}$, satisfying the following commutation relations:

$$\left[a_i^\alpha\,,\,a_j^\beta\right] = \epsilon^{\alpha\beta}(\delta_{ij} + \nu_0(\delta_{ij}\sum_{l=1,\,l\neq i}^{N} K_{il} - \delta_{i\neq j}K_{ij}) + \nu_1\delta_{ij}R_i + \nu_2(\delta_{ij}\sum_{l\neq i} K_{il}R_iR_l + \delta_{i\neq j}K_{ij}R_iR_j))\quad, \tag{1}$$

where $\epsilon^{\alpha\beta} = -\epsilon^{\beta\alpha}$, $\epsilon^{01} = 1$. Here repeated Latin indices do not imply summation.

The operators K_{ij}, $K_{ij}R_iR_j$ and R_i are the reflections connected with the vectors of the classical root systems $a_i^\alpha - a_j^\alpha$, $a_i^\alpha + a_j^\alpha$ and a_i^α correspondingly. They act on a_i^α as follows

$$R_i a_i^\alpha = -a_i^\alpha R_i, \qquad R_i a_j^\alpha = a_j^\alpha R_i \text{ for } i \neq j\ , \tag{2}$$

$$K_{ij}a_j^\alpha = a_i^\alpha K_{ij}, \qquad K_{ij}a_k^\alpha = a_k^\alpha K_{ij} \text{ for } i\neq j\neq k\neq i\ , \tag{3}$$

and generate the group algebra $\mathfrak{G} \subset H_G(\nu)$ of the group G.

The commutational relations (1) and associativity of $H_{W(\mathbf{R})}(\nu)$ are consistent when one of the following conditions takes place:
A) $\nu_1 = \nu_2 = 0$, and reflections $K_{ij}R_iR_j$ and R_i are excluded,
B,C) $\nu_0 = \nu_2$,
D) $\nu_0 = \nu_2$, $\nu_1 = 0$, and every monomial in $H_{W(\mathbf{R})}(\nu)$ contains even number of reflections R_i.

The parity π: $\pi(a_i^\alpha) = 1$, $\pi(K_{ij}) = \pi(R_i) = 0$ gives the structure of superalgebra on $H_{W(\mathbf{R})}(\nu)$ with superbrackets $[f, g\} = fg-(-1)^{\pi(f)\pi(g)}gf$ $\forall f, g \in H_{W(\mathbf{R})}(\nu)$.

It is obvious that every element $u \in H_G(\nu)$ can be presented in the form $u = \sum_{\sigma\in G} P_\sigma(a_i^\alpha)\sigma$, where P_σ are some polynomials of a_i^α.

2. When one of conditions A), B,C), D) takes place, the oscillators a_i^α admit the following presentation $a_i^\alpha = (x_i + (-1)^\alpha D_i(x))/\sqrt{2}$ where $D_i(x)$ are Dunkl's differential-difference operators (Dunkl (1989)) connected with the corresponding root systems,

$$D_i = \tfrac{\partial}{\partial x_i} + \nu_1 \tfrac{1}{x_i}(1 - R_i) +$$

$$\sum_{l\neq i}^{N} \left(\nu_0 \tfrac{1}{x_i - x_l}(1 - K_{il}) + \nu_2 \tfrac{1}{x_i + x_l}(1 - K_{il}R_iR_l)\right)$$

3. An important property of $H_{W(\mathbf{R})}(\nu)$ which allows one to solve the Calogero model (Calogero (1969, 1971)) is that this algebra possesses inner sl_2 automorphisms with the generators

$$T^{\alpha\beta} = \frac{1}{2}\sum_{i=1}^{N}(a_i^\alpha a_i^\beta + a_i^\beta a_i^\alpha) \tag{4}$$

which act on the generating elements a_i^α as on sl_2 vectors

$$[T^{\alpha\beta}, a_i^\gamma] = \epsilon^{\alpha\gamma} a_i^\beta + \epsilon^{\beta\gamma} a_i^\alpha \ . \tag{5}$$

The Hamiltonian of Calogero model associated with root system (Olshanetsky, Perelomov (1983)) is identified with second-order differential operator $H = T^{01} = \frac{1}{2}\sum_{i=1}^{N}\{a_i^0, a_i^1\}$. The operators a_i^α serve as generalized oscillators underlying the Calogero problem and allow one (Polychronakos (1992), Brink, Hansson, Vasiliev (1992)) to construct wave functions via the standard Fock procedure with the Fock vacuum $|0\rangle$ such that $a_i^0|0\rangle=0$.

4. Define the supertrace as a linear complex-valued function str$(\cdot)$ on the algebra $H_{W(\mathbf{R})}(\nu)$ such that

$$\mathrm{str}([u, v\}) = 0 \quad \forall u, v \in H_{W(\mathbf{R})}(\nu) \ . \tag{6}$$

To know the supertraces is useful in various respects. One of the most important is that that they define polylinear invariant forms

$$\mathrm{str}(u_1 u_2 \dots u_n) \tag{7}$$

what allows for example to construct the lagrangians for dynamical theories based on these algebras. Another useful property is that since null vectors of invariant bylinear form $\mathrm{str}(u_1 u_2)$ span a both-side ideal of the algebra, this gives a powerful device for investigating ideals which decouple from everything under the supertrace operation as it happens in $H_{W(A_1)}(\nu)$ for half-integer ν (Vasiliev (1989, 1991)).

An important motivation for the analysis of the supertraces of $H_{W(\mathbf{R})}(\nu)$ is due to its deep relationship with the analysis of the representations of this algebra, which in its turn gets applications to the analysis of the wave functions of the Calogero models. For example, given representation of $H_{W(\mathbf{R})}(\nu)$, one can speculate that it induces some supertrace on this algebra as (appropriately regularized) supertrace of (infinite) representation matrices. When the corresponding bylinear form $\mathrm{str}(u_1 u_2)$ degenerates this would imply that the representation becomes reducible.

5. In (Konstein, Vasiliev (1995, 1996), Konstein (1996)) it is shown that there exist the supertraces and that the number of independent supertraces is equal to

- the number of partitions of N into a sum of odd positive integers for the case A_{N-1},
- the number of partitions of N into a sum of positive integers for the cases B_N and C_N,
- the number of partitions of N into a sum of positive integers with even number of even integers for the case A_{N-1}.

2 Recurrent Equations for Supertraces

1. Consider the defining relations (6) as the equations for finding the supertrace.

One can show that if $\mathrm{str}(u) \neq 0$ for some supertrace $\mathrm{str}(\cdot)$ and some element $u \in H_G(\nu)$ then u is singlet under action of sl_2, *i.e.* $[T^{\alpha\beta}, u\} = 0$ and as a consequence $\pi(u) = 0$. So, if $\pi(u) = 1$ then $\mathrm{str}(u) = 0$.

To write down the equations for $\mathrm{str}(P(a_i^\alpha)\sigma)$ with even polynomial P it is convenient to introduce the new basis of oscillators $\mathfrak{B}_\sigma=\{b^I\}$, $I = 1, ..., 2N$, (linear combinations of vectors a_i^α) instead of $\{a_i^\alpha\}$. Namely let b^I be the eigenvectors of operator σ: $\sigma b^I = \lambda_I b^I \sigma$. If there exist such I that $\lambda_I = -1$ let us reorder the indices I in such a way that $\lambda_I = -1$ if $1 \leq I \leq 2E(\sigma)$ and $\lambda_I \neq -1$ if $2E(\sigma \leq I \leq 2N$. The function $E(\sigma)$ defined in such a way is integer valued function on $\mathfrak{G}$.

Polynomial $P(b^I)$ is called special if it does not depend on b^I with $\lambda_I \neq -1$.

2. Let us introduce together with $\mathfrak{B}_\sigma$ the symplectic form $\mathcal{C}^{IJ}=$ $[b^I, b^J]\,|_{\nu_0=\nu_1=\nu_2=0}$ which serves in the following consideration for raising and lowering of indices I. Let $f^{IJ} = [b^I, b^J] - \mathcal{C}^{IJ}$.

The next lemma proved in (Konstein, Vasiliev (1995, 1996), Vasiliev (1989, 1991)) ensures the existence of supertraces.

Lemma If $\lambda_I = \lambda_J = -1$ then $E(f^{IJ}\sigma) = E(\sigma) - 1$.

3. Let us deduce the equation for $\mathrm{str}(P(b^I)\sigma)$ where P is non-special monomial of even degree. Without loss of generality one can assume that $P = b^I Q$ where $\lambda_I \neq -1$ and monomial Q has an odd degree. Then (6) gives

$$\mathrm{str}(b^I Q\sigma) = -\mathrm{str}(Q\sigma b^I) = -\lambda_I \mathrm{str}(Qb^I\sigma)$$

and as a consequence

$$\mathrm{str}(b^I Q\sigma) = \frac{\lambda_I}{1+\lambda_I}\mathrm{str}([b^I, Q]\sigma) \ . \tag{8}$$

Relation (8) reduces the computation of supertrace of non-special monomials to the computation of the supertrace of polynomials of lower degree.

Introduce the following two sets of generating functions:

$$\Psi_\sigma(\mu) \stackrel{\mathrm{def}}{=} \mathrm{str}\left(\mathrm{e}^S\sigma\right) , \qquad S = \sum_{L=1}^{2N}(\mu_L b^L) \tag{9}$$

and

$$\Phi_\sigma(\mu) \stackrel{\mathrm{def}}{=} \mathrm{str}\left(\mathrm{e}^{S'}\sigma\right) , \qquad S' = \sum_{L=1}^{2E(\sigma)}(\mu_L b^L) , \tag{10}$$

$$\Phi_\sigma(\mu) = \Psi_\sigma(\mu)\Big|_{(\mu_I=0 \ \forall I:\lambda_I\neq -1)} ,$$

where σ is some fixed element of G, $b^L \in \mathfrak{B}_\sigma$ and $\mu_L \in \mathbb{C}$ are independent parameters. By differentiating over μ_L one can obtain an arbitrary polynomial of b^L in front of σ.

In terms of Ψ_σ and Φ_σ the recursion (8) takes the form (Konstein, Vasiliev (1995, 1996))

$$\begin{aligned} \Psi_\sigma &= \Phi_\sigma(\mu) \\ &+ \sum_{L:\,\lambda_L\neq -1}\int_0^1 \frac{\mu_L \mathrm{d}\tau}{1+\lambda_L}\int_0^1 \mathrm{d}t\,(\lambda_L t + t - 1)\times \\ &\times \mathrm{str}\Big(\mathrm{e}^{t(\tau S''+S')}[b^L, (\tau S''+S')]\,\mathrm{e}^{(1-t)(\tau S''+S')}\sigma\Big) , \end{aligned} \tag{11}$$

where $S'' = S - S'$.

4. Now consider $\mathrm{str}(b^I Q\sigma)$ where $\lambda_I = -1$ and Q is odd-degree special monomial. From (6) it follows that $\mathrm{str}(b^I Q\sigma) = \mathrm{str}(Qb^I\sigma)$ and hence

$$\mathrm{str}([b^I, Q]\sigma) = 0 \ . \tag{12}$$

But since $\mathrm{str}([b^I, Q]\sigma) = \sum_{J=1}^{2E(\sigma)} \mathrm{str}(Q_J(C^{IJ} + f^{IJ})\sigma)$ with some special monomials Q_J the equation (12) takes the form

$$\mathrm{str}(Q^I\sigma) = -\sum_{J=1}^{2E(\sigma)} \mathrm{str}(Q_J f^{IJ}\sigma) \ . \tag{13}$$

The degree of polynomial Q^I is equal to the degrees of monomials Q_J, but due to lemma $E(f^{IJ}\sigma) = E(\sigma) - 1$. So the equations (13) are recurrent on the grading $E(\sigma)$.

In terms of the generating functions Φ_σ the recursion (13) have more explicit form

$$\Phi_\sigma(\mu) = \Phi_\sigma(0) + \frac{\nu}{8E(\sigma)} \sum_{I,J=1}^{2E(\sigma)} \int_0^1 \frac{dt}{t}(1 - t^{2E(\sigma)})(L_{IJ}R^{IJ})(t\mu) \ , \tag{14}$$

where

$$R^{IJ}(\mu) = \sum_{M=1}^{2E(\sigma)} \mathrm{str}\Big(\exp(S')\{b^J, f^{IM}\}\mu_M\sigma\Big) + (I \leftrightarrow J) \tag{15}$$

and

$$L^{IJ} = \frac{\partial}{\partial\mu_J}\mu^I + \frac{\partial}{\partial\mu_I}\mu^J \ . \tag{16}$$

It should be noticed that (14) is less restrictive than (13), namely it does not contain the equations for $\Phi_\sigma(0)$ while (13) does.

In terms of Ψ_σ and Φ_σ the analysis of the equations (6) (or equivalently (8) and (13)) becomes more convenient and is carried out in (Konstein, Vasiliev (1995, 1996)) where it is shown, that (8) and (13) are consistent and the solution has the recurrent form describing by (11) and (14).

5. To complete the finding of supertrace with the aid of (14) it remains to find $\Phi_\sigma(0) = \mathrm{str}(\sigma)$. The relations (13) give

$$C^{IJ}\mathrm{str}(\sigma) = \mathrm{str}(f^{IJ}\sigma) \tag{17}$$

if $\lambda_I = \lambda_J = -1$. Due to lemma it means that the supertrace of some element σ of Weyl group is expressed via the supertraces of the elements with smaller value of grading E. In (Konstein, Vasiliev (1995, 1996), Konstein (1996)) it is proved that (17) are consistent and that there is no restrictions on the $\mathrm{str}(\sigma)$ with such σ that $E(\sigma) = 0$. So due to G-invariance of supertrace the number of independent supertraces is equal to the number of conjugacy classes with $E = 0$ in Weyl group G. These numbers are listed in 1.5.

3 Examples

1. Consider $N=2$ rational Calogero model with harmonic term in potential. After separation of center mass its algebra of observables $H_{W(A_1)}(\nu)=H_{S_2}(\nu)$ is generated by the oscillators $a^\alpha=\frac{1}{\sqrt{2}}(a_1^\alpha-a_2^\alpha)$ and by the reflection $K\stackrel{\text{def}}{=}K_{12}$ with the commutational relations:

$$Ka^\alpha=-a^\alpha K,\qquad [a^0,a^1]=1-2\nu K,\qquad K^2=1\,. \tag{18}$$

There exists only one supertrace on this algebra with the generating functions

$$\Psi_n(\mu)\stackrel{\text{def}}{=}\mathrm{str}(e^{\mu_0 a^0+\mu_1 a^1}K^n),\quad n=1,2 \tag{19}$$

It is shown in (Vasiliev (1989, 1991)) that $\Psi_n(\mu)=\Psi_n(0)=\mathrm{str}(K^n)$ and $\mathrm{str}(K)=-2\nu\mathrm{str}(1)$.
2. The algebra $H_{W(A_2)}(\nu)$ corresponding to $N=3$ Calogero model has two independent supertraces. To write down the expressions for the generating functions of these supertraces it is necessary to introduce some notations.

Let $\lambda=\exp(2\pi i/3)$. After separation of the center of mass the algebra is generated by

$$\begin{aligned}
x^\alpha &= a_1^\alpha+\lambda a_2^\alpha+\lambda^2 a_3^\alpha\ ,\\
y^\alpha &= a_1^\alpha+\lambda^2 a_2^\alpha+\lambda a_3^\alpha\ ,\\
L_0 &= 1/3(K_{12}+K_{23}+K_{31})\ ,\\
L_{-1} &= 1/3(\lambda^2 K_{12}+K_{23}+\lambda K_{31})\ ,\\
L_{+1} &= 1/3(\lambda K_{12}+K_{23}+\lambda^2 K_{31})\ ,\\
Q_0 &= 1/3(1+c+c^2)\ ,\\
Q_{+1} &= 1/3(1+\lambda c+\lambda^2 c^2)\ ,\\
Q_{-1} &= 1/3(1+\lambda^2 c+\lambda c^2)\ ,
\end{aligned}$$

where $c=K_{12}K_{13}$, $c^2=K_{12}K_{23}$. The set of elements L_μ and Q_μ is the basis of the group algebra $\mathfrak{G}$ of symmetric group $S_3=W(A_2)$.

It is quite evident that all singlets under sl_2 (1.3.) have the form $f(m)g$ where $g\in\mathfrak{G}$ and $m=\frac{1}{2}\{x^\alpha,y^\beta\}\epsilon_{\alpha\beta}$. The supertrace of singlet element $m^k g$ can be found by differentiating of the following generating functions over ξ

$$\begin{aligned}
\mathcal{P}_\pm &\stackrel{\text{def}}{=} \mathrm{str}(e^{2/3\xi m}Q_{\pm1})=\\
&=\frac{2\nu S_1}{\Delta}\left(\frac{\mathrm{ch}(3\nu\xi)}{3\nu}\left(-e^{\pm2\xi}+2e^{\mp\xi}\right)\pm\mathrm{sh}(3\nu\xi)\left(e^{\pm2\xi}+e^{\mp\xi}\right)\right)+\\
&+\frac{S_2}{2\Delta}\left(e^{\mp2\xi}-2e^{\pm\xi}+3\right)\ ,\\
\mathcal{P}_0 &\stackrel{\text{def}}{=} \mathrm{str}(\mathrm{ch}(\xi\sqrt{4/9m^2+9\nu^2}Q_0)=\\
&=\frac{2\nu S_1}{\Delta}\left(-3\frac{\mathrm{ch}(3\nu\xi)}{3\nu}+\frac{\mathrm{sh}(3\nu\xi)}{2}\left(-e^{3\xi}+e^{-3\xi}\right)\right)+
\end{aligned}$$

$$+\frac{S_2}{2\Delta}\left(e^{2\xi}+e^{-2\xi}-2e^{\xi}-2e^{-\xi}\right)\right) ,$$

$$\mathcal{R}_0 \stackrel{\text{def}}{=} \operatorname{str}(\operatorname{sh}(\xi m)Q_0)=0) ,$$

$$\mathcal{L}_\pm \stackrel{\text{def}}{=} \operatorname{str}(e^{2/3\xi m}L_{\pm 1})=0 ,$$

$$\mathcal{L}_0 \stackrel{\text{def}}{=} \operatorname{str}(e^{2/3\xi m}L_0)=\nu S_1 ,$$

where $\Delta = \exp(-3\xi)(\exp(3\xi)+1)^2$ and S_1, S_2 are arbitrary parameters, defining 2-dimensional space of supertraces.

Acknowledgments. Author is very grateful to the Organizing Committee of International Seminar dedicated to the memory of Dmitrij Volkov for invitation and hospitality. Author thanks also M. A. Vasiliev and V. Sh. Burd for useful discussions. This work was supported in part by the Russian Basic Research Foundation, Grant 96-02-17314, and INTAS Grant 93-0633.

References

Brink, L., Hansson, H., and Vasiliev, M.A. (1992): Phys. Lett. **B286**, 109.
Calogero, F. (1969): J. Math. Phys., **10**, 2191, 2197;
Calogero, F. (1971): *ibid* **12**, 419.
Dunkl, C.F. (1989): Trans. Am. Math. Soc. **311**, 167
Konstein, S.E. (1996): hep-th/9612253; Theor. Math. Phys. (in press).
Konstein, S.E. and Vasiliev, M.A. (1995): hep-th/9512038;
Konstein, S.E. and Vasiliev, M.A. (1996): J. Math. Phys. **37**, 2872.
Olshanetsky, M.A. and Perelomov, A.M. (1983): Phys. Rep., **94**, 313.
Polychronakos, A. (1992): Phys. Rev. Lett. **69**, 703.
Vasiliev, M.A. (1989): JETP Letters, **50**, 344-347;
Vasiliev, M.A. (1991): Int. J. Mod. Phys. **A6**, 1115.

Harish–Chandra Embedding and q-Analogues of Bounded Symmetric Domains

S. Sinel'shchikov and L. Vaksman

Institute for Low Temperature Physics & Engineering
National Academy of Sciences of Ukraine

1. This work is devoted to study of a very restricted class of homogeneous spaces associated to quantum groups (Drinfeld (1987)), Jantzen (1996). We follow (Sinelshchikov, Vaksman (1997)) in describing here the construction of algebras of functions and differential forms on these quantum homogeneous spaces.

We hope to extend to the above context a great deal of the results of function theory and harmonic analysis in bounded symmetric domains (Hua (1963)). This is shown here to be available for the simplest one among such domains, the quantum disc (Sinel'shchikov, Shklyarov, Vaksman (to be published)).

Our subsequent constructions are q-analogues of the corresponding Harish-Chandra's constructions which allow one to embed a Hermitian symmetric space of non-compact type into $\mathbb{C}^N$ (Helgason (1962)).

Let A be a Hopf algebra, ε its counit, and S its antipode. Consider an algebra F equipped also with a structure of A-module. F is said to be an A-module (covariant) algebra if

i) the multiplication $m: F \otimes F \to F$, $m: f_1 \otimes f_2 \mapsto f_1 \cdot f_2$; $f_1, f_2 \in F$ is a morphism of A-modules;

ii) the unit $1 \in F$ is an invariant: $\xi 1 = \varepsilon(\xi)1$, $\xi \in A$.

If A is a Hopf $*$-algebra, and F is also equipped with an involution, then the definition of covariance should include the following compatibility condition for involutions:

$$\forall \xi \in A, f \in F \quad (\xi f)^* = (S(\xi))^* f^* \ .$$

In the sequel all the algebras of "functions" ($\mathbb{C}[\mathfrak{g}_{-1}]_q$, $\mathbb{C}[\overline{\mathfrak{g}}_{-1}]_q$, $\mathrm{Pol}(\mathfrak{g}_{-1})_q$) and "differential forms" are covariant algebras.

2. Let $\mathfrak{g}$ be a simple complex Lie algebra, $\mathfrak{h} \subset \mathfrak{g}$ a Cartan subalgebra, $\alpha_j \in \mathfrak{h}^*, j = 1, \ldots, l$, a system of simple roots with α_{j_0} being one of those roots. Consider the $\mathbb{Z}$-grading $\mathfrak{g} = \bigoplus_m \mathfrak{g}_m$ given by

$$\mathfrak{g}_m = \{\xi \in \mathfrak{g} |\, [H_0, \xi] = 2m\xi\} \ ,$$

with $H_0 \in \mathfrak{h}$ such that

$$\alpha_j(H_0) = 0,\ j \neq j_0;\ \alpha_{j_0}(H_0) = 2\ .$$

If this grading terminates,

$$\mathfrak{g} = \mathfrak{g}_{-1} + \mathfrak{g}_0 + \mathfrak{g}_{+1}\ ,$$

then clearly $\mathfrak{g}_{\pm 1}$ are Abelian Lie subalgebras. This is just the case when Harish-Chandra's construction presents a bounded symmetric domain U in the vector space $\mathfrak{g}_{-1}$.

Note also that in the case $\mathfrak{g} = \mathfrak{sl}_{m+n}$, $\alpha_{j_0} = \alpha_m$, U is the matrix ball in the space of $m \times n$ matrices:

$$\mathfrak{g}_{-1} \simeq \mathrm{Mat}(m,n); \quad U = \{T \in \mathrm{Mat}(m,n) |\ \| T \| < 1\}\ .$$

3. Turn to the quantum case and fix $q \in (0,1)$.

Remind that the Hopf algebra $U_q\mathfrak{sl}_2$ is determined by its generators $K^{\pm 1}, E, F$ and the relations

$$K \cdot K^{-1} = K^{-1} \cdot K = 1, \quad K^{\pm 1} \cdot E = q^{\pm 2} E K^{\pm 1}\ ,$$

$$K^{\pm 1} \cdot F = q^{\mp 2} F K^{\pm 1}, \quad EF - FE = (K - K^{-1})/(q - q^{-1})\ .$$

Comultiplication $\Delta : U_q\mathfrak{sl}_2 \to U_q\mathfrak{sl}_2 \otimes U_q\mathfrak{sl}_2$ is given by $\Delta(E) = E \otimes 1 + K \otimes E$, $\Delta(F) = F \otimes K^{-1} + 1 \otimes F$, $\Delta(K^{\pm 1}) = K^{\pm} \otimes K^{\pm 1}$.

This Hopf algebra was introduced by E. Sklyanin, and its generalization $U_q\mathfrak{g}$ to the case of an arbitrary simple Lie algebra $\mathfrak{g}$ in the works of V. Drinfeld and M. Jimbo (Drinfeld (1987)). $U_q\mathfrak{g}$ is determined by the generators $\{K_j^{\pm 1}, E_j, F_j\}_{j=1,\dots,l}$ and the well known relations (Jantzen (1996)). In this setting, every simple root $\alpha_j, j = 1, \dots, l$, generates an embedding $\varphi_j : U_{q_j}\mathfrak{sl}_2 \to U_q\mathfrak{g}$ given by

$$\varphi_j : K^{\pm 1} \mapsto K_j^{\pm 1},\ \varphi_j : E \mapsto E_j,\ \varphi_j : F \mapsto F_j\ .$$

Here $q_j = q^{d_j}$ with $d_j > 0$ such that $d_i a_{ij} = a_{ji} d_j$ for all i, j.

4. Equip $U_q\mathfrak{g}$ with a structure of graded algebra:

$$\deg K_j \ = \ \deg E_j \ = \ \deg F_j \ = \ 0, \quad j \neq j_0$$

$$\deg K_{j_0} \ = \ 0, \quad \deg E_{j_0} \ = \ 1, \quad \deg F_{j_0} \ = \ -1\ .$$

The embedding $\mathfrak{g}_{-1} \subset U\mathfrak{g}$ has no good q-analog. This forces us to use the generalized Verma modules instead of $U\mathfrak{g}$.

Let V be a graded $U_q\mathfrak{g}$-module determined by its generator $v \in V$ and the relations

$$E_i v = 0, \quad K_i^{\pm 1} v = v, \quad i = 1, \dots, l,$$

$$F_j v = 0, \quad j \neq j_0\ .$$

In the classic limit $q \to 1$ there is an embedding $\mathfrak{g}_{-1} \hookrightarrow V$, $\xi \mapsto \xi v$. This allows one to treat the homogeneous component $V_{-1} = \{v \in V \,|\, \deg v = -1\}$ as a q-analog of the vector space $\mathfrak{g}_{-1}$.

We need also the graded $U_q\mathfrak{g}$-module V' given by its generator v' and the relations

$$E_i v' = 0, \quad K_i^{\pm 1} v' = q^{\mp a_{ij_0}} v', \quad i = 1, \ldots, l \,,$$

$$F_j^{-a_{ij_0}+1} v' = 0, \quad j \neq j_0; \quad \deg v' = -1 \,.$$

5. Introduce the notation $\mathbb{C}[\mathfrak{g}_{-1}]_q = \bigoplus_m (V_m)^*$, $\bigwedge^1(\mathfrak{g}_{-1})_q = \bigoplus_m (V'_m)^*$ for the dual to the $U_q\mathfrak{g}$-modules V and V' respectively graded $U_q\mathfrak{g}$-modules. The elements $f \in \mathbb{C}[\mathfrak{g}_{-1}]_q$ are be called holomorphic polynomials, and the elements $\omega \in \bigwedge^1(\mathfrak{g}_{-1})_q$ differential 1-forms. The linear operator $d : \mathbb{C}[\mathfrak{g}_{-1}]_q \to \bigwedge^1(\mathfrak{g}_{-1})_q$ is defined via the adjoint operator $d^* : V' \to V$. In turn, d^* is defined as the unique $U_q\mathfrak{g}$-module morphism with $d^* : v' \mapsto F_{j_0} v$. Evidently, the differential d is a morphism of $U_q\mathfrak{g}^{op}$-modules.

6. Besides the comultiplication $\Delta : U_q\mathfrak{g} \to U_q\mathfrak{g} \otimes U_q\mathfrak{g}^{op}$ we need also the opposite comultiplication Δ^{op}. It is used to equip the vector spaces $V \otimes V$, $V \otimes V'$, $V' \otimes V$ with a structure of $U_q\mathfrak{g}$-modules.

The maps $v \mapsto v \otimes v$, $v' \mapsto v \otimes v'$, $v' \mapsto v' \otimes v$ admit the unique extensions to morphisms of $U_q\mathfrak{g}$-modules

$$V \to V \otimes V, \quad V' \to V \otimes V', \quad V' \to V' \otimes V \,.$$

The adjoint operator to the comultiplication $V \to V \otimes V$ equips $\mathbb{C}[\mathfrak{g}_{-1}]_q$ with a structure of associative algebra. Similarly, the operators dual to the above morphisms $V' \to V \otimes V'$, $V' \to V' \otimes V$ equip $\bigwedge^1(\mathfrak{g}_{-1})_q$ with a structure of a bimodule over $\mathbb{C}[\mathfrak{g}_{-1}]_q$.

It is easy to show that $\mathrm{d}(f_1 f_2) = \mathrm{d}f_1 \cdot f_2 + f_1 \cdot \mathrm{d}f_2$ for all $f_1, f_2 \in \mathbb{C}[\mathfrak{g}_{-1}]_q$. This allows one to pass from the 1-forms to the higher differential forms (see, for instance, the construction of G. Maltsiniotis (Maltsiniotis (1993))).

7. It is possible to describe the above algebras by their generators and relations. Even in the simplest case $\mathfrak{g} = \mathfrak{sl}_N$ our approach yields the profound results (Sinel'shchikov, Vaksman (to be published)).

It should be noted that our approach to the construction of algebras of differential forms is completely analogous to that of V. G. Drinfeld to the construction of the algebra of functions on a formal quantum group (Drinfeld (1987)).

8. If we replace in the above construction the $U_q\mathfrak{g}$-module V with a highest weight vector by the $U_q\mathfrak{g}$-module with a lowest weight vector, we obtain the algebra $\mathbb{C}[\overline{\mathfrak{g}}_{-1}]_q$ of antiholomorphic polynomials on the quantum vector space $\mathfrak{g}_{-1}$.

The tensor product

$$\mathrm{Pol}(\mathfrak{g}_{-1})_q = \mathbb{C}[\mathfrak{g}_{-1}]_q \otimes \mathbb{C}[\overline{\mathfrak{g}}_{-1}]_q$$

is equipped with a structure of algebra by means of the universal R-matrix together with the corresponding "commutativity morphism" (Drinfeld (1987)):

$$\check{R} : \mathbb{C}[\overline{\mathfrak{g}}_{-1}]_q \otimes \mathbb{C}[\mathfrak{g}_{-1}]_q \to \mathbb{C}[\mathfrak{g}_{-1}]_q \otimes \mathbb{C}[\overline{\mathfrak{g}}_{-1}]_q \ .$$

9. For the sake of passage from "complex quantum Lie groups to real ones" equip the Hopf algebra $U_q\mathfrak{g}$ with an involution:

$$E_j^* = \begin{cases} K_jF_j \big| j \neq j_0 \\ -K_jF_j \big| j = j_0 \end{cases}, \qquad F_j^* = \begin{cases} E_jK_j^{-1} \big| j \neq j_0 \\ -E_jK_j^{-1} \big| j = j_0 \end{cases},$$

$(K_j^{\pm 1})^* = K_j^{\pm 1}, \quad i,j \in \{1,\dots,l\}$.

The involution in $\mathrm{Pol}(\mathfrak{g}_{-1})_q$ presented in (Sinelshchikov, Vaksman (1997)) possess the property $\forall \xi \in U_q\mathfrak{g},\ f \in \mathrm{Pol}(\mathfrak{g}_{-1})_q \qquad (\xi f)^* = (S(\xi))^* f^*$.

In the simplest case $\mathfrak{g} = \mathfrak{sl}_2$ one obtains the well known $*$-algebra given by the generators z, z^* and the relation $z^*z - q^2zz^* = 1 - q^2$ (Sinelshchikov, Vaksman (1997)).

The passage from the polynomial algebra to the algebra of continuous functions in the closure of a bounded symmetric domain is made by means of a C*-completion. In the special case of quantum disc this argument was used, in particular, in (Nagy, Nica (1994)).

In (Sinel'shchikov, Shklyarov, Vaksman (to be published)) the q-analogs for the basic integral representations of the function theory in the quantum disc were obtained. Besides, there are several results for the quantum ball (Vaksman (1995)).

10. Finally, we express our gratitude to V. P. Akulov for helpful discussions of the results of this work.

References

Drinfeld, V.G. (1987): *Quantum groups*, in Proceedings of the International Congress of Mathematicians, Berkeley, 1986, A. M. Gleason (ed.), American Mathematical Society, Providence, R. I., 798 - 820.

Helgason, S. (1962): Differential Geometry and Symmetric Spaces, Acad. Press, N.-Y. - London.

Hua L.-K. (1963): *Harmonic analysis of functions of several complex variables in the classical domains*, Transl. Math. Mono, Vol. 6, Amer. Math. Soc., Providence.

Jantzen, J. C. (1996): Lectures on Quantum Groups. Graduate studies in Mathematics, Vol. 6, Amer. Math. Soc.

Maltsiniotis, G. (1993): *Le langage des espaces et des groupes quantiques*, Commun. Math. Phys., **151**, 275 - 302.

Nagy, G., Nica, A. (1994): *On the "quantum disc" and a "non-commutative circle"*, in: Algebraic Methods on Operator Theory, R. E. Curto, P. E. T. Jorgensen (eds.), Birkhauser, Boston, p. 276 - 290.
Sinel'shchikov, S., and Vaksman L. (1997): On q-analogues of bounded symmetric domains and Dolbeault complexes. Preprint 1997, q-alg/9703005.
Sinel'shchikov, S., Shklyarov, D., and Vaksman, L. (to be published): On function theory in quantum disc: integral representations.(to appear in q-alg)
Sinel'shchikov, S. and Vaksman, L. (to be published): *Hidden symmetry of the differential calculus on the quantum matrix space*, to appear in J. Phys. A.
Vaksman, L.L. (1995): *Integral intertwinning operators and quantum homogeneous spaces*, Theoretical and Mathematical Physics, **105**, 3.

q-Differential Calculus and Deformed Light–Cone

V. P. Akulov[1], V.V. Chitov[2] and S. Duplij[3]

[1] Theory Division, Kharkov Institute of Physics and Technology, Kharkov 310108, Ukraine
[2] Department of Physics and Technology, Kharkov State University, Kharkov 310077, Ukraine
[3] Alexander von Humboldt Fellow.
On leave of absence from Theory Division, Nuclear Physics Laboratory, Kharkov State University,
Kharkov 310077, Ukraine,
E-mail: duplij@physik.uni-kl.de
Internet: http://gluon.physik.uni-kl.de/~duplij;
Physics Department, University of Kaiserslautern, Postfach 3049, D-67653 Kaiserslautern, Germany

Abstract. We propose a "short" version of q-deformed differential calculus on the light-cone using twistor representation. The commutation relations between coordinates and momenta are obtained. The quasi-classical limit introduced gives an exact shape of the off-shell shifting.

The deformed differential calculus (Wess and Zumino (1990)) plays an important role in physical applications. In this connection the light-cone approach attracts the special interest, because it is employed for the description of massless particles, null-strings and null-membranes (Green, Schwarz and Witten (1987)). The usual formalism of the bicovariant differential calculi (Faddeev and Pyatov (1994)) and right (left) invariant differential calculi (Akulov and Gershun (1995)) are not applicable on the light-cone due to vanishing of the corresponding quantum determinant. That is the reason why the Maurer-Cartan forms cannot be defined here at all, and therefore it is not possible to build the deformed differential calculus on the light-cone in the standard way.

From general mathematical viewpoint some objects under consideration belong to the class of quantum semigroups (Demidov (1993)) which consist of the standard quantum groups and ideals (for abstract semigroup theory see (Clifford and Preston (1961)) and for application to supersymmetry see (Duplij (1996),Duplij (1997))).

Here we construct a version of q-deformed differential calculus on the light-cone which is mostly close to the standard one (Zupnik (1993)). We therefore hope that the obtained calculus can be directly used in physical applications.

Let us remind the deformed differential calculi on the quantum plane $E_q(2)$ (Wess and Zumino (1990))

$$xy = qyx. \tag{1}$$

In the standard case we have two solutions for $E_q\,(2)$

a)

$$\begin{cases} \delta x\; x = q^2 x\; \delta x\;, \\ \delta x\; y = qy\; \delta x + (q^2-1)\; x\; \delta y\;, \\ \delta y\; x = qx\; \delta y\;, \\ \delta y\; y = q^{-2} y\; \delta y\;; \end{cases} \tag{2}$$

b)

$$\begin{cases} \delta x\; x = q^2 x\; \delta x\;, \\ \delta x\; y = q^{-1} y\; \delta x\;, \\ \delta y\; x = q^{-1} x\; \delta y\; + (q^{-2}-1)\; y\; \delta x\;, \\ \delta y\; y = q^{-2} y\; \delta y\;. \end{cases} \tag{3}$$

It is well-known Faddev, Reshetikhin, Takhtajan (1990) that there exists the automorphism of the quantum plane (1)

$$\begin{pmatrix} x \to y \\ y \to x \\ q \to q^{-1} \end{pmatrix} \tag{4}$$

which is stipulated by the following property of the R-matrix

$$R_q \to R^{-1}_{q^{-1}}\;. \tag{5}$$

If we introduce the evolution parameter and consider the dynamics on $E_q\,(2)$ where (2) are commutation relations between momentum and coordinate, then we come to very complicated and inappropriate conditions for the phase space. The reason lays in the second term of the second formula in (2). Such undesirable terms (so called "long" solutions) result in difficulties while deriving and exploiting of self-consistent Poisson bracket.

In search of "short" solutions we found the following ones on $E_q\,(2)$ additionally to (2) and (3)

$$\begin{cases} \delta x\; x = x\; \delta x\;, \\ \delta x\; y = qy\; \delta x\;, \\ \delta y\; x = q^{-1} x\; \delta y\;, \\ \delta y\; y = y\; \delta y\;. \end{cases} \tag{6}$$

In formulas (2), (3) and (6) δ is the standard exterior differential satisfying the classical Leibniz rule, lemma Poincare and having the properties $\delta x\; \delta y\; = -q\delta y\; \delta x\;$, $(\delta x\;)^2 = 0$, $(\delta y\;)^2 = 0$.

In contrast to the "long" solutions (2) and (3) which are transformed one to another by the automorphism (4), the "short" solution (6) is its fixed point.

Let us consider q-deformed null-vector in 4-dimensional Minkowski space which is described by the following (2×2) q-matrix

$$X = \begin{pmatrix} x^{1\dot{1}} & x^{1\dot{2}} \\ x^{2\dot{1}} & x^{2\dot{2}} \end{pmatrix} . \tag{7}$$

We define the q-deformed light-cone as

$$\det_{q^2} X = x^{1\dot{1}} x^{2\dot{2}} - q^2 x^{1\dot{2}} x^{2\dot{1}} = 0 . \tag{8}$$

Due to multiplicativity of the quantum determinant the Lorentz transformations (represented by $GL_{q^2,\frac{1}{q^2}}(2,C)$ matrices Schirrmacher, Wess and Zumino (1991)) do not change the light-cone condition (8).

The q-deformed components have the following commutation relations

$$\begin{cases} x^{1\dot{1}} x^{1\dot{2}} = q^2 x^{1\dot{2}} x^{1\dot{1}} , \\ x^{1\dot{1}} x^{2\dot{1}} = q^2 x^{2\dot{1}} x^{1\dot{1}} , \\ x^{1\dot{2}} x^{2\dot{1}} = x^{2\dot{1}} x^{1\dot{2}} , \\ x^{1\dot{2}} x^{2\dot{2}} = q^2 x^{2\dot{2}} x^{1\dot{2}} , \\ x^{2\dot{1}} x^{2\dot{2}} = q^2 x^{2\dot{2}} x^{2\dot{1}} , \\ x^{1\dot{1}} x^{2\dot{2}} - x^{2\dot{2}} x^{1\dot{1}} = \left(q^2 - q^{-2}\right) x^{1\dot{2}} x^{2\dot{1}} . \end{cases} \tag{9}$$

As usual a null vector has a twistor representation. In q-deformed case we introduce the twistors having q-deformed components

$$\varphi_q = \left(\varphi_q^A\right) = \begin{pmatrix} x \\ y \end{pmatrix} , \ A = 1, 2 \tag{10}$$

where x and y satisfy (1).

The q-deformed Levi-Chivita tensor is

$$\epsilon_q = \left(\epsilon_q^{AB}\right) = \begin{pmatrix} 0 & q^{1/2} \\ -q^{-1/2} & 0 \end{pmatrix} \tag{11}$$

and has the property

$$\epsilon_q^{AB} \epsilon_{q,BA} = q + q^{-1} . \tag{12}$$

The q-antisymmetry of ϵ_q leads to

$$\varphi_q^A \varphi_q^B \epsilon_{q,AB} = \varphi_q^A \varphi_{q,A} = 0 .$$

The complex conjugated q-deformed twistor is

$$\overline{\varphi}_q = \left(\overline{\varphi}_q^{\dot{A}}\right) = (\overline{x}, \overline{y}) .$$

We define here the involution $x \to \bar{x}$ as the standard Hermitian conjugation.

The q-null-vector in terms of q-twistor components has the standard form

$$X^{A\dot{A}} = \varphi_q^A \overline{\varphi}_q^{\dot{A}} = \begin{pmatrix} x\overline{x} & x\overline{y} \\ y\overline{x} & y\overline{y} \end{pmatrix} . \tag{13}$$

In classical case $\varphi_q^A \to \varphi^A$ where φ^A is a commuting Majorana-Weil spinor for $D = 10$, Weil spinor for $D = 6$ and Majorana spinor for $D = 3, 4$, in correspondence to the division algebras $\mathcal{O}, \mathcal{H}, \mathcal{C}, \mathcal{R}$.

The light-cone condition (8) in terms of q-twistor components is

$$\det_{q^2} X = x\overline{x}y\overline{y} - q^2 x\overline{y}y\overline{x} = 0 . \tag{14}$$

Here we conisider the complex case $D = 4$ only, but all the results can be obviously extended on other division algebras.

So we have two copies of the quantum plane (x, y) and $(\overline{x}, \overline{y})$. Now we need such commutation relations between components of the q-twistor and its conjugate which satisfy (9) and the standard involution.

From (1) and its conjugate we easily obtain the condition

$$\overline{q} = q^{-1} = \exp(-ih) \tag{15}$$

or $|q| = 1$.

Let other commutation relations have the general form

$$\begin{cases} x\overline{x} = q^n \overline{x}x, \\ x\overline{y} = q^m \overline{y}x, \\ y\overline{x} = q^k \overline{x}y, \\ y\overline{y} = q^l \overline{y}y, \end{cases} \tag{16}$$

where n, m, k, l are arbitrary constants which should be determined from (9) and involution. So we have the following independent equations

$$\begin{cases} m - n = 1, \\ n - k = 1, \\ l - k = 1. \end{cases}$$

Then

$$\begin{cases} x\overline{x} = q^n \overline{x}x, \\ x\overline{y} = q^{n+1} \overline{y}x, \\ y\overline{x} = q^{n-1} \overline{x}y, \\ y\overline{y} = q^n \overline{y}y. \end{cases} \tag{17}$$

Now we consider the reality condition, as the consequence of the possibility to reduce the dimension of q-deformed null-vector up to three where it is real. So we derive $n = 0$, and the commutation relations between twistor components become

$$\begin{cases} xy = qyx, \\ x\overline{y} = q\overline{y}x, \\ \overline{x}y = qy\overline{x}, \\ \overline{x}\overline{y} = q\overline{y}\overline{x}, \\ \overline{x}x = x\overline{x}, \\ \overline{y}y = y\overline{y}. \end{cases} \tag{18}$$

Then we can find the "short" version of q-deformed differential calculus on twistor components

$$\begin{cases} \delta x\; x = x\; \delta x\ , \\ \delta x\; \bar{x} = \bar{x}\; \delta x\ , \\ \delta x\; y = qy\; \delta x\ , \\ \delta x\; \bar{y} = q\bar{y}\; \delta x\ , \end{cases} \quad \begin{cases} \delta \bar{x}\; x = x\; \delta \bar{x}\ , \\ \delta \bar{x}\; \bar{x} = \bar{x}\; \delta \bar{x}\ , \\ \delta \bar{x}\; y = qy\; \delta \bar{x}\ , \\ \delta \bar{x}\; \bar{y} = q\bar{y}\; \delta \bar{x}\ , \end{cases} \tag{19}$$

$$\begin{cases} \delta y\; x = q^{-1}x\; \delta y\ , \\ \delta y\; \bar{x} = q^{-1}\bar{x}\; \delta y\ , \\ \delta y\; y = y\; \delta y\ , \\ \delta y\; \bar{y} = \bar{y}\; \delta y\ , \end{cases} \quad \begin{cases} \delta \bar{y}\; x = q^{-1}x\; \delta \bar{y}\ , \\ \delta \bar{y}\; \bar{x} = q^{-1}\bar{x}\; \delta \bar{y}\ , \\ \delta \bar{y}\; y = y\; \delta \bar{y}\ , \\ \delta \bar{y}\; \bar{y} = \bar{y}\; \delta \bar{y}\ , \end{cases} \tag{20}$$

where $(\delta x\,)^2 = (\delta \bar{x}\,)^2 = (\delta y\,)^2 = (\delta \bar{y}\,)^2 = 0$.

The commutation relations between the differentials themselves take the form

$$\begin{cases} \delta x\; \delta \bar{x} &= -\delta \bar{x}\; \delta x\ , \\ \delta x\; \delta y &= -q\delta y\; \delta x\ , \\ \delta x\; \delta \bar{y} &= -q\delta \bar{y}\; \delta x\ , \\ \delta y\; \delta x &= -q^{-1}\; \delta x\; \delta y\ , \\ \delta y\; \delta \bar{x} &= -q^{-1}\; \delta \bar{x}\; \delta y\ , \\ \delta y\; \delta \bar{y} &= -\delta \bar{y}\; \delta y\ . \end{cases} \tag{21}$$

Using the obtained q-deformed differential calculus on q-twistors we can build q-deformed differential calculus on any composite objects, as q-deformed null-vectors and tensors.

So for q-deformed null-vector we obtain

$$\begin{cases} \delta x^{1\dot{1}} x^{1\dot{1}} = x^{1\dot{1}}\; \delta x^{1\dot{1}}, \\ \delta x^{1\dot{1}} x^{1\dot{2}} = q^2 x^{1\dot{2}}\; \delta x^{1\dot{1}}, \\ \delta x^{1\dot{1}} x^{2\dot{1}} = q^2 x^{2\dot{1}}\; \delta x^{1\dot{1}}, \\ \delta x^{1\dot{1}} x^{2\dot{2}} = q^4 x^{2\dot{2}}\; \delta x^{1\dot{1}}, \end{cases} \quad \begin{cases} \delta x^{1\dot{2}} x^{1\dot{2}} = x^{1\dot{2}}\; \delta x^{1\dot{2}}, \\ \delta x^{1\dot{2}} x^{2\dot{1}} = x^{2\dot{1}}\; \delta x^{1\dot{2}}, \\ \delta x^{1\dot{2}} x^{1\dot{1}} = q^{-2} x^{1\dot{1}}\; \delta x^{1\dot{2}}, \\ \delta x^{1\dot{2}} x^{2\dot{2}} = q^2 x^{2\dot{2}}\; \delta x^{1\dot{2}}, \end{cases} \tag{22}$$

$$\begin{cases} \delta x^{2\dot{1}} x^{2\dot{1}} = x^{2\dot{1}}\; \delta x^{2\dot{1}}, \\ \delta x^{2\dot{1}} x^{1\dot{1}} = q^{-2} x^{1\dot{1}}\; \delta x^{2\dot{1}}, \\ \delta x^{2\dot{1}} x^{1\dot{2}} = x^{1\dot{2}}\; \delta x^{2\dot{1}}, \\ \delta x^{2\dot{1}} x^{2\dot{2}} = q^2 x^{2\dot{2}}\; \delta x^{2\dot{1}}, \end{cases} \quad \begin{cases} \delta x^{2\dot{2}} x^{2\dot{2}} = x^{2\dot{2}}\; \delta x^{2\dot{2}}, \\ \delta x^{2\dot{2}} x^{1\dot{1}} = q^{-4} x^{1\dot{1}}\; \delta x^{2\dot{2}}, \\ \delta x^{2\dot{2}} x^{1\dot{2}} = q^{-2} x^{1\dot{2}}\; \delta x^{2\dot{2}}, \\ \delta x^{2\dot{2}} x^{2\dot{1}} = q^{-2} x^{2\dot{1}}\; \delta x^{2\dot{2}}, \end{cases} \tag{23}$$

and

$$\begin{aligned} \delta x^{1\dot{1}}\; \delta x^{1\dot{2}} &= -q^2 \delta x^{1\dot{2}}\; \delta x^{1\dot{1}}, \\ \delta x^{1\dot{1}}\; \delta x^{2\dot{1}} &= -q^2 \delta x^{2\dot{1}}\; \delta x^{1\dot{1}}, \\ \delta x^{1\dot{1}}\; \delta x^{2\dot{2}} &= -q^4 \delta x^{2\dot{2}}\; \delta x^{1\dot{1}}, \\ \delta x^{1\dot{2}}\; \delta x^{2\dot{1}} &= \ -\delta x^{2\dot{1}}\; \delta x^{1\dot{2}}, \\ \delta x^{1\dot{2}}\; \delta x^{2\dot{2}} &= -q^2 \delta x^{2\dot{2}}\; \delta x^{1\dot{2}}, \\ \delta x^{2\dot{1}}\; \delta x^{2\dot{2}} &= -q^2 \delta x^{2\dot{2}}\; \delta x^{2\dot{1}}. \end{aligned} \tag{24}$$

From the above formulas we can find the commutation relations between the derivatives

$$\begin{aligned}
\frac{\partial}{\partial x^{1\dot{1}}}\frac{\partial}{\partial x^{1\dot{2}}} &= q^2\frac{\partial}{\partial x^{1\dot{2}}}\frac{\partial}{\partial x^{1\dot{1}}},\\
\frac{\partial}{\partial x^{1\dot{1}}}\frac{\partial}{\partial x^{2\dot{1}}} &= q^2\frac{\partial}{\partial x^{2\dot{1}}}\frac{\partial}{\partial x^{1\dot{1}}},\\
\frac{\partial}{\partial x^{1\dot{1}}}\frac{\partial}{\partial x^{2\dot{2}}} &= q^4\frac{\partial}{\partial x^{2\dot{2}}}\frac{\partial}{\partial x^{1\dot{1}}},\\
\frac{\partial}{\partial x^{1\dot{2}}}\frac{\partial}{\partial x^{2\dot{1}}} &= \frac{\partial}{\partial x^{2\dot{1}}}\frac{\partial}{\partial x^{1\dot{2}}},\\
\frac{\partial}{\partial x^{1\dot{2}}}\frac{\partial}{\partial x^{2\dot{2}}} &= q^2\frac{\partial}{\partial x^{2\dot{2}}}\frac{\partial}{\partial x^{1\dot{2}}},\\
\frac{\partial}{\partial x^{2\dot{1}}}\frac{\partial}{\partial x^{2\dot{2}}} &= q^2\frac{\partial}{\partial x^{2\dot{2}}}\frac{\partial}{\partial x^{2\dot{1}}}.
\end{aligned} \tag{25}$$

By analogy we can write the commutation relation between q-deformed differentials and derivatives.

The above relations are consistent with the following condition

$$\delta\left(\det_{q^2} X\right) = 0. \tag{26}$$

Let us introduce q-matrix for momenta

$$P_q = \begin{pmatrix} P_{1\dot{1}} & P_{1\dot{2}} \\ P_{2\dot{1}} & P_{2\dot{2}} \end{pmatrix} = \begin{pmatrix} -i\frac{\partial}{\partial x^{1\dot{1}}} & -i\frac{\partial}{\partial x^{1\dot{2}}} \\ -i\frac{\partial}{\partial x^{2\dot{1}}} & -i\frac{\partial}{\partial x^{2\dot{2}}} \end{pmatrix} \tag{27}$$

with commutation relations determined by (25).

The q-D'Alembertian is defined by

$$\Box_{q^2} = -\det_{q^2} P_q = \frac{\partial}{\partial x^{1\dot{1}}}\frac{\partial}{\partial x^{2\dot{2}}} - q^2\frac{\partial}{\partial x^{1\dot{2}}}\frac{\partial}{\partial x^{2\dot{1}}} \ . \tag{28}$$

The light-cone condition (8) leads to the similar condition for momenta

$$\Box_{q^2} = 0 \ . \tag{29}$$

In the classical limit $q = \exp(ih) \to 1$ we decompose the q-D'Alembertian as follows

$$\begin{aligned}
\Box_{q^2} &= \Box - \left(q^2 - 1\right)\frac{\partial}{\partial x^{1\dot{2}}}\frac{\partial}{\partial x^{2\dot{1}}}\\
&= \Box - 2ihP_{1\dot{2}}P_{2\dot{1}} = 0 \ ,
\end{aligned} \tag{30}$$

where $\Box$ is the ordinary D'Alembertian.

The second term in (30) gives us a new way of the off-shell approximation and is responsible for its exact shape. In contrast to the standard picture Azcarraga, Kulish and Rodenas (1994) the momenta entering into the additional off-shell term in (30) commute.

We have proposed a version of the differential calculus on q-deformed light-cone which can be applied to description of the dynamics of the massless quantum particles, q-deformed null-strings and null-membranes.

Using the obtained q-deformed differential calculus on q-twistors we have the possibility to construct the corresponding calculi on q-tensors of any rank.

The work is partially supported by INTAS-(93/0493ext and 93/0127ext).

References

Akulov, V. P. and Gerschun, V. D. (1995): Matched reduction of differential calculus on quantum groups $GL_q(2,c)$, $SL_q(2,C)$ and $E_q(2)$., preprint q-alg/9509030.

Azcarraga, J. A., Kulish, P. P., and Rodenas, F. (1994): Quantum groups and deformed special relativity, preprint FTUV 94-21, hep-th/9405161.

Clifford, A. H. and Preston, G. B. (1961): *The Algebraic Theory of Semigroups*, Vol. 1, Amer. Math. Soc., Providence.

Demidov, E. E. (1993): Some aspects of the theory of quantum groups, *Russian Math. Surv.* **48**, 41–79.

Duplij, S. (1996): On an alternative supermatrix reduction, *Lett. Math. Phys.* **37**, 385–396.

Duplij, S. (1997): Some abstract properties of semigroups appearing in superconformal theories, *Semigroup Forum* **54**, 253–260.

Faddeev, L. D. and Pyatov, P. N. (1984): The differential calculus on linear quantum groups, preprint hep-th/9402070.

Faddeev, L. D., Reshetikhin, N. Y., and Takhtajan, I. A. (1990): Quantum lie groups and lie algebras, *Leningrad Math. J.* **1**, 193–236.

Green, M. B., Schwarz, J. H., and Witten, E. (1987): *Superstring Theory*, Vol. 1,2, Cambridge Univ. Press, Cambridge, 1987.

Schirrmacher, A., Wess, J., and Zumino, B. (1991): The two parameter deformation of $GL(2)$ its differential calculus, and Lie algebra, *Z. Phys.* **49**, 317–321.

Wess, J. and Zumino, B. (1990): Covariant differential calculus on the quantum hyperplane, *Nucl. Phys. (Proc. Suppl.)* **B18**, 302–312.

Zupnik, B. M. (1993): Minimal deformations of the commutative algebra and the linear group $GL(n)$, *Theor. Math. Phys.* **95**, 403–415.

σ-Models on the Quantum Group Manifolds $SL_q(2,R)$, $SL_q(2,R)/U_h(1)$, $C_q(2|0)$ and Infinitesimal Transformations

V.D. Gershun

Kharkov Institute of Physics and Technology, 310108, Kharkov, Ukraine

Abstract. The differential and variational calculus on the $SL_q(2,R)$ group is constructed. The spontaneous breaking symmetry in the WZNW model with $SL_q(2,R)$ quantum group symmetry and in the σ-models with $SL_q(2,R)/U_h(1)$,$C_q(2|0)$ quantum group symmetry is considered. The Lagrangian formalism over the quantum group manifolds is discussed. The classical solution of $C_q(2|0)$ σ-model is obtained.

1 Differential Calculus on the $SL_q(2,R)$ Group

The matrix quantum group (Faddeev, Reshetikhin, Takhtajan (1989)) $G = SL_q(2,R)$ is defined by the q-commutation relations (C.R.) of its group parameters. Let

$$g = \begin{pmatrix} a^1 & a^2 \\ a^3 & a^4 \end{pmatrix},$$

$$\begin{array}{lll} a^1a^2 = qa^2a^1, & a^2a^4 = qa^4a^2, & a^2a^3 = a^3a^4 \\ a^1a^3 = qa^3a^1, & a^3a^4 = qa^4a^3, & a^1a^4 = a^4a^1 + (q-q^{-1})a^2a^3 \end{array} \tag{1}$$

a^k - hermitian, $|q| = 1, Det_q g = a^1a^4 - qa^2a^3 = 1$.

For any elements g, $g' \in SL_q(2,R)$ element $g'' = g'g$ will belong $SL_q(2,R)$ if $a^{k'}a^l = a^la^{k'}$. In the Gauss decomposition (Akulov et al. (1992))

$$g = \begin{pmatrix} 1 & \varphi_- \\ 0 & 1 \end{pmatrix} \begin{pmatrix} 1 & 0 \\ \varphi_+ & 1 \end{pmatrix} \begin{pmatrix} \rho & 0 \\ 0 & \rho^{-1} \end{pmatrix} = \begin{pmatrix} \rho + \varphi_-\varphi_+\rho & \varphi_-\rho^{-1} \\ \varphi_+\rho & \rho^{-1} \end{pmatrix} \tag{2}$$

the C.R. are:

$$\rho\varphi_\pm = q\varphi_\pm\rho, \quad \varphi_-\varphi_+ = q^2\varphi_+\varphi_- \tag{3}$$

Let the quantum group is a manifolds of any possible transformations $g' = gg_0$. There are two kinds of the variation: the variation in the neighborhood of the arbitrary point of the group space $g' = g + \mathrm{d}g$ and variation in the neighborhood of the unit of the group $g = 1 + \delta g$. First variation defines the group invariants: element of the distance between two neighboring points, element of the volume around the point. Second variation defines the group symmetry of this invariants. The C.R. between variation $\mathrm{d}g$ and g define the type of the differential calculus.

The left-invariant differential calculus (Akulov et al (1993)) on the $GL_q(2,C)$ group, matched with the differential calculi on the $SL_q(2,C)$ subgroup and on the Borel subgroups $B_L(C), B_U(C)$, was constructed in (Akulov, Gershun (1995), Akulov, Gershun (1995a)). Let $\omega = g^{-1}dg$ is the left differential Kartan 1-form

$$\omega = \begin{pmatrix} \omega^1 & \omega^2 \\ \omega^3 & \omega^4 \end{pmatrix}, \quad \mathrm{Tr}_q\omega = q^2\omega^1 + \omega^4 = 0 \tag{4}$$

The differential calculus on the $SL_q(2,R)$ group is defined by the C.R.

$$\begin{array}{lll} \omega^1\rho = \frac{1}{q^2}\rho\omega^1, & \omega^2\rho = \frac{1}{q}\rho\omega^2, & \omega^3\rho = \frac{1}{q}\rho\omega^3 \\ \omega^1\varphi_\pm = \varphi_\pm\omega^1, & \omega^2\varphi_\pm = \varphi_\pm\omega^2, & \omega^3\varphi_\pm = \varphi_\pm\omega^3 \end{array} \tag{5}$$

The C.R. between the group parameters and their differentials are more complicated:

$$\begin{array}{ll} \mathrm{d}\rho\rho = \frac{1}{q^2}\rho\mathrm{d}\rho, & \mathrm{d}\varphi_+\varphi_+ = \frac{1}{q^2}\varphi_+\mathrm{d}\varphi_+ + (q^4-1)\varphi_+^3\mathrm{d}\varphi_- \\ \mathrm{d}\varphi_-\varphi_- = q^2\varphi_-\mathrm{d}\varphi_-, & \mathrm{d}\rho\varphi_- = q\varphi_-\mathrm{d}\rho \end{array}$$

$$\begin{array}{ll} \mathrm{d}\varphi_-\varphi_+ = q^2\varphi_+\mathrm{d}\varphi_-, & \mathrm{d}\varphi_+\varphi_- = \frac{1}{q^2}\varphi_-\mathrm{d}\varphi_+ \\ \mathrm{d}\rho\mathrm{d}\varphi_- = -q\mathrm{d}\varphi_-\mathrm{d}\rho & \mathrm{d}\varphi_-\mathrm{d}\varphi_+ = -q^2\mathrm{d}\varphi_+\mathrm{d}\varphi_- \\ \mathrm{d}\varphi_-\rho = \frac{1}{q}\rho\mathrm{d}\varphi_- & \\ & \mathrm{d}\varphi_+\rho = \frac{1}{q}\rho\mathrm{d}\varphi_+ - q(q^2-1)\varphi_+^2\rho\mathrm{d}\varphi_- \\ & \mathrm{d}\rho\varphi_+ = q\varphi_+\mathrm{d}\rho - q^2(q^2-1)\varphi_+^2\rho\mathrm{d}\varphi_- \end{array} \tag{6}$$

$$\mathrm{d}\rho\mathrm{d}\varphi_+ + q\mathrm{d}\varphi_+\mathrm{d}\rho + q^3(q^2-1)\varphi_+^2\mathrm{d}\varphi_-\mathrm{d}\rho - \frac{(q^4-1)}{q^3}\varphi_+\rho\mathrm{d}\varphi_-\mathrm{d}\varphi_+ = 0$$

The Kartan 1-forms are:

$$\begin{array}{ll} \omega^1 = \rho^{-1}\mathrm{d}\rho + \varphi_+\mathrm{d}\varphi_-, & \omega^3 = \frac{1}{q}\rho^2\mathrm{d}\varphi_+ - q^5\varphi_+^2\rho^2\mathrm{d}\varphi_- \\ \omega^2 = q\rho^{-2}\mathrm{d}\varphi_- & \end{array} \tag{7}$$

$$\begin{array}{ll} (\omega^1)^2 = (\omega^2)^2 = (\omega^3)^2 = 0, & \omega^4 = -q^2\omega^1 \\ \omega^1\omega^2 + q^4\omega^2\omega^1 = 0 & \omega^1\omega^3 + q^{-4}\omega^3\omega^1 = 0 \\ \omega^2\omega^3 + q^{-2}\omega^3\omega^2 = 0 & \end{array}$$

The left vector fields ∇_k can be obtained from the applying the left differential to an arbitrary function on the quantum group $df = (f\frac{\partial}{\partial a^k})da^k = (f\nabla_k)\omega^k$.

$$\nabla = \begin{pmatrix} \nabla_1 & \nabla_2 \\ \nabla_3 & \nabla_4 \end{pmatrix} \quad \hat{\nabla}_1 = \nabla_1 - q^2\nabla_4 \quad \hat{\nabla}_4 = \nabla_1 + \nabla_4 \tag{8}$$

The C.R. for vector fields have following form

$$\begin{array}{lll} \rho\hat{\nabla}_1 = \frac{1}{q^2}\hat{\nabla}_1\rho + \rho & \varphi_-\hat{\nabla}_1 = \hat{\nabla}_1\varphi_- & \varphi_+\hat{\nabla}_1 = \hat{\nabla}_1\varphi_+ \\ \rho\nabla_2 = \frac{1}{q}\nabla_2\rho - \varphi_+\rho^3 & \varphi_-\nabla_2 = \nabla_2\varphi_- + \frac{1}{q}\rho^2 & \varphi_+\nabla_2 = \nabla_2\varphi_+ + q\varphi_+^2\rho^2 \\ \rho\nabla_3 = \frac{1}{q}\nabla_3\rho & \varphi_-\nabla_3 = \nabla_3\varphi_- & \varphi_+\nabla_3 = \nabla_3\varphi_+ + q\rho^{-2} \end{array} \tag{9}$$

$$q^2\hat{\nabla}_1\nabla_3 - \tfrac{1}{q^2}\nabla_3\hat{\nabla}_1 = (q^2+1)\nabla_3, \quad \nabla_3\nabla_2 - \tfrac{1}{q^2}\nabla_2\nabla_3 = \hat{\nabla}_1$$
$$q^2\nabla_2\hat{\nabla}_1 - \tfrac{1}{q^2}\hat{\nabla}_1\nabla_2 = (q^2+1)\nabla_2 \tag{10}$$

and ∇_k have the form

$$\hat{\nabla}_1 = \frac{\partial}{\partial\rho}\rho, \quad \nabla_2 = \frac{1}{q}\frac{\partial}{\partial\varphi_-}\rho^2 - \frac{\partial}{\partial\rho}\varphi_+\rho^3 + q\frac{\partial}{\partial\varphi_+}\varphi_+^2\rho^2, \quad \nabla_3 = q\frac{\partial}{\partial\varphi_+}\rho^{-2}$$

The left vector fields and the left derivatives act on the any function of the group parameters from the right side.

2 WZNW model on the $SL_q(2,R)$ Group

The existing of the quantum group structure in the WZNW model was shown in (Faddeev (1990), Alekseev, Shatashvili (1990)). The σ-models with a quantum group symmetry was considered in (Akulov et al. (1992), Arefeva, Volovich (1991), Frishman, Lukierski, Zakrzewski (1993), Gershun (1996)). To construct the WZNW model with $SL_q(2,R)$ group symmetry, we consider the space $M^{1,1} \oplus SL_q(2,R)$, where $M^{1,1}$ is the commutative (undeformed) space. The element of the volume in $M^{1,1}$ space, which is the invariant of $SL_q(2,R)$, is

$$\frac{\mathrm{Tr_d}[\omega(\mathrm{d})\wedge \mathrm{d}z^\mu][\omega(\mathrm{d})\wedge \mathrm{d}z^\mu]}{2\epsilon^{\lambda\rho}\mathrm{d}z^\lambda\wedge\mathrm{d}z^\rho} = \mathrm{Tr}_q(\omega_\mu\omega^\mu)\mathrm{d}^2z, \tag{11}$$

where $\omega(\mathrm{d}) = \omega_\mu \mathrm{d}z^\mu, z^\mu \epsilon M^{1,1}, \mu = 1,2$. For any 2×2 matrix $A, \mathrm{Tr}_q A = q^2A^1 + A^4$. As a result we have

$$\mathrm{Tr}_q(\omega_\mu\omega^\mu) = q^5[2]_q\rho^{-2}\partial_\mu\rho\partial^\mu\rho + q^5[2]_q\rho^{-1}\varphi_+(\partial_\mu\varphi_-\partial^\mu\rho + \frac{1}{q}\partial_\mu\rho\partial^\mu\varphi_-) +$$
$$(\partial_\mu\varphi_-\partial^\mu\varphi_+ + q^2\partial_\mu\varphi_+\partial^\mu\varphi_-) - q^2(q^4-1)\varphi_+^2\partial_\mu\varphi_-\partial^\mu\varphi_- \tag{12}$$

The C.R. are now in the same space-time point, $\mathrm{d}\rho = \partial_\mu\rho \mathrm{d}z^\mu, \mathrm{d}\varphi_\pm = \partial_\mu\varphi_\pm\mathrm{d}z^\mu$ and $[n]_q = \frac{q^n - q^{-n}}{q-q^{-1}}$. The Wess-Zumino term

$$\mathrm{Tr}_q(\omega(\mathrm{d})\wedge\omega(\mathrm{d})\wedge\omega(\mathrm{d})) = \frac{q[2]_q[3]_q}{6}\varepsilon^{\mu\nu\lambda}\partial_\lambda(\rho^{-1}\partial_\mu\rho\partial_\nu\varphi_-\varphi_+)\mathrm{d}^3z \tag{13}$$

is the total derivative.

Finally, the WZNW-action with the $SL_q(2,R)$ quantum group symmetry describes the 2-dimensional relativistic string in the background gravity and antisymmetric fields

$$S[\rho,\varphi_-,\varphi_+] = \frac{k}{4\pi}\int \mathrm{d}^2z(G_{AB}\partial_\mu X^A\partial^\mu X^B + B_{AB}\varepsilon_{\mu\nu}\partial^\mu X^A\partial^\nu X^B), \tag{14}$$

where $X^A = (\rho, \varphi_-, \varphi_+)$ and the background gravity and antisymmetric fields have the following form:

$$G_{AB} = \begin{pmatrix} q^5[2]_q\rho^{-2} & q^4[2]_q\rho^{-1}\varphi_+ & 0 \\ q^5[2]_q\rho^{-1}\varphi_+ & -q^2(q^4-1)\varphi_+^2 & 1 \\ 0 & q^2 & 0 \end{pmatrix}$$

$$B_{AB} = \frac{q^3[2]_q[3]_q}{6}\varphi_+\rho^{-1}\begin{pmatrix} 0 & 1 & 0 \\ -1 & 0 & 0 \\ 0 & 0 & 0 \end{pmatrix}$$

The group symmetry of this model is $\mathrm{SL}_q(2,R) \otimes \mathrm{SL}_q(2,R)$, because under the left multiplication on the group $g' = g_0 g$ the differential forms of Kartan are invariant, $\omega' = \omega$, and under the right multiplication $g' = g g_0$ the differential forms are covariant, $\omega' = g_0^{-1}\omega g_0$. But $\mathrm{Tr}_q A$ is invariant of the transformation $A' = g_0^{-1} A g_0$, because the elements of matrix A commute with the elements of matrix g_0, by definition of the quantum group. Therefore, this model describes the spontaneous breaking of the $\mathrm{SL}_q(2,R) \otimes \mathrm{SL}_q(2,R)$ symmetry to the $\mathrm{SL}_q(2,R)$ one.

3 σ-Model on the $\mathrm{SL}_q(2,R)/\mathrm{U}_h(1)$ Group

Let us consider the spontaneous breaking symmetry in the σ−model with the $\mathrm{SL}_q(2,R)/\mathrm{U}_h(1)$ group symmetry. Let $G = KH$, K-coset, H-subgroup. The Kartan 1-forms

$$k^{-1}\mathrm{d}k = \begin{pmatrix} q^2\varphi_+\mathrm{d}\varphi_- & \mathrm{d}\varphi_- \\ \mathrm{d}\varphi_+ - q^2\varphi_+^2\mathrm{d}\varphi_- & -\varphi_+\mathrm{d}\varphi_- \end{pmatrix} = \omega + \theta \ , \tag{15}$$

where $\omega\epsilon K, \theta\epsilon H$ and the coset elements $\varphi_\pm$ commute with the subgroup parameter ρ and satisfy to C.R. of $\mathrm{SL}_q(2,R)$ group among themselves. There is a question:how do coset and subgroup separate from $k^{-1}\mathrm{d}k$? In opposite to the classical case, there is the 3-parametric family of the $U(1)$ subgroups. The Lagrangian has the following form:

$$\begin{aligned} L_n &= \tfrac{1}{2}\mathrm{Tr}_q(\omega_\mu\omega^\mu) = \\ &\tfrac{(q^4+1)}{4q^4}(\partial_\mu\varphi_-\partial^\mu\varphi_+ + q^2\partial_\mu\varphi_+\partial^\mu\varphi_-) - c_n(q)\varphi_+^2\partial_\mu\varphi_-\partial^\mu\varphi_- \ , \end{aligned} \tag{16}$$

where $c_n(q)$ depends on the choice of a subgroup. There are three most interesting examples.

1)Undeformed $U(1)$ subgroup: $c_1 = \frac{2q^4-q^2+1}{2}$

$$\omega = \begin{pmatrix} (q^2-1)\varphi_+\mathrm{d}\varphi_- & \mathrm{d}\varphi_- \\ \mathrm{d}\varphi_+ - q^2\varphi_+^2\mathrm{d}\varphi_- & 0 \end{pmatrix}, \quad \theta = \varphi_+\mathrm{d}\varphi_-\begin{pmatrix} 1 & 0 \\ 0 & -1 \end{pmatrix} \tag{17}$$

The algebra symmetry of this Lagrangian is defined by the Maurer-Kartan equations:

$$d\theta = -\begin{pmatrix} q^{-2} & 0 \\ 0 & 1 \end{pmatrix} \omega\omega + (q^2-1)\omega\theta, \quad d\omega = -\begin{pmatrix} \frac{q^2-1}{q^2} & 0 \\ 0 & 0 \end{pmatrix} \omega\omega - q^3[2]_q\omega\theta$$

$$\theta\omega = q^4\omega\theta$$

The C.R. between the coset and the subgroup forms are common for all of the examples

$$\begin{aligned} \omega^1\omega^3 + q^4\omega^3\omega^1 = 0, \, \omega^2\omega^3 + q^2\omega^3\omega^2 = 0 \\ \omega^1\omega^3 + q^4\omega^3\omega^1 = 0, \, \omega^4\omega^3 + q^4\omega^3\omega^4 = 0 \end{aligned} \tag{18}$$

2) Classical coset structure: $c_2 = \frac{q^6+1}{4}$

$$\omega = \begin{pmatrix} 0 & d\varphi_- \\ d\varphi_+ - q^2\varphi_+^2 d\varphi_- & 0 \end{pmatrix}, \quad \theta = \varphi_+ d\varphi_- \begin{pmatrix} q^2 & 0 \\ 0 & -1 \end{pmatrix} \tag{19}$$

$$d\theta = -\omega\omega, \quad d\omega = -\omega\theta - \theta\omega, \; \theta\omega = q^2\omega\theta$$

3) There is one of the examples of the 2- parametric family $U_q(1)$ subgroups:

$$c_3 = \frac{2q^4 - q^2 + 1}{2q^2}$$

$$\omega = \begin{pmatrix} \frac{(q^2-1)}{q^2}\varphi_+ d\varphi_- & d\varphi_- \\ d\varphi_+ - q^2\varphi_+^2 d\varphi_- & 0 \end{pmatrix}, \quad \theta = \frac{1}{q^2}\varphi_+ d\varphi_- \begin{pmatrix} 1 & 0 \\ 0 & -q^2 \end{pmatrix} \tag{20}$$

$$d\theta = -\begin{pmatrix} q^{-4} & 0 \\ 0 & 1 \end{pmatrix} \omega\omega + (q^4-1)\omega\theta, \, d\omega = -\begin{pmatrix} \frac{q^4-1}{q^4} & 0 \\ 0 & 0 \end{pmatrix} \omega\omega - q^4[2]_q\omega\theta$$

$$\theta\omega = q^6\omega\theta$$

Why we have obtained different algebras of a symmetry for the same subgroup? That is possible because we can use the different map from the algebra to the group, for example:

$$g = \exp(\varphi_-\tau_+)\exp(\varphi_+\tau_-)\exp(\ln\rho\tau_3) \;, \tag{21}$$

where τ are the Pauli matrices – the fundamental representation of the $U_q(SL(2,R)$ algebra. The group stability of the vacuum is $U(1)$. In the another parametrization

$$g = \exp(\varphi_-\tau_+)\exp(\varphi_+\tau_-)(1 - \frac{(q^2-1)}{q^2}\nabla_3)^{\frac{\ln\rho}{\ln q^{-2}}}, \; \nabla_3 = \begin{pmatrix} 1 & 0 \\ 0 & -q^2 \end{pmatrix} \tag{22}$$

the group stability of the vacuum is $U_q(1)$.

The group symmetry of this Lagrangians is $SL_q(2,R)$ spontaneously broken to $U_h(1)$, $h = q^{\pm 2n}, n = 0, 1....$ Under the left multiplication on the group $G' = G_0 G$, the differential form $G^{-1}dG = H^{-1}(\omega+\theta)H = G'^{-1}dG'$. Therefore, $\omega' + \theta' = H'H^{-1}(\omega+\theta)HH'^{-1}$.

Again, the decomposition on the coset and the subgroup forms is not unique after transformation. The group transformation can transform the Lagrangian with the $U_{h_1}(1)$ subgroup of the vacuum stability to the Lagrangian with the $U_{h_2}(1)$ subgroup.

4 Variational Calculus on the $\mathbf{SL}_q(2, \boldsymbol{R})$ Group

It is possible to obtain the variational calculus on the group by two ways: from the C.R. between the left vector fields and group parameters and from the infinitesimal transformations on the group. Let us multiply the C.R. (8) between ∇_n and group parameters on the parameters of transformation R^n. The form of the infinitesimal transformations of the group parameters is obtained under the requirement

$$[X_A, \nabla_n R^n] = X_A \delta_{R^n}, \quad X_A = (\rho, \varphi_-, \varphi_+), \quad [A, B] = AB - BA \tag{23}$$

By imposing the C.R. between the parameters of infinitesimal transformations and group parameters

$$\begin{array}{lll} \rho R^1 = q^2 R^1 \rho & \varphi_- R^1 = R^1 \varphi_- & \varphi_+ R^1 = R^1 \varphi_+ \\ \rho R^2 = q R^2 \rho & \varphi_- R^2 = R^2 \varphi_- & \varphi_+ R^2 = R^2 \varphi_+ \\ \rho R^3 = q R^3 \rho & \varphi_- R^3 = R^3 \varphi_- & \varphi_+ R^3 = R^3 \varphi_+ \end{array} \tag{24}$$

we obtain the infinitesimal transformation of the group parameters

$$\rho\delta = \rho R^1 - \varphi_+ \rho^3 R^2 \quad \varphi_- \delta = \tfrac{1}{q} \rho^2 R^2 \quad \varphi_+ \delta = q \varphi_+^2 \rho^2 R^2 + q \rho^{-2} R^3 \tag{25}$$

In the terms of the components δ_{R_n} the infinitesimal transformations have the following form:

$$\begin{array}{lll} \rho \delta_{R^1} = \rho, & \rho \delta_{R^2} = -\varphi_+ \rho^3, & \rho \delta_{R^3} = 0 \\ \varphi_- \delta_{R^1} = 0, & \varphi_- \delta_{R^2} = \frac{1}{q} \rho^2, & \varphi_- \delta_{R^3} = 0 \\ \varphi_+ \delta_{R^1} = 0, & \varphi_+ \delta_{R^2} = q \varphi +^2 \rho^2 & \varphi_+ \delta_{R^3} = q \rho^{-2} \end{array} \tag{26}$$

We postulate the $U_q SL(2, R)$ algebra of the vector fields in the form, which is common for a boxon theory and a supersymmetric theory

$$\begin{array}{c} [\nabla_1 R^1, \nabla_2 R^2] = (q^2 + 1) \nabla_2 R^1 R^2, \\ [\nabla_3 R^3, \nabla_1 R^1] = (q^2 + 1) \nabla_3 R^3 R^1, \\ [\nabla_2 R^2, \nabla_3 R^3] = \nabla_1 R^2 R^3 \end{array} \tag{27}$$

The same result we can obtain from the right infinitesimal multiplication on the group $g' = g g_0$, where $g_0 = 1 + \delta g_0$. For

$$\delta g_0 = \begin{pmatrix} R^1 & R^2 \\ R^3 & -q^2 R^1 \end{pmatrix} \tag{28}$$

we see, that $\mathrm{d}g = g \delta g_0$ and C.R. for δg_0 are the same as for left forms ω simultaneously with condition $R^4 = -q^2 R^1$. The C.R. of the variational calculus

$$\begin{array}{ll} (\rho\delta)\rho = \frac{1}{q^2} \rho (\rho\delta) & (\varphi_- \delta)\rho = \frac{1}{q} \rho (\varphi_- \delta) \\ (\varphi_+ \delta)\rho = \frac{1}{q} \rho (\varphi_+ \delta) - \frac{(q^2-1)}{q} \varphi_+ (\rho\delta) & (\rho\delta)\varphi_- = \varphi_- (\rho\delta) \\ (\varphi_- \delta)\varphi_- = \varphi_- (\varphi_- \delta) & (\varphi_+ \delta)\varphi_- = \varphi_- (\varphi_+ \delta) \\ (\rho\delta)\varphi_+ = \varphi_+ (\rho\delta) & (\varphi_- \delta)\varphi_+ = \varphi_+ (\varphi_- \delta) \\ (\varphi_+ \delta)\varphi_+ = \varphi_+ (\varphi_+ \delta) & \end{array} \tag{29}$$

are consistent with the C.R. (3) and are simpler than the C.R. of the differential calculus (6). The $U_q(SL(2,R))$ algebra is the condition of the compatibility of the relations (25)

$$\begin{aligned} X^A(q^2\delta_{R^1}\delta_{R^3} - q^{-2}\delta_{R^3}\delta_{R^1} = (q^2+1)\delta_{R^3}) \\ X^A(q^2\delta_{R^2}\delta_{R^1} - q^{-2}\delta_{R^1}\delta_{R^2} = (q^2+1)\delta_{R^2}) \\ X^A(\delta_{R^3}\delta_{R^2} - q^2\delta_{R^2}\delta_{R^3} = \delta_{R^1}) \end{aligned} \tag{30}$$

5 Equations of Motion

We use the extremum principle of the action to obtain the equations of motion and we must to commute the variations of fields and their derivatives on the right or on the left side. We can use both variation $\mathrm{d}X^A$ and δX^A to do this. The C.R. of the differential calculus on the $SL_q(2,R)$ group are insufficient to do this. Therefore, we need in the differential calculus on the Lagrangian manifolds $(\rho, \varphi_\pm, \dot\rho, \dot\varphi_\pm, \acute\rho, \acute\varphi_\pm)$. This is not the quantum group manifold and we can not use the formalism of 1-forms. We can require, that the Lagrangian equation of motion be coincident with the conservation law $\partial_\mu\omega^\mu = 0$ for Lagrangian with $SL_q(2,R)$ group symmetry. At last, we can investigate the 1- dimensional σ- models. The variational calculus is more suitable to obtain the equations of motion. The C.R. between the X^A, $\dot X^A$, $\acute X^A$ and R^n, $\dot R^n$, $\acute R^n$ can obtain by differentiating the relations (24).

$$\begin{aligned} \dot\rho R^1 = q^2 R^1\dot\rho \quad \dot\rho\dot R^1 = q^2\dot R^1\dot\rho \quad \rho\dot R^1 = \dot R^1\rho \\ \dot\rho R^2 = qR^2\dot\rho \quad \dot\rho\dot R^2 = q\dot R^2\dot\rho \quad \rho\dot R^2 = q\dot R^2\rho \\ \dot\rho R^3 = qR^3\dot\rho \quad \dot\rho\dot R^3 = q\dot R^3\dot\rho \quad \rho\dot R^3 = q\dot R^3\rho \end{aligned} \tag{31}$$

The derivatives of $\varphi_\pm$ commute with the derivatives of R^n.

6 One Dimensional σ-Model on the Quantum Plane $C_q(2|0)$

The differential calculus on the $C_q(2|0)$ is coincide with the differential calculus on the Borel subgroup of $SL_q(2,C)$ and can be obtained from the differential calculus on the $SL_q(2,C)$ by surjection: π:$SL_q(2,C) \to B_L$ such that $\pi(b) = 0$.

$$g = \begin{pmatrix} x & 0 \\ y & x^{-1} \end{pmatrix}, \quad \begin{aligned} xy = qyx \quad \dot y y = q^{-2} y\dot y \\ \dot x x = q^{-2} x\dot x \; \dot x y = q^{-1} y\dot x \end{aligned} \tag{32}$$

$$\omega = \begin{pmatrix} x^{-1}\mathrm{d}x & 0 \\ x\mathrm{d}y - qy\mathrm{d}x & -q^2x^{-1}\mathrm{d}x \end{pmatrix}, \quad \dot y x = q^{-1}x\dot y - \frac{(q^2-1)}{q^2} y\dot x$$

In term of the variables $\rho, \varphi_\pm$

$$g = \begin{pmatrix} \rho & 0 \\ \varphi_+\rho & \rho^{-1} \end{pmatrix}; \quad \omega = \begin{pmatrix} \rho^{-1}\mathrm{d}\rho & 0 \\ \frac{1}{q}\rho^2\mathrm{d}\varphi_+ & -q^2\rho^{-1}\mathrm{d}\rho \end{pmatrix} \tag{33}$$

$$L = \frac{1}{2}\mathrm{Tr}_q(\omega_\mu \omega^\mu) = \frac{q^4(q^2+1)}{2}\rho^{-2}\dot{\rho}^2$$

The equation of motion $\dot{\omega}^1 = \rho^{-1}\ddot{\rho} - q^2\rho^{-1}\dot{\rho}^2 = 0$ will coincide with Lagrangian equation, if we impose the C.R. $\delta\dot{\rho}\dot{\rho} = \frac{1}{q^2}\dot{\rho}\delta\dot{\rho}$. The classical solution of this equation is

$$\rho = \alpha \exp(\beta t), \quad \alpha\beta = q^2\beta\alpha \tag{34}$$

and C.R.

$$\rho(t)\rho(t') = \rho(q^2 t')\rho(\frac{1}{q^2}t), \quad \rho(t)\rho(t') = \exp[q^2(q^2-1)\beta(t-t')]\rho(t')\rho(t) \tag{35}$$

There are 4×4 matrix representations of α, β such, that $det_q\alpha = 0$ or $det_q\beta = 0$. Therefore, we can rewrite this Lagrangian as a 4×4 matrix model for the commuting fields. In conclusion, note that 2-dimensional σ-model on the guantum plane

$$L = \frac{q^4(q^2+1)}{2}\rho^{-2}\partial_\mu\rho\partial^\mu\rho \tag{36}$$

leads to the C.R. $\delta\acute{\rho}\,\acute{\rho} = \frac{1}{q^2}\acute{\rho}\,\delta\acute{\rho}$ and the equation of motion $\partial_\mu\partial^\mu\rho - q^2\rho^{-1}\partial_\mu\rho\partial^\mu\rho = 0, \quad \mu = 1, 2.$

I would like to thank J.Wess, V.Dobrev, J, Lukierski, A.Isaev, P.Pyatov, B.Zupnik, V.Lyakhovsky, A.Akulov for stimulating discussions.

This work was supported in part by the Ukrainian Ministry of Science and technology, grant 2.5.1/54, by grants INTAS 93-633 (Extension) and INTAS 93-127 (Extension).

References

Alekseev, A. and Shatashvili, S. (1990): *Commun. Math. Phys.*, **133**, p.353

Akulov, V., Gershun, V. (1995): *q-alg/9509030*

Akulov, V.P., Gershun, V.D. (1995a): *Proc. Int. conf. on math, phys. (Rahov, Ukraine, 1995)*, (in press)

Akulov, V.P., Gershun, V.D., Gumenchuk, A.I. (1992): *JETP Lett.*, **56**, p.177

Akulov, V.P., Gershun, V.D., Gumenchuk, A.I. (1993): *JETP Lett.*, **58**, p.474

Arefeva, I.Y. and Volovich, I.V. (1991): *Phys. Lett.*, **264B**, p.62

Fadeev, L.D. (1990): *Commun. Math, Phys.*, **132**, p.131

Faddeev, L., Reshetikhin, N., Takhtajan, L. (1989): *Algebra i Analiz* **1**, p.178

Frishman, Y., Lukierski, J., Zakrzewski, W.J. (1993): *J.Phys.A:Math.Gen*, **26**, p.301

Gershun, V.D. (1996): *Proc. Xth Int. conf. "Problems of Quantum Field Theory" (Alushta, Ukraine, 1996)*, JINR, Dubna, p.119

Integrating a Generic Algebra

R. Casalbuoni

Dipartimento di Fisica, Universita' di Firenze
I.N.F.N., Sezione di Firenze
e-mail: CASALBUONI@FI.INFN.IT

During the last years there has been a lot of interest in generalized classical theories. The most typical examples are theories involving Grassmann variables (Berezin, Marinov (1975, 1977), Casalbuoni (1976)) (this last paper was largely inspired by the work in (Volkov, Akulov (1973))). The corresponding path-integral quantization requires the notion of integration over the phase-space variables. This procedure is very well known for the particular case mentioned above (Berezin (1966)). The problem of defining the path-integral in the general case is too much complicated and we have limited ourselves to the first necessary step, that is to define an integration procedure over an arbitrary algebra. This approach is described more completely in paper (Casalbuoni (1997)), to which we refer for all the details. Here we will outline only the most important steps. We want to define the integral as a linear mapping between the given algebra and the real numbers, but we need to specify further the properties of such a mapping. We do this by requiring the physical principle of the combination law for the probability amplitudes. In ordinary quantum mechanics this is mathematically expressed through the completeness of the eigenstates of the position operator. In order to extend this idea to the general case we use the same approach followed in the study of non-commutative geometry (Connes (1994)) and of quantum groups (Drinfeld (1986)). The approach starts from the observation that in the normal case one can reconstruct a space from the algebra of its functions. Giving this fact, one lifts all the necessary properties in the function space. In this way one is able to deal with cases in which no concrete realization of the space itself exists.

In order to see how we can lift up the completeness from the base space to the space of functions, let us suppose that this admits an orthonormal set of functions. Then, any function on the base space can be expanded in terms of the complete set $\{\psi_n(x)\}$. It turns out convenient to define a generalized Fick space, $\mathcal{F}$,and the following special vector in it

$$|\psi\rangle = \begin{pmatrix} \psi_0(x) \\ \psi_1(x) \\ \cdots \\ \psi_n(x) \\ \cdots \end{pmatrix} \tag{1}$$

Then, a function $f(x) = \sum_n a_n \psi_n(x)$ can be represented as $f(x) = \langle a|\psi\rangle$ where $\langle a| = (a_0, a_1, \cdots, a_n, \cdots)$. To write the orthogonality relation in terms of this new formalism it is convenient to realize the complex conjugation as a linear operation on $\mathcal{F}$. In fact, $\psi_n^*(x)$ itself can be expanded in terms of $\psi_n(x)$, $\psi_n^*(x) = \sum_n \psi_m(x) C_{mn}$ or $|\psi^*\rangle = C^T|\psi\rangle$. Defining a bra in $\mathcal{F}$ as the transposed of the ket $|\psi\rangle$

$$\langle\psi| = (\psi_0(x), \psi_1(x), \cdots (x), \psi_n(x), \cdots) \tag{2}$$

the orthogonality relation becomes

$$\int |\psi\rangle\langle\psi^*|\; \mathrm{d}x = \int |\psi\rangle\langle\psi| C\; \mathrm{d}x = 1 \tag{3}$$

Another important observation is that the orthonormal functions define an algebra. In fact we can expand the product of two eigenfunctions in terms of the eigenfunctions

$$\psi_m(x)\psi_n(x) = \sum_p c_{nmp}\psi_p(x) \tag{4}$$

with

$$c_{nmp} = \int \psi_n(x)\psi_m(x)\psi_p^*(x)\; \mathrm{d}x \tag{5}$$

The relation (3) makes reference only to the elements of the algebra of functions that we have organized in the space $\mathcal{F}$, and it is the key element in order to define the integration rules on the algebra. In fact, we can now use the algebra product to reduce the expression (3) to a linear form. If the resulting expression has a solution for $\int \psi_p(x)\; \mathrm{d}x$, then we are able to define the integration over all the algebra of functions, by linearity. Notice that a solution always exists, if the constant function is in the set $\{\psi_n(x)\}$.

The procedure we have outlined here is the one that we will generalize to arbitrary algebras. Before doing that we will consider the possibility of a further generalization. In the usual path-integral formalism sometimes one makes use of the coherent states instead of the position operator eigenstates. In this case the basis in which one considers the wave functions is a basis of eigenfunctions of a non-hermitian operator $\psi(z) = \langle\psi|z\rangle$ with $a|z\rangle = |z\rangle z$. The wave functions of this type close an algebra, as $\langle z^*|\psi\rangle$ do. But this time the two types of eigenfunctions are not connected by any linear operation. In fact, the completeness relation is defined on an algebra which is the direct product of the two algebras

$$\int \frac{\mathrm{d}z^*\mathrm{d}z}{2\pi i} \exp(-z^*z)|z\rangle\langle z^*| = 1 \tag{6}$$

Therefore, in similar situations, we will not define the integration over the original algebra, but rather on the algebra obtained by the tensor product of the algebra times a copy. The copy corresponds to the complex conjugated functions of the previous example.

Let us start with a generic algebra $\mathcal{A}$ with $n+1$ elements x_i, with $i = 0, 1, \cdots n$. We do this for simplicity, but there are no problems in letting $n \to \infty$, or in taking a continuous index. We assume the multiplication rules

$$x_i x_j = f_{ijk} x_k \tag{7}$$

with the usual convention of sum over the repeated indices. The structure constants f_{ijk} define uniquely the algebraic structure. Consider for instance the case of an abelian algebra. In this case

$$x_i x_j = x_j x_i \longrightarrow f_{ijk} = f_{jik} \tag{8}$$

Or, for an associative algebra, from $x_i(x_j x_k) = (x_i x_j) x_k$, one gets

$$f_{ilm} f_{jkl} = f_{ijl} f_{lkm} \tag{9}$$

We introduce now the space $\mathcal{F}$, and the special vector

$$|x\rangle = \begin{pmatrix} x_0 \\ x_1 \\ \cdot \\ \cdot \\ x_n \end{pmatrix}, \qquad |x\rangle \in \mathcal{F} \tag{10}$$

In order to be able to generalize properly the discussion made for the functions, it will be of fundamental importance to look for linear operators having the vector $|x\rangle$ as eigenvector and the algebra elements x_i as eigenvalues. This notion is strictly related to the mathematical concept of **right and left multiplication algebras** associated to a given algebra (Casalbuoni (1997)). The linear operators we are looking for are defined by the relation

$$X_i|x\rangle = |x\rangle x_i \tag{11}$$

that is

$$(X_i)_{jk} x_k = x_j x_i = f_{jik} x_k \tag{12}$$

or

$$(X_i)_{jk} = f_{jik} \tag{13}$$

In a complete analogous way we can consider a bra $\langle\tilde{x}|$, defined as the transposed of the ket $|x\rangle$ and we define left multiplication through the equation

$$\langle\tilde{x}|\Pi_i = x_i\langle\tilde{x}| \tag{14}$$

implying

$$(\Pi_i)_{kj} = f_{ijk} \tag{15}$$

The two matrices X_i and Π_i corresponding to right and left multiplication are generally different. For instance, consider the abelian case. It follows from eq. (8)

$$X_i = \Pi_i^T \tag{16}$$

If the algebra is associative, then from eq. (9) the following three relations can be shown to be equivalent:

$$X_i X_j = f_{ijk} X_k, \quad \Pi_i \Pi_j = f_{ijk} \Pi_k, \quad [X_i, \Pi_j^T] = 0 \tag{17}$$

The first two say that X_i and Π_i are linear representations of the algebra. The third that the right and left multiplication commute for associative algebras.

Recalling the discussion made for the functions we would like first consider the case of a basis originating from hermitian operators. Notice that the generators x_i play here the role of generalized dynamical variables. It is then natural to look for the case in which the operators X_i admit both eigenkets and eigenbras. This will be the case if

$$\Pi_i = C X_i C^{-1} \tag{18}$$

that is if Π_i and X_i are connected by a non-singular C matrix. This matrix is exactly the analogue of the matrix C defined in the case of functions. From eq. (14), we get

$$\langle \tilde{x} | C X_i C^{-1} = x_i \langle \tilde{x} | \tag{19}$$

By putting

$$\langle x | = \langle \tilde{x} | C \tag{20}$$

we have

$$\langle x | X_i = x_i \langle x | \tag{21}$$

In this case, the equations (11) and (21) show that X_i is the analogue of an hermitian operator. We will define now the integration over the algebra by requiring that

$$\int_{(x)} |x\rangle\langle x| = 1 \tag{22}$$

where 1 is the identity matrix on the $(n+1) \times (n+1)$ dimensional linear space of the linear mappings on the algebra. In more explicit terms we get

$$\int_{(x)} x_i x_j = \int_{(x)} f_{ijk} x_k = (C^{-1})_{ij} \tag{23}$$

If we can invert this relation in terms of $\int_{(x)} x_i$, we can say to have defined the integration over the algebra, because we can extend the operation by linearity. In particular, if $\mathcal{A}$ is an algebra with identity, let us say $x_0 = 1$, then, by using (23), we get

$$\int_{(x)} x_i = (C^{-1})_{0i} = (C^{-1})_{i0} \tag{24}$$

and it is always possible to define the integral.

We will discuss now the transformation properties of the integration measure with respect to an automorphism of the algebra. In particular, we will restrict our analysis to the case of a simple algebra (that is an algebra having as ideals only the algebra itself and the null element). Let us consider an invertible linear transformation on the basis of the algebra leaving invariant the multiplication rules (that is an automorphism) $x'_i = S_{ij}x_j$ with $x'_i x'_j = f_{ijk}x'_k$. For a simple algebra, one can show that (Casalbuoni (1997))

$$C^{-1}S^T C = kS^{-1} \tag{25}$$

where k is a constant. It follows that the measure transforms as

$$\int_{(x')} = \frac{1}{k}\int_{(x)} \tag{26}$$

Let us consider now the case in which the automorphism S can be exponentiated in the form $S = \exp(\alpha D)$. Then D is a derivation of the algebra. If it happens that for this particular automorphism S, one has $k = 1$, the integration measure is invariant, and the integral satisfies

$$\int_{(x)} D(f(x)) = 0 \tag{27}$$

for any function $f(x)$ on the algebra. On the contrary, a derivation always defines an automorphism of the algebra by exponentiation. So, if the corresponding k is equal to one, the equation (27) is always valid.

Of course it may happen that the C matrix does not exist. This would correspond to the case of non-hermitian operators as discussed before. So we look for a copy $\mathcal{A}^*$ of the algebra. By calling x^* the elements of $\mathcal{A}^*$, the corresponding generators will satisfy $x_i^* x_j^* = f_{ijk}x_k^*$. It follows

$$\langle \tilde{x}^*|\Pi_i = x_i^*\langle \tilde{x}^*| \tag{28}$$

Then, we define the integration rules on the tensor product of $\mathcal{A}$ and $\mathcal{A}^*$ in such a way that the completeness relation holds

$$\int_{(x,x^*)} |x\rangle\langle \tilde{x}^*| = 1 \tag{29}$$

This second type of integration is invariant under orthogonal transformation or unitary transformations, according to the way in which the * operation acts on the transformation matrix S. If * acts on complex numbers as the ordinary conjugation, then we have invariance under unitary transformations, otherwise if * leaves complex numbers invariant, then the invariance is under orthogonal transformations. Notice that the invariance property does not depend on S being an automorphism of the original algebra or not.

The two cases considered here are not mutually exclusive. In fact, there are situations that can be analyzed from both points of view (Casalbuoni

(1997)). We want also to emphasize that this approach does not pretend to be complete and that we are not going to give any theorem about the classification of the algebras with respect to the integration. What we are giving is rather a set of rules that one can try to apply in order to define an integration over an algebra. As argued before, there are algebras that do not admit the integration as we have defined in (23). Consider, for instance, a simple Lie algebra. In this case we have the relation $f_{ijk} = f_{jki}$ which implies $X_i = \Pi_i$ or $C = 1$. Then the eq. (23) requires

$$\delta_{ij} = \int_{(x)} x_i x_j = \int_{(x)} f_{ijk} x_k \tag{30}$$

which cannot be satisfied due to the antisymmetry of the structure constants. Therefore, we can say that, according to our integration rules, there are algebras with a complete set of states and algebras which are not complete. On the contrary there are many examples in which our rules allow the definition of an integration. We recall here, bosonic and fermionic integration, the q-oscillator and the paraGrassmann cases, and finally the integration over the algebras of quaternions and octonions (all these examples are discussed in (Casalbuoni (1997))).

The work presented here is only a first approach to the problem of quantizing a theory defined on a configuration space made up of non-commuting variables, the simplest example being the case of supersymmetry. In order to build up the functional integral, a second step would be necessary. In fact, one needs a different copy of the given algebra to each different time along the path-integration. This should be done by taking convenient tensor products of copies of the algebra. Given this limitation, we think, however, that the step realized in this work is a necessary one in order to solve the problem of quantizing the general theories discussed here.

References

Berezin, F.A. (1966): *The method of second quantization*, Academic Press (1966).

Berezin, F.A. and Marinov, M.S. (1975, 1977): JETP Lett. **21**, 321, *ibidem* Ann. of Phys. **104**, 336.

Casalbuoni, R. (1976): Il Nuovo Cimento, **33A**, 115 and *ibidem* 389.

Casalbuoni, R. (1977): Florence preprint DFF-270/02/1997, physics/9702019.

Connes, A. (1994): *Noncommutative geometry*, Academic Press (1994).

Drinfeld, V.G. (1986): *Quantum Groups*, in Proceedings of the International Congress of Mathematicians, Berkeley 1986, pp. 798-820, AMS, Providence, RI.

Volkov, D. and Akulov, V.P. (1973): Phys. Lett. **B46**, 109.

Part III

Selected Works and List of Main Publications of D.V. Volkov

On the Quantization of Half-Integer Spin Fields

D.V. Volkov

Physico–Technical Institute, Academy of Sciences, Ukrainian S.S.R.

Submitted to JETP editor December 8, 1958
J. Exptl. Theoret. Phys. (U.S.S.R.) **36**, 1560–1566 (May, 1959)

Abstract. A method of quantization for half–integer spin fields is considered which is different from the usual one involving anticommutators and is consistent with the principle of relativistic causality, positive definiteness of the energy (for non–interacting fields), the Lagrangian formalism in Schwinger's formulation (Schwinger (1956)), and with invariance under TCP transformations (Pauli (1955)). The main difference between the proposed method and the usual one is that the maximal occupation number is two.

1 Introduction

It is a well–known fact that nonrelativistic quantum mechanics does not explain the connection between spin and statistics. Moreover, the equations of nonrelativistic quantum mechanics admit of solutions which are neither completely summetric nor completely antisymmetric, and which transform according to different irreducible representations of the permutation group if the particles are interchanged. In the relativistic quantum theory, the existing methods of quantization lead to a unique connection between spin and statistics (Pauli(1947)); however, from the very beginning only two alternatives are considered in this case: either we quantize with commutators, or with anti-commutators. In this connection it is of interest to investigate whether other possibilities, which are admissible in nonrelativistic quantum mechanics, are consistent with the basic principles of the relativistic theory.

In the present paper we consider, on the simplest example of the Dirac equation, the possibility of constucting an algebra of operator wave functions with the following properties: it leads to a new statistics with the maximal occupation number two for each individual state,[1] and is at the same time consistent with the principle of relativistic causality, the positive definiteness of the enery (for non–interacting fields), and with the Lagrangian formalism. In setting up the Lagrangian formalism we make use of the variational principle of (Schwinger (1956)). We show that this method of quantization

[1] This method of quantization can be generalized for the case of arbitrary maximal occupational numbers.

is a consequence of the variational principle of Schwinger based on a class of admissible variations of the operator wave function which is different from that used in the usual quantization scheme.

2 The Condition of Relativistic Causality

The requirement of relativistic causality is usually formulatied as the condition that commutators of physical operators reduce to zero for points which are separated by a space–like interval (outside the light cone). With the condition that operators corresponding to measurable quantities are bilinear combinations of the opoerator wave functions (as in the case of half–integer spin fields), the requirement of relativistic causality will be fulfilled if the commutators or anticommutators of the operator wave functions reduce to zero outside the light cone. These are the only two cased usually considered. It is, however, possible to satisfy the requirement of relativistic causlity using a different algebra for the operator wave functions.

As an example, we consider the field satisfying the free Dirac equation,

$$\left(\gamma_\mu \frac{\partial}{\partial x_\mu} + m\right)\psi = 0 \ , \tag{1}$$

where γ_μ (μ=1,2,3,4) are matrices defined by the relation $\gamma_\mu\gamma_\nu+\gamma_\nu\gamma_\mu = 2\delta_{\mu\nu}$.

We define the operator properties of the wave functions $\psi(c)$ and $\bar{\psi}(x) = \psi^+(x)\gamma_4$ with the help of the commutation relations

$$\begin{aligned}
&\psi_\alpha(x)\psi_\beta(x')\psi_\gamma(x'') + \psi_\gamma(x'')\psi_\beta(x')\psi_\alpha(x) = 0,\\
&\psi_\alpha(x)\bar{\psi}_\beta(x')\psi_\gamma(x'') + \psi_\gamma(x'')\bar{\psi}_\beta(x')\psi_\alpha(x) =\\
&\qquad = -\mathrm{i}S_{\alpha\beta}(x-x')\psi_\gamma(x'') - \mathrm{i}S_{\gamma\beta}(x''-x')\psi_\alpha(x),\\
&\bar{\psi}_\alpha(x)\bar{\psi}_\beta(x')\psi_\gamma(x'') + \psi_\gamma(x'')\bar{\psi}_\beta(x')\psi_\alpha(x) =\\
&\qquad = -\mathrm{i}S_{\gamma\beta}(x''-x')\bar{\psi}_\alpha(x) \ ,
\end{aligned} \tag{2}$$

where $S(x)$ is the known commutator function (Akhiezer and Berestetskii (1953))

$$S_{\alpha\beta}(x) = -\left(\gamma_\mu \frac{\partial}{\partial x_\mu} - m\right)_{\alpha\beta} \Delta(x),$$

$$\Delta(x) = \mathrm{i}(2\pi)^{-3}\int \mathrm{e}^{ipx}\varepsilon(p)\delta\left(p^2+m^2\right)\mathrm{d}^4p \ . \tag{3}$$

It follows from the properties of the function $S(x)$ that the relations (2) are consistent with Eq. (1). We verify that all commutators of the form $\left[\psi_\alpha(x)\psi_\beta(x),\ \psi_\gamma(x')\psi_\delta(x')\right]$ are zero outside the light cone. For the proof we write the relations (2) in the following compact form:

$$\psi_\alpha(x)\psi_\beta(x')\psi_\gamma(x'') + \psi_\gamma(x'')\psi_\beta(x')\psi_\alpha(x) =$$
$$= \left\{\psi_\alpha(x), \psi_\beta(x')\right\}_F \psi_\gamma(x'') + \left\{\psi_\gamma(x''), \psi_\beta(x')\right\}_F \psi_\alpha(x), \qquad (2')$$

where ψ now stands for the usual as well as the Dirac conjugate spinor, and the brackets with the index F are given by

$$\left\{\psi_\alpha(x), \bar{\psi}_\beta(x')\right\}_F \equiv -\mathrm{i}S_{\alpha\beta}(x - x'),$$
$$\left\{\psi_\alpha(x), \psi_\beta(x')\right\}_F \equiv 0 \ , \qquad (4)$$

i.e., by the usual values for the anticommutators in the quantization according to Fermi–Dirac statistics.[2]

Using the relations (2'), we can transform this commutator to the form

$$\psi_\alpha(x)\psi_\beta(x)\psi_\gamma(x')\psi_\delta(x') - \psi_\gamma(x')\psi_\delta(x'\psi_\alpha(x)\psi_\beta(x) =$$
$$- \left\{\psi_\alpha(x), \psi_\delta(x')\right\}_F \psi_\gamma(x')\psi_\beta(x) + \left\{\psi_\beta(x), \psi_\gamma(x')\right\}_F \psi_\alpha(x)\psi_\delta(x') \ .$$

It follows from the properties of the function $S(x)$ that this expression is zero outside thew light cone.

It is easily shown that the commutators of physical quantities also reduce to zero outside the light cone for commutation relations of the type

$$\psi_\alpha(x)\psi_\beta(x')\psi_\gamma(x'') + \psi_\gamma(x'')\psi_\beta(x')\psi_\alpha(x) =$$
$$= \left\{\psi_\alpha(x), \psi_beta(x')\right\}_F \psi_\gamma(x'') + \left\{\psi_\gamma(x''), \psi_\beta(x')\right\}_F \psi_\alpha(x)$$
$$+ \rho\left\{\psi_\alpha(x), \psi_\gamma(x'')\right\}_F \psi_\beta(x'); \qquad (5)$$

$$\psi_\alpha(x)\psi_\beta(x')\psi_\gamma(x'') - \psi_\gamma(x'')\psi_\beta(x'\psi_\alpha(x) =$$
$$\left\{\psi_\alpha(x), \psi_\beta(x')\right\}_F \psi_\gamma(x'') - \left\{\psi_\gamma(x''), \psi_\beta(x')\right\}_F \psi_\alpha(x) \ , \qquad (6)$$

where ρ is an arbitrary number.

The commiutation relations (5) have nono-zero solutions only for ρ=0 and ρ=-1. In the last case the algebra for the operator wave functions correspons to th usual method of quantization using anticommutators.

The commutation relations (6) differ from the commutation relations (2') by a change of sign. This difference (as in the case of commutators and anticommutators) leads to an energy for halfinteger spin fields which is not positive definite.

In principle, the relations (6) can be used for the quantization of fields with integer spin, with the requirement that the interaction Lagrangian contains an even number of field operators.

---

[2] We note that in this quantization scheme the anticommutator $\left\{\psi_\alpha(x), \psi_\beta(x')\right\} \neq \left\{\psi_\alpha(x), \psi_\beta(x')\right\}_F$.

3 Momentum Representation

The algebra for the field operators is conveniently realized in the momentum representation. We make the transition to the momentum representation by expanding the operator wave functions and the commutator function $S(x)$ into Fourier series (Akhiezer and Berestetskii (1953)):

$$\psi_\alpha(x) = V^{-\frac{1}{2}} \sum_p \sum_{r=1}^{2} \{a_{pr} u^r_\alpha(\mathrm{p}) \mathrm{e}^{\mathrm{i}px} + b^+_{pr} v^r_\alpha(\mathrm{p}) \mathrm{e}^{-\mathrm{i}px}\} ,$$

$$\bar{\psi}_\alpha(x) = V^{-\frac{1}{2}} \sum_p \sum_{r=1}^{2} \{a^+_{pr} \bar{u}^r_\alpha(\mathrm{p}) \mathrm{e}^{-\mathrm{i}px} + b_{pr} v^r_\alpha(\mathrm{p}) \mathrm{e}^{\mathrm{i}px}\} , \tag{7}$$

$$-\mathrm{i}S_{\alpha\beta}(x) = \frac{1}{V} \sum_p \sum_r^{2} \{u^r_\alpha(\mathrm{p}) \bar{u}^r_\beta(\mathrm{p}) \mathrm{e}^{\mathrm{i}px} + v^r_\alpha(\mathrm{p}) \bar{v}^r_\beta(\mathrm{p}) \mathrm{e}^{-\mathrm{i}px}\} \ , \tag{8}$$

where V is the normalization volume; the sum over r implies summation over the states with different polarization; u^r_α and v^r_α are constant spinors subject to the orthonormality conditions

$$\sum_\alpha u^{r\,*}_\alpha u^s_\alpha = \delta_{rs}, \quad \sum_\alpha v^{r\,*}_\alpha v^s_\alpha = \delta_{rs} \ ,$$

Substituting the expansions (7) and (8) in (2), we obtain the following commutation relations for the operators a and b:

$$\begin{aligned}
&a_k a^+_l a_m + a_m a^+_l a_k = \delta_{kl} a_m + \delta_{ml} a_k, \\
&a_k a^+_l a^+_m + a^+_m a^+_l a_k = \delta_{kl} a^+_m, \\
&b_k b^+_l b_m + b_m b^+_l b_k = \delta_{kl} b_m + \delta_{ml} b_k, \\
&b_k b^+_l b^+_m + b^+_m b^+_l b_k = \delta_{kl} b^+_m, \\
&a_k a^+_l b_m + b_m a^+_l a_k = \delta_{kl} b_m, \\
&b_k b^+_l a_m + a_m b^+_l b_k = \delta_{kl} a_m \ ;
\end{aligned} \tag{9}$$

the indices k, l, and m define the momentum and the polarization.

All the remaining commutation relations of the same type, except those which derive from (9) by Hermitian conjugation, are equal to zero.[3]

The operators corresponding to the basic physical quantities can be simply expressed in terms of the operators $N_k = a^+_k a_k - a_k a^+_k + 1$ [cf. Eqs. (17') to (19') below]. We show that this operator can be interpreted as the opeeeerator

[3] With $a_k = \alpha_k + \mathrm{i}\beta_k$ and $b_k = \gamma_k + \mathrm{i}\delta_k$, where α_k, β_k, γ_k and δ_k are Hermitian matrices, relations (9) go over into the Duffin–Kemmer relations. The algebra of the Duffin–Kemmer matrices for an arbitrary number of matrices was considered by (Fujiware (1953)).

corresponding to the number of particles in the state k. To determine the eigenvalues of the operator N_k, we use the relation

$$\left(a_k^+ a_k - a_k a_k^+\right)^3 = a_k^+ a_k - a_k a_k^+ \ , \tag{10}$$

which is readily proven with the help of the commutation relations (0. It follows from formula (10) that the eigenvalues of the operator $a_k^+ a_k - a_k a_k^+$ are equal to -1, 0, or 1, i.e., the corresponding eigenvalues of the operator N_k are equal to 0,1,or 2.

We consider the commutation relations of the operator N_k with the operators a_l and a_l^+:

$$\left[a_k^+ a_k - a_k a_k^+, a_l\right] = a_k^+ a_k a_l - a_k a a_k^+ a_l - a_l a_k^+ a_k + a_l a_k a_k^+ \ . \tag{11}$$

From the relations (9) we have

$$\begin{aligned} a_k^+ a_k a_l + a_l a_k a_k^+ &= a_l, \\ a_k a_k^+ a_l + a_l a_k^+ a_k &= a_l + \delta_{lk} a_k \ . \end{aligned} \tag{12}$$

Substituting (12) in (11), we obtain

$$[N_k, a_l] = -\delta_{kl} a_l \ . \tag{13}$$

Similarly, we have for the operator a_l^+:

$$\left[N_k, a_l^+\right] = \delta_{kl} a_l^+ \ . \tag{14}$$

The relations (13) and (14) are analogous to the corresponding relations in the usual quantization scheme. In particular, they inmply that the operators a_l^+ and a_l can be interpreted as the cr4eation and annihilation operators for particles in the stae l.

Indeed, if $\Psi_{n_1,n_2,\ldots}$ is the simultaneous eigenvector of the operators N_k (the operators N_k commute with one another) with the eigenvalues n_k, then $a_k\Psi_{n_1,\ldots}$ and $a_k^+\Psi_{n_1,\ldots}$ are also eigenvectors of these operators, where the eigenvalue n_k is loweered or raised by unity, respectively.

We define the vacuum state as the statew in which all occupation numbers n_k are equal to zero, i.e.,

$$\left(a_k^+ a_k - a_k a_k^+ + 1\right)\Psi_0 = 0 \ . \tag{15}$$

It follows from the definition (115) and the relations (13) and (14) that

$$\begin{aligned} &a_k\Psi_0 = 0, \quad a_k a_l^+ \Psi_0 = 0 \quad (k \neq l); \\ &a_k a_k^+ \Psi_0 = \Psi_0 \ . \end{aligned} \tag{16}$$

We can generate a complete set of basis vectors by successively acting on the vacuum vector with the creation operators a_k^+:

$$\Psi_0,\ a_k^+\Psi_0,\ a_k^+ a_l^+\Psi_0,\ \text{etc.}$$

We note that, in contrast to the quantization scheme using anticommutators,, we now have $a_k^+ a_k^+ \Psi_0 \neq 0$; futhermore, basis vectors differing in the order of the operators a_k^+ can be independent. For example, in the case of two particles the vectors $a_k^+ a_l^+ \Psi_0$ and $a_l^+ a_k^+ \Psi_0$ are independent; for three particles, we have the following independent vectors:

$$a_k^+ a_l^+ a_m^+ \Psi_0, \quad a_k^+ a_m^+ a_l^+ \Psi_0, \quad a_m^+ a_k^+ a_l^+ \Psi_0$$

etc. This difference manifests itself in configuration space through the appearance of partially symmetric wave functions.

The commutation relations (9) together with the relations (16) permit the calculation of the result of operating with the operators a_k and a_k^+ on an arbitrary basis vector; thus we can determine the explicit form of these operators in the representation under consideration.

4 Operators Corresponding to Physical Quantities

The operators of energy, momentum, and charge for the free Dirac field are given by the following expressions, which are antisymmetric in the operator wave functions:

$$E = \mathrm{i} \int \left(\bar{\psi} \gamma_4 \frac{\partial}{\partial t} \psi - \frac{\partial}{\partial t} \psi \gamma_4^T \bar{\psi} \right) \mathrm{d}V; \tag{17}$$

$$\mathbf{P} = -\mathrm{i} \int \left(\bar{\psi} \gamma_4 \nabla \psi - \nabla \psi \gamma_4^T \bar{\psi} \right) \mathrm{d}V \tag{18}$$

$$Q = e \int \left(\bar{\psi} \gamma_4 \psi - \psi \gamma_4^T \bar{\psi} \right) \mathrm{d}V \ . \tag{19}$$

The conventional expression for the energy, momentum, and charge differ from the expressions (17) to (19) by the factor $\frac{1}{2}$. This difference is connected with the normalizations of the commutation relations for the operator wave functions; it can be removed by changing the function $S(x - x')$ in the commutation relations to $2S(x - x')$.

In the momentum representation the operators of energy, momentum, and charge have the form

$$\begin{aligned} E &= \sum_{pr} |p_0| \left(a_{pr}^+ a_{pr} - a_{pr} a_{pr}^+ - b_{pr} b_{pr}^+ + b_{pr}^+ b_{pr} \right) \\ &= \sum_{pr} |p_0| \left(N_{pr}^{(+)} + N_{pr}^{(-)} - 2 \right); \end{aligned} \tag{20}$$

$$\mathbf{P} = \sum_{pr} \mathbf{p} \left(a_{pr}^+ a_{pr} - a_{pr} a_{pr}^+ - b_{pr} b^+ pr + b_{pr}^+ b_{pr} \right)$$

$$= \sum_{pr} \mathbf{p} \left(N_{pr}^{(+)} + N_{pr}^{(-)} \right) ; \tag{21}$$

$$Q = e \sum_{pr} \left(a_{pr}^{+} a_{pr} - a_{pr} a_{pr}^{+} + b_{pr} b_{pr}^{+} - b_{pr}^{+} b_{pr} \right)$$
$$= e \sum_{pr} \left(N_{pr}^{(+)} - N_{pr}^{(-)} \right) , \tag{22}$$

where $N_{pr}^{(+)}$ and $N_{pr}^{(-)}$ are the number operators for particles and antiparicles.

The infinite term $\sum 2|p_0|$ in expression (20) does not contain any operators and can therefore be omitted, as in the usual theory. As a result, the energy becomes a positive definite quantity. The spectum of the operators E, $\mathbf{P}$, and Q admits of the usual interpretation in terms of the number of particles occupying the individual states, with the only difference that now the maximal occupation number for each state is equal to two.

We now consider the commutators of operators for physical quantities with the operator wave functions. From the relations (13) and (14) we have

$$[E, \psi(x)] = -\mathrm{i}\frac{\partial \psi(x)}{\partial t},$$
$$[\mathbf{P}, \psi(x)] = \mathrm{i}\nabla\psi(x),$$
$$[Q, \psi(x)] = -\psi(x) \ . \tag{23}$$

The relations (23) give the usual connection between the operators of energy, momentum, and charge and the infinitesimal canonical transformations. Similar relations can also be obtained for other physical operators.

5 The Variational Principle of Schwinger TCP Invariance

The variational principle of (Schwinger (1956)) contains the most consistent formulation of the basic postulates of the quantum theory of localized fields. We show that our method of quantization is contained in the variational principle of Schwinger as a special solution.

Since the detailed exposition of Schwinger's variational principle can be found in the literature, we shall deal only with those aspects which change as we make the transition to our method of quantization.

We take the Lagrangian in the form

$$L = -\frac{1}{2}\left[\bar{\psi}, \gamma_\mu \frac{\partial}{\partial x_\mu}\psi + m\psi\right] - \frac{1}{2}\left[-\frac{\partial}{\partial x_\mu}\bar{\psi}\gamma_\mu + m\bar{\psi}, \psi\right] + \dots \ , \tag{24}$$

... stands for any arbitrary interaction terms.

We shall assume that the terms describing the interaciton are antisymmetrized with respect to the operator functions ψ and $\bar{\psi}$. The class of admissible variations is restricted by the conditions

$$\psi_\alpha(x)\delta\psi_\beta(x)\psi_\gamma(x) + \psi_\gamma(x)\delta\psi_\beta(x)\psi_\alpha(x) = 0; \tag{25}$$

$$\begin{aligned}\delta\psi_\alpha(x)\left(\psi_\beta(x)\psi_\gamma(x) - \psi_\gamma(x)\psi_\beta(x)\right)\\ + \left(\psi_\gamma(x)\psi_\beta(x) - \psi_\beta(x)\psi_\gamma(x)\right)\delta\psi_\alpha(x) = 0\ .\end{aligned} \tag{26}$$

where $\psi(x)$ stands for the usual as well as the Dirac conjugate spinor. We shall see later on that we have to supplement the definition of the class of admissible variations to obtain the commutation relations.

Condition (26) is sufficient for the derivation of the equations of motion. Indeed, owing to the antisymmety of the Lagrangian, condition (26) permits us to move the variations either completely to the right or to the left, depending on whether the variation is in an even or an odd position in the formula. Again we see from the antisymmetry of the Lagrangian that the coefficients of the variations standing to the left or to the right in the formula are equal, and can be set equal to zero simultaneously.

To obtain the commutation relations, we consider the operators $G(\psi)$ and $G(\bar{\psi})$ which generate an infinitesimal transformation of the functions $\psi(x)$ and $\bar{\psi}(x)$ (Schwinger (1956));

$$G(\psi) = \mathrm{i}\int \mathrm{d}V\ [\bar{\psi}(x), \gamma_4\delta\psi(x)]\ ; \tag{27}$$

$$G(\bar{\psi}) = -\mathrm{i}\int \mathrm{d}V\ [\delta\bar{\psi}(x), \gamma_4\psi(x)]\ \ ; \tag{28}$$

the time t is assumed to be the same in both operators and is not indicated ecplicitly.

The commutators of G with ψ and $\bar{\psi}$ are equal to

$$[\psi(x), G(\psi)] = \mathrm{i}\delta\psi(x); \tag{29}$$

$$[\bar{\psi}(x), G(\psi)] = \mathrm{i}\delta\psi(x)\ . \tag{30}$$

The other commutators are equal to zero.

Substituting (27) in (29) and using (25), we find

$$\begin{aligned}\int \mathrm{d}V \Big\{\psi_\mu(x')\bar{\psi}_\nu(x)(\gamma_4)_{\nu\rho}\delta\psi_\rho(x)+\\ \delta\psi_\rho(x)(\gamma_4)_{\nu\rho}\bar{\psi}_\nu(x)\psi_\mu(x')\Big\} = \delta\psi_\mu(x')\ .\end{aligned} \tag{31}$$

Hence

$$\begin{aligned}\psi_\mu(x')\bar{\psi}_\mu(x)(\gamma_4)_{\nu\rho}\delta\psi_\rho(x) + \delta\psi\rho(x)(\gamma_4)_{\nu\rho}\bar{\psi}_\nu(x)\psi_\mu(x')\\ = \delta(x' - x)\delta\psi_\mu(x)\ .\end{aligned} \tag{32}$$

Analogous relations are obtained for the other commutators of G with ψ and $\bar{\psi}$.

The relations (32) further delimits the class of admissible variations. However, this delimitation is not sufficiently comlete to obtain the commutation relations in explicit form.

We note that the more general relations of the type

$$\psi_\mu(x)\bar{\psi}_\nu(x')\delta\psi_\rho(x'') + \delta\psi_\rho(x'')\bar{\psi}_n u(x')\psi_\mu(x)$$
$$= (\gamma_4)_{\mu\nu}\delta(x - x')\delta\psi_\rho(x''); \qquad (33)$$

$$\psi_\mu(x)\delta\bar{\psi}_\nu(x')\psi_\rho(x'') + \psi_\rho(x'')\delta\bar{\psi}_\nu(x')\psi_\mu(x) = 0 \ , \qquad (34)$$

are also valid. They are consistent with (25), (26), and (32).

Applying formula (33) to the relations

$$\left[\bar{\psi}(x)\psi(x'), G(\psi)\right] = \mathrm{i}\bar{\psi}(x)\delta\psi(x') \qquad (35)$$

we obtain the following expression for the commutation relations:

$$\psi_\mu(x)\bar{\psi}_n u(x')\psi_\rho(x'') + \psi_\rho(x'')\bar{\psi}_\nu(x')\psi_\mu(x)$$
$$= (\gamma_4)_{\mu\nu}\delta(x - x')\psi_\rho(x'') + (\gamma_4)_{\rho\nu}\delta(x'' - x')\psi_\mu(x) \qquad (36)$$

etc., in agreement with (2').

All otherapplications of the variational principle of Schwinger remain practically unaltered in changing the quantization method.

In concluding this section, we note that our method of quantization preserves the TCP invariance (Pauli (1955)). Indeed, the TCP invariance for the case of spin $\frac{1}{2}$ fields is a consequence of the antisymmetrization of the equations of motion with respect to the operator wave functions. But this antisymmetrization also lies at the basis of our method of quantization.

The author expresses his gratitude to A.I.Akhiezer, E.V.Inopin, I.M.Lifshitz, S.V.Peletminskii, and P.I.Fomin for valuable discussions of the results of this paper.

References

Akhiezer, A.I. and Berestetskii, V.B. (1953): Quantum Electrodymanics, Moscow. [Engl. Transl. by U.S. Dept. Comm.]

Fujiware, I. (1953): Progr. Theoret. Phys. **10**, 589.

Pauli, W. (1955): Article in "Niels Bohr and the Development of Phyusics", McGraw–Hill, N.Y.

Pauli, W. (1947): Relativistic Theory of Elementary Particles (Russ. Transl.), Moscow.

Schwinger, J. (1956): Theory of Quantized Fields (Russ. Transl.), Moscow; [Phys. Rev. **91**, 713, 728; **92**, 1283; **93**, 615; **94**, 1362.]

Translated by R. Lipperheide

S-Matrix in the Generalized Quantization Method

D.V. Volkov

Physico-technical Institute, Academy of Sciences, Ukrainian S.S.R.

J. Exptl. Theoret. Phys. (USSR), **38**, 518–523 (February, 1960)

Abstract. The formalism of the S–matrix for interacting electromagnetic field and half–spin particle field is considered. Particle field quantization is carried out according to a scheme suggested in the works of (Green (1953)) and the author (Volkov (1959)). It is shown that the basic concepts of the conventional theory of S–matrices (N–product, Wick's theorem, Feynman graphs) allow a simple generalization within the framework of the quantization scheme considered.

1. The customary methods of quantization of wave fields use as commutation relations commutators of anticommutators based on a choice of completely symmetric or completely antisymmetric wave functions in the configuration space of many identical particles. The confinement to symmetric of antisymmetric wave functions corresponds to the experimental data known at present as regards the statistics of elementary particles, but is evidently not rigorously established from the theoretical point of view. The problem as to why other possibilities are not realized in nature, "equally valid in the sense of the correspondence principle, in which " lies the essence of this limited choice of nature" (Pauli (1947)), has been discussed in lively fashion in the literature in the period of the development of quantum mechanics (see, for example, reference (Pauli (1947)).

With the development of methods of quantum theory, great progress has been achieved in the understanding of the connection of symmetric and antisymmetric wave functions with the value of the spin of particles (Pauli (1957)) and with the TCP invariance (Schwinger (1951), Pauli (1958)). However, consideration of the problems mentioned has always been carried out within the framework of the following alternative: either symmetric or antisymmetric wave functions; all other possibilities have been entirely neglected.

In this connection it is of interest to attempt to formulate this old problem, which arises in nonrelativistic quantum mechanics, in terms of the theory of wave fields.

The generalization of the existing methods of quantum field theory, which takes into consideration the presence not only of symmetric and antisymmetric wave functions, but which is also compatible with the fundamental premises of relativistic quantum theory, was carried out in the work of (Green (1953)) and later in a research of the author (Volkov (1959)) [1].

In references 1 and 2, however, questions connected with interaction were not considered. At the same time the possibility was not excluded that precisely the interaction between fields could be decisive for explanation of the separation of the existing methods of quantization [2].

In the present article we consider the formalism of the scattering matrix (S–matrix) for interacting electromagnetic field and the field of half–spin charged particles. Quantization of the field of the particles is carried out on the basis of transformed commutation relations (see below, Eq. (3)). It is shown that, in spite of the change of the quantization rules, there exists a unique procedure of expansion of the S–matrix in a series of normal derivatives (analogous to the usual technique of (Wick (1950))), which makes it possible to isolate the vacuum effects in the S–matrix. The results obtained without any essentioal change are applicable also to other local variants of interacting fields.

2. The scattering matrix for the case under consideration has the form

$$S = T(\exp(-\mathrm{i}\int H(x)\ \mathrm{d}^1 x)) \tag{1}$$

where $H(x)$ is the Hamiltonian density in the interaction representation

$$H(x) = \mathrm{i}\ e[\bar{\psi}(x), \gamma_\mu \psi(x)] A_\mu(x), \tag{2}$$

$\psi(x)$ and $\bar{\psi}(x) = \psi^+(x)\gamma_4$ are the field operators of particles satisfying the Dirac equation without interaction in the commutation representation[3]

$$\begin{aligned}
&\psi_\alpha(x)\psi_\beta(x')\psi_\gamma(x'') + \psi_\gamma(x'')\psi_\beta(x')\psi_\alpha(x) = 0\\
&\psi_\alpha(x)\bar{\psi}_\beta(x')\psi_\gamma(x'') + \psi_\gamma(x'')\bar{\psi}_\beta(x')\psi_\alpha(x) = -\mathrm{i}S_{\alpha\beta}(x-x')\psi_\gamma(x'')\\
&\qquad\qquad\qquad\qquad\qquad\qquad\qquad\qquad - \mathrm{i}S_{\gamma\beta}(x''-x')\psi_\alpha(x)\\
&\bar{\psi}_\alpha(x)\bar{\psi}_\beta(x')\psi_\gamma(x'') + \psi_\gamma(x'')\bar{\psi}_\beta(x')\bar{\psi}_\alpha(x) = -\mathrm{i}S_{\gamma\beta}(x''-x')\bar{\psi}_\alpha(x).
\end{aligned} \tag{3}$$

$A_\mu(x)$ are the operators of the electromagnetic field, which satisfy the usual rules of commutation.

Thanks to the commutability of the operators $H(x)$ and $H(x')$, the T–product in Eq.(1) outside the light cone is determined in a unique, relativistically–invariant fashion.

---

[1] The work of Green was not known to the author during preparation of reference 2 for publication.

[2] The possible connection of the symmetry of a wave function with a definite type of interaction in nonrelativistic quantum mechanics has been investigated by (Yaffe (1930)).

[3] We use the notation of reference 2, which is cited below as I.

The operators of the electromagnetic field and the field of particles commute with one another; therefore the T–product in Eq.(1) can be represented in the form of the products of two independent one of which contains only the field operators of the particles, while the other contains only the operators of the electromagnetic field. The latter of these T–products will not be considered, since it has the same form as in ordinary theory.

The absence in the quantization method under consideration of simple commutation rules between the two operators makes difficult the separation of the vacuum effects in the T–product, which depend on the field operators of the particles, and requires a generalization of the concept of normal product.

In order to make clear the idea of such a generalization, let us look first at the simplest case, in which there are two operators: a_k is the destruction of a particle in the state k and a_l^+ that of the creation of a particle in the state l [or, similarly the operators b_k (b_k^+) of destruction (creation) of antiparticles].

The fundamental properties of these vectors are defined by the relations (9), (13), and (14) of I.

Let us determine the normal product $N(a_l^+ a_k)$ of the operators a_l^+ and a_k by the direct action of the N–product on the arbitrary basis vector[4]:

$$\begin{aligned}N(a_i^+ a_k)a_1^+ a_2^+ \dots a_n^+ \Phi_0 &= \delta_{k1} a_l^+ a_2^+ \dots a_n^+ \Phi_0 + \delta_{k2} a_1^+ a_l^+ \dots a_n^+ \Phi_0 + \dots \\ &\quad + \delta_{kn} a_1^+ a_2^+ \dots a_l^+ \Phi_0 + \dots \\ &= \sum_{j=1}^{n} \delta_{kj} a_1^+ \dots a_{j-1}^+ a_l^+ a_{j+1}^+ \dots a_n^+ \Phi_0, \end{aligned} \tag{4}$$

where Φ_0 is the vector of the vacuum state for nonintersecting fields and the indices $1, 2, \dots n$ determine the state of the particle.

As is seen directly from the definition (4), the N–product in the first place preserves the symmetry of the wave function, which is important in the establishment of the connection with nonrelativistic theory, and, in the second place, does not contain the vacuum effects, which are connected with the possibility of the destruction by the operator a_k of a particle previously created by the operator a_l^+.

We note that in the quantization with anticommutators, the determination just considered of the normal derivative coincides with the usual one.

Making use of the commutation relations for the operators a and a^+ (9, I), it is easy to find an explicit expression for the normal product $N(a_l^+ a_k)$ in terms of the operators a_l^+ and a_k:

$$N(a_l^+ a_k) = a_l^+ a_k - a_k a_l^+ - \delta_{lk} \tag{5}$$

[4] In Eq.(4) [and in the subsequent formula (6)] the operators b^+ (or a^+) which can enter into the determination of the basis vector are omitted. Such operators, if there are any, do not affect the action of the N–products considered in (4) and (6), and without change in their position go over into the righthand parts of the corresponding equations.

The normal product of the operators b_k^+ and b_l is determined in similar fashion:

$$N(b_l b_k^+) b_1^+ b_2^+ \dots b_n^+ \Phi_0 = -\sum_{j=1}^{n} \delta_{jl} b_1^+ \dots b_{j-1}^+ b_k^+ b_{j+1}^+ \dots b_n^+ \Phi_0 \tag{6}$$

where

$$N(b_l b_k^+) = b_l b_k^+ - b_k^+ b_l + \delta_{kl} \tag{7}$$

For the case of two particle and antiparticle creation operators, and correspondingly for two destruction operators, we determine the normal product with the aid of the following relations:

$$N(a_l^+ b_k^+) = a_l^+ b_k^+ - b_k^+ a_l^+, \tag{8}$$

$$N(b_l a_k) = b_l a_k - a_k b_l \tag{9}$$

The relations (5), (7)—(9) make it possible to write down the current operator in the form of a normal product. Actually, if the wave functions of the particle and antiparticle in the state k are connected by the relation $v_k = C\bar{u}_k$, where C is the charge–conjugation matrix, u_k and v_k are the coefficients in the expansion (7,I), then $\bar{u}_k \gamma_\mu u_k = \bar{v}_k \gamma_\mu v_k$, as a consequence of which,

$$ie[\bar{\psi}(x), \gamma_\mu \psi(x)] = ieN[\bar{\psi}(x) \gamma_\mu \psi(x)]. \tag{10}$$

In the general case, the N–product depends on an arbitrary number of pairs of operators[5] and is determined by the following relations:

$$N(a_1^+ a_2; \dots; a_{1'}^+ b_{2'}^+; \dots; b_{1''} a_{2''}; \dots; b_{1'''} b_{2'''}^+; \dots) =$$
$$N(a_{1'}^+ b_{2'}^+) \dots N(a_1^+ a_2; \dots) N(b_{1'''} b_{2'''}^+; \dots) N(b_{1''} a_{2''}) \dots \tag{11}$$

The order of arrangement of pairs of operators under the sign of the N–product in this formula is arbitrary.

The normal products $N(a_1^+ a_2; \dots)$ and $N(b_1 b_2^+; \dots)$ in Eq. (11) depend only on pairs of operators of the form $a^+ a$ and bb^+, respectively. The N–products of such a type are determined, similarly to (4) and (6), by the action of these products on the basis vectors:

$$N(a_k^+ a_l; a_m^+ a_r; \dots) a_1^+ a_2^+ \dots a_n^+ \Phi_0 =$$
$$\sum_{i,j\dots=1}^{n} \delta_{li} \delta_{rj} \dots a_1^+ \dots a_{i-1}^+ a_k^+ a_{i+1}^+ \dots a_{j-1}^+ a_m^+ a_{j+1}^+ \dots a_n^+ \Phi_0 \tag{12}$$

$$N(b_k b_l^+; b_m b_r^+; \dots) b_1^+ b_2^+ \dots b_n^+ \Phi_0 = (-1)^P \sum_{i,\dots=1}^{n} \delta_{ki} \delta_{mj}$$
$$\dots b_1^+ \dots b_{i-1}^+ b_l^+ b_{i+1}^+ \dots b_{j-1}^+ b_r^+ b_{j+1}^+ \dots b_n^+ \Phi_0 \tag{13}$$

[5] We limit ourselves her to a consideration of the N–products only of an even number of field operators of the particles. Such a limitation is not essential in what follows, since an even number of particle field operators always enters into the S–matrix and into all observable physical quantities.

summation in (12) and (13) is carried out over all non–coinciding indices; P is the number of pairs of operators of the form bb^+ (see the last footnote but one).

Equations (8),(9),and (11)–(13) determine the N–product for an arbitrary even number of operators and make it possible to represent any product of N–products (including the T–product) in the form of a sum of normal products.

As an example, let us consider the product $N[\bar{\psi}(1)\psi(2)]\ N[\bar{\psi}(3)\psi(4)]$ (the numbers 1,2,3,4 indicate the dependence of the operators on the coordinates and spinor indices). Making use of the commutation relations for the operators a, a^+, b and b^+ [Eqs. (3) and (9,I)] and the determination of the normal products, we obtain

$$\begin{aligned} &N(\bar{\psi}(1)\psi(2))N(\bar{\psi}(3)\psi(4)) = \\ &\quad N(\bar{\psi}(1)\psi(2);\bar{\psi}(3)\psi(4) - \mathrm{i}S^+(2,3)N(\bar{\psi}(1)\psi(4)) \\ &\quad + \mathrm{i}S^-(4,1)N(\bar{\psi}(3)\psi(2)) - 2S^+(2,3)S^-(4,1), \end{aligned} \tag{14}$$

where S^+ and S^- are the usual (+)- and (−)- fold commutation functions:

$$\begin{aligned} S^+(x) &= -\frac{\mathrm{i}}{(2\pi)^3}(\gamma\frac{\partial}{\partial x} - m)\int_{p_0>0}\delta(p^2+m^2)\mathrm{e}^{\mathrm{i}px}\mathrm{d}^4p, \\ S^-(x) &= \frac{\mathrm{i}}{(2\pi)^3}(\gamma\frac{\partial}{\partial x} - m)\int_{p_0<0}\delta(p^2+m^2)\mathrm{e}^{\mathrm{i}px}\mathrm{d}^4p. \end{aligned} \tag{15}$$

A similar formula holds for the T–product

$$\begin{aligned} &T[N(\bar{\psi}(1)\psi(1'))N(\bar{\psi}(2)\psi(2'))] = \\ &\quad N(\bar{\psi}(1)\psi(1');\bar{\psi}(2)\psi(2')) + S^F(1',2)N(\bar{\psi}(1)\psi(2')) + \\ &\quad S^F(2',1)N(\bar{\psi}(2)\psi(1')) - 2S^F(1',2)S^F(2',1), \end{aligned} \tag{16}$$

where

$$S^F(x) = \frac{\mathrm{i}}{(2\pi)^4}(\gamma\frac{\partial}{\partial x} - m)\int_{p_0<0}\delta(p^2+m^2-\mathrm{i}\varepsilon)\mathrm{e}^{\mathrm{i}px}\mathrm{d}^4p;\ \varepsilon\to 0. \tag{17}$$

The prime indicates the possible difference of spinor indices in the corresponding operators.

In the general case of a T—product from an arbitrary number of N–products, the following rule holds, similar to the rule of Wick in the ordinary quantization theory.

In order to expand a T–product of the form

$$T[N(\bar{\psi}(1)\bar{(}1'))N(\bar{\psi}(2)\psi(2'))\ldots N(\bar{\psi}(n)\psi(n'))$$

in a sum of normal products, it is necessary to consider all possible couplings of operators $\psi(a'($ and $\bar{\psi}(b)$, which do not enter into the composition of one and the same normal product, and to substitute these couplings in the functions $S^F(a',b)$. As a result of the superposition of the couplings, all the

N–products located under the sign of the T–product are united in groups which are uncoupled among themselves, and which either contain no operators (closed loop) or contain two operators $\bar{\psi}(a)$ and $\psi(b')$ (open lines). In the latter case, it is necessary to join the two disconnected operators in a pair and to put under the sign N–products of the form $N(\bar{\psi}(a)\psi(b')\dots)$. In the presence of closed loops, each of them must be multiplied by an additional factor of 2[6].

We note that the rule formulated above has the usual graphical interpretation i terms of a Feynman diagram.

3. The relations considered in the preceding section make it possible to investigate in a simple fashion the matrix element of the scattering matrix corresponding to some particular process. As an illustration, we consider the process of scattering of two particles.

In order to determine completely the state of the two particles (in the given system of quantization), it is necessary, in addition to the quantities that characterize the individual states (spin, momentum), also to give the symmetry of the wave function (in the case of pure states) or the relative weights[7] of the symmetric and antisymmetric functions (for mixed states).

The basic orthonormalized vectors of states for different types of symmetry have the form

$$\frac{1}{\sqrt{2}}(a_k^+ a_l^+ \pm a_l^+ a_k^+)\Phi_0;\ k \neq l,$$

k and l are indices characterizing the spin and angular momentum of the individual states.

To determine the probability amplitudes of the scattering process under consideration, we compute the matrix elements of the N–product between the different basis vectors.

Separating in the N–products the terms giving non-vanishing contributions to the matrix element

$$N(\bar{\psi}(1)\gamma_\mu\psi(1); \bar{\psi}(2)\gamma_\mu\psi(2)) =$$

[6] The fundamental difference between the ordinary technique of (Wick (1950)) and its generalization considered here consists in the appearance of this factor takes place not only for virtual processes, but also for processes which occur with the creation of pairs of real particles and antiparticles (as a consequence of the normalization of the operator wave function). In the case of more complicated schemes of quantization (Green (1953)), which lead in the general case to statistics of particles with maximal occupation number m for each of the individual states, the expansion of the T–product in a sum of N–products takes place in precisely the same fashion, but in this case each closed loop acquires an additional factor of m.

[7] A more detailed realization of the state (furnishing of coefficients in the expansion of the wave function over symmetric and antisymmetric states) has no meaning because of the identity of the particles.

$$2N(a_{k'}^{+}a_k; a_{l'}^{+}a_l)\bar{u}_{k'}(1)\gamma_\mu u_k(1)\bar{u}_{l'}(2)\gamma_\mu u_l(2) +$$
$$2N(a_{l'}^{+}a_k; a_{k'}^{+}a_l)\bar{u}_{l'}(1)\gamma_\mu u_k(1)\bar{u}_{k'}(2)\gamma_\mu u_l(2), \qquad (18)$$

where the primed indices characterize the state of the particles in the final states, while $u_k(1)$ etc are wave function of single particle states, and noting that as a consequence of (12),

$$N(a_{k'}^{+}a_k; a_{l'}^{+}a_l)\frac{1}{\sqrt{2}}(a_k^{+}a_l^{+} \pm a_l^{+}a_k^{+})\Phi)0 = \frac{1}{\sqrt{2}}(a_{k'}^{+}a_{l'}^{+} \pm a_{l'}^{+}a_{k'}^{+})\Phi)0$$

$$N(a_{l'}^{+}a_k; a_{k'}^{+}a_l)\frac{1}{\sqrt{2}}(a_k^{+}a_l^{+} \pm a_l^{+}a_k^{+})\Phi)0 = \frac{1}{\sqrt{2}}(a_{l'}^{+}a_{k'}^{+} \pm a_{k'}^{+}a_{l'}^{+})\Phi)0 \qquad (19)$$

we get the following expression for the non–vanishing matrix elements:

$$\Phi_0^*\frac{1}{\sqrt{2}}(a_{l'}a_{k'} \pm a_{k'}a_{l'})N(\bar{\psi}(1)\gamma_\mu\psi(1); \bar{\psi}(2)\gamma_\nu\psi(2))\frac{1}{\sqrt{2}}(a_k^{+}a_l^{+} \pm a_l^{+}a_k^{+})\Phi_0 =$$
$$2(\bar{u}_{k'}(1)\gamma_m u u_k(1)\bar{u}_{l'}(2)\gamma_\nu u_l(2) \pm \bar{u}_{l'}(1)\gamma_\mu u_k(1)\bar{u}_{k'}(2)\gamma_\nu u_l(2)). \qquad (20)$$

Taking into account the sign (−) in Eq. (20), we obtain the well–known formula of Møller. The sign (+) in Eq. (20) leads to the following expression for the scattering cross section (in the center–of–mass system):

$$\mathrm{d}\sigma_{(+)} = \frac{r_0^2}{\varepsilon^2(\varepsilon^2-1)^2}\left\{\frac{(2\varepsilon^2-1)^2}{\sin^4\theta} - \frac{4\varepsilon^4 - 5\varepsilon^2 + \frac{5}{4}}{\sin^2\theta} + \frac{(\varepsilon^2-1)^2}{4}\right\}.$$

For the case of a mixed state, the scattering cross sections $\mathrm{d}\sigma^{+}$ and $\mathrm{d}\sigma^{-}$ are averaged with the corresponding weighting factors.

In conclusion the author expresses his gratitude to A. I. Akhiezer, S.V. Peletminskii and P.I.Fomin for valuable discussions.

References

Green, M. (1953): Phys. Rev., **90**, 270.
Pauli, W. 1947): General Principles of Wave Mechanics (Russian translation, IIL).
Pauli, W. (1957): Relativistic Theory of Elementary Particles (Russian translation, IIL).
Pauli, W. (1958): article in the volume, Niels Bohr and the Development of Physics (Russian translation, IIL).
Schwinger, J. (1951): Phys. Rev. **82**, 914.
Volkov, D.V. (1959): JETP **36**, 1560; Soviet Phys. JETP **9** ,1107 .
Wick, G.C. (1950): Phys. Rev. **80**, 268.
Yaffe, G. (1930): Z. Physik **66**, 748.

Translated by R.T. Beyer

Regge Poles in Nucleon–Nucleon and Nucleon–Antinucleon Scattering Amplitudes

D.V. Volkov[1] and V.N. Gribov[2]

[1] Physico–technical Institute, Academy of Sciences, Ukr. S.S.R., Khar'kov,
[2] A.F.Ioffe Physico–technical Institute, Academy of Sciences, U.S.S.R.

Submitted to JETP editor October 24,1962
J.Exptl. Theoret. Phys. (U.S.S.R.) **44**, 1068–1077 (March, 1963)

Abstract. The spin structure and the character of the kinematic singularities of the contributions of Regge poles with various quantum numbers to nucleon–nucleon and nucleon–antinucleon scattering amlitudes are cinsidered. It is shown that the presence of kinematic singularities at $t = 0$ in the contributions from certain Regge poles together with the requirement of analyticity of the whole amplitude leads to simple relations, at $t = 0$, between the positions of Regge poles belonging to different trajectories. The dependence of the spin structure of the forward scattering amplitude on the character of these relations is discussed.

1 Introduction

Regge poles (Regge (1959, 1960)) in nn scattering amplitudes have been discussed in a number of recent papers (Regge (1959, 1960), Gribov (1961, 1962), Chew, Frautschi (1961), Frautschi, Gell–Mann, Zachariasen (1962), Udgaonkar (1962), Gell–Mann (1962), Gribov, Pomeranchuk (1962), Gribov, Pomeranchuk (1962a)). In particular, a detailed invesrigation has been made in (Gribov, Pomeranchuk (1962a)) of the spin stucture on the nn scatterung amplitude as determined by the dominant vacuum pole – the Pomeranchuk pole.

In the present paper we consider the spin structure of the nn and na scattering amplitudes thaking into account r with other quantum numbers as well. It will be shown that for fixed signature (Frautschi, Gell–Mann, Zachariasen (1962)) and isotopic spin there are, besides the poles whose remaining quantum numbers are those the vacuum, three further kinds of poles which give a contribution to these amplitudes.

It turns out that the analyticity requirement on the amplitude implies that two of the three families of poles must coincide at $t = 0$(t is the momentum transfer), whereas the poles of the third family have angular momenta which differ from the amgular momenta of the first two families by ± 1. The r at $t = 0$ correspond to particles with zero mass. The coincidence of the furst two families implies that there is a degeneracy in parity and angular momentum for zero mass particles not belonging to the vacuum family.

The situation is of a similar nature as the one noted by one of us (Gribov (1962b, 1963)) in discussing the 180° meson-nucleon scattering amplitudes. In that case the poles of amplitudes with different parity became identical for vanishing mass.

In both cases the degeneracy is due to the fact that there exists an additional symmetry in forwand or backward scattering which is connected with the presence of only one distinguished direction.

The essential difference between the boson r considered in this paper and the fermion poles considered in (Gribov (1962b, 1963)) consists in the fact that in our case poles with different quantum numbers do not become complex conjugated for $t, 0$. As was shown in (Gribov, Pomeranchuk (1962a)), the contribution from the vacuum poles at $t = 0$ to the amplitudes is indeprndent of the spin. The other poles considered in this paper lead to a spin dependence of the forward scattering amplitude. The spin structure is strongly dependent on whether the angular momentum of the dominant pole of the third family differs from that of the dominant pole of the two other families by $+1$ or -1. In the first case the additional contribution to the forward scattering amplitude has axial vector character and leads to a correlation of the longitudinal polarizations of the nucleons. In the second case it has tensor character and leads to a correlation of the transverse polarizations.

2 Classification of the Poles

Let us consider a na system in the channel where t is the energy. We characterize the state of the system by the total angular momentum j and the projections of the particle spins on their directions of motion. Let us call these staes $|j, \pm, \pm >$. The states $|j, \lambda p, \lambda >$ zre related to the states with definite total spin and parity in the following way (Goldberger, Grisaru, MacDowell, Wong (1960)):

$$|j, 0, - >= |j, +, + > - |j, -, - > \ , \quad \text{(1a)}$$

$$|j, 1, - >= |j, +, - > - |j, -, + > \ , \quad \text{(1b)}$$

$$|j, 0, + >= |j, +, + > - |j, -, - > \ , \quad \text{(1c)}$$

$$|j, 1, + >= |j, +, - > - |j, -, + > \ , \quad \text{(1d)}$$

where the figures 0 andl denote the absolute valie of the spin projection of the relative direction of motion and the signs $\pm$ define the symmetry under interchange of the spin projections of the particles.

The states $|j, 0, - >$ are singlets ($s = 0$) with parity $(-1)^{j+1}$ the state $|j, 0, - >$ is a triplet with parity $(-1)^{j+1}$, and the states $|j, 0, + >$ are triplet states with parity $(-1)^j$.

The first two states are conserved. The amplitudes in these states will be denoted by f_0^j and f_1^j. The last two stated are not conserved. The amplitudes in these two states will be denoted by f_{00}^j and f_{11}^j, and the amplitude for transition between them by f_{01}^j . In what follows we shall consider the amplitudes $f_1^j, f_1^j, f_{00}^j, f_{11}^j$, and f_{01}^j for complex j and their moving poles.

As is known (Gribov (1961a, 1962a), Chew, Frautschi (1961), Frautschi, Gell–Mann, Zachariasen (1962)), it is necessary for this to consider separately the amplitudes withdifferent signature (Frautschi, Gell–Mann, Zachariasen (1962)), i.e., coinciding with the physical partial waves for even and odd j. In order to classify the different Regge trajectories it is convenient to introduce the following quantum numbers: the signature of parity of $j(P_j)$, the parity P, the G parity, and the isospin T. The relation between the trajectories with quantum numbers $P-j, P, G$, and T and the states $|j, 0, \pm>$ is given in the table.

P_j	P	G	T	*States*
$+$	$+$	$+$ $-$	0 1	$\|j,0,+>$, $\|j,1,->$
$+$	$+$	$-$ $+$	0 1	*none*
$+$	$-$	$+$ $-$	0 1	$\|j,0,->$
$+$	$-$	$-$ $+$	0 1	$\|j,1,+>$

The simultaneous change of the signes of P_j, P, and G with fixed T does not change the states.

The states $|j, 0, +>$ and $|j, 1, +>$ are not conserved and have a common family of poles. However, it is convenient for the following to divide this system of poles into two families according to whether they contribute to one or the other eigenvalue of the matrix

$$\begin{pmatrix} f_{00}^j & f_{01}^j \\ f_{10}^j & f_{11}^j \end{pmatrix} \tag{2}$$

for continuous changes of t. It will be shown below that the transition matrix element $f_{01}^j = 0$ for $t = 0$. As a consequence, the amplitudes f_{00}^j and f_{11}^j and the corresponding two systems of poles become independent at $t = 0$. In the following wee shall call these systems of poles of the type α and the type β.

Since the states of the nucleon–antinucleon system which transform the matrix (2) into diagonal form do not go over into one another duiring the scattering, only one of them can go over into two mesons. Otherwise they would transform into one another via the two–meson state. Therefore the two–meson annihilation and meson–meson scattering amplitudes contain poles of only one type: either α or β. For $t = 0$ the state $|j, 1, +>$ cannot go into two mesons, since two mesons always have a vanishing spin projection on the relative direction of motion. Thus the meson–meson and meson–nucleon scattering amplitudes have only poles of type α. The vacuum poles usually considered, and in particular, the Pomeranchuk pole, belong to the type α.

3 Partial Wave Expansion of the Amplitude

We write the nucleon–antinucleon scattering amplitude as a sum of the five fermion variants:

$$F = \sum_{i=1}^{5} H_i(t,s)\left(\bar{u}(p_2)O^i_\alpha v(p_2')\right)\left(\bar{v}(p_1')O^i_\alpha u(p_1)\right) \quad , \tag{3}$$

$$Q^1_\alpha = 1,\ \ Q^2_\alpha = \gamma_\mu,\ \ O^3_\alpha = \sigma_{\mu\nu}\ \ (\mu > \nu)\ ,$$
$$O^4_\alpha = i\gamma_5\gamma_\mu,\ \ O^5_\alpha = \gamma_5\ ; \tag{4}$$

p_1, p_1', p_2, p_2' are the momenta of the nucleons and antinucleons in the initial and final states, respectively;

$$t = (p_1 + p_1')^2,\quad s = (p_1 - p_1')^2 .$$

We use the Feynman choice of matrices and the summation convention $p_\mu x_\mu = p_0 x_0 - \mathbf{px}$.

It was shown by Goldberger et al (Goldberger, Grisaru, MacDowell, Wong (1960)) that the invariant functions $H_i(t,s)$ have no kinematic singularities. This will be essential for the following discussion.

In order to calculate the contributions of the different Regge poles to F it is necessary to use the partial wave expansion of the amplitudes. It is convenient to do this in terms of the helicity amplitudes (Jacob, Wick (1959))

$$\varphi(\lambda_2', \lambda_2, \lambda_1', \lambda_1) =< \lambda_2', \lambda_2 | F | \lambda_1', \lambda_1 >,$$

where λ is the spin projection of the particle on its momentum. Choosing the spinors u and v of the form

$$u_\lambda(p) = \begin{pmatrix} \sqrt{p_0 + m} \\ 2\lambda\sqrt{p_0 - m} \end{pmatrix} \varphi_\lambda,\quad v_\lambda(p') = C\bar{u}_{\lambda'}(p') \ , \tag{5}$$

where φ_λ is the spinor of the state with projection λ on the direction of the momentum p, we easily obtain a relation between the amplitudes and the invariant functions $H_i(t,s)$ (Goldberger, Grisaru, MacDowell, Wong (1960)):

$$\begin{aligned}
\varphi_1 &= \varphi(\lambda\lambda\lambda\lambda) = 4p^2 H_1 - 4m^2 z H_2 - 4m^2 z H_3 - 4m^2 H_4 - 4p_0^2 H_5, \\
\varphi_2 &= \varphi(\lambda\lambda - \lambda - \lambda) = -4p^2 H_1 + 4m^2 z H_2 - 4(p_0^2 + p^2) z H_3 \\
&\qquad - 4m^2 H_4 - 4p_0^2 H_5, \\
\varphi_3 &= \varphi(\lambda - \lambda\lambda - \lambda) = -4p_0^2(1+z) H_2 - 4m^2(1+z) H_3 + 4p^2(1+z) H_4, \\
\varphi_4 &= \varphi(\lambda - \lambda - \lambda\lambda) = 4p_0^2(1-z) H_2 + 4m^2(1-z) H_3 + 4p^2(1-z) H_4, \\
\varphi_5 &= \varphi(\lambda\lambda\lambda - \lambda) = -2\lambda \sin\theta \left[4p_0 m H_2 + 4p_0 m H_3\right] \ .
\end{aligned} \tag{6}$$

Here $p = \frac{1}{2}\sqrt{t - 4m^2}$ and $p_0 = \frac{1}{2}\sqrt{t}$ are the momentum and the energy of the particles in the center of mass system (c.m.s.) and

$$s = -2p^2(1+z), \quad z = \cos\theta, \quad \theta = \theta_{p_1 p_2}.$$

The partial wave expansion of the amplitude $\varphi(\lambda_2'\lambda_2;\ \lambda_1'\lambda_1)$ is of the form (Jacob, Wick (1959))

$$4\lambda_2'\lambda_1'\varphi(\lambda_2'\lambda_2;\lambda_1'\lambda_1) = \sum_j (2j+1) < \lambda_2'\lambda_2|F^j|\lambda_1'\lambda_1 > d^j_{\mu_1\mu_2}(z)\ , \tag{7}$$

where $d^j_{\mu_1\mu_2}(z)$ is the reduced rotation matrix and $\mu_1 = \lambda_1' - \lambda_1$, $\mu_2 = \lambda_2' - \lambda_2$. The factor $4\lambda_2'\lambda_1'$ is due to our choice of phase for the state with negative energy.

Following (Goldberger, Grisaru, MacDowell, Wong (1960)) it is convenient to introduce instead of the amplitudes $\varphi(\lambda_2'\lambda_2;\ \lambda_1'\lambda_1)$ their linear combinations

$$\begin{aligned} &f_1 = \varphi_1 + \varphi_2, \quad f_2 = \varphi_1 - \varphi_2, \\ &f_3 = \frac{1}{1+z}\varphi_3 + \frac{1}{1-z}\varphi_4, \quad f_4 = \frac{1}{1+z}\varphi_3 - \frac{1}{1-z}\varphi_4, \\ &f_5 = -\frac{m}{\lambda_0 \sin\theta}\varphi_5\ . \end{aligned} \tag{8}$$

The partial wave expansion of the amplitudes $f_i(t,s)$ has the form

$$f_1 = \sum_j (2j+1) f_0^j(t) P_j(z), \tag{9a}$$

$$f_2 = \sum_j (2j+1) f_{00}^j(t) P_j(z), \tag{9b}$$

$$f_3 = \sum_j \frac{2j+1}{j(j+1)} \left\{ f_1^j(t) \left[P_j'(z) + zP_j''(z) \right] - f_{11}^j(t) P_j''(z) \right\} \tag{9c}$$

$$f_4 = \sum_j \frac{2j+1}{j(j+1)} \left\{ -f_1^j(t) P_j''(z) + f_{11}^j(t) \left[P_j'(z) + zP_j''(z) \right] \right\} \tag{9d}$$

$$f_5 = \sum_j \frac{2j+1}{\sqrt{j(j+1)}} \frac{m}{p_0} f_{01}^j(t) P_j'(z), \tag{9e}$$

where the quantities $f^j(t)$ are defined in the Introduction. In deriving (9a) to (9e) we have made use of the explicit form of $d^j_{\mu_1\mu_2}(z)$ (Jacob, Wick (1959)).

The invariant functions $H_i(t,s)$ are related, according to (7) and (8), to the functions $f_j(t,s)$ by

$$H_1 = \frac{1}{8p^2}\left[f_2 + zf_4 + z\frac{p_0^2+m^2}{m^2}f_5\right], \quad H_2 = -\frac{1}{8p^2}[f_4+f_5],$$

$$H_3 = \frac{1}{8p^2}\left[f_4 + \frac{p_0^2}{m^2}f_5\right], \quad H_4 = \frac{1}{8p^2}f_3,$$

$$H_5 = -\frac{1}{8p_0^2}\left[f_1 - zf_4 + \frac{m^2}{p^2}f_3 - z\frac{p_0^2}{m^2}f_5\right] . \tag{10}$$

Using (9) and (10), we may now compute the contributions of the different Regge poles to the scattering ampitude. It follows from (10) that the functions $f_i(t,s)$ as well as the functions $H_i(t,s)$ satisfy ordinary dispersion relations in the momentum transfer. This permits us, with the help of (9), to introduce amplitudes $f_0^j(t)$, $f_1^j(t)$, $f_{00}^j(t)$, $f_{11}^j(t)$, and $f_{01}^j(t)$ with complex j in the same manner as in the case of spinless particles (Frautschi, Gell–Mann, Zachariasen (1962), Gribov (1961a, 1962a)). Considering separately the symmetric and antisymmetric parts of the functions $f_i(t,s)$ and changing the sum into an integral, we easily find the contribution from the pole with the largest $\mathrm{Re}\, j$ of the amplitudes f_0^j, f_1^j, f_{00}^j, and f_{01}^j to the amplitudes $f_i(t,s)$ and hence to F.

4 Contributions to the Scattering Amplitude from Poles with Different Quantum Numbers

A. Contribution from the pole with $P_j = \pm 1$, $P = \mp 1$, $G = \pm 1$ for $T = 0$ and $G = \mp 1$ for $T = 1$. This pole occurs only in the amplitude f_0^j. According to (9) and (10), it gives a contribution only to the amplitudes $f_1(t,s)$ and $H_5(t,s)$. Leaving out the spinors, we find that in this case

$$F_0 = \frac{\pi}{4t}\frac{2j_0+1}{\sin \pi j_0}\alpha_{j_0} r_0^{\pm}(t)\left[(-z)^{j_0} \pm z^{j_0}\right]\gamma_5^{(1)} \times \gamma_5^{(2)} , \tag{11}$$

where $j_0 = j_0(t)$ is the position of the pole and $r_0^{\pm}(t)$ the residue of $f_0^j(t)$. The signs $\pm$ refer to the signature, and $\alpha_j = \frac{\Gamma(2j+1)}{2^j \Gamma^2(j+1)}$.

B. Contribution from the pole with $P_j = \pm 1$, $P = \mp 1$, $G = \mp 1$ for $T = 0$ and $G = \pm 1$ for $T = 1$. This pole occurs in the amplitude f_1^j and gives a contribution to the functions $f_3(t,s)$ and $f_4(t,s)$. However, f_3 and zf_4 have the same order of magnitude for large z because of the presence of the term $P_j''(z)$ in (9d).

Taking account of the fact that for large $z(s)$ the scalar and pseudoscalar variants give a contribution one power of a smaller than the rest, it can be seen that this pole gives a contribution only to the pseudovector variant for large s. We thus obtain

$$F_1 = -\frac{\pi}{4(t-4m^2)}\frac{2j_1+1}{\sin\pi j_1}r_1^{\pm}(t)\frac{j_1}{j_1+1}\alpha_{j_1}\left[-(-s)^{j_1-1} \pm z^{j_1-1}\right]\cdot$$
$$\cdot i\gamma_5^{(1)}\gamma_\mu^{(1)} \times i\gamma_5^{(2)}\gamma_\mu^{(2)} \ . \tag{12}$$

C. Contibution from the poles with $P_j = \pm 1$, $P = \pm 1$, $G = \pm 1$ for $T = 0$ and $G = \mp 1$ for $T = 1$. This family of poles includes, in particular, the vacuum pole considered in (Gribov, Pomeranchuk (1962a)). Poles with these quantum numbers occur in the amplitudes f_{00}^j, f_{11}^j, and f_{01}^j and give contributions to all functions $f_i(t,s)$ except $f_1(t,s)$. However, for large $z(s)$ the function $f_3(t,s)$ can be omitted by the same arguments as were used to leave out the function $f_4(t,s)$ in the case B. The expressions for the remaining functions are of the form

$$f_2 = -\frac{\pi}{2}\frac{2j+1}{\sin\pi j}\alpha_j r_{00}^{\pm}(t)\left[(-z)^j \pm z^j\right],$$

$$f_4 = -\frac{\pi}{2}\frac{2j+1}{\sin\pi j}\alpha_j\frac{j}{j+1}r_{11}^{\pm}(t)\left[-(-z)^j \pm z^j\right],$$

$$f_5 = -\frac{\pi}{2}\frac{2j+1}{\sin\pi j}\frac{m}{p_0}\alpha_j\sqrt{\frac{j}{j+1}}r_{01}^{\pm}(t)\left[-(-z)^j \pm z^j\right] \ . \tag{13}$$

At first sight the formulas (13) xontain three independent parametes r_{00}, r_{11}, and r_{01}. But the residues of different amplitudes at one and the same pole factorize by virtue of the unitarity condition (Gell–Mann (1962), Gribov, Pomeranchuk (1962), Charap, Squires (1962), Gribov, Ioffe, Pomeranchuk, Rudik (1962)), so that

$$r_{00}r_{11} = r_{01}^2 \tag{14}$$

This relation implies the factorization of the whole amplitude into a part referring to the initial state and a part referring to the finel state. Substituting (13) in (10), we obtain an expression containing four Fermion variants which do not manifestly factorize even if (14) is taken into account. In order to obtain an explicitly, factorized expression, we proceed in the following fashion. Instead of the tensor variant we introduce a variant of the form

$$\Delta = \gamma_\mu^{(1)}(p_2 - p_2')_\mu + \gamma_\mu^{(2)}(p_1 - p_1')_\mu \ . \tag{15}$$

It is easy to show that the following relation is satisfied:

$$tT = 4m^2V + (s-u)(S+P) - 2m\Delta \ , \tag{16}$$

where

$$V = \gamma_\mu^{(1)} \times \gamma_\mu^{(2)}, \quad S = 1 \times 1, \quad P = \gamma_5^{(1)} \times \gamma_5^{(2)},$$
$$T = \frac{1}{2}\sigma_{\mu\nu}^{(1)} \times \sigma_{\mu\nu}^{(2)}, \quad u = (p_1 - p_2)^2, s - u = -4p'^2 z \ .$$

Using this relation, we can write the contibution from the poles in the form

$$F_{\text{vac}} = H'_1(p_1 - p'_1)_\mu(p_2 - p'_2)_\mu S + H'_2 V + H'_3 \Delta \;, \tag{17}$$

where

$$H'_1 = -\frac{1}{32p^4 z}\left(f_2 + z f_4 \frac{m^2}{p_0^2} + 2z f_5\right),$$

$$H'_2 = -\frac{1}{8p_0^2} f_4, \quad H'_3 = -\frac{m}{16p^2 p_0^2}\left(f_4 + \frac{p_0^2}{m^2} f_5\right) . \tag{18}$$

We easily see with the help of (18) and (14) that

$$H'_1 H'_2 = H'^2_3 \; . \tag{19}$$

This implies that the amplitude F_{vac} factorizes and can ve written in the form

$$\begin{aligned} F_{\text{vac}} = & \frac{\pi}{4(t-4m^2)^2} \frac{2j+1}{\sin \pi j} \alpha_j \left[-(-z)^{j-1} \pm z^{j-1}\right] \cdot \\ & \cdot \left[(p_1 - p'_1)_\mu(\rho_0 + \rho_1) + \frac{t-4m^2}{2m} \rho_1 \gamma^{(1)}_\mu\right] \cdot \\ & \cdot \left[(p_2 - p'_2)_\mu(\rho_0 + \rho_1) + \frac{t-4m^2}{2m} \rho_1 \gamma^{(2)}_\mu\right] \end{aligned} \tag{20}$$

where

$$\rho_0^2 = r_{00}, \quad \rho_1^2 = \frac{4m^2}{t} \frac{j}{j+1} r_{11} \; . \tag{21}$$

The form (20) of the contribution of the vacuum pole to the scattering amplitude found in (Gribov, Pomeranchuk (1962a)) was proposed by L.B.Okun'. It provided the basis for the derivation given above.

In concluding this section, we note that the structure of (20) does not depend, for $t \neq 0$, on whether we are dealing with a pole of type α or type β. The difference between the two cases for $t = 0$ shows up in the following way.

It is clear from (9e) and the fact that $f_5(t,s)$ has no singularity at $t = 0$ ($p_0 = 0$) that $f^j_{01} = 0$ for $t = 0$. This means that r_{01} for $t = 0$. Then it follows from (14) that at $t = 0$ either $r_{11} = 0$ or $r_{00} = 0$. In the first case we are dealing with a pole of type α. Here the quantities ρ_0 and ρ_1 remain finite of $t = 0$ [see (21)] and both appear in the expression for F_{vac} for $t = 0$. In the second case we have a pole of type β. It then follows from (21) that $\rho_0 = 0$ and $\rho_1 \to \infty$. This case will be discussed in detail in the netxt section.

Up to this point we have discussed the nucleon–antinucleon scattering amplitude in the channel in which t is the energy and (11), (12), and (20) refer to the region of large unphysical momentum transfers. It is clear that for $t \leq 0$ these formulas give the asymptotic form for $s \to \pm\infty$ of the nucleon–nucleon and nucleon–antinucleon scattering amplitudes, respectively, where s or u are the energy.

5 Asymptotic Form of the Forward Scattering Amplitude

It is well known that the forward scattering amplitude has an additional symmety, as there is only one distinguished direction. Because of this the forward scattering amplitude is defined by three invariant functions instead of five. The lower number of independent functions is connected with the circumstance that in forward scattering there is , first, no polarization (the expression for the amplitude in the c.m.s. does not contain $\sigma_1 \cdot n + \sigma_2 \cdot n$; n is the normal to the scattering plane) and, second, the spin correlations can be defined by expressions of the type $\sigma_{1x}\sigma_{2x} + \sigma_{1y}\sigma_{2y}$ and $\sigma_{1z}\sigma_{2z}$ alone (z the relative direction of motion). The vanishing of the polarization in the relativistic treatment is guaranteed by the fact that the invariant functions H_i have no root singularity at $t = 0$. The necessary form of the correlation terms is guaranteed by the vanishing of the pseudoscalar variant (H_5 has no pole at $t = 0$).

It is obvious that the number of independent amplitudes in the nucleon–antinucleon scattering channel must be the same for $t = 0$. This implies that there must be two relations between the five partial wave amplitudes $f_0^j, f_1^j, f_{00}^j, f_{11}^j$, and f_{01}^j.

As already noted, the partial wave amplitude f_{01}^j is equal to zero for $t = 0$ because of the absence of a $\frac{1}{\sqrt{t}}$ type singularity in the amplitude f_5. The second relation is obtained with the help of the last of Eqs. (10), by requiring that $H_5(t, s)$ does not become infinite at $t = 0$ ($P_0 = 0$). This requirement leads to the condition

$$f_1 - z f_4 - f_3 = 0 \ . \tag{22}$$

The relation (22) can easily be transformed into a relation between the partial wave amplitues, using (9a) and (9b):

$$\begin{aligned} & f_0^{j-1} - f_0^{j+1} - \frac{j-1}{j} f_{11}^{j-1} + \\ & \frac{j+2}{j+1} f_{11}^{j+1} - \frac{2j+1}{j(j+1)} f_1 j = 0 \end{aligned} \tag{23}$$

Relations (22) and (23) have been used for different purposes in (Goldberger, Grisaru, MacDowell, Wong (1960)). It is clear that (23), which is valid for integer j, will also hold true for arbitrary complex j. Interpreting the relation (23) in this sense, we find that the positions of the singularities fo different amplitudes are related at $t = 0$.

We thus obtain relations between the positions of the poles of the amplitudes f_0^j and f_1^j and the positions of the poles of type β which enter in the expression for the amplitude f_{11}^j at $t = 0$. We note that the poles of the type α are distinguised in the sense that there are no relations between their positions and the positions of other poles. Since the above-mentioned relation is a finite difference equation for the amplitudes f_0^j and f_{11}^j, there will in general be, for each pole of f_1^j, an infinite number of poles of f_0^j and f_{11}^j which are displaced from one another by $\Delta j = 2$. On the other hand, if we arbitrarily select a pole at $j = j_0$ in the amplitude f_0^j or f_{11}^j, we find from (23) that the amplitude f_1^j will have two poles at $j = j_0 \pm 1$.

According to the physical interpretation of Regge poles as the analytic continuations in the angular momentum of possible states of a dymanical system, both these situations are completely unreasonable. In particular, there is no reason whatsoever that an arbitrary interaction should have an additional symmetry at $t = 0$ which leads to integral numberd intervals between the angular momenta of the poles of one and the same amplitude. The coincidence of poles of different amplitudes, on the othe hand, is ietirely natural, since there is an additional space symmetry for $t = 0$.

If we require that there be no simple integral numbered relations between the positions of the poles of one and the same amplitude, then we have two possibilities of preserving the validity of (23); the poles of the anplitues f_0^j and f_{11}^j coincide and th angular momentum of the pole of f_1^j differs by ± 1. This implies that it is impossible to ascribe a definite spin and parity to particles with vanishing mass which do not belong to family α.

1. If f_0^j and f_{11}^j have a pole at $j = j_0$ and f_1^j has one at $j = j_0 + 1$, we obtain [by setting $j = j_0 \pm 1$ in (23)] the following relations between the residues of the amplitudes at these poles:

$$r_{11} = r_0 \frac{j_0}{j_0 + 1}, \quad r_1 = r_0 \frac{(2j_0 + 1)(j_0 + 2)}{2j_0 + 3)(j_0 + 1)} \ . \tag{24}$$

2. If f_0^j and f_{11}^j have a pole at $j = j_0$ and f_1^j has one at $j = j_0 - 1$, we have

$$r_{11} = r_0 \frac{j_0 + 1}{j_0}, \quad r_1 = r_0 \frac{(2j_0 + 1)(j_0 - 1)}{(2j_0 - 1)j_0} \ . \tag{25}$$

Let us consider, in correspondence with these two possibilities, the forward scattering amplitude in the chammels where s or u are the energy.

In case 1, where the pole of f_1^j is to the right of the poles of f_0^j and f_{11}^j, we can neglect the contribution of the latter for large z. The forward scattering amplitude will be a sum of contributions from all vacuum poles of the type α which do not lead to a spin dependence and of the expression (12).

In two-component form the amplitude can be written as

$$F = A + B\sigma_{1z}\sigma_{2z} \ . \tag{26}$$

In case 2, where the pole of f_1^j is to the left of the poles of f_0^j and f_{11}^j, we can neglect the contribution from the pole of f_1^j. The scattering amplitude has the form

$$F = \sum_i F_{\alpha_i} + F_0 + f_\beta \ , \tag{27}$$

where F_0 is defined by(11) and F_β is easily obtained from (20) by setting $\rho_0 = 0$:

$$F_\beta = \frac{\pi m^2}{(t-4m^2)^2} \frac{2j_0+1}{\sin \pi j_0} \alpha_{j_0} \left[-(-z)^{j_0-1} \pm z^{j_0-1}\right] \frac{r_{11}}{t} \frac{j_0}{j_0+1} \cdot$$

$$\cdot \left\{ (s-u)S + \frac{(t-4m^2)^2}{4m^2} V + \frac{t-4m^2}{2m} \Delta \right\} . \tag{28}$$

Keeping the terms linear in t in the curly brackets and using (16) and the first of Eqs. (25) we easily find

$$F_0 + F_\beta = -\frac{\pi}{4m^2} \frac{2j_0+1}{\sin \pi j_0} \alpha_{j_0} \left[-\left(-\frac{s}{4m^2}\right)^{j_0-1} \pm \left(\frac{s}{4m^2}\right)^{j_0-1} \right] \cdot$$

$$\cdot \frac{r_0}{2} \sigma^{(1)}_{\mu\nu} \times \sigma^{(2)}_{\mu\nu} . \tag{29}$$

Here the amplitude has the following two-component form:

$$F = A + B(\sigma_{1x}\sigma_{2x} + \sigma_{1y}\sigma_{2y}) . \tag{30}$$

We note that we have not made full use of (24) and (25) in deriving (12), (26), and (29). Therfore, a comparison of these formulas with experiment does not verify our picture of the coincidence of poles at $t = 0$. However, the mere appearance, in (29), of the tensor variant in which each vertex does not have a definite parity, implies already that the "particle" whose exchange gives rise to such an amplitude has no definite parity.

In conclusion we should like to thank I.Ya.Pomeranchyuk, L.B.Okun, V.B.Berestetskii,, and I.M.Shmushkevich for very useful discussions.

References

Charap, I.M. and Squires, E.J. (1962): preprint UCRL-10115.
Chew, G.F. and Frautschi, S.C. (1961): Phys. Rev. Lett. **7**, 394.
Frautschi, Gell–Mann, and Zachariasen (1962): Phys. Rev. **126**, 2204.
Gell–Mann, M. (1962): Phys. REv. Lett. **8**, 263.
Goldberger, Grisaru, MacDowell, and Wong (1960): Phys. Rev. **120**, 2250.
Gribov, V.N. (1961, 1962): JETP **41**, 667, Soviet Phys. JETP **41**, 478.
Gribov, V.N. (1961a, 1962a): JETP **41**, 1962, Soviet Phys. JETP **14**, 1395.
Gribov, V.N. (1962b, 1963): JETP **43**, 1529, Soviet Phys. JETP **16**, 1080.
Gribov, Ioffe, Pomeranchuk, and Rudik (1962): JETP 42, 1419, Soviet Phys. JETP **15**, 984.
Gribov, V.N. and Pomeranchuk, I.Ya. (1962): JETP **42**, 1141, Soviet Phys. JETP **15**, 788.
Gribov, V.N. and Pomeranchuk, I.Ya. (1962a): Phys. Rev. Lett. **8**, 412.
Jacob, M. and Wick, G.C. (1959): Ann. of Physics **7**, 404.
Regge, T. (1959, 1960): Nuovo cimento **14**, 951 and **18**, 947.
Udgaonkar, B.M. (1962): Phys. Rev. Lett. **8**, 142.

Translated by R. Lipperheide

SU(3) × SU(3) Symmetry and the Baryon–Meson Coupling Constants

D.V. Volkov

Institute of Physics & Technology
Kharkov, USSR

Submitted to JETP editor August 10, 1959
J. Exptl. Theoret. Phys. **38**, 518–523, (February, 1960)

Abstract. In this paper we consider the relations between the coupling constants of a unitary octet of baryons with octets of pseudoscalar and singlet vector mesons. The analysis is based in the assumption that the breaking of the SU(6) symmetry for a vertex function with three external lines has a kinematic nature and is due to the presence of two independent four–dimensional energy–momentum vectors in lieu of one, as is the case for the self–energy of the particles when SU(6) symmetry is satisfied.

The requirement that the configuration of the system 4–momenta be invariant leads to the reduction of the SU(6) group to the group SU(3) × SU(3) × U, where the two SU(3) groups correspond to unitary transformations of quarks with different polarization directions along a preferred axis, while the group U corresponds to the usual spatial rotations about this axis. The 35- and 56–plet representations of the group SU(6), corresponding to mason and baryon supermultiplets, contain the following irreducible representations of the group SU(3) × SU(3):

$$(35) \rightarrow (3, 3^*); (3^*, 3); (8, 1); (1, 8); (1, 1)$$

$$(56) \rightarrow (1, 10); (10, 1); (6, 3); (3, 6) \ . \qquad (1)$$

It follows from (1) that the vertex (56*)(56)(35), when is invariant against SU(3)×SU(3) transformations, contains in the general case eight independent parity–conserving interaction constants.

If we confine ourselves to vertices which do not contain the (10), 1) (1, 10) (10, 1), and (10, 1) multiplets, which are classified as baryon resonances because the spin projection in these states is equal to 3/2, then the vertices of the interaction between the baryons proper and the mesons is characterized by four independent coupling constants in place of the eight constants in the case of SU(3) invariance.

The relations obtained by us for the coupling constants, which are valid for arbitrary values of the meson mass and off the mass shell, have the following form

$$G_C^D = \frac{\mu}{2m}(\frac{2}{3}G^D - G^F) \ , \qquad (2)$$

$$G_C^F = -\frac{\mu}{2m}\left(\frac{5}{9}G^D + \frac{2}{3}G^F\right) , \tag{3}$$

$$G_M^D : G_M^F : G_M = 3 : 2 : 1^1 . \tag{4}$$

The upper index refers here to the type of coupling (F or D) of the meson octet, while the power index determines the character of the interaction of the vector mesons with the baryons (C — electric–charge, M — magnetic moment). In determining the coupling constants we started from the following normalization of the interaction variants;

$$\bar{U}\gamma_C U \qquad \text{pseudoscalar mesons}$$

$$\left.\begin{array}{ll} \frac{2m}{4m^2\mu^2}(eq)\bar{U}U & C - \text{interaction} \\ \bar{U}[e\gamma) - \frac{2m}{4m^2-\mu^2}(eq)]U & M - \text{interaction} \end{array}\right\} \text{vector mesons} \tag{5}$$

(e — polarization vector of the vector meson; $q = p_1 + p_2$; p_a and p_2 — four–dimensional baryon momenta) and from the usual definition of the F and D coupling.

Relations (2)–(4) are compatible with the relations obtained by (Gursey, Radicati and Pais (1964)) for the coupling constants only when the coupling constant G_C^D coincides with the analogous constant in the Dirac variant, i.e., when the meson mass or all the constants of the M–interaction are equal to zero.

It is interesting to note that in the case when $\mu = 0$ there is no C – interaction of the vector meson octet with baryons at all. As a result, the variants of the dynamic theory of strong interactions, in which, in analogy with electrodynamics, the minimum interaction of the vector mesons of zero mass with baryons is regarded as the main bare interaction, are not SU(3) × SU(3)–invariant[2]. An analogous paradox concerning the incompatibility of the minimal electromagnetic interaction and SU(6) invariance was noted by (Beg, Lee and Pais (1964)).

References

Beg, Lee and Pais (1964): Phys.Rev.Lett., **13**, 514.
Gursey, Radicati and Pais (1964): Phys.Rev.Lett., **13**, 299.
Ruhl, W. (1965): Phys.Lett. **14**, 350.

[1] Relationship (4) was recently obtained by (Ruhl (1965)) on the basis of a relativistic generalization of SU(6) transformation which he proposed.

[2] Relationship (4) was recently obtained by (Ruhl (1965)) on the basis of a relativistic generalization of SU(6) transformation which he proposed.

Phenomenological Lagrangian for Spin Waves

D.V. Volkov, A.A. Zheltukhin, and Yu.P. Bliokh

Kharkov Institute of Physics and Technology

Fizika Tverdogo Tela, **13**, (1971), 1668

Abstract. A phenomenological Lagrangian describing the interaction of long-wavelength spin waves is obtained. The derivation of the Lagrangian is based on nonlinear realizations of the symmetry group of the Heisenberg Hamiltonian and on the assumption that the spin-wave excitation mechanism on ferromagnets, ferrites, and antiferromagnets is of the Goldstone type.

1. At low temperatures, the excitation spectrum of solids is governed by collective excitations, such as phonons, spin waves, etc. The collective excitations have been extensively studied using various methods. In our view, it is of great interest to establish rigorously the existence of a special type of collective excitation, i.e., Goldstone's particles (or quasiparticles Goldstone (1961)). The energy spectrum of such excitations has a special form (the excitation frequency $\omega(k)$ vanishes when the quasimomentum k tends to zero[1]), and the excitations are related to the symmetry properties of systems with a large number of degrees of freedom.

The fact that spin waves in ferromagnets, antiferromagnets, and ferrites correspond to Goldstone excitations is well known (see, for example, Hugenholtz (1967)). The following qualitative excitation mechanism of spin waves was proposed: the existence of a preferred direction of the magnetization breaks the symmetry of the original microscopic Hamiltonian of the system and, as a result, collective excitation (i.e., spin waves) are created, which tend to restore the broken symmetry.

A similar excitation mechanism is assumed in the case of general Goldstone particles and they are regarded as a specific reaction of the system to the symmetry breaking of the ground state. In connection with the discovery of the approximate "chiral" symmetries in elementary particle physics and of the properties of π mesons (pseudoscalar meson octet) with the properties of Goldstone particles of the $SU(2)\times SU(2)$ $[SU(3)\times SU(3)]$ symmetry group, a new phenomenological method of description of the interaction of Goldstone particles was recently proposed (Nambu and Jona–Lasinio (1961); Weinberg (1967); Schwinger (1967)). This method is based on the assumption that the symmetry of the system of Goldstone particles is fully restored.

[1] It will be shown that this relationship between the frequency and the quasimomentum $[\omega(k) \to 0$ if $|k| \to 0]$, which is obeyed by relativistic Goldstone particles, may not be satisfied in the case of essentially nonrelativistic particles (see Sec.3).

The mathematical formulation of this assumption is equivalent to the requirement that the phenomenological Lagrangian describing the interaction of Goldstone particles is strictly invariant under the transformations of the symmetry group in question. In fact, the Lagrangian can be expressed in terms of variables with a nonlinear transformation law (under the transformations of the group in question) which correspond to the Goldstone fields. The most general formulation of the Lagrangian for relativistic Goldstone particles with a dispersion law $\omega^2 - k^2 = 0$, which is valid for antisymmetry group, has been given by Coleman et al. (1969); Callan et al. (1969), and by Volkov (1969).

Our aim is to describe the interaction of an arbitrary number of "soft" spin waves, using the phenomenological Lagrangian method. This is equivalent to the approximation in which only the lowest powers of the magnon energy and momentum are retained in all the interaction matrix elements. In Sec. 2, the case of antiferromagnetic spin waves is discussed. There is a complete correspondence between this simplest case and the situation encountered in the elementary particle physics. In Secs. 4 and 5, the cases of more complicated excitation spectra are considered. The most general case is discussed in Sec. 5.

We shall try to make our treatment essentially independent of other approaches but, for brevity, we have to omit the proofs of important concepts in the phenomenological Lagrangian method. For example, we shall omit the proof that the matrix elements of the S matrix on the mass surface are independent of the actual choice of the parametrization of the spin-wave operators. Likewise we omit the justification of the term which describes the interaction of spin waves with other quasiparticles, etc, (see, for example, Coleman et al. (1969); Callan et al. (1969), Volkov (1969)).

2. To illustrate the general method of phenomenological Lagrangians, we shall make use of the well-known boson creation and annihilation operators of spin waves, which were introduced by Holstein and Primakoff (1941) and by Dyson (1956). We shall consider the Heisenberg Hamiltonian describing the exchange interaction

$$H_H = \sum \mathcal{J}_{ik} \mathbf{S}_i \mathbf{S}_k \ . \tag{1}$$

The transition from the Hamiltonian defined by Eq. (1) to the Holstein–Primakoff-Dyson Hamiltonian, which describes the spin–wave interaction, can be accomplished by expressing the spin operators $\mathbf{S}_i$ in terms of the creation and the annihilation operators a_i^+, a_i. In the phenomenological Lagrangian method, the actual form of the representation of the operators $\mathbf{S}_i$ in terms of the operators a_i^+, a_i is not relevant. The crucial fact is that, irrespective of the representation in question, the rotational symmetry of the Hamiltonian in the spin space is conserved, i.e., the Hamiltonian is invariant under the transformations of the SO(3) symmetry group. The only property which depends on the actual form of the representation of $\mathbf{S}_i$ is the form of the transformations of the operators a_i, a_i^+, which realize the transformations of the SO(3) group. Since the relationship between the operators $\mathbf{S}_i$ and the operators a_i^+, a_i is nonlinear, the transformation of the operators a_i^+, a_i under the action of the SO(3) group is also nonlinear.

However, the direct determination of the symmetry properties of the spin-wave interaction Hamiltonian in the Holstein-Primakoff-Dyson form is very complicated and, as far as we know, has not been carried out successfully.

Therefore, we shall try to deduce the transformation properties of the spin–wave operators under the action of the SO(3) group independently of the form of the Hamiltonian defined by Eq. (1) and of the actual form of the representation of the operators $\mathbf{S}_i$ in terms of a_i, a_i^+. On the basis of the transformation properties of the spin–wave operators and the invariance of the spin–wave interaction Lagrangian under the transformations of SO(3) group, we shall obtain, in the "soft–magnon' limit, the interaction Lagrangian, which depends only on several phenomenological parameters.

It should be noted that there is an important difference between the Holstein-Primakoff-Dyson approach and the phenomenological Lagrangian method. In fact, in the former case, the operators a_i, a_i^+ correspond to the nonrenormalized magnon operators. Therefore, in the calculation of the interaction between real magnons, it is necessary to take into account all the Feynman diagrams, including the closed loops. In the long–wavelength limit, these diagrams lead to a renormalization of the energy spectrum and of the interaction constants. On the other hand, in the phenomenological Lagrangian method, all such effects are assumed to have been taken into account, i.e., the local fields in question correspond to renormalized phenomenological magnon fields. On other words, the phenomenological Lagrangian method assumes that, in the "soft–magnon" limit, the matrix elements of the S matrix are smooth functions of the real–magnon momenta (with the exception of poles which are due to magnon exchange). The terms containing poles, which can be deduced from the Lagrangian derived solely on the basis of symmetry considerations, correspond to "tree–like" diagrams (diagrams without loops).

The proposed program constitutes the basis of the phenomenological Lagrangian method.

3. We shall demonstrate the derivation of the phenomenological Lagrangian in the simplest case of antiferromagnetic spin waves. The spin–wave spectrum in an antiferromagnet is doubly degenerate and, for small k, the magnon frequency is a linear function of k . Therefore, noninteracting antiferromagnetic spin waves can be described by two local–field operators $A_i(\mathbf{x}, t)$ $(i = 1, 2)$ which satisfy the following equation:

$$\partial^2 A_i(\mathbf{x}, t) - c^2 \nabla^2 A_i(\mathbf{x}, t) = o, \tag{2}$$

where ∂ and ∇ denote, respectively, the derivatives with respect to time and space. Equation (2) can be derived from the Lagrangian

$$L = \frac{1}{2}\{\partial A_i(\mathbf{x},t)\partial A_i(\mathbf{x},t) - c^2\nabla A_i(\mathbf{x},t)\nabla A_i(\mathbf{x},t)\} \ . \tag{3}$$

To obtain the Lagrangian which describes the interaction of spin waves, we shall assume that the transformation of the local fields $A_i(\mathbf{x}, t)$ under the action of the SO(3) group is governed by the following equations:

$$\delta A_i(\mathbf{x},t) = \varepsilon_3 \varepsilon_{ik3} A_i(\mathbf{x},t) \ , \tag{4}$$

$$\delta A_i(\mathbf{x},t) = \varepsilon_i(1 - f^2 A^2(\mathbf{x},t)) + f^2 \cdot 2(\varepsilon_k A_k) A_i \ , \tag{5}$$

where $f^2 \neq 0$. Here, ε_3 and ε_i are the parameters of infinitesimal transformations of the SO(3) group and $\varepsilon_3\varepsilon_{ik3}$ is an antisymmetric tensor. Equations (4) and (5) describe the infinitesimal transformations of the three–parameter SO(3) group . These equations have the following simple geometrical meaning: the fields $A_k(\mathbf{x}, t)$ represent the coordinates in a plane corresponding to the stereographic projection of a sphere of radius $1/f$ on the plane; the coordinates in the plane transform according to Eqs. (4) and (5) under infinitesimal rotations of the sphere. Therefore, Eqs. (4) and (5) correspond to the introduction of a coordinate system on a sphere which has the property that each coordinate corresponds to a local field A_i; they also describe the transformations of these fields under infinitesimally small rotations of the sphere. For $f^2 = 0$, the transformations (4) and (5) correspond to the transformation group of a plane. Clearly, the Lagrangian defined by Eq. (3) is invariant under such transformations.

We shall require the interaction Lagrangian to be invariant under the transformations (4) and (5) for $f^2 \neq 0$. In the "soft–magnon" limit, the invariant Lagrangian is defined uniquely as follows:

$$L = \frac{1}{2}\frac{\partial A_i \partial A_i - c^2 \nabla\partial A_i \nabla\partial A_i}{(1 + f^2 A_k^2)^2} \ . \tag{6}$$

The constant f^2 which appears in Eq. (6) is a phenomenological constant which governs the interaction strength of spin waves.

The interaction of an arbitrary number of "soft" magnons can be described by means of an S matrix, corresponding to the Lagrangian defined by Eq. (6), in the approximation of "tree–like" diagrams.

For two–magnon scattering processes, the matrix element is governed by the first term in the expansion of Eq. (6) in powers of f^2 :

$$L' = -f^2(\partial A_i \partial A_i - c^2 \nabla\partial A_i \nabla\partial A_i) A_k^2 \tag{7}$$

and is identical with the matrix element which was discussed in (Cartan (1910)).

The many–magnon scattering processes are governed by higher–order terms in the expansion of the S matrix in powers of f^2. Because of the nonlinearity of the transformations (4-5), the symmetry properties make it possible to obtain relationships between the matrix elements corresponding to different numbers of magnons.

It should be noted that the form of the Lagrangian defined by Eq. (6) depends on the form of the realization of the SO(3) group transformations. If we perform in Eqs. (4) and (5) the following transformation to new local fields $A'(x)$ (which corresponds to the choice of new coordinates on a sphere):

$$A'x = A(x) + \eta\ (A(x)),$$

where η is an arbitrary analytic function such that $\eta(0) = 0$, Eqs. (4) and (5) and the Lagrangian defined by Eq. (6) will have a different form in these new fields. Nevertheless, the form of the S matrix on the mass surface will be the same for the original and the transformed Lagrangians (see Coleman et al. (1969); Callan et al. (1969), Volkov (1969)).

We shall write the Lagrangian in a form which is convenient in many problems and which corresponds to normal coordinates on a sphere, i.e.,

$$L = \frac{1}{2} g_{\alpha\beta}(A_\gamma)(\partial A_\alpha \partial A_\beta - c^2 \nabla A_\alpha \nabla A_\beta)\ , \tag{8}$$

where

$$g_{\alpha\beta}(A_i) = \delta_{\alpha\beta} + \sum_{k=1}^{\infty} \frac{(-1)^k 2^{2k+1}}{(2k+2)!} (m^k)_{\alpha\beta} \tag{9}$$

and

$$m_{\alpha\beta} = \tilde{J}^2 (A_k^2 \delta_{\alpha\beta} - A_\alpha A_\beta)\ ; \tag{10}$$

$(m^k)_{\alpha\beta}$ in Eq. (9) denotes the k-th power of the matrix $m_{\alpha\beta}$ defined by Eq. (10).

Equation (9) can be also written in the form

$$g_{\alpha\beta}(A) = \delta_{\alpha\beta} + \frac{m_{\alpha\beta}}{A^4 \tilde{f}^4} \left\{ \frac{1 - \cos 2\sqrt{\tilde{f}^2 A_k^2}}{2} - \tilde{J}^2 A_k^2 \right\}\ . \tag{11}$$

It should be noted that Eq. (8) is invariant under the transformations (4) and also under the transformations

$$\delta A_\alpha = (\sqrt{m}\ \mathrm{ctg} \sqrt{m})_{\alpha\beta}\, \varepsilon_\beta\ , \tag{12}$$

which realize the transformations of the SO(3) group. The quantities ε_β in Eq. (12) are the parameters of an infinitesimal transformation.

For a two–magnon process, we obtain ($\tilde{f} \equiv 2f$)

$$L = -\frac{(2f)^2}{3!} \{A_\alpha^2 [(\partial A_\alpha)^2 - c^2 (\nabla A_\alpha)^2] - [(A_\alpha \partial A_\alpha)^2 - c^2 (A_\alpha \nabla A_\alpha)^2]\}\ . \tag{13}$$

Let us compare this matrix element with that defined by Eq. (7). To simplify Eq. (13), we shall make use of the laws of conservation of energy and momentum:

$$\partial[A_\alpha^2(A_\beta \partial A_\beta)] = c^2\nabla[A_\alpha^2(A_\beta \nabla A_\beta)] = 0 \tag{14}$$

which implies

$$2[(A_\alpha\partial A_\alpha)^2 - c^2((A_\alpha\nabla A_\alpha)^2] + A_\alpha^2[(\partial A_\alpha)^2 - c^2(\nabla A_\alpha)^2] =$$
$$= -A_\alpha^2[(A_\alpha\partial^2 A_\alpha) - c^2(A_\alpha\nabla^2 A_\alpha)] \ . \tag{15}$$

Unless stated otherwise, we shall always use the abbreviations $(A_\alpha \partial A_\alpha)$, $(A_\alpha \nabla A_\alpha)$, etc., for $\sum_{\alpha=1}^{2}(A_\alpha\partial A_\alpha), \sum_{\alpha=1}^{2}(A_\alpha\partial A_\alpha)$, etc.

Therefore, on the mass surface, Eq. (13) becomes identical with L':

$$L' = L = -f^2 A_\alpha^2[(\partial A_\alpha)^2 - c^2(\nabla A_\alpha)^2]. \tag{16}$$

However, it should be noted that this fact is trivial since, for the Lagrangian defined be Eq. (8), where $g_{\alpha\beta}(A)$ is an arbitrary function, the matrix element of the scattering on the mass surface can be always brought to the form (16) by the transformation (15), irrespective of the symmetry properties of the Lagrangian.

4. The case discussed in Sec. 3 concerns spin waves with a linear dispersion law. Let us now investigate more complicated dispersion laws. It should be noted that expressions (6) and (8) are the only invariants under the transformations (4-5) which do not contain higher than second-order derivatives. Nevertheless, it will be shown later that terms containing the first derivative of arbitrary fields with respect to time can be added to expressions (6) and (8). Such terms are not invariant under the transformations of the SO(3) group but, since the resulting contribution represents the total time derivative, these terms do not influence the physical properties of the system. On the other hand, the presence in the Lagrangian of the terms which contain first derivatives with respect to time changes the form of the magnon spectrum, which makes it possible to use the phenomenological Lagrangian method to describe the interaction of spin waves in ferrites and ferromagnets.

Let us consider the following expression, which is linear in the first derivative with respect to the field A_α:

$$\omega_3 = 2\,\frac{A_1\partial A_2 - A_2\partial A_1}{1 + f^2 A_\alpha^2} \ . \tag{17}$$

Using the explicit form of the transformation (4), it can be shown that Eq. (17) is invariant under this transformation.

The change in ω_3 as a result of the transformation (5) has the form

$$\delta\omega_3 = 2\partial(\epsilon_1 A_2 - \epsilon_2 A_1) \ , \tag{18}$$

i.e., it is a total time derivative.

Adding to the Lagrangian defined by Eq. (6) the expression (17) multiplied by an arbitrary phenomenological constant, we obtain a modified expression for the Lagrangian whose invariance properties are defined up to terms containing total time derivatives.

It can be shown that, in the case of two spin wave branches and for a given choice of the coordinates on a sphere, the term (17) is defined uniquely. This problem will be discussed in detail in the next section.

Let us investigate the form of the spectrum generated by the Lagrangian defined by Eq. (6) with the additional term (17). Retaining in Eqs. (6) and (17) only the terms bilinear in the fields, we obtain the following expressions in the momentum representation:

$$\left.\begin{aligned}(\omega^2 - c^2\mathbf{k}^2)A_1 + 2ib\omega A_2 = 0,\\ (\omega^2 - c^2\mathbf{k}^2)A_2 - 2ib\omega A_1 = 0,\end{aligned}\right\} \tag{19}$$

where $1/2\, b$ is the phenomenological constant in front of the term (17) in the total Lagrangian.

It follows from Eq. (19) that the spectrum of spin waves has the form

$$\omega = \sqrt{b^2 + c^2k^2} \pm b\ , \tag{20}$$

which is identical with the spectrum of magnons in ferrites and, for large b, with the spectrum of ferromagnetic magnons (Akhiezer et al. (1968)). It should be noted that the parameter b is proportional to the internal field in the system. In fact, this follows either from a comparison of Eq. (20) with the standard expressions for the spin–wave spectra in ferrites, or directly from the transformation properties of Eq. (17) under the reflection of the spatial coordinates. The Lagrangian of a ferromagnet corresponds to the sum of Eq. (17) and the second term in Eq. (6).

As in the preceding section, the expansion of the Lagrangian defined by Eqs. (6), (17) in powers of f^2 makes it possible to obtain the matrix elements of the S matrix for an arbitrary number of "soft magnons." By analogy with the relativistic theory, it is assumed in the discussion of the Goldstone excitations in solids that ω vanishes in the limit $\mathbf{K} \to 0$. It follows from our discussion that, as far as the breaking and the subsequent restoration of the symmetry is concerned, the two branches in Eq. (2) are completely equivalent.

Furthermore, only in the limit $b \to \infty$ (ferromagnetic case) can the restoration of the symmetry be achieved by considering only one excitation branch. As already noted, this special feature of the present model (in contrast to the relativistic case) is due to the fact that additional terms linear in the field derivatives can be added to the Lagrangian.

As before, the S matrix elements describing many-magnon scattering processes can be obtained by expanding the phenomenological Lagrangian in powers of the field A_i and using the standard perturbation theory in which only the "tree–like" diagrams are considered. As a result of such an expansion of Eqs. (17) or (19), the following products appear in the S matrix:

$$\epsilon_{ikl} H_i A_k \partial A_l f(A_i^2) \ . \tag{21}$$

These terms depend explicitly on the internal magnetic field, and, therefore, define a direction in the spin–wave space. In the derivation of Eq. (21) from Eqs. (17) or (19), we have taken into account the fact that the parameter b is proportional to the internal magnetic field and we have used the tensor notation (assuming that the magnetic field is in the direction of the third axis).

On the other hand, we assume that the parameter b and the corresponding magnetic field are invariant under the transformations (4),(15) of the SO(3) group. This last requirement is compatible with Eq. (24), provided the magnetic field is orthogonal to the surface of the sphere at every point. In this case, the magnetic field is in the direction of the third axis only if the expansion in powers of A_i is carried out at the origin. If, in the derivation of the S matrix, the Lagrangian is expanded in powers of $A_i - A_{0i}$, where A_{0i} is a fixed point on the sphere, the magnetic field will be orthogonal to the sphere at this point. Physically, this ambiguity in the direction of the magnetic field is related to the fact that the vacuum state of the system is infinitely degenerate. Before a certain vacuum state is chosen, the system is completely symmetric. The apparent symmetry breaking occurs only in the derivation of the S matrix when the states involved in the scattering are considered, i.e., a well-defined vacuum state is chosen. As already discussed, such a choice of the vacuum state corresponds to the expansion of the Lagrangian in powers of the field in the neighborhood of a fixed point. For any scattering process involving a finite number of magnons, the expansion in question is determined by differential surface elements of finite order, and, therefore, the states corresponding to different vacuum states (i.e., expansions at different points) are independent. Therefore, the phenomenological Lagrangian under study can be used to demonstrate the relationship between the Goldstone particles and the symmetry of the system as well as the vacuum degeneracy.

We shall conclude this section by giving the explicit form of the additional term in the Lagrangian defined by Eq. (8) and its variation under the transformations (12):

$$\omega_3 = \frac{A_1 \partial A_2 - A_2 \partial A_1}{f^2 A^2} \{1 - \cos\sqrt{f^2 A^2}\} \ , \tag{22}$$

$$\delta\omega_3 = \frac{\partial}{\partial t} \left\{ (e_1 A_2 - e_2 A_1) \frac{\mathrm{tg}\frac{\sqrt{f^2 A^2}}{2}}{\sqrt{A^2 f^2}} \right. \ . \tag{23}$$

Here, $A^2 = A_1^2 + A_2^2$. The matrix elements of the S matrix on the mass surface corresponding to the Lagrangians defined by Eqs. (6),(17) and (8),(22) are identical for any number of interacting magnons.

5. Let us now consider the most general expressions for the phenomenological spin–wave Lagrangian. We shall base our discussion solely on the symmetry properties of the Lagrangian under the transformation of the SO(3) group and make no assumptions about the spin–wave spectrum. The discussion presented in Secs. 4 and 5 concerns the special cases corresponding to a special choice of phenomenological parameters. The method of the Cartan differential forms (see, for example, Cartan (1910)) can used to take into account the symmetry requirements. To introduce the Cartan forms, we shall consider an arbitrary parametrization of the SO(3) group elements

$$g = g(\alpha_i), \quad g \in \mathrm{SO}(3); \qquad i = 1,2,3 \tag{24}$$

and define three Cartan forms $\omega_i(\alpha_i, \mathrm{d}\alpha_k)$ by the following equation:

$$g^{-1}(\alpha)\mathrm{d}g(\alpha) = \mathrm{i}\omega_k(\alpha, \mathrm{d}\alpha)\frac{\sigma_k}{2} , \tag{25}$$

where $\sigma_k/2$ are the generators of the SO(3) group , i.e., the Pauli matrices. Clearly, the Cartan forms are invariant under the action of the elements of the SO(3) group on the left, i.e.,

$$g(\tilde{\alpha}_i)g(\alpha) = g(\alpha') \tag{26}$$

where $g(\tilde{\alpha}_i)$ is an arbitrary element of the SO(3) group.

In fact, we obtain

$$\mathrm{i}\omega_i(\alpha', \mathrm{d}\alpha')\frac{\sigma_i}{2} = g^{-1}(\alpha')\mathrm{d}g(\alpha') = g^{-1}(\alpha)g^{-1}(\tilde{\alpha})g(z)\mathrm{d}g(\alpha) = \mathrm{i}\omega_i(\alpha, \mathrm{d}\alpha)\frac{\sigma_i}{2} . \tag{27}$$

Using the linear independence and completeness of the differential forms $\omega_i(\alpha, \mathrm{d}\alpha)$, it can be shown that any invariant form $\Omega(\alpha, \mathrm{d}\alpha)$ can be represented in the form

$$\Omega(\alpha, \mathrm{d}\alpha) = d_i\omega_i(\alpha, \mathrm{d}\alpha) . \tag{28}$$

where the coefficients d_i are constant, i.e., the forms $\omega_i(\alpha, \mathrm{d}\alpha)$ represent a complete system of the invariants of the SO(3) group. To derive the most general expression for the phenomenological Lagrangian of spin waves, we shall assume that to each parameter α_i there corresponds a local field $A_i(\mathbf{x}, t)$ and that the differentials $\mathrm{d}\alpha_i$ correspond either to the time or the space derivatives of the field. The Lagrangian which contains no higher than the second–order derivatives with respect to the field and is invariant under the thansformations of the SO(3) group [assuming that the local fields $A_i(\mathbf{x}, t)$ transform as the group parameters] has the form

$$L = \frac{1}{2}a_{ik}\omega_i(A, \partial A)\omega_k(A, \partial A) + 2b_i\omega_i(A, \partial A)$$
$$-\frac{1}{2}C_{ik;\, lm}\omega_i(A, \nabla_i A)\omega_k(A, \nabla_m A) + 2d_{i;\, l}\omega_i(A, \nabla_l A) . \tag{29}$$

Let us first determine the values of the phenomenological parameters and the choice of the coordinates in the group space which correspond to the examples discussed in the previous sections.

The following parametrization of the SO(3) group elements corresponds to the Lagrangian defined by Eq. (6):

$$g(\alpha) = \left[\frac{1+\mathrm{i}f(\alpha_1\sigma_1+\alpha_2\sigma_2)}{1-\mathrm{i}f(\alpha_1\sigma_1+\alpha_2\sigma_2)}\right]^{1/2} \mathrm{e}^{\mathrm{i}\alpha_3\frac{\sigma_3}{2}} \ . \tag{30}$$

The parametrization (30) and Eq. (25) imply that ω_i have the form

$$\left.\begin{aligned} \omega_1 &= \tfrac{2}{1+f^2\alpha_i^2}[\partial\alpha_1\cos\alpha_3 - \partial\alpha_2\sin\alpha_3] \\ \omega_2 &= \tfrac{2}{1+f^2\alpha_i^2}[\partial\alpha_1\sin\alpha_3 - \partial\alpha_2\cos\alpha_3] \end{aligned}\right\} \tag{31}$$

and

$$\omega_3 = \frac{2(\alpha_1\partial\alpha_2 - \alpha_2\partial\alpha_1)}{1+f^2\alpha_i^2} + \partial\alpha_3 \ . \tag{32}$$

Here, $\alpha_i^2 = \alpha_1^2 + \alpha_2^2$.

Equations (30-32) contain three group parameters $\alpha_1, \alpha_2, \alpha_3$, whereas Eq. (6) involves only two local fields. To eliminate one of the components, we shall choose the parameters a_{ik} and $c_{ik,\ lm}$ so that the Lagrangian is independent of A_3. This requirement can be satisfied if

$$a_{11} = a_{22}; \qquad c_{11,\ lm} = c_{22,\ lm} \tag{33}$$

and all the other constants vanish. The Lagrangian defined by Eq. (6) satisfies the conditions (33) and the additional requirement

$$c_{11,\ lm} = c_{11}\cdot\delta_{lm} \ , \tag{34}$$

which is equivalent to the requirement that the dependence of the frequencies on the wave vector is anisotropic. It is impossible to eliminate the components A_3 from the terms in Eq. (29) which are linear in the derivatives. However, since ω_3 only as a factor under the total derivative sign, we can retain this term in the Lagrangian defined by Eq. (29), i.e.,

$$b_3 \neq 0 \ , \tag{35}$$

which corresponds, in the isotropic case $d_{3l} \neq 0$, to the addition of the expression (17) to the Lagrangian defined by Eq. (6). The Lagrangian defined by Eqs. (8-10) and the expression (22) corresponds to the following parametrization of the group elements:

$$g(\alpha) = \mathrm{e}^{\mathrm{i}(\frac{\sigma_1}{2}\alpha_1+\frac{\sigma_2}{2}\alpha_2)f}\mathrm{e}^{\mathrm{i}\frac{\sigma_3}{2}\alpha_3} \ , \tag{36}$$

which leads to the following differential forms:

$$\left.\begin{aligned} \omega_1 &= F_1 \cos\alpha_3 - F_2 \sin\alpha_3, \\ \omega_2 &= F_1 \sin\alpha_3 + F_2 \cos\alpha_3, \\ F_i &= \partial\alpha_i \frac{\sin\sqrt{\alpha^2 f^2}}{\sqrt{\alpha^2 f^2}} + \alpha_i \frac{(\alpha_k \partial\alpha_k)}{f^2\alpha^2}\left[1 - \frac{\sin\sqrt{\alpha^2 f^2}}{\sqrt{\alpha^2 f^2}}\right]. \end{aligned}\right\} . \tag{37}$$

Here, $\alpha^2 = \alpha_1^2 + \alpha_2^2 (\alpha_k \partial\alpha_k) = \alpha_1 \partial\alpha_1 + \alpha_2 \partial\alpha_2$ and

$$\omega_3 = \frac{(\alpha_1 \partial\alpha_2 - \alpha_2 \partial\alpha_1)}{\alpha^2 f^2}(1 - \cos\sqrt{\alpha^2 f^2}) + \partial\alpha_3 \ . \tag{38}$$

In the isotropic case, the elimination of the third component corresponds to Eqs. (33-35). Assuming that the frequency depends only on $\mathbf{k}^2$ (isotopic case), we shall discuss the form of the excitation spectrum, which follows from the Lagrangian defined by Eq. (29). It is convenient to consider the highest–symmetry parametrization of the SO(3) group elements, i.e.,

$$g(\alpha) = \mathrm{e}^{\mathrm{i}\frac{\sigma}{2}\alpha}; \quad f = 1 \ . \tag{39}$$

Using eq .(25), we obtain, with the accuracy up to terms quadratic in the group parameters, the following expression for the forms $\omega_i (i = 1, 2, 3)$:

$$\omega_i = \mathrm{d}\alpha_i + \frac{1}{2}\varepsilon_{ikl}\alpha_k \mathrm{d}\alpha_l + \dots \ . \tag{40}$$

Substituting eq/(40) in Eq. (25) and retaining in Eq. (25) only the terms quadratic in the field, we obtain the Lagrangian of noninteracting spin waves,

$$L = \frac{1}{2} a_{ik} \partial A_i \partial A_k + b_i \varepsilon_{ikl} A_k \partial A_l - \frac{1}{2} c_{ik} \nabla A_i \nabla A_k \ . \tag{41}$$

The Lagrangian defined by Eq. (41) leads to the following equations of motion in the momentum representation:

$$\omega^2 a_{ik} A_k + 2 b_i \mathrm{i} \varepsilon_{ikl} A_k \omega - \mathbf{k}^2 c_{ik} A_k = 0 \ . \tag{42}$$

The condition that the determinant of the system of equations (42) vanishes yields the dispersion equation

$$\alpha\omega^6 + 3\beta\omega^4 + 3\gamma\omega^2 + \delta = 0 \ , \tag{43}$$

where the determinants α, β, γ, and δ are defined by

$$\left.\begin{aligned} &\alpha = M_1, \ = - (M_5 + M_2 \mathbf{k}^2), \\ &\gamma = (M_6 \mathbf{k}^2 + M_3 \mathbf{k}^4), \quad \delta = -M_4 \mathbf{k}^6 \end{aligned}\right\} \tag{44}$$

and

$$\left.\begin{aligned} M_1 &= |a_{ik}|, \ \ M_4 = |c_{ik}|, \\ M_2 &= \tfrac{1}{3\,!}\ \varepsilon_{ikl}\ \varepsilon_{prs}\ a_{ip}\ a_{kr}\ c_{ls}, \\ M_3 &= \tfrac{1}{3\,!}\ \varepsilon_{ikl}\ \varepsilon_{prs}\ a_{ip}\ c_{kr}\ c_{ls}, \\ M_5 &= \tfrac{1}{3}\ b_i\ a_{ik}\ b_k, \quad M_6 = \tfrac{1}{3}\ b_i\ c_{ik}\ b_k. \end{aligned}\right\} \tag{45}$$

Since Eq. (43) is bicubic, for every root ω_i of Eq. (43), there is also a root $-\omega_i$. Therefore, using the standard method, we can separate in the local–field operators the terms corresponding to the creation and the annihilation of magnons.

The dependence of the coefficients in Eq. (43) on $\mathbf{k}^2$ [Eq. (44)] yields the frequency spectrum for small $\mathbf{k}^2$. If $\mathbf{k}^2 = 0$, we find that $\gamma = \delta = 0$ and Eq. (43) has only one nonvanishing root. The product of the roots which tend to zero in the limit of small $\mathbf{k}^2$ is proportional to δ, and, therefore, to $\mathbf{k}^6$; it follows from the form of the dependence of the coefficient γ on $\mathbf{k}^2$ that the sum of the roots is proportional to $\mathbf{k}^2$. Therefore, for small $\mathbf{k}^2$, the roots of Eq. (43) have the form

$$\left.\begin{aligned} \omega_1^2 &= A + B\mathbf{k}^2, \\ \omega_2^2 &= C\mathbf{k}^2, \\ \omega_3^2 &= D\mathbf{k}^4. \end{aligned}\right\} \tag{46}$$

Substituting Eq. (46) in Eq. (43), we obtain

$$\left.\begin{aligned} A &= \tfrac{M_5}{M_1}\ , B = \tfrac{M_2}{M_1} - \tfrac{M_6}{M_5}\ , \\ C &= \tfrac{M_6}{M_5}\ , D = \tfrac{M_4}{M_6}\ . \end{aligned}\right\} \tag{47}$$

Since $\omega_i^2 > 0$, the coefficients A, B, and D should be positive, which constrains the possible values of phenomenological constants appearing in the Lagrangian defined by Eq. (29).

If the coefficients b_i in the Lagrangian defined by Eq. (29) and in Eq. (42) vanish, then $M_5 = M_6 = 0$, which implies that all the frequencies ω_i are proportional to $\mathbf{k}^2$.

It should be noted that nine out of fifteen phenomenological constants a_{ik}, c_{ik}, and b_i are related to the structure of the equation describing noninteracting spin waves, and the remaining six govern the interaction of spin waves. In fact, the transformation in the free and in the interaction Lagrangian from the fields A_k to fields A'_k ,

$$A_k = (a^{-1/3})_{kl} A_l \ , \tag{48}$$

leads to a free Lagrangian of the type defined by Eq. (41), in which the coefficient of the first term is proportional to the unit matrix; this also represents the standard form of the free Lagrangian. In the expansion of the interaction Lagrangian in powers of the field A' , the six coefficients of the matrix $(a^{-1/2})_{kl}$ correspond to the phenomenological coupling constants and describe the interaction of an arbitrary number of "soft magnons."

The derivation of the S matrix, which is based on the expansion of the Lagrangian defined by Eqs. (6), (8) in powers of the field A_i, corresponds to a special choice of the vacuum state.

It follows from Eq. (42) that the apparent symmetry breaking of the Lagrangian is due to the fact that the quantities a_{ik}, c_{ik} are tensors and b_i

is a vector (they all depend on the macroscopic parameters of the system). However, such a symmetry violation does not contradict the invariance of Eq. (29), in which all the tensor quantities are defined in a local basis which is closely related to the properties of the SO(3) group.

The authors are grateful to V.G.Bar'yakhtar and S.V.Peletminskii for their helpful discussions.

References

Akhiezer, A.I., Bar'yakhtar, V.G. and Peletminskii, S.V. (1968): Spin Waves, North–Holland, Amsterdam.

Bar'yakhtar, V.G., Zarochintsev, R.V. and Popov, V.A. (1968): Fiz. Metal. Metalloved., **25**, 3.

Cartan, E. (1910): Bull. Soc. Math. Fr., **34**, 250

Coleman, S., Wess, J. and Zimino, B. (1969): Phys.Rev., **177**, 2239;
Callan, C.G. Jr., Coleman, S., Wess, J. and Zimino, B. (1969): **177**, 2247.

Dyson, F.J. (1956): Phys.Rev., **102**, 1217, 1230.

Holstein, T.D. and Primakoff, H. (1941): Phys.Rev., **58**, 1048.

Hugenholtz, N.M. (1967): Quantum Theory of Many–Body Systems [Russian translation], Mir, Moscow.

Goldstone, J. (1961): Nuovo Cimento, **19**, 154.

Nambu, Y. and Jona–Lasinio, G. (1961): Phys.Rev., **122**, 345;
Weinberg, S. (1967): Phys.Rev.Lett., **18**, 507 ;
Schwinger, J. (1967): Phys.Lett., **24B**, 473.

Volkov, D.V. (1969): Preprint ITF-69-75 [in Russian], Kiev.

Possible Universal Neutrino Interaction

D.V. Volkov, and V.P. Akulov

Kharkov

JETP Lett., **16**, (1972), 367

Much attention is being paid of late in elementary–particle physics to the possible degeneracy of vacuum, and the ensuing spontaneous breaking of one symmetry or another. The most direct consequence of vacuum degeneracy is the occurrence of the zero–mass particles called Goldstone particles (Goldstone (1961)).

Of all the presently known elementary particles, only the neutrino, photon, and graviton have zero mass. The last two, however, correspond to gauge fields and apparently do not require vacuum degeneracy for their description. The neutrino is therefore the only particle whose existence may be directly connected with vacuum degeneracy.

We wish to point out here thet the hypothesis that the neutrino is a Goldstone particle leads to a definite type of neutrino interaction both with other neutrinos and with all other particles. The interaction is fully defined by a single phenomenological coupling constant and is in this sense universal.

To determine the type of symmetry whose spontaneous violation causes the degeneracy of vacuum and the corresponding properties of the neutrino as as Goldstone particle, let us consider the symmetry properties of the equation for the free neutrino

$$\mathrm{i}\sigma_\mu \frac{\partial}{\partial x_\mu}\psi = 0 \ . \tag{1}$$

This equation is invariant both with respect to the Poincare group in chiral transformations, and with respect to shifts in spinor space, i.e., with respect to transformations of the type

$$\begin{aligned} \psi &\to \psi' = \psi + \zeta, \\ x &\to x' = x \ , \end{aligned} \tag{2}$$

where ζ is a constant spinor that anticommutes with ψ .

We retain the character of the transformations of x_μ and ψ in the transformations of the Poincare group, and replace the transformations (2) by the transformations

$$\begin{aligned} \psi &\to \psi' = \psi + \zeta, \\ \psi^+ &\to \psi^{+\prime} = \psi^+ + \zeta^+, \\ x_\mu &\to x'_\mu = x_\mu - \frac{a}{2\mathrm{i}}(\zeta^+\sigma_\mu\psi - \psi^+\sigma_\mu\zeta) \ . \end{aligned} \tag{3}$$

The resultant structure with ten commuting and four anticommuting parameters has the structure of a group[1] and is the only possible generalization of (2) without introducing additional group parameters.

The constant a in the transformations (3) is arbitrary and has the dimension of length raised to the fourth power. We postulate that the equations for the neutrino with allowance for the interaction are invariant against the transformations (3) . We assume also that the terms of interaction contain the minimum number, compatible with the invariance requirement, of field derivatives.

To construct the phenomenological action integral under the foregoing assumptions, it suffices to use the following differential forms, which are invariant against the transformations (3)

$$\omega_\mu = \mathrm{d}x_\mu + \frac{a}{2\mathrm{i}}(\psi^+\sigma_\mu \mathrm{d}\psi - \mathrm{d}\psi^+\sigma_\mu\psi) \ . \tag{4}$$

The action integral invariant against transformations (3) and the transformations of the Poincare group is given by

$$S = \frac{1}{\sigma}\int \omega_0 \wedge \omega_1 \wedge \omega_2 \wedge \omega_3 \ , \tag{5}$$

where $\wedge$ denotes an external product. The expression (5) corresponds to certain four–dimensional volume in the group–parameter space.

If the 4–volumes are defined by specifying the function $\psi = \psi(x)$, the action integral (5) can be written in the following more usual form:

$$S = \frac{1}{a}\int |W| \mathrm{d}^4x \ , \tag{6}$$

where $|W|$ is the determinant of the matrix W

$$W_{\mu\nu} = \delta_{\mu\nu} + aT_{\mu\nu} \ ,$$

$$T_{\mu\nu} = \frac{1}{2\mathrm{i}}(\psi^+\sigma_\mu\partial_\nu\psi - \partial_\nu\psi^+\sigma_\mu\psi) \ . \tag{7}$$

It follows from (6) and (7) that the action integral as a function of the tensor T takes the form

$$S = \int \Bigg[\frac{1}{a} + T_{\mu\mu} + \frac{a}{2}(T_{\mu\mu}T_{\nu\nu} - T_{\mu\nu}T_{\nu\mu}) +$$

$$+ \frac{a^2}{3\,!}\sum_p (-1)^p \, T_{\mu\mu}T_{\nu\nu}T_{\rho\rho} + \frac{\sigma^3}{4\,!}\sum_p (-1)^p T_{\mu\mu}T_{\nu\nu}T_{\rho\rho}T_{\sigma\sigma}\Bigg] \mathrm{d}^4x \ , \tag{8}$$

where the summation over p corresponds to a sum over all the permutations of the second indices in the products of the tensors T.

[1] Lie groups with commuting and anticommuting parameters were recently considered by Berezin and Kats (1970).

The term with $T_{\mu\mu}$ corresponds to the kinetic terms, and the terms with products of two, three, and four tensors T describe interactions in which four, six, and eight fields, respectively, participate. The degrees of the field derivatives in the interaction terms are determined by the number of factors T.

The neutrino interaction with other fields can be determined in manner that is invariant with respect to the transformations (3).

Thus, for example, the action integral for a Dirac particle is given by

$$S = \int \left[R_{\mu\mu} + a(R_{\mu\mu}T_{\nu\nu} - R_{\mu\nu}T_{\nu\mu}) + \frac{a^2}{2} \sum_p (-1)^p R_{\mu\mu}T_{\nu\nu}T_{\rho\rho} + \right.$$

$$\left. + \frac{a^3}{3\,!} \sum_p (-1)^p R_{\mu\mu}T_{\nu\nu}T_{\rho\rho}T_{\sigma\sigma} + m\bar{\phi}\phi|W| \right] \mathrm{d}^4x \ , \qquad (9)$$

where

$$R_{\mu\nu} = \frac{1}{2\mathrm{i}}(\bar{\phi}\gamma_\mu\partial_\nu\phi - \partial_\nu\bar{\phi}\gamma_\mu\phi) \qquad (10)$$

and the tensor T and the determinant $|W|$ are defined in (7).

Weak interactions can be included in the scheme under consideration by introducing gauge fields for the approximate unitary symmetry group of the neutrino and other leptons. The electromagnetic interaction for charged leptons is introduced simultaneously. The weak and electromagnetic interactions are turned on simultaneously by the well–known mechanism of spontaneous breaking of the unitary–group symmetry (Weinberg (1967)). In the zero–lepton–mass limit, the unitary symmetry is exact. To obtain the action integral for the leptons in the unitary–symmetry limit, it suffices to regard in (3), (4) ,and (7) the spinor products as invariant products of unitary multiplets. It is also possible to add to formulas (3), (4), and (7) the terms for leptons with opposite chirality. The unitary groups for states with different chirality need not necessarily coincide in this case.

The gauge fields can be introduced into the so–generalized action integral in a manner covariant with respect to the transformations (3).

The gravitational interaction can be introduced into the scheme in analogous fashion, by introducing gauge fields corresponding to the Poincare group.

We note that if we introduce also gauge fields corresponding to the transformations (3), then, as a consequence of the Higgs effect (Higgs (1966)), a massive gauge field with spin 3/2 arises, and the Goldstone particles with spin 1/2 vanish.

References

Berezin, F.A. and Kats, G.I. (1970): Matematicheskii sbornik **82**, 343.
Higgs, P.W. (1966): Phys.Rev. **145**, 1156.
Goldstone, J. (1961): Nuovo Cim. **19**, 154.
Weinberg, S. (1967): Phys.Rev.Lett. **19**, 1264.

Higgs Effect for Goldstone Particles with Spin 1/2

D.V. Volkov and V.A. Soroka

Institute of Physics & Technology, Kharkov, USSR

JETP Lett., **18**, (1973), 529

Abstract. We consider the introduction of gauge fields and the Higgs effect for Goldstone particles with spin 1/2

Higgs and others (Higgs (1966), Migdal and Polyakov (1966), Kibble (1967)) have noted that when interactions are turned on between Goldstone particles and gauge fields, the Goldstone particles vanish and those gauge fields whose quantum numbers coincide with the quantum numbers of the Goldstone particles acquire mass (the Higgs effect). We consider in this article a variant of the Higgs effect, in which the "absorption" of the Goldstone particles is effected by a "foreign" gauge field, i.e., by a gauge fields whose quantum numbers differ from those of the Goldstone particles.

We consider as the symmetry group G the direct product of a Poincare group with a certain group of internal symmetry, supplemented by the following transformations:

$$\begin{aligned} &\psi \to \psi' = \psi + \xi, \\ &x_\mu \to x'_\mu = x_\mu - \frac{a}{2\,!}(\xi^+ \sigma_\mu \psi - \psi^+ \sigma_\mu \xi) \ . \end{aligned} \tag{1}$$

We use the definitions of Volkov and Akulov (1973).

The spontaneous violation of the considered symmetry group under the assumption that the direct product of the Poincare group and of the internal–symmetry group leaves the vacuum invariant, leads to the appearance of Goldstone particle with spin 1/2/ (Volkov and Akulov (1972), Volkov and Akulov (1973)).

Let us consider the gauge transformations of the group G with parameters l, u, t, and ξ, which depend on x_μ and gauge fields that are coefficient functions of the differentials $\mathrm{d}x_\mu$ and $\mathrm{d}\psi$ of the differential form $A(\mathrm{d})$ with the following transformation law:

$$A'(\mathrm{d}) = G(t,\xi;l,u)\ A(\mathrm{d})G^{-1}(t,\xi;l,u) + \frac{1}{f}G(t,\xi;l,u)\mathrm{d}G^{-1}(t,\xi;l,u)\ , \tag{2}$$

where $A(\mathrm{d}) = A'(\mathrm{d})Z_i$ and Z_i are the generators of group G.

The differential forms

$$\bar{\omega}(\mathrm{d}) = H^{-1}(l,u)\omega(\mathrm{d})H(l,u) \tag{3}$$
$$= G^{-1}(x,\psi;l,u)\mathrm{d}G(x,\psi;l,u) + fG^{-1}(x,\psi;l,u)A(\mathrm{d})G(x,\psi;l,u) \;,$$

which are invariant to the translation (2) and correspond to the transformation generators in the transformations (1), take the following form as functions of the gauge fields:

$$\begin{aligned}
\omega_\mu(\mathrm{d}) &= Dx_\mu + \frac{\sigma}{2\mathrm{i}}[\psi^+\sigma_\mu(D\psi + 2f\Phi(\mathrm{d})) - \\
&\qquad (D\psi + 2f\Phi(\mathrm{d}))^+\sigma_\mu\psi] + fW_\mu(\mathrm{d}) \;, \\
\omega_a^\alpha(\mathrm{d}) &= (D\psi + f\Phi(\mathrm{d}))_a^\alpha \;, \quad \omega^{a\dot{\alpha}}(\mathrm{d}) = \omega_a^\alpha(\mathrm{d})^+, \\
Dx_\mu &= \mathrm{d}x_\mu + 2\Omega_{\mu\nu}(\mathrm{d})x^\nu \;, \\
D\psi &= \mathrm{d}\psi + f\Omega^{\mu\nu}(\mathrm{d})I_{\mu\nu}\psi - \mathrm{i}fV^a(\mathrm{d})I_a\psi \;,
\end{aligned} \tag{4}$$

where $I_{\mu\nu} = (1/4)(\tilde{\sigma}_\mu\sigma_\nu - \tilde{\sigma}_\nu\sigma_\mu)$, I_a are the generators of the internal–symmetry group, and the gauge fields $\Omega_{\mu\nu}(\mathrm{d}), W(\mathrm{d}), V^a(\mathrm{d})$, and $\Phi_a^\alpha(\mathrm{d})$ are the coefficients of the expansion of the form $A(\mathrm{d})$ in terms of generators of the considered group.

The forms $\omega(\mathrm{d})$, which correspond to the generators of the Lorentz group, as well as the internal–symmetry group, depend on the parameters l and u . There is no such dependence in the second–order differential forms

$$\begin{aligned}
R^{\mu\nu}(\delta,\mathrm{d}) &= \delta\Omega^{\mu\nu}(\mathrm{d}) - \mathrm{d}\Omega^{\mu\nu}(\delta) + 2f[\Omega_\gamma^\mu(\delta)\Omega^{\gamma\nu}(\mathrm{d}) - \Omega_\gamma^\nu(\delta)\Omega^{\gamma\mu}(\mathrm{d})]; \\
F^a(\delta,\mathrm{d}) &= \delta V^a(\mathrm{d}) - dV^a(\delta) + fC_{bc}^a V^b(\delta)V^c(\mathrm{d}) \;,
\end{aligned} \tag{5}$$

where C_{bc}^a are the structure constants of the internal symmetry group.

The invariant action integral is made up in the form of outer products of fourth order, which are invariant with respect to the Lorentz group and the internal–symmetry group, of the forms (4) and (5) (or in the general case in the form of a homogeneous first–order function of such products (see Volkov and Akulov (1973), Volkov (1973))).

The simplest permissible invariant combinations are

$$\mathrm{i}\omega_\alpha^{\dot\beta}(\mathrm{d}_0)\Lambda\omega_{\dot\beta}{}^\gamma(\mathrm{d}_1)\Lambda\omega_\gamma^{\dot\delta}(\mathrm{d}_2)\Lambda\omega_{\dot\delta}{}^\alpha(\mathrm{d}_3) \;, \tag{6}$$

$$D_0\Lambda\omega_a^\alpha(\mathrm{d}_1)\Lambda\omega_{\alpha\dot\beta}(\mathrm{d}_2)\Lambda\omega^{a\dot\beta}(\mathrm{d}_3) - D_0\Lambda\omega^{a\dot\alpha}(\mathrm{d}_1)\Lambda\omega_{\dot\alpha\beta}(\mathrm{d}_2)\Lambda\omega_a^\beta(\mathrm{d}_3) \;, \tag{7}$$

$$\mathrm{i}[R_{\dot\alpha}^{\dot\beta}(\mathrm{d}_0,\mathrm{d}_1)\Lambda\omega_{\dot\beta\gamma}(\mathrm{d}_2)\Lambda\omega^{\gamma\dot\alpha}(\mathrm{d}_3) - R_\beta^\alpha(\mathrm{d}_0,\mathrm{d}_1)\Lambda\omega^{\beta\dot\gamma}(\mathrm{d}_2)\Lambda\omega_{\dot\gamma\alpha}(\mathrm{d}_3)] \;, \tag{8}$$

$$\frac{[F^a(\mathrm{d}_0,\mathrm{d}_1)\Lambda\omega_\alpha^{\dot\beta}(\mathrm{d}_2)\Lambda\omega_\gamma^{\dot\delta}(\mathrm{d}_3)][F^a(\mathrm{d}_0,\mathrm{d}_1)\Lambda\omega_{\dot\beta}^\alpha(\mathrm{d}_2)\Lambda\omega_{\dot\delta}^\gamma(\mathrm{d}_3)]}{\mathrm{i}\omega_\alpha^{\dot\beta}(\mathrm{d}_0)\Lambda\omega_{\dot\beta}{}^\gamma(\mathrm{d}_1)\Lambda\omega_\gamma^{\dot\delta}(\mathrm{d}_2)\Lambda\omega_{\dot\delta}^\alpha(\mathrm{d}_3)} \;, \tag{9}$$

where Λ is the outer–product symbol.

Expressions (7-9) contain kinetic terms for gauge fields with spins $3/2, 2$ and 1, respectively. expression (6) contains a kinetic term for Goldstone particles.

All the field variables in the products (6-9) can be parametrized in the form of functions of x_μ, so that the gauge form can be reduced, by redefining the fields, to a form containing only the differentials x_μ and the redefined gauge fields corresponding to the coefficients of the differentials.

To consider the Higgs effect it suffices to examine the structure of forms (4,5). In the form $\omega_a^\alpha(\mathrm{d})$ the gauge form $\Phi(\mathrm{d})$ for spin 3/2 enters together with $D\psi$, so that we can define as a new gauge form

$$f\Phi'(\mathrm{d}) = D\psi + f\Phi(\mathrm{d}) \ . \tag{10}$$

In this case, however, the Goldstone field ψ in the form $\omega_\mu(\mathrm{d})$ is not eliminated completely, so that there is no Higgs effect for this field. Moreover, by imposing the invariant condition[1]

$$\omega^\alpha(\mathrm{d}) = 0 \tag{11}$$

we can exclude from consideration the gauge field with spin 3/2.

The gauge form $W_\mu(\mathrm{d})$ enters in the form $\omega_\mu(\mathrm{d})$ together with other terms that contain, in particular, kinetic terms of Goldstone particles. A change to a new gauge form

$$fW'(\mathrm{d}) = \omega_\mu(\mathrm{d}) \tag{12}$$

eliminates the variables connected with the Goldstone fields. This circumstance corresponds to the Higgs effect in which the Goldstone particles vanish as a result of a redefinition of the metric tensor.

Owing to redefinition of the gauge field $W_\mu(\mathrm{d})$, the invariants (7-9) correspond to interacting Yang–Mills fields, to a field with spin 3/2, and to gravitational field with a cosmological term (6).

A Goldstone effect field with spin 1/2 can be retained only by violating the gauge group. Therefore the program proposed in [4], of including the weak and electromagnetic interactions in the scheme of Goldstone particles, can be carried only with allowance for this violation.

References

Higgs, P.W. (1966): Phys.Rev. **145**, 1156.

Kibble, T.W. (1967): Phys.Rev. **155**, 1557.

Migdal, A.A. and Polyakov, A.M. (1966): Zh.Eksp.Teor.Fiz. **51**, 135 [Sov.Phys.-JETP **24**, 91 (1967)].

Volkov, D.V. and Akulov, V.P. (1972): ZhETF Pis.Red. **16**, 621 [JETP Lett. **16**, 438 (1972)].

Volkov, D.V. and Akulov, V.P. (1973): Preprint ITF–73–51R, Kiev.

Volkov, D.V. (1973): Fiz.Elem.Chastits At.Yad. **4**, 3 [Sov.J.Part.Nuclei **4**, 1, (1973)]

[1] The possibility of using conditions such as (11) to eliminate some of the considered fields was pointed out to us by V.I.Ogievetskii.

Gauge Fields on Superspaces with Different Holonomy Groups

V.P. Akulov, D.V. Volkov, and V.A. Soroka

Kharkov Institute of Physics and Technology

JETP Lett., v. 22, No 7, p. 187

Abstract. We discuss variants of a generally–covariant theory of superfields with nonzero values of the torsion tensors and the curvature tensor.

1. A generally-covariant theory of superfields was recently proposed (Arnowitt, Nath, Zumino (1975), Arnowitt and Nath (1975)) for a space with coordinates $Z^A = (\chi^\mu, \phi^\alpha, \phi^{\dot\alpha})$, ($\chi^\mu$ are the usual spatial coordinates, and ϕ^α and $\phi^{\dot\alpha}$ are anticommuting spinor coordinates).[1]

The generalized Einstein equation for the superspace takes the form (Arnowitt and Nath (1975))

$$R_{AB} = 0 \tag{1}$$

where $R_{AB} = R^C_{AC;B}$ and $R^D_{AB;C}$ is the curvature tensor of the superspace.

In the expansion of the superfield of the metric tensor g_{AB} in terms of ordinary fields, Eq. (1), leads to second–order equations both for the fields with integer spins and for fields with half–integer spin. The latter circumstance is due to the absence from the structure constants of a maximal holonomy group[2] of Riemannian superspace satisfying Eq. (1), of quantities that can play the role of the matrices γ_μ in the equations for fields with half–integer spin, and is a shortcoming of the theory.

We wish to call attention in this paper to the existence of generally–covariant theories free of the foregoing shortcoming.

2. The Cartan equation for a superspace with coordinates z_A can be written in the form (Volkov, Soroka (1972))

$$\mathrm{d}\omega^A(\delta) + \omega^B(\delta) \wedge \Gamma_B{}^A(\mathrm{d}) = \frac{1}{2}\omega^B(\delta) \wedge \omega^C(\mathrm{d}) T_{CB}{}^A \ , \tag{2}$$

$$\mathrm{d}\Gamma_A{}^B(\delta) + \Gamma_A{}^C(\delta) \wedge \Gamma_C{}^B(\mathrm{d}) = \frac{1}{2}\omega^C(\delta) \wedge \omega^D(\mathrm{d}) R_{DC;A}{}^B \ , \tag{3}$$

where T^A_{CB} and $R^B_{DC;A}$ are the torsion and curvature tensors. The differentiations and the products of the forms in expressions (2) and (3) are external

---

[1] For a detailed bibliography on supersymmetry see Zumino (1974), where no reference is made, however, to the pioneering paper Golfand and Likhtman (1971).

[2] The holonomy group is a group of transformations of a reference frame in parallel transfer of the latter along an infinitesimal closed contour.

and are determined, just for ordinary spaces, by alternating the differentials d and δ.

We choose as the holonomy group of the considered superspace the Poincare group supplemented by translation of the spinor variables. In this case only the components $\Gamma^{\beta}_{\alpha}(\mathrm{d}), \Gamma^{\dot\beta}_{\dot\alpha}(\mathrm{d})$, and $\Gamma^{\nu}_{\mu}(\mathrm{d})$ of the differential form of connectivity differ from zero and satisfy the relations

$$2g^{\mu\rho}\Gamma^{\nu}_{\rho}(\mathrm{d}) = \Gamma^{\alpha}_{\beta}(\mathrm{d})(\sigma^{\mu\nu})^{\beta}_{\alpha} + \Gamma^{\dot\alpha}_{\dot\beta}(\mathrm{d})(\sigma^{\mu\nu})^{\dot\beta}_{\dot\alpha} \ , \tag{4}$$

where

$$\sigma^{\mu\nu} = \frac{1}{2}(\sigma^{\mu}\sigma^{\nu} - \sigma^{\nu}\sigma^{\mu}) \ .$$

The invariant action integral for the superfields that determine the differential forms $\omega^{A}(\mathrm{d})$ and $\Gamma^{A}_{B}(\mathrm{d})$, can be represented in the form

$$\int L(R,T)W\prod_{A}\mathrm{d}z^{A} \tag{5}$$

where L is an invariant function of the curvature and torsion tensors, and

$$W = \det|\omega_{A}{}^{B}| \ , \tag{6}$$

where the matrix ω^{B}_{A} is determined by the form coefficients[3] $\omega^{B}(\mathrm{d}) = \mathrm{d}z^{A}\omega^{B}_{A}$.

In the simplest case, L is a function of the following invariant quantities

$$R_1 = \mathrm{i}[R_{\mu\nu;\alpha}{}^{\beta}(\sigma^{\mu\nu})^{\alpha}_{\beta} - R_{\mu\nu;\dot\alpha}{}^{\dot\beta}(\sigma^{\mu\nu})^{\dot\alpha}_{\dot\beta}] \ , \tag{7a}$$

$$R_2 = \mathrm{i}(R_{\alpha\beta;\gamma}{}^{\beta}\,\epsilon^{\alpha\gamma} - R_{\dot\alpha\dot\beta;\dot\gamma}{}^{\dot\beta}\,\epsilon^{\dot\alpha\dot\gamma}) \ , \tag{7b}$$

$$T_1 = \mathrm{i}T_{\alpha\dot\beta;}{}^{\mu}\,(\sigma_{\mu})^{\dot\beta\alpha} \ , \tag{7c}$$

$$T_2 = \mathrm{i}[T_{\mu\dot\alpha;}{}^{\beta}\,(\sigma^{\mu})^{\dot\alpha}_{\beta} - T_{\mu\alpha;}{}^{\dot\beta}\,(\sigma^{\mu})^{\alpha}_{\dot\beta}] \ . \tag{7d}$$

3. We consider the case when

$$L = a_1R_1 + a_2R_2 \ . \tag{8}$$

The variation of the superfields for individual terms of the action integral (5) and (8) takes the form

$$(\widetilde{R_{DC;A}{}^{B}W}) = \Big\{\Big[-\omega^{F}_{C}R_{FD;A}{}^{B} + \frac{1}{2}(-)^{F}\tilde\omega^{F}_{F}R_{CD;A}{}^{B} - (-)^{F}T_{CF;}{}^{F}\tilde\Gamma^{A}_{DB} + \frac{1}{2}T_{CD}{}^{F}\tilde\Gamma^{B}_{FA}\Big] - (-)^{CD}(C\leftrightarrow D)\Big\}W \ , \tag{9}$$

[3] The holonomy group is a group of transformations of a reference frame in parallel transfer of the latter along an infinitesimal closed contour.

where the tilde over a quantity denotes its variation, and the factors of the form $(-)^F$ determine the signs of the corresponding terms, depending on their graduation.

The equations for the superfields are established by substituting the expression (9) in (7a,b) and by equating to zero the coefficients of the independent variations $\tilde{\omega}_A^B, \tilde{\Gamma}_{A\alpha}^B$, and $\tilde{\Gamma}_{A\alpha}^{\dot{B}}$.

An essential difference between the equations obtained from the Lagrangian (8) and the generalized Einstein equations considered in (Arnowitt, Nath, Zumino (1975), Arnowitt and Nath (1975)) is the fact that as a result of the weakening of the holonomy group not all the components of the torsion tensor are equal to zero[4]. In particular, the Lagrangian (8) contains a solution corresponding to superspace with constant torsion tensor

$$T^{\mu}_{\alpha\dot{\beta}} = \mathrm{i}t(\sigma^{\mu})_{\alpha\dot{\beta}} \tag{10}$$

for which

$$\Gamma^{\beta}_{A\alpha} = \Gamma^{\dot{\beta}}_{A\dot{\alpha}} = 0, \quad \omega^{\alpha} = d\phi^{\dot{\alpha}}, \quad \omega^{\dot{\alpha}} = \mathrm{d}\phi^{\dot{\alpha}}$$

$$\omega^{\mu}(\mathrm{d}) = \mathrm{d}x^{\mu} + \frac{t}{2\mathrm{i}}(\phi\sigma^{\mu}\mathrm{d}\phi^{+} - \mathrm{d}\phi\sigma^{\mu}\phi^{+}) \ . \tag{11}$$

Such a superspace, first introduced in (Volkov, Akulov (1972, 1974), Salam and Strathdee (1974)), constitutes the basis of different variants of supersymmetry theory.

For solutions that differ little from (11), the Lagrangian leads to typically supersymmetric structure of the superfields.

4. In the presence of invariant combinations of the torsion tensor in the Lagrangian, the variations of such combinations can be expressed in terms of the quantity

$$\begin{aligned}\widetilde{(T^A_{BC;}W)} = \Bigg\{\Bigg[&-\tilde{\omega}^F_B T^A_{FC;} + \frac{1}{2}(-)^F\tilde{\omega}^F_F T^A_{BC;} + (-)^{AC+F}\omega^A_B T^F_{FC} \\ &+\frac{1}{2}(-)^{(A+F)(B+C+F)}\tilde{\omega}^A_F T^F_{BC;} + \tilde{\Gamma}^A_{BC:}\Bigg] - (-)^{BC}(B \leftrightarrow C)\Bigg\} W\end{aligned}$$

$$\tilde{W} = (-)^F\tilde{\omega}^F_F W \ . \tag{12}$$

The requirement that the space with a constant vector torsion be edmitted by the equations of motion leads in this case to definite relations between the constants at different degrees of the torsion tensor (7c) in the Lagrangian.

---

[4] We note that if we require in addition that all the components of the torsion tensor be equal to zero, then the latter holds for the Lagrangian (8) only under the condition that all the components of the curvature tensor also vanish.

References

Arnowitt, R. and Nath, P. (1975): Phys. Lett., **56B**, 117.
Arnowitt, R., Nath, P. and Zumino, B. (1975): Phys. Lett., **56B**, 81.
Golfand, Yu. A. and Likhtman, E.P. (1971): ZhETF Pis. Red., **13**, 452. [JETP Lett., **13**, 323]
Salam, A. and Strathdee, J. (1974): Nucl. Phys. **B76**, 477.
Volkov, D.V., Soroka, V.A. (1972): Teor. Mat. Fiz., **20**, 291.
Volkov, D.V., Akulov, V.P. (1972, 1974): ZhETP Pis. Red., **16**, 621 [JETP Lett., **16**, 438];
Teor. Mat. Fiz., **18**, 39.
Zumino, B. (1974): CERN Preprint TH.1901.

Spontaneous Compactification of Subspace Due to Interaction of the Einstein Fields with the Gauge Fields

D.V. Volkov and V.I. Tkach

Institute of Physics & Technology
Kharkov, USSR

Pis'ma Zh. Eksp. Theoret. Phys., **32**, No 11, 681, (1980)

Abstract. The mechanism for compactification of spaces due to interaction of the Einstein fields with the gauge fields, which is different from the mechanism previously analyzed by Cremmer and Scherk, is determined.

The approximate method of compactification of additional space–time measurements is now used extensively in constructing dual models with internal symmetries and Lagrangians for O(N) broadened supergravitation (Cremmer and Scherk (1976), Cremmer and Scherk (1977), Cremmer and Julia (1979)). It was shown by Cremmer and Scherk (1976, 1977, 1979) and later by Luciani (1978) in a more general way that compactification of additional measurements under certain conditions may be due to the original Lagrangian, i.e., it can have a spontaneous nature. In this paper we examine the mechanism of spontaneous compactification of subspaces, which is different from the mechanism examined in Ref. 2.

Assuming that the vacuum state of the fields in the subspace subject to compactification is determined by zero values of all the fields except the Einstein fields and gauge fields, we examine the Lagrangian corresponding to them

$$L = \alpha\sqrt{g}R + \beta\sqrt{g}F^{\alpha\beta}_{lm}F^{\alpha\beta\ lm} \quad . \tag{1}$$

The indices $l, m = 1, 2, ...n$ and $\alpha, \beta = 1, 2, ...n$ pertain to the space and intrinsic variables, respectively. $(F^{\alpha\beta})_{lm} = -(F^{\beta\alpha})_{lm}$ are the strengths of the O(n) gauge group.

The ratio α/β can be larger of smaller than zero, depending on whether the examined subspace is spacelike or timilake with respect to the signature of the metric tensor of the total space on which the real space and time are included with its usual signature.

The field equations have the usual form,

$$\alpha(R_{lm} - \frac{1}{2}\, G_{lm}\, R) = -T_{lm} \quad , \tag{2}$$

$$T_{lm} = \beta(F^{\alpha\beta}_{l\ s}\, F^{\alpha\beta\ s}_{m} - \frac{1}{4} g_{lm}\, F^{\alpha\beta}_{p\ s} = F^{\alpha\beta\ ps}) \quad , \tag{3}$$

and

$$(D_l\ F_l^m)^{\alpha\beta} = 0 \tag{4}$$

where D_l represents a covariant derivative that contains both a metric connectivity defined by Christoffel symbols Γ^l_{mn} and the gauge fields.

To find a nontrivial vacuum solution, the Riemanniam space, which is determined by Eq. (2), must be a space of constant, positive curvature, i.e.,

$$R_{lm\ np} R(g_{mn} g_{lp} - g_{lp} g_{mn})\ ,\ \ K > 0 \tag{5}$$

and the gauge fields must satisfy the additional condition

$$(D_l\ Fmn)^{\alpha\beta} = 0\ , \tag{6}$$

which identically satisfies Eqs. (4).

If we take Eq. (5) into account, then Eq. (6) will have the following solution for the strengths

$$(F_{lm})^{\alpha\beta} = -K(\chi_l^\alpha \chi_m^\beta - \chi_m^\alpha \chi_l^\beta)\ , \tag{7}$$

and for the gauge fields

$$(A_m)^{\alpha\beta} = -\chi^{\alpha\ p}\ \Gamma^s_{mp}\ \chi_s^\beta + (\partial_m \chi_s^\alpha)\chi^{s\ \beta}\ , \tag{8}$$

where χ_m^α are continuous functions that satisfy the orthogonality conditions

$$\chi_m^\alpha\ = \chi_m^\alpha\ g_{lm},\ \ g^{lm}\ \chi_l^\alpha\ \chi_m^\beta\ = \delta_{\alpha\beta}\ . \tag{9}$$

The relation (8) establishes a connection between the gauge connectivity and the metric connectivity.

For the specified functions χ_m^α , which corresponds to the selection of specific gauge, the expressions (7) and (8) are invariants of the global $O(n+1)$group that coincides with the metric–invariance group.

Substituting the strengths (7) in Eq. (2) and taking Eq. (3) into account, we obtain the following equation:

$$\alpha(n-2)K = \beta(n-4)K^2\ , \tag{10}$$

witch determines the dependence of the Gaussian curvature on the ratio α/β. Equation (10) has two solutions for n unequal to two $-K = 0$, which corresponds to the trivial case of the flat space, and

$$K = \frac{\alpha(n-2)}{\beta(n-4)}\ , \tag{11}$$

for a space of constant curvature.

Thus, as a consequence of the nonlinear equation (10), which is in a certain sense analogous to the equation for spontaneous symmetry breaking by Higgs bosons, the Lagrangian (1) contains a spontaneous transition in which the flat space in transformed into spheres.

It follows from the relation (11) that $\alpha/\beta < 0$ for $n = 3$ and $\alpha/\beta > 0$ for $n > 4$. As result, compactification can occur only if the space is timelike for $n = 3$ and spacelike for $n > 4$.

In conclusion, we note that the investigation performed above can be extended, without substantial changes, to compactification of spaces for which

$$R_{lm;\ np,\ r} = 0 \tag{12}$$

i.e., to the general case of symmetrical spaces (Khelgason 1964). The gauge group for the fields $F^{\alpha\beta}_{mn}$ in this case must coincide with the holonomy group of symmetrical space.

References

Cremmer, T. and Scherk, J.(1976): Nucl.Phys. **103**, 339.
Cremmer, T. and Scherk, J. (1977): Nucl.Phys. **118**, 61.
Cremmer, T. and Julia, B. (1979): Nucl.Phys. **159**, 141.
Khelgason, S. (1964): Differentsial'naya geometriya i simmetricheskie prostranstva (Differential Giometry and Symmetrical Spaces), Moscow.
Luciani, T.F. (1978): Nucl.Phys. **135**, 111.

Hamiltonian Systems with Even and Odd Poisson Brackets: Duality of Their Conservation Laws

D.V. Volkov, A.I. Pashnev, V.A. Soroka and V.I. Tkach

Kharkov

Sov. J. Nucl. Phys., **44**, 810 (1986)

Abstract. The existence of Hamiltonian systems with even and odd Poisson brackets is proved in the example of Witten's supersymmetric mechanics.

Of the three known versions of Poisson brackets, two of which are even and one odd with respect to the Grassmann gauge of canonical variables (Berezin (1983); Leites (1983)), the odd Poisson brackets, whose canonical variables have the opposite Grassmann gauge if nontrivial (Volkov (1983); Volkov et al. (1984)). It is definitely worthwhile to study the various physical applications of odd Poisson brackets.

We wish to call attention to the circumstance that a Hamiltonian system with equal numbers of even and odd canonical variables allows the simultaneous introduction of even and odd Poisson brackets. When bracket operations of different gauges are used, the equations for the canonical variables do not change, but the integrals of motion with the opposite Grassmann gauge become duals, converting into each other upon the transformation to the Poisson brackets with the opposite gauge.

We require that the same equations of a dynamic system containing even (^{a} and odd x^α) canonical variables be reproduced by even Poisson brackets $\{,\}_0$ with an even Hamiltonian H and by odd Poisson brackets $\{,\}_1$ with an odd Poisson brackets $\bar{H}$. In other words, we require

$$\dot{X}^A = \{X^A, H\}_0 = \{X^A, \bar{H}\}_1 \tag{1}$$

where $X^A - (x^a, a^\alpha)$. Relation (10 is equivalent to the equations

$$\bar{\omega}_{AB}(X)\omega^{BC}\partial_C H = \partial_A \bar{H} \ , \tag{2}$$

which determine $\bar{H}$ and the coefficients of the closed odd external form

$$\bar{\omega}_{AB}(X) = \partial_A \varphi_B - (-1)^{AB}\partial_B \varphi_A, \tag{3}$$

which corresponds to odd brackets for the given H and to the even canonical form ω^{AB}, where φ_A are coefficients of an odd Liouville form.

To illustrate the point, we seek the solution of Eqs. (2) for the case of Witten's supersymmetric mechanics (Witten (1981)) with the Hamiltonian

$$H = H_0 + i\eta^1\eta^2 W^1(q) \ , \tag{4}$$

where $x^a = (q,p)$ and $a^\alpha = (\eta^1, \eta^2)$, and $H_0 = [p^2 + W^2(q)]/2$. For Hamiltonian (4), the fermion charge $F = i\eta^2\eta^2$ and the supercharges $Q_1 = p\eta^1 - W\eta^2$, $Q_2 = \eta^2 - W\eta^1$, which form a superalgebra with even Poisson brackets

$$\{Q_\alpha, Q_\beta\}_0 = -2i\delta_{\alpha\beta}H, \quad QQ\{F, Q_\alpha\}_0 = \epsilon_{\alpha\beta}Q_\beta \ . \tag{5}$$

are also conserved quantities. By virtue of equations of motion (1), the quantities H, F, Q_1 and Q_2 and also arbitrary functions of them are also integrals of motion with respect to the odd brackets $\{,\}_1$ with the Hamiltonian $\bar{H}$. Equations (2) with Hamiltonian (4) determine $\bar{H}$ and $\bar{\omega}_{AB}$ within six arbitrary functions that depend of H_0. Making use of this arbitrariness, we can require

$$\bar{H} = Q_1 \tag{6}$$

and that the three other independent quantities, which are conserved with respect to $\bar{H}$ in odd brackets, i.e., the quantities $\bar{F}, \bar{Q}_1$ and $\bar{Q}_2$, must be linear in the integrals H, F, Q_1, and Q_2 and must form with odd brackets the superalgebra

$$\{\bar{Q}_\alpha, \bar{Q}_\beta\}_1 = -2\delta_{\alpha\beta}\bar{H}, \quad \{\bar{F}, \bar{Q}_\alpha\}_1 = \varphi_{\alpha\beta}\bar{Q}_\beta \ ,$$

which is the same as (5). The integrals of motion $\bar{F}, \bar{Q}_1$ and $\bar{Q}_2$, are then related in the following way to the conserved quantities in Witten's mechanics with Hamiltonian (4):

$$\bar{F} = \frac{1}{2i}Q_2, \quad \bar{Q}_1 = H, \quad \bar{Q}_2 = i(2F - H) \ . \tag{7}$$

Upon the transformation to odd brackets, the supercharges Q_1 and Q_2 acquire the meaning of an of Hamiltonian $\bar{H}$ and an odd fermion charge $\bar{F}$, while the role of the supercharges $\bar{Q}_1$ amd $\bar{Q}_2$ is played by linear combinations of Witten's Hamiltonian (4) and the fermion charge F. By virtue of the symmetry between the charges $Q_1$and Q_2 in Witten's mechanics, we could have used Q_2 as the odd Hamiltonian $\bar{H}$.

Under additional condition (6) and (7), Eqs. (2) have the following solution for the coefficients φ_A of the odd Louiville form;

$$\varphi_q = (q - \alpha_2 W)\eta^2 \ ,$$

$$\varphi_p = \frac{1}{W'}[(1 + \alpha_1 p)\eta^2 - \alpha_2 p\eta^1] \ ,$$

$$\varphi_{\eta^1} = i\eta^1\eta^2\alpha_2 \ , \qquad \varphi_{\eta^2} = i\eta^1\eta^2\alpha_1$$

where

$$\alpha_1 = W' \left[WJ(H_0, q) + \frac{p}{H_0} \right] - \frac{1}{p} \ , \alpha_2 = W' \left[pJ(H_0, q) + \frac{W}{H_0} \right] \ ,$$

$$J(H_0, q) = \int_{q_0}^{q} [2H_0 - W^2(q')]^{-3/2} dq' \ .$$

We have thus proved that there is and odd Louiville form, determined internally, for Witten's Hamiltonian systems, and we have proved that duality relations (6) and (7) hold between the even and odd integrals of motion; specifically, they hold between the Hamiltonian and the supercharge. The duality relations between the Hamiltonian and the supercharge are of particular interest for relativistic systems, which will be analyzed in a separate paper.

References

Berezin, F.A.(1983): Vvedenie v algebru i analiz s antikommutiruyushchimi pepemennymi (Introduction to Algebra and Analysis with Anticommuting Variables), MGU, Moscow;
Leites, D.A. (1983): Theoria sujpermnogoobrazii (Theory of Supermanifolds), Petrozavodsk.

Volkov, D.V. (1983): Pis'ma Zh.Eksp.Teor.Fiz. **38**, 508 [JETP Lett. **38**, 615];
Volkov, D.V., Soroka, V.A. and Tkach, V.I. (1984): in Problemy fiziki vysokikh energii i kvntovou tiorii polya (Problems of High-Energy Physics and Quantum Field Theory), Vol. 1, Protvino, p.48.

Witten, T. (1981): Nucl.Phys. **B188**, 513.

List of Main Publications of D.V. Volkov

1. D. V. Volkov, A. I. Akhieser and V. F. Aleksin, On some effects caused by the interaction of electromagnetic field with vacuum. *Docladi Akademii Nauk USSR* **104** (1955) 830.
2. D. V. Volkov and V. F. Aleksin, Radiation corrections to particle scattering in external field and to Compton effect in scalar quantum electrodynamics. *Sov. Phys. JETP* **33** (1957) 1044.
3. D. V. Volkov, On the connection of Pauli theorem with the inversion of time. *Uchenie Zapiski Kharkovskogo Universiteta* **7** (1958) 75.
4. D. V. Volkov, One-photon mass operator in scalar quantum electrodynamics. Ibid. p. 79.
5. D. V. Volkov, Mass operator of scalar particle in constant magnetic field. Ibid. p. 91.
6. D. V. Volkov and V. N. Oraevskij, On the role of the form factor in the decay of $\pi^0 \to \gamma + e^- + e^-$. *Ukrainian Phys. Journal* **4** (1959) 804.
7. D. V. Volkov, On the possibility of separating relativistic charged particles with respect to their masses by means of waveguids of running waves. *Soviet J. of Technical Physics* **29** 1959 804.
8. D. V. Volkov, On the spin one-half field quantization. *Sov. Phys. JETP* **36** (1959) 1560
9. D. V. Volkov and S. V. Peletminskij, On a Lagrangian formalism for spin variables. *Sov. Phys. JETP* **37** (1959) 170.
10. D. V. Volkov, S–matrix in a generalized quantization procedure. *Sov. Phys. JETP* **38** (1960) 518.
11. D. V. Volkov and E. V. Inopin, The motion of nucleons in the anisotropic oscillator potential with taking into account the spin-orbit interaction. *Sov. Phys. JETP* **38** (1960) 1765.
12. D. V. Volkov and M. P. Rekalo, On the possibility of determining the sign of the decay $\pi^0 \to 2\gamma$ amplitude. *Nucl. Phys.* **37** (1962) 172.
13. D. V. Volkov, On the Coulomb exitation of the Λ particle. *Sov. Phys. JETP* **43** (1962) 1112.
14. D. V. Volkov and A. S. Bakai, On quantum electrodynamics with a Reggized photon. *Phys. Lett.* **5** (1963) 223.
15. D. V. Volkov and V. N. Gribov, Regge poles in the N-N and N-aN scattering amplitudes. *Sov. Phys. JETP* **44** (1963) 1068.
16. D. V. Volkov, On a factorization of particle scattering amplitudes in the Regge pole. *Sov. Phys. JETP* **45** (1963) 742.
17. D. V. Volkov and E. Kuraev, $K \to 2\pi$ decays and SU(3). *Sov. J. Nucl. Phys.* **2** (1965) 272.

18. D. V. Volkov and M. Chernishev, On propagators and vortex operators for particles with arbitrary spin. *Ukrainian Phys. Journal* **10** (1965) 51.
19. D. V. Volkov, SU(6)–symmetry and electromagnetic splitting of mass. *Sov. Phys. JETP Lett.* **1** (1965) 51.
20. D. V. Volkov, $SU(3) \times SU(3)$ symmetry and baryon–meson coupling constants. *Sov. Phys. JETP Lett.* **1** (1965) 129.
21. D. V. Volkov, Baryon electromagnetic formfactors and SU(6)–symmetry. *Sov. Phys. JETP Lett.* **2** (1965) 284.
22. D. V. Volkov, Negative parity barion resonances and broken $SU(6) \times SU(6)$ symmetry. *Nuovo Cimento* **40** (1965) 281.
23. D. V. Volkov, U(6,6) symmetry and high baryon resonances. *Nuovo Cimento* **40** (1965) 284.
24. D. V. Volkov, $U(3) \times U(3)$ symmetry for collinear processes. *Nuovo Cimento* **43** (1966) 84.
25. D. V. Volkov, Interaction of baryons with mesons and electromagnetic fields and SU(6) symmetry. *Sov. J. Nucl. Phys.* **3** (1966) 526.
26. D. V. Volkov and V. N. Guriev, Coupling constants of S6–S6 35–multiplets in the representation of SU(6) and $SU(3) \times SU(3)$ symmetry. *Sov. J. Nucl. Phys.* **3** (1966) 359.
27. D. V. Volkov, E. Kuraev and V. I. Tkach, Relations between coupling constants of 70, 56 and 35 vortices. *Ukrainian Phys. Journal* **11** (1966) 1296.
28. D. V. Volkov, E. Kuraev and V. I. Tkach, Meson resonance decays in the U(6,6) symmetry scheme. *Sov. J. Nucl. Phys.* **4** (1966) 601.
29. D. V. Volkov and V. N. Guriev, Locality of interaction and the parametrization of spiral amplitudes. *Sov. J. Nucl. Phys.* **5** (1967) 1290.
30. D. V. Volkov, On the possible confluence of bosonic Regge trajectories at $\tau \to 0$. *Sov. Phys. JETP Lett.* **5** (1967) 341.
31. D. V. Volkov and V. L. Lazarenko, Electroproduction of higher baryon resonances and SU(6) symmetry. *Sov. J. Nucl. Phys.* **7** (1968) 843.
32. D. V. Volkov, Phenomenological Lagrangian for baryon–meson interaction invariant under chiral group. *Sov. Phys. JETP Lett.* **7** (1968) 385.
33. D. V. Volkov and V. N. Guriev, Generalization of the notion of signature and asymptotic features of the scattering amplitude. *Ukrainian Phys. Journal* **13** (1969) 863.
34. D. V. Volkov, Phenomenological Lagrangians for Goldstone particle interaction. Preprint ITP-69-75, KIev, 1969.
35. D. V. Volkov, Current structure of phenomenological Lagrangians. *Theor. Mat. Phys.* **5** (1970) 321.
36. D. V. Volkov, M. Bliokh and A. A. Zheltukhin, Phenomenological Lagrangian for spin waves. *Fisika Tverdogo Tela* **13** (1971) 1668.
37. D. V. Volkov and V. I. Radchenko, On the fermion resonances in the Veneziano model. *Phys. Lett.* **36B** (1971) 83.

38. D. V. Volkov, Phenomenological Lagrangians invariant under symmetry groups containing the Poincare group as a subgroup. Preprint FIAN N 141/1971, Moscow, 1971.
39. D. V. Volkov, V. D. Gershun, A. I. Pashnev and A. A. Zheltukhin. Duality and Adler princile. Composite and dual models. Kiev, 1971.
40. D. V. Volkov, V. I. Tkach and A. A. Zheltukhin. On the minimal interaction of mesons. *Theor. Mat. Phys.* **10** (1972) 329.
41. D. V. Volkov, Broken chiral symmetries and holonomy groups. *Theor. Mat. Phys.* **11** (1972) 173.
42. D. V. Volkov and V. P. Akulov, U(4) symmetry of quarks and leptons. *Sov. J. Nucl. Phys.* **15** (1972) 827.
43. D. V. Volkov, Phenomenological Lagrangians. *Soviet J. of Elementary Particles and Atomic Nuclear* **4** (1973) 3.
44. D. V. Volkov, V. D. Gershun and A. A. Zheltukhin, Adler principle and algebraic duality. *Theor. Mat. Phys.* **15** (1973) 245.
45. D. V. Volkov and V. P. Akulov, On a possible universal interaction of the neutrino. *Sov. Phys. JETP Lett.* **16** (1972) 367.
46. D. V. Volkov and V. P. Akulov, Interaction of Goldstone neutrino with electromagnetic field. *Sov. Phys. JETP Lett.* **17** (1973) 367.
47. D. V. Volkov and V. P. Akulov, Is the neutrino a Goldstone particle. *Phys. Lett.* **46B** (1973) 109.
48. D. V. Volkov and V. A. Soroka, Higgs effect for Goldstone particles with spin one half. *Sov. Phys. JETP Lett.* **18** (1973) 529.
49. D. V. Volkov and V. I. Tkach, On gauge fields for symmetry $ISL(6, C)$ group. *Theor. Mat. Phys.* **20** (1974) 324.
50. D. V. Volkov and V. A. Soroka, Gauge fields for symmetry groups with anticommuting parameters. *Theor. Mat. Phys.* **20** (1974) 291.
51. D. V. Volkov, A. I. Pashnev and A. A. Zheltukhin, On vacuum states in dual models. 1. *Sov. J. Nucl. Phys.* **18** (1973) 902.
52. D. V. Volkov and V. P. Akulov, Goldstone particles with spin one half. *Theor. Mat. Phys.* **18** (1974) 39.
53. D. V. Volkov, A. I. Pashnev and A. A. Zheltukhin, Spontaneous vacuum transitions in dual models. *Sov. Phys. JETP Lett.* **20** (1974) 488.
54. D. V. Volkov, A. I. Pashnev and A. A. Zheltukhin, Quark structure of resonances as consequence of spontaneous vacuum transitions in dual models. *Sov. Phys. JETP Lett.* **21** (1975) 454.
55. D. V. Volkov, A. I. Pashnev and A. A. Zheltukhin, On spontaneous vacuum transitions in dual models. *Sov. J. Nucl. Phys.* **21** (1975) 1104.
56. D. V. Volkov, A. I. Pashnev and A. A. Zheltukhin, On the quark structure of resonance states in dual models. *Sov. J. Nucl. Phys.* **22** (1975) 1225.
57. D. V. Volkov, V. P. Akulov and V. A. Soroka, On gauge fields in superspaces with different holonomy groups. *Sov. Phys. JETP Lett.* **22** (1975) 396.

58. D. V. Volkov, V. P. Akulov and V. A. Soroka, On general covariant theories of gauge fields in superspaces. *Theor. Mat. Phys.* **31** (1977) 12.
59. D. V. Volkov, Vacuum transitions in Neveu–Schvarz dual model. *Sov. J. Nucl. Phys.* **27** (1978) 243.
60. D. V. Volkov and V. P. Akulov, Einstein superspaces of minimal dimensions. *Theor. Mat. Phys.* **41** (1979) 147.
61. D. V. Volkov and V. P. Akulov, On supersymmetric equations in (1,2) spaces. *Theor. Mat. Phys.* **42** (1980) 16.
62. D. V. Volkov and A. I. Pashnev, On a supersymmetric Lagrangian for particles in proper time. *Theor. Mat. Phys.* **44** (1980) 321.
63. D. V. Volkov and A. A. Zheltukhin, Phenomenological Lagrangian of spin waves in space–distorted media. *Physics of Low Temperatures* **5** (1979) 1359.
64. D. V. Volkov and A. A. Zheltukhin, On spin wave in space disordered magnetic media. *Solid State Communications* **36** (1980) 733.
65. D. V. Volkov and A. A. Zheltukhin, On spin wave propagation in spatially distorted media. *Izv. Academii Nauk USSR, Phys. Ser.* **44** (1980) 1487; *Sov. Phys. JETP* **78** (1980) 1867.
66. D. V. Volkov and V. I. Tkach, On the spontaneous compactification of subspaces by the interaction of Einstein and gauge fields. *Sov. Phys. JETP Lett.* **32** (1980) 681.
67. D. V. Volkov and V. I. Tkach, Spontaneous compactification of subspaces. *Theor. Mat. Phys.* **51** (1982) 171.
68. D. V. Volkov, Space–time, physical fields and quantum statistics. In "Problems of modern theoretical physics", p. 92. Naukova Dumka, Kiev, 1982.
69. D. V. Volkov, V. D. Gershun and V. I. Tkach, On the algebra of nonlocal charges in two–dimensional chiral models. *Ukrainian Phys. Journal* **38** (1983) 641.
70. D. V. Volkov, D. P. Sorokin and V. I. Tkach, Gauge fields in mechanisms of spontaneous compactification. *Teor. Mat. Fiz.* **56** (1983) 177 (Sov. J. Theor. Math. Phys)
71. D. V. Volkov, D. P. Sorokin and V. I. Tkach, Mechanisms of spontaneous compactification of N=2, d=10 supergravity. *JETP Letters* **38** (1983) 481.
72. D. V. Volkov, D. P. Sorokin and V. I. Tkach, Spontaneous compactification in d=10,11 supergravities. *Sov. J. Nucl. Phys.* **39** (1984) 823.
73. D. V. Volkov, D. P. Sorokin and V. I. Tkach, Supersymmetric vacuum configurations in d=11 supergravity, *JETP Lett.* **40** (1984) 1162.
74. D. V. Volkov, D. P. Sorokin and V. I. Tkach, Spontaneous compactification into symmetric spaces with non-semisimple holonomy group. *Teor. Mat. Fiz.* **61** (1984) 241 (Sov. J. Theor. Math. Phys.)

75. D. V. Volkov, D. P. Sorokin and V. I. Tkach, On geometrical structure of compactified subspaces in d=11 supergravity. *Sov. J. Nucl. Phys.* **41** (1985) 872.
76. D. V. Volkov, D. P. Sorokin and V. I. Tkach, Kaluza-Klein theories and spontaneous compactification of extra space dimensions. Proceed. of 3rd Seminar on Quantum Gravity. World Scientific. P. Co. 1985.
77. D. V. Volkov, D. P. Sorokin and V. I. Tkach, On the relationship of compactified vacua in D=11 and D=10 supergravity. *Phys. Lett.* **161B** (1985) 301.
78. D. V. Volkov and A. A. Zheltukhin, Off shell superfield description of particles with spin 1/2. *Ukrainian Phys. Journal* **30** (1985) 72.
79. D. V. Volkov and A. A. Zheltukhin, On a description of strings in space and superspace. *Theor. Mat. Phys.* **30** (1985) 809.
80. D. V. Volkov, V. P. Akulov and V. I. Pashnev, Quantum oscillator with spontaneously broken supersymmetry. *Ukrainian Phys. Journal* **30** (1985) 1263.
81. D. V. Volkov, V. A. Soroka and V. I. Tkach, On classical and quantum Hamiltonian systems with the odd Poisson bracket. *Sov. J. Nucl. Phys.* **44** (1986) 810.
82. D. V. Volkov, A. I. Pashnev, V. A. Soroka and V. I. Tkach, On Hamiltonian systems with the even and odd Poisson bracket and the duality of their conservation laws. *Sov. Phys. JETP Lett.* **44** (1986) 55.
83. D. V. Volkov, D. P. Sorokin and V. I. Tkach, Structure of N=3,1 SUSY vacua in d=11 supergravity. *Sov. J. Nucl. Phys.* **43** (1986) 142.
84. D. V. Volkov and V. A. Soroka, On the quantization of dynamical systems with the odd Poisson bracket. *Sov. J. Nucl. Phys.* **46** (1987) 110.
85. D. V. Volkov, V. A. Soroka and V. I. Tkach, An odd Poisson bracket and spinor structure of space-time. *Ukrainian Phys. Journal* **32** (1987) 1622.
86. D. V. Volkov, Trends of developing supersymmetric theories. *Ukrainian Phys. Journal* **32** (1987) 1782.
87. D. V. Volkov, A. I. Pashnev, V. A. Soroka and V.I. Tkach, On the Hamilton dynamical systems with the even and odd Poisson brackets. *Theor. Mat. Phys.* **79** (1989) 424.
88. D. V. Volkov, D. P. Sorokin and V. I. Tkach, Superparticles, twistors and Siegel symmetry. *Mod. Phys. Lett.* **A4** (1989) 901.
89. D. V. Volkov and A. A. Zheltukhin, An extension of Penrose representation and its application to the description of supersymmetric models. *Sov. Phys. JETP Lett.* **48** (1988) 61.
90. D. V. Volkov and A. A. Zheltukhin, On the equivalence of the Lagrangians of massless Dirac and supersymmetrical particles. *Lett.in Math. Phys.* **17** (1989) 141.
91. D. V. Volkov, D. P. Sorokin, V. I. Tkach, and A. A. Zheltukhin, From the superparticle Siegel symmetry to the spinning particle proper time supersymmetry. *Phys. Lett.***216B** (1989) 302.

92. D. V. Volkov and A. A. Zheltukhin , Lagrangian for massless particles and strings with local and global supersymmetry. *Nucl. Phys.* **B335** (1990) 723.
93. D. V. Volkov, D. P. Sorokin and V. I. Tkach, Spinning superparticles and extended supersymmetry *Yad. Fiz.* **49** (1990) 844 (Sov.J.Nucl.Phys.).
94. D. V. Volkov. Quartions in relativistic field theory. *Sov. Phys. JETP Lett.* **49** (1989) 473.
95. D. V. Volkov, D. P. Sorokin and V. I. Tkach, On relativistic field theories with fractional statistics and spin in D=2+1, 3+1. in "Problems of Modern Quantum Field Theory". Eds. A. A. Belavin, A. U. Klimyk and A.A.Zamolodchikov, Springer-Verlag, 1989.
96. D. V. Volkov, D. P. Sorokin and V. I. Tkach, Twistor displacement in the equations of motion of relativistic particles and strings, *JETP Lett.* **52**(1990) 1124.
97. D. V. Volkov, V. A. Soroka, D. P. Sorokin and V. I. Tkach, Generalized twistor dynamics of relativistic particles and strings. *Int. J. Mod. Phys.* **A7** (1992) N24, 5977.
98. D. V. Volkov and D. P. Sorokin, (Anti)commuting spinors and supersymmetric dynamics of semions. ICTP Preprint IC/92/121, 1992; *Nucl. Phys.* **B** (1993) (in press).
99. D. V. Volkov, Twistors and Supersymmetry. in "Spinors, Twistors, Clifford Algebras and Quantum Deformations" p. 109. Edit. Z. Oziewicz et al. Kluwer Academic Publishers. The Netherlands, 1993.
100. D. V. Volkov and D. P. Sorokin, Tracing an analogy between the Dirac-Maxwell-Einstein theory and a field model for semions. *JETP Lett.***57** (1993) 343.
101. D. V. Volkov and D. P. Sorokin, $D = (0|2)$ Dirac-Maxwell-Einstein theory as a way for describing supersymmetric quartions. Padova preprint DFPD/93/TH/46, 1993; *Int. J. Mod. Phys.* **A** (in press).
102. D. V. Volkov, I. A. Bandos, D. P. Sorokin, and M. Tonin, Doubly supersymmetric null strings and string tension generation. Padova preprint DFPD/93/TH/48, 1993. *Phys.Lett.* **B319**(1993)445-450
103. I.A. Bandos, M. Cederwall, D.P. Sorokin and D.V. Volkov, Towards a complete twistorization of the heterotic string, *Mod.Phys. Lett.* **A9** (1994) No. 32, 2987–2997. **(hep-th/9403181)**
104. I.A. Bandos, A.Yu. Nurmagambetov, D.P. Sorokin and D.V. Volkov, On another version of the twistor–like approach to superparticles, *JETP.Lett.* **60** (1994) 613–618. **(hep-th/9409439)**
105. I. Bandos, P. Pasti, D. Sorokin, M. Tonin and D. Volkov, Superstrings and supermembranes in the doubly supersymmetric geometric approach, *Prerpint* **DFPD 95/TH/02**, January 1995, **hep-th/9501113**, *Nucl.Phys.* **B446** (1995) 79–119.
106. I. Bandos, D. Sorokin and D. Volkov, The generalized action principle for superstrings and supermembranes.

Preprint **DFPD 95/TH/06**, February 1995, (**hep-th/9502141**). *Phys.Lett.* **B352** (1995) n.3,4; pp.269–275.

107. I.A. Bandos, A.Yu. Nurmagambetov, D.P. Sorokin and D.V. Volkov, Twistor–like approach to superparticles revisited, *Preprint* **DFPD 95/TH/08 (hep-th/9502143)**, *Class.Quantum Grav.*.12 (1995) 1881–1891.
108. I. Bandos, D. Sorokin and D. Volkov, New supersymmetric generalizations of the Liouville equation. *Preprint* **DFPD 95/TH/58**, October 1995, **hep-th/9510220**. (to be published in *Phys.Lett.* **B**.)
109. D.V. Volkov, Generalized Action Principle for superstrings and supermembranes. *Preprint* December 1995 **hep-th/9512103**. (to be published in Proceedings of International Workshop "SUSY-95", Palaiseau, France, 15-19 May 1995).

Lecture Notes in Physics

For information about Vols. 1–469
please contact your bookseller or Springer-Verlag

Vol. 475: J. Klamut, B. W. Veal, B. M. Dabrowski, P. W. Klamut, M. Kazimierski (Eds.), Recent Developments in High Temperature Superconductivity. Proceedings, 1995. XIII, 362 pages. 1996.

Vol. 476: J. Parisi, S. C. Müller, W. Zimmermann (Eds.), Nonlinear Physics of Complex Systems. Current Status and Future Trends. XIII, 388 pages. 1996.

Vol. 477: Z. Petru, J. Przystawa, K. Rapcewicz (Eds.), From Quantum Mechanics to Technology. Proceedings, 1996. IX, 379 pages. 1996.

Vol. 478: G. Sierra, M. A. Martín-Delgado (Eds.), Strongly Correlated Magnetic and Superconducting Systems. Proceedings, 1996. VIII, 323 pages. 1997.

Vol. 479: H. Latal, W. Schweiger (Eds.), Perturbative and Nonperturbative Aspects of Quantum Field Theory. Proceedings, 1996. X, 430 pages. 1997.

Vol. 480: H. Flyvbjerg, J. Hertz, M. H. Jensen, O. G. Mouritsen, K. Sneppen (Eds.), Physics of Biological Systems. From Molecules to Species. X, 364 pages. 1997.

Vol. 481: F. Lenz, H. Grießhammer, D. Stoll (Eds.), Lectures on QCD. VII, 276 pages. 1997.

Vol. 482: X.-W. Pan, D. H. Feng, M. Vallières (Eds.), Contemporary Nuclear Shell Models. Proceedings, 1996. XII, 309 pages. 1997.

Vol. 483: G. Trottet (Ed.), Coronal Physics from Radio and Space Observations. Proceedings, 1996. XVII, 226 pages. 1997.

Vol. 484: L. Schimansky-Geier, T. Pöschel (Eds.), Stochastic Dynamics. XVIII, 386 pages. 1997.

Vol. 485: H. Friedrich, B. Eckhardt (Eds.), Classical, Semiclassical and Quantum Dynamics in Atoms. VIII, 341 pages. 1997.

Vol. 486: G. Chavent, P. C. Sabatier (Eds.), Inverse Problems of Wave Propagation and Diffraction. Proceedings, 1996. XV, 379 pages. 1997.

Vol. 487: E. Meyer-Hofmeister, H. Spruit (Eds.), Accretion Disks – New Aspects. Proceedings, 1996. XIII, 356 pages. 1997.

Vol. 488: B. Apagyi, G. Endrédi, P. Lévay (Eds.), Inverse and Algebraic Quantum Scattering Theory. Proceedings, 1996. XV, 385 pages. 1997.

Vol. 489: G. M. Simnett, C. E. Alissandrakis, L. Vlahos (Eds.), Solar and Heliospheric Plasma Physics. Proceedings, 1996. VIII, 278 pages. 1997.

Vol. 490: P. Kutler, J. Flores, J.-J. Chattot (Eds.), Fifteenth International Conference on Numerical Methods in Fluid Dynamics. Proceedings, 1996. XIV, 653 pages. 1997.

Vol. 491: O. Boratav, A. Eden, A. Erzan (Eds.), Turbulence Modeling and Vortex Dynamics. Proceedings, 1996. XII, 245 pages. 1997.

Vol. 492: M. Rubí, C. Pérez-Vicente (Eds.), Complex Behaviour of Glassy Systems. Proceedings, 1996. IX, 467 pages. 1997.

Vol. 493: P. L. Garrido, J. Marro (Eds.), Fourth Granada Lectures in Computational Physics. XIV, 316 pages. 1997.

Vol. 494: J. W. Clark, M. L. Ristig (Eds.), Theory of Spin Lattices and Lattice Gauge Models. Proceedings, 1996. XI, 194 pages. 1997.

Vol. 495: Y. Kosmann-Schwarzbach, B. Grammaticos, K. M. Tamizhmani (Eds.), Integrability of Nonlinear Systems. XX, 404 pages. 1997.

Vol. 496: F. Lenz, H. Grießhammer, D. Stoll (Eds.), Lectures on QCD. VII, 483 pages. 1997.

Vol. 497: J.P. De Greve, R. Blomme, H. Hensberge (Eds.), Stellar Atmospheres: Theory and Observations. Proceedings, 1996. VIII, 352 pages. 1997.

Vol. 498: Z. Horváth, L. Palla (Eds.), Conformal Field Theories and Integrable Models. Proceedings, 1996. X, 251 pages. 1997.

Vol. 499: K. Jungmann, J. Kowalski, I. Reinhard, F. Träger (Eds.), Atomic Physics Methods in Modern Research. IX, 448 pages. 1997.

Vol. 500: D. Joubert (Ed.), Density Functionals: Theory and Applications. XVI, 194 pages. 1998.

Vol. 501: J. Kertész, I. Kondor (Eds.), Advances in Computer Simulation. VIII, 166 pages. 1998.

Vol. 502: H. Aratyn, T. D. Imbo, W.-Y. Keung, U. Sukhatme (Eds.), Supersymmetry and Integrable Models. XI, 379 pages. 1998.

Vol. 503: J. Parisi, S. C. Müller, W. Zimmermann (Eds.), A Perspective Look at Nonlinear Media. From Physics to Biology and Social Sciences. VIII, 372 pages. 1998.

Vol. 504: A. Bohm, H.-D. Doebner, P. Kielanowski (Eds.), Irreversibility and Causality. Semigroups and Rigged Hilbert Space. XIX, 385 pages. 1998.

Vol. 505: D. Benest, C. Froeschlé (Eds.), Impacts on Earth. XVII, 223 pages. 1998.

Vol. 506: D. Breitschwerdt, M. J. Freyberg, J. Trümper (Eds.), The Local Bubble and Beyond. Proceedings, 1997. XXVIII, 603 pages. 1998.

Vol. 507: J. C. Vial, K. Bocchialini, P. Boumier (Eds.), Space Solar Physics. Proceedings, 1997. XIII, 296 pages. 1998.

Vol. 508: H. Meyer-Ortmanns, A. Klümper (Eds.), Field Theoretical Tools for Polymer and Particle Physics. XVI, 258 pages. 1998.

Vol. 509: J. Wess, V. P. Akulov (Eds.), Supersymmetry and Quantum Field Theory. Proceedings, 1997. XV, 405 pages. 1998.

Vol. 510: J. Navarro, A. Polls (Eds), Microscopic Quantum Many-Body Theories and Their Applications. Proceedings, 1997. XIII, 379 pages. 1998.

New Series m: Monographs

Vol. m 2: P. Busch, P. J. Lahti, P. Mittelstaedt, The Quantum Theory of Measurement. XIII, 165 pages. 1991. Second Revised Edition: XIII, 181 pages. 1996.

Vol. m 3: A. Heck, J. M. Perdang (Eds.), Applying Fractals in Astronomy. IX, 210 pages. 1991.

Vol. m 4: R. K. Zeytounian, Mécanique des fluides fondamentale. XV, 615 pages, 1991.

Vol. m 5: R. K. Zeytounian, Meteorological Fluid Dynamics. XI, 346 pages. 1991.

Vol. m 6: N. M. J. Woodhouse, Special Relativity. VIII, 86 pages. 1992.

Vol. m 7: G. Morandi, The Role of Topology in Classical and Quantum Physics. XIII, 239 pages. 1992.

Vol. m 8: D. Funaro, Polynomial Approximation of Differential Equations. X, 305 pages. 1992.

Vol. m 9: M. Namiki, Stochastic Quantization. X, 217 pages. 1992.

Vol. m 10: J. Hoppe, Lectures on Integrable Systems. VII, 111 pages. 1992.

Vol. m 11: A. D. Yaghjian, Relativistic Dynamics of a Charged Sphere. XII, 115 pages. 1992.

Vol. m 12: G. Esposito, Quantum Gravity, Quantum Cosmology and Lorentzian Geometries. Second Corrected and Enlarged Edition. XVIII, 349 pages. 1994.

Vol. m 13: M. Klein, A. Knauf, Classical Planar Scattering by Coulombic Potentials. V, 142 pages. 1992.

Vol. m 14: A. Lerda, Anyons. XI, 138 pages. 1992.

Vol. m 15: N. Peters, B. Rogg (Eds.), Reduced Kinetic Mechanisms for Applications in Combustion Systems. X, 360 pages. 1993.

Vol. m 16: P. Christe, M. Henkel, Introduction to Conformal Invariance and Its Applications to Critical Phenomena. XV, 260 pages. 1993.

Vol. m 17: M. Schoen, Computer Simulation of Condensed Phases in Complex Geometries. X, 136 pages. 1993.

Vol. m 18: H. Carmichael, An Open Systems Approach to Quantum Optics. X, 179 pages. 1993.

Vol. m 19: S. D. Bogan, M. K. Hinders, Interface Effects in Elastic Wave Scattering. XII, 182 pages. 1994.

Vol. m 20: E. Abdalla, M. C. B. Abdalla, D. Dalmazi, A. Zadra, 2D-Gravity in Non-Critical Strings. IX, 319 pages. 1994.

Vol. m 21: G. P. Berman, E. N. Bulgakov, D. D. Holm, Crossover-Time in Quantum Boson and Spin Systems. XI, 268 pages. 1994.

Vol. m 22: M.-O. Hongler, Chaotic and Stochastic Behaviour in Automatic Production Lines. V, 85 pages. 1994.

Vol. m 23: V. S. Viswanath, G. Müller, The Recursion Method. X, 259 pages. 1994.

Vol. m 24: A. Ern, V. Giovangigli, Multicomponent Transport Algorithms. XIV, 427 pages. 1994.

Vol. m 25: A. V. Bogdanov, G. V. Dubrovskiy, M. P. Krutikov, D. V. Kulginov, V. M. Strelchenya, Interaction of Gases with Surfaces. XIV, 132 pages. 1995.

Vol. m 26: M. Dineykhan, G. V. Efimov, G. Ganbold, S. N. Nedelko, Oscillator Representation in Quantum Physics. IX, 279 pages. 1995.

Vol. m 27: J. T. Ottesen, Infinite Dimensional Groups and Algebras in Quantum Physics. IX, 218 pages. 1995.

Vol. m 28: O. Piguet, S. P. Sorella, Algebraic Renormalization. IX, 134 pages. 1995.

Vol. m 29: C. Bendjaballah, Introduction to Photon Communication. VII, 193 pages. 1995.

Vol. m 30: A. J. Greer, W. J. Kossler, Low Magnetic Fields in Anisotropic Superconductors. VII, 161 pages. 1995.

Vol. m 31: P. Busch, M. Grabowski, P. J. Lahti, Operational Quantum Physics. XI, 230 pages. 1995.

Vol. m 32: L. de Broglie, Diverses questions de mécanique et de thermodynamique classiques et relativistes. XII, 198 pages. 1995.

Vol. m 33: R. Alkofer, H. Reinhardt, Chiral Quark Dynamics. VIII, 115 pages. 1995.

Vol. m 34: R. Jost, Das Märchen vom Elfenbeinernen Turm. VIII, 286 pages. 1995.

Vol. m 35: E. Elizalde, Ten Physical Applications of Spectral Zeta Functions. XIV, 228 pages. 1995.

Vol. m 36: G. Dunne, Self-Dual Chern-Simons Theories. X, 217 pages. 1995.

Vol. m 37: S. Childress, A.D. Gilbert, Stretch, Twist, Fold: The Fast Dynamo. XI, 410 pages. 1995.

Vol. m 38: J. González, M. A. Martín-Delgado, G. Sierra, A. H. Vozmediano, Quantum Electron Liquids and High-T_c Superconductivity. X, 299 pages. 1995.

Vol. m 39: L. Pittner, Algebraic Foundations of Non-Commutative Differential Geometry and Quantum Groups. XII, 469 pages. 1996.

Vol. m 40: H.-J. Borchers, Translation Group and Particle Representations in Quantum Field Theory. VII, 131 pages. 1996.

Vol. m 41: B. K. Chakrabarti, A. Dutta, P. Sen, Quantum Ising Phases and Transitions in Transverse Ising Models. X, 204 pages. 1996.

Vol. m 42: P. Bouwknegt, J. McCarthy, K. Pilch, The W_3 Algebra. Modules, Semi-infinite Cohomology and BV Algebras. XI, 204 pages. 1996.

Vol. m 43: M. Schottenloher, A Mathematical Introduction to Conformal Field Theory. VIII, 142 pages. 1997.

Vol. m 44: A. Bach, Indistinguishable Classical Particles. VIII, 157 pages. 1997.

Vol. m 45: M. Ferrari, V. T. Granik, A. Imam, J. C. Nadeau (Eds.), Advances in Doublet Mechanics. XVI, 214 pages. 1997.

Vol. m 46: M. Camenzind, Les noyaux actifs de galaxies. XVIII, 218 pages. 1997.

Vol. m 47: L. M. Zubov, Nonlinear Theory of Dislocations and Disclinations in Elastic Body. VI, 205 pages. 1997.

Vol. m 48: P. Kopietz, Bosonization of Interacting Fermions in Arbitrary Dimensions. XII, 259 pages. 1997.

Vol. m 49: M. Zak, J. B. Zbilut, R. E. Meyers, From Instability to Intelligence. Complexity and Predictability in Nonlinear Dynamics. XIV, 552 pages. 1997.

Vol. m 50: J. Ambjørn, M. Carfora, A. Marzuoli, The Geometry of Dynamical Triangulations. VI, 197 pages. 1997.

Vol. m 51: G. Landi, An Introduction to Noncommutative Spaces and Their Geometries. XI, 200 pages. 1997.

Vol. m 52: M. Hénon, Generating Families in the Restricted Three-Body Problem. XI, 278 pages. 1997.

Vol. m53: M. Gad-el-Hak, A. Pollard, J.-P. Bonnet (Eds.), Flow Control. Fundamentals and Practices. XII, 527 pages. 1998.